T0135288

Advances in Intelligent Systems and Computing

Volume 1039

The series "Advances in Intelligent Systems and Computing" contains publications on theory, applications, and design methods of Intelligent Systems and Intelligent Computing. Virtually all disciplines such as engineering, natural sciences, computer and information science, ICT, economics, business, e-commerce, environment, healthcare, life science are covered. The list of topics spans all the areas of modern intelligent systems and computing such as: computational intelligence, soft computing including neural networks, fuzzy systems, evolutionary computing and the fusion of these paradigms, social intelligence, ambient intelligence, computational neuroscience, artificial life, virtual worlds and society, cognitive science and systems, Perception and Vision, DNA and immune based systems, self-organizing and adaptive systems, e-Learning and teaching, human-centered and human-centric computing, recommender systems, intelligent control, robotics and mechatronics including human-machine teaming, knowledge-based paradigms, learning paradigms, machine ethics, intelligent data analysis, knowledge management, intelligent agents, intelligent decision making and support, intelligent network security, trust management, interactive entertainment, Web intelligence and multimedia.

The publications within "Advances in Intelligent Systems and Computing" are primarily proceedings of important conferences, symposia and congresses. They cover significant recent developments in the field, both of a foundational and applicable character. An important characteristic feature of the series is the short publication time and world-wide distribution. This permits a rapid and broad dissemination of research results.

** Indexing: The books of this series are submitted to ISI Proceedings, EI-Compendex, DBLP, SCOPUS, Google Scholar and Springerlink **

More information about this series at http://www.springer.com/series/11156

A. Pasumpon Pandian · Klimis Ntalianis ·
Ram Palanisamy
Editors

Intelligent Computing, Information and Control Systems

ICICCS 2019

 Springer

Editors
A. Pasumpon Pandian
Department of Computer Science
and Engineering
Vaigai College of Engineering
Madurai, Tamil Nadu, India

Klimis Ntalianis
Electrical and Computer Engineering
University of Applied Sciences
Egaleo, Attiki, Greece

Ram Palanisamy
Department of Business Administration
The Gerald Schwartz School of Business
Antigonish, NS, Canada

ISSN 2194-5357   ISSN 2194-5365   (electronic)
Advances in Intelligent Systems and Computing
ISBN 978-3-030-30464-5   ISBN 978-3-030-30465-2   (eBook)
https://doi.org/10.1007/978-3-030-30465-2

This Springer imprint is published by the registered company Springer Nature Switzerland AG
The registered company address is: Gewerbestrasse 11, 6330 Cham, Switzerland

We are honored to dedicate the proceedings of International Conference on Intelligent Computing, Information and Control Systems (ICICCS 2019) to all the participants and editors of ICICCS 2019.

Foreword

It is with deep satisfaction that I write this foreword to the proceedings of the ICICCS 2019 held in Malla Reddy College of Engineering, Dhulapally, Secunderabad, at June 27–28, 2019.

This conference was bringing together researchers, academics, and professionals from all over the world, experts in Intelligent Control Technologies, Intelligent Computing Technologies, and Intelligent Information Systems.

This conference particularly encouraged the interaction of research students and developing academics with the more established academic community in an informal setting to present and to discuss new and current work. The papers contributed the most recent scientific knowledge known in the fields of Intelligent Computing Models and Systems, Intelligent Information Systems, fuzzy sets, Intelligent Computing Technologies Control Applications on automotive, energy, autonomous systems, big data, and machine learning. Their contributions helped to make the conference as outstanding as it has been. The Local Organizing Committee members and their helpers put much effort into ensuring the success of the day-to-day operation of the meeting.

We hope that this program will further stimulate research in Intelligent Systems and Computing, and provide practitioners with better techniques, algorithms, and tools for deployment. We feel honored and privileged to serve the best recent developments in the fields of Intelligent Control and automation, knowledge-based systems, computational and communication constraints, network intelligence and network control, and fuzzy logic control through this exciting program.

We thank all authors and participants for their contributions.

A. Pasumpon Pandian
Guest Editor

Preface

This conference proceedings volume contains the written versions of most of the contributions presented during the conference of ICICCS 2019. The conference provided a setting for discussing recent developments in a wide variety of topics including Intelligent Control Technologies, artificial intelligence [AI], machine learning, Intelligent Information Retrieval, Intelligent Agents, fuzzy logic control, neuro-fuzzy control, and evolutionary computing. The conference has been a good opportunity for participants coming from various destinations to present and discuss topics in their respective research areas.

ICICCS 2019 conference tends to collect the latest research results and applications on Intelligent Data Communication Technologies and Internet of things. It includes a selection of 79 papers from 312 papers submitted to the conference from universities and industries all over the world. All of accepted papers were subjected to strict peer-reviewing by 2–4 expert referees. The papers have been selected for this volume because of quality and the relevance to the conference.

ICICCS 2019 would like to express our sincere appreciation to all authors for their contributions to this book. We would like to extend our thanks to all the referees for their constructive comments on all papers, and especially, we would like to thank Organizing Committee for their hard working. Finally, we would like to thank the Springer publications for producing this volume.

P. John Paul
Conference Chair

Acknowledgements

ICICCS 2019 would like to acknowledge the excellent work of our conference Organizing Committee, keynote speakers for their presentation on June 27–28, 2019. The organizers also wish to acknowledge publicly the valuable services provided by the reviewers.

On behalf of the editors, organizers, authors, and readers of this conference, we wish to thank the keynote speakers and the reviewers for their time, hard work, and dedication to this conference. The organizers wish to acknowledge Dr. A. Pasumpon Pandian, Dr. Klimis Ntalianis, and Dr. Ram Palanisamy for the discussion and suggestion, and finalize the paper of this conference. The organizers also wish to acknowledge the speakers and participants who attend this conference. Many thanks are given for all persons who help and support this conference. ICICCS 2019 would like to acknowledge the contribution made to the organization by its many volunteers. Members contribute their time, energy, and knowledge at local, regional, and international levels.

We also thank all the chairpersons and Conference Committee members for their support.

Contents

CNN and Raspberry PI for Fruit Tree Disease Detection

Fanghong Hu, Zhu Li$^{(\boxtimes)}$, and Liping Yan

School of Electronic Information, Hangzhou Dianzi University,
Hangzhou 310018, Zhejiang, China
hjwwy66@163.com, lzl126@hdu.edu.cn, lpyan@hotmail.com

Abstract. Fruit tree disease detection system is very essential to reduce the production pressure of fruit farmers. The disease of fruit trees can be detected through the appearance recognition of leaves. CNN is widely used for object detection and recognition. In order to save cost, we proposed a fruit tree disease detection scheme based on CNN. By cutting and compressing the traditional neural network model, we designed a lightweight neural network model, which can be run on Raspberry PI and can be used to detect fruit tree diseases in an offline environment. The Plant Village dataset was used in this article to verify its validity.

Keywords: CNN · Raspberry PI · Plant Village · Lightweight neural network

1 Introduction

Fruit tree disease detection by optical observation of symptoms on the leaves of fruit trees has obvious high complexity. Because of the complexity of this plant pathology problem, even plant pathologists and agronomists with extensive experience often fail to correctly diagnose specific diseases, leading to erroneous conclusions and treatments [1]. In recent years, deep learning has achieved great success in image classification, image segmentation, object detection and other computer vision fields. Lots of excellent convolution network models have emerged [2]. However, the scale of these network models is very large. The performance requirements of the running platform are also very high [3]. So we compressed these giant CNN [4] models to design and implement a lightweight running on the Raspberry PI by pruning neural network model [5]. This builded automatic detection can be used for offline environment system of fruit tree disease [6], for effective prevention and control of fruit diseases, minimizing losses from diseases, and helps to improve the fruit production yield.

2 CNN Model Compression Method

There are many methods for model compression. In this paper, the CNN model will be pruned [7]. The main idea is to pick out and remove the unimportant weights or convolution kernels in the model [8]. After removing the unimportant weights or convolution kernels, the performance of the model can be restored through retraining.

© Springer Nature Switzerland AG 2020
A. P. Pandian et al. (Eds.): ICICCS 2019, AISC 1039, pp. 1–8, 2020.
https://doi.org/10.1007/978-3-030-30465-2_1

So that the model parameters can be compressed to the maximum extent and the model acceleration can be realized under the premise of guaranteeing the model performance [9]. In this research, how to find an effective standard to measure the weights or importance of convolution kernels is the key point. We put forward the CNN pruning steps as stated in the following. Based on an original network [10], the convolution kernels of low importance are removed to conduct the pruning of convolution kernels firstly [11]; Then, heavy pruning is used to achieve further compression. Compared with the original CNN, the CNN after pruning has less convolution kernels and weights [12]. The framework of the pruning is shown in Fig. 1.

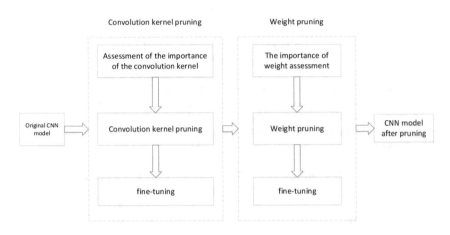

Fig. 1. Framework of network pruning

Convolution kernel pruning consists of the following four steps, as shown in Fig. 2:

(1) Evaluate the importance of convolution kernels. The main idea is to use the statistics of i + 1 layer to guide the pruning of i layer, to approximate the output of i + 1 layer by using the subset of the i + 1 input [13].
(2) Cut off unimportant channels and their corresponding convolution kernels. The channel in i + 1 layer input is generated by the convolution kernel in i layer, so the unimportant channel of the i + 1 layer and the convolution kernels of the corresponding i layer can be safely removed.
(3) After pruning, the model generalization ability will be damaged partly. The entire network should be fine-tuned for more one or two times to restore its performance.
(4) Finally, the decision is made to end the pruning. If the pruning stops, you can fine tune multiple times to get a more accurate model [14]. Otherwise, steps (1)–(3) are repeated to continue pruning the next layer. The linear process of its output is simulated by use of the input of the i + 1 layer, the output of the i layer. If the optimized processing is found, the linear combination in the input of the i + 1 layer is equal to zero [15].

After partly discarding the corresponding convolution kernels of the i layer output, we can get a similar output with less input. At length, the ideal compression model is achieved without affecting the model effect.

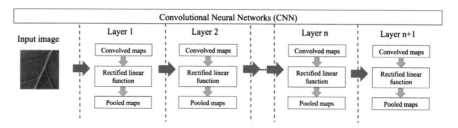

Fig. 2. CNN study on fruit tree leaves

Weight pruning consists of 3 steps:

(1) The importance of weights should be assessed firstly. The weights from normal training are decided to be relatively important and unimportant weights [16].

(2) A threshold is picked out for seting the weight in the range of the threshold to zero. Then the network becomes a sparsely connected network.

(3) Retrain this sparse network to get the final model. Due to limited data, too many training parameters and excessive training may cause over-fitting problems of convolutional neural networks. Currently, regularization and dropout are two common methods to prevent over-fitting. After the normal training network, the relatively important and unimportant weights are obtained respectively. In the weight pruning, the L_1/L_2 regularization method, the convolutional layer and the fully connected layer have different sensitivities.

By increasing the depth of the network, you can make it better to approximate the loss function, increase the nonlinearity, and obtain a better feature representation. However, the complexity of the network will also increase with the increase of the depth, making the network more difficult to be optimized and fitted. To train large and dense networks, we use L_1/L_2 regularization and dropout to generate sparse weights, and then delete some redundant connections with reducing the weights of the network that provide less contribution. In addition, the most straightforward way to prevent overfitting is to enlarge the used data set and narrow down the used network structure. However, expanding the data set is not implemented straightforward. Reducing the network structure can effectively reduce the number of parameters. But the deeper the network, the stronger the expressive ability. Therefore, the use of regularization and dropout is designed to solve the problem of overfitting. L_2 regularization is the most commonly used regularization method and can be expressed as

$$C = C_0 + \frac{\lambda}{2n} \sum_w w^2 \tag{1}$$

Where C represents the regularized loss function and C_0 represents the original loss function. λ is a regularization factor. Therefore, the commonly used cross entropy loss function can be expressed as:

$$C = -\frac{1}{n}\sum_x [y \ln a + (1 - y)\ln(1 - a)] + \frac{\lambda}{2n}\sum_w w^2 \tag{2}$$

Where x represents the sample, n represents the total amount of the sample, y represents the actual value, and a represents the output value. From the above formula, regularization is achieved by adding weights to the loss function [17]. The loss value can be decreased by reducing the weight. Regularization is adjusted using a regularization factor: the larger λ, the more inclined to reduce the weight; The smaller λ, the more inclined to reduce the original loss function [18]. In the first training of the pruning process, we determine whether the weight is important or not. Positiving degree could affect the size of the weight, which resolves which connections in the network would be pruned.

Unlike L_2 regularization, dropout can offset the problem of the combination during training by changing the network structure. İn detail, dropout randomizes certain weights with a certain probability. The network sparseness acceleration convergence is increased according to formulas (3) and (4):

$$C_i = N_i N_{i-1} \tag{3}$$

$$D_r = D_o \sqrt{\frac{C_{ir}}{C_{io}}} \tag{4}$$

Where C_i represents the number of connections in the i-th layer, N_i represents the number of neurons in the i-th layer, C_{io} represents the original number of connections, C_{ir} represents the number of connections during retraining, and D_o represents the original dropout [19]. Different types of network levels and different regularization methods have different sensitivities in pruning. Therefore, their different trim thresholds will also have different effects on their accuracy. Sensitivity analysis of the neural network is done layer by layer by deleting the part that has little influence on the accuracy of the neural network after resetting the weight to zero. Then the weights are sorted and a threshold is set. At the same time, the weight below the threshold is reset to zero. Meanwhile, the other weights remain unchanged. The training is run continuously until the model accuracy is restored. The above process is repeated for times. Then the threshold is increased with increasing the threshold.

3   Experimental Results and Analysis

Based on the open source dataset Plant Village, the CNN architecture uses the classic VGG-16. Plantvillage dataset contains 54306 plant images containing 14 diseases. These images are assigned to 38 categories of labels. VGG-16 is a 16-layer CNN with 13 convolutional layers and 3 FC layers [20]. VGG-16 is used to evaluate the weight performance on the Plant Village dataset, creating the same architecture as VGG-16, with masks and threshold variables added to the layer that needs to be pruned. The weight tensor and variable mask of the network layer have the same shape and determine which weight participation graphs [21] are executed in the forward direction. The operation is then added to the training map. It correctly calculates layer thresholds and monitors the distribution of weights in the layer masking the weight below the threshold to the level of sparsity required for the current training step.

Table 1. The changes of the VGG-16 model pruning.

Method	Accuracy	Compression
Original	87.68%	1.00 X
Weight pruning	86.33%	13.32 X
Convolution nuclear pruning	86.67%	16.65 X
Mixed pruning	86.13%	19.21 X

The VGG-16 network is used to evaluate the performance of the convolution kernel pruning method based on the Plant Village dataset. The Botanical Village dataset containing 38 categories from which 50,000 training images are extracted for the training set for channel selection. Then, 10 images are selected from each category in the training set to form our evaluation set. There are a total of 1000 training sets for finding the optimal subset of channels through the greedy algorithm. During fine-tuning, the network is trimmed each time. When all layers have been pruned, we fine tune to restore the original accuracy for 15 times.

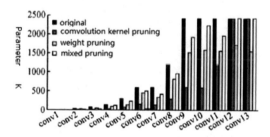

Fig. 3. Parameters reduction statistics for VGG-16 on Plant Village

The performance of the hybrid pruning method was also evaluated on the Plant Village dataset using the VGG-16 network. Firstly the unimportant convolution kernels are removed. Then weights reduction is performed on the pruned model to achieve further compression. As shown in Fig. 3, VGG-16 optimizes the number of pruning layers by different pruning methods on the Plant Village dataset. The results show that the proposed hybrid pruning method significantly reduces the number of parameters compared to the use of only weighted pruning or convolutional nuclear pruning. The convolutional layer of the VGG-16 network occupies approximately 90% of floating-point operations, while the FC layer has 89.36% of the parameters. The Raspberry PI development board used in this experiment is shown in Fig. 4. For model acceleration, it is easier and more efficient to replace the FC layer with the GAP layer. The performance changes of the VGG-16 model after pruning are shown in Table 1. Compared with other benchmarks, the hybrid pruning method can realize better compression and acceleration. For the VGG-16 model, the proposed hybrid pruning method can be used up to 19.21 × compression. Figure 4. The trained cropping network is run on this Raspberry PI to get the test results as shown in Table 2.

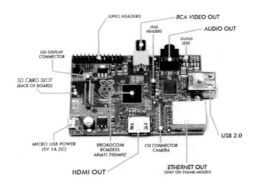

Fig. 4. The Raspberry Pi board

Table 2. The fruit tree disease detection confidence

Plant name	Disease name	Confidence
Apple	Apple scab	90.8%
Banana	Banana Speckle	89.5%
Blueberry	Black rot	92.2%
Cherry	Powdery mildew	91.4%
Grape	Leaf blight	89%
Watermelon	Black rot	90.4%
Orange	Huanglongbing	87%
Peach	Bacterial sport	88.5%
Strawberry	Leaf scorch	86.3%
Tomato	Early blight	90.8%

4 Conclusion

In model compression, weight pruning or convolution kernel pruning is carried out. The compressed convolutional neural network still has parameter redundancy problem. Therefore, a hybrid pruning method combining convolution kernel pruning and weight pruning is proposed. By deleting the less important convolution kernels in the VGG-16 model, the initial purpose of the compression network is achieved. The pruned model is further weighted and pruned to achieve further model compression. The recalibration is used to restore the model accuracy during the pruning process. The results show that this hybrid pruning method can effectively reduce the parameter redundancy existing in CNN, realize network acceleration, and reduce the scale of the network model running on Raspberry PI to complete the detection.

References

1. Li, Z.Q., Zeng, S.M.: Wheat Rusts in China, pp. 1–379. China Agriculture Press, Beijing (2002)
2. Li, H.Y., Chen, Z.Y., Zhou, J.C., Zhang, J.X.: Occurrence and damage of the main diseases of grape in greenhouse during growthperiod and shelf life. Mod. Agric. Sci. Technol. **39**, 187–189 (2010)
3. Wang, H.G., Ma, Z.H., Zhang, M.R., Shi, S.D.: Application of computer technology in plant pathology. Agric. Netw. Inform. **19**, 31–34 (2004)
4. Wang, N., Wang, K.R., Xie, R.Z., Lai, J.C., Ming, B., Li, S.K.: Maize leaf disease identification based on fisher discrimination
5. Ma, X.D., Qi, G.Y.: Investigation and recognition on diseased spots of soybean laminae based on neural network. J. Heilongjiang Aug. First Land Reclamat. Univ. **18**, 84–87 (2006)
6. Song, K., Sun, X.Y., Ji, J.W.: Corn leaf disease recognition based on support vector machine method. Trans. CSAE **23**, 155–157 (2007)
7. Bock, C.H., Poole, G.H., Parker, P.E., Gottwald, T.R.: Plant disease severity estimated visually, by digital photography and image analysis, and by hyperspectral imaging. Crit. Rev. Plant Sci. **29**, 59–107 (2010)
8. Ahmad, I.S., Reid, J.F., Paulsen, M.R., Sinclair, J.B.: Color classifier for symptomatic soybean seeds using image processing. Plant Dis. **83**, 320–327 (1999)
9. Al-Hiary, H., Bani-Ahmad, S., Reyalat, M., Braik, M., ALRahamneh, Z.: Fast and accurate detection and classification of plant diseases. Int. J. Comput. Appl. **17**, 31–38 (2011)
10. Li, Z.R., He, D.J.: Research on identify technologies of apple's disease based on mobile photograph image analysis. Comput. Eng. Design **31**, 3051–3053 (2010)
11. Wang, Y.P., Dai, X.P., Huang, H., Zhang, R.: Identification of tomatoes with diseases based on mathematical morphology and neural network. J. Hum. Agric. Univ. (Nat. Sci.) **32**, 344–346 (2006)
12. Tian, Y.W., Li, T.L., Li, C.H., Piao, Z.L., Sun, G.K., Wang, B.: Method for recognition of grape disease based on support vectormachine. Trans. CSAE **23**, 175–180 (2007)
13. Ren, D., Yu, H.Y., Wang, J.H.: Research on plant disease recognition based on linear combination of the kernel function support vector machine. J. Agric. Mech. Res. **29**, 41–43 (2007)
14. Camargo, A., Smith, J.S.: Image pattern classification for the identification of disease causing agents in plants. Comput. Electron. Agric. **66**, 121–125 (2009)
15. Li, G.L., Ma, Z.H., Huang, C., Chi, Y.W., Wang, H.G.: Segmentation of color images of grape diseases using K_means clustering algorithm. Trans. CSAE **26**, 32–37 (2010)
16. Li, G.L.: Preliminary study on automatic diagnosis and classification method of plant diseases based on image recognition technology, pp. 1–64. China Agricultural University, Beijing (2011)
17. Strange, R.N., Scott, P.R.: Plant disease: a threat to global food security. Annu. Rev. Phytopathol. **43**, 83–116 (2005)

18. Harvey, C.A., et al.: Extreme vulnerability of smallholder farmers to agricultural risks and climate change in madagascar. Philos. Trans. Roy. Soc. London B: Biol. Sci. **369**, 20130089 (2014)
19. Ehler, L.E.: Integrated pest management (IPM): definition, historical development and implementation, and the other IPM. Pest Manag. Sci. **62**(9), 787–789 (2006)
20. Hughes, D.P., Salathé, M.: An open access repository of images on plant health to enable the development of mobile disease diagnostics through machine learning and crowdsourcing. CoRR abs/1511.08060 (2015)
21. Everingham, M., Van Gool, L., Williams, C.K., Winn, J., Zisserman, A.: The pascal visual object classes challenge. Int. J. Comput. Vision **88**(2), 303–338 (2010)

Stock Market Trend Prediction in Sub-Saharan Africa Using Generalized Additive Models (GAMs)

Dennis Murekachiro[1(✉)], Thabang M. Mokoteli[1],
and Hima Vadapalli[2]

[1] Wits Business School, University of the Witwatersrand,
Johannesburg, South Africa
dtmureka@gmail.com,
thabang-mokoaleli-mokoteli@wits.ac.za
[2] School of Computer Science and Applied Mathematics,
University of the Witwatersrand, Johannesburg, South Africa
hima.vadapalli@wits.ac.za

Abstract. Pattern discovery emerges as a significant factor to identify the direction of the market. This study sought to test the usefulness of GAMs in predicting the frontier and emerging stock markets in Africa for pattern discovery by comparing its prediction capability to deep neural models namely Long Short Term Memory (LSTM), Recurrent Neural Networks (RNNs), Gated Recurrent Units (GRUs), Bidirectional LSTM, Bidirectional RNN and Bidirectional GRU. Using daily stock market index, the data from Bloomberg database for the period 2012 to 2018, and this study aims to predict daily closing prices for the next 365 days as well as determining the direction of the stock markets. Prediction accuracies were 99.76%, 97.55%, 100%, 99.21%, 99.50%, 99.32%, 99.58%, 99.88%, 99.59% and 99.52% for Botswana, Egypt, Kenya, Mauritius, Morocco, Nigeria, South Africa, Tunisia, Zambia and Zimbabwe stock markets respectively. The GAM model outperformed the deep neural models and it can be used for enhancing investment decision making in Africa.

1 Introduction

The field of stock market forecasting has and is still receiving substantial notice from both practitioners as well as researchers as it is an essential matter for stock fund managers, individual investors and financial analysts amongst other players in the stock markets [1]. The ability to correctly predict future market trends according to [2] is a prerequisite for successful financial market trading. In [3], the authors believe that the success of market trading strategies depends upon the accurate predictions of Stock price movements. Those who will win in today's business world are those with the ability to predict the future, or at least having some future information upon which they can support their decisions [4]. Generally, Stock markets are described in [5] as complex, evolutionary and nonlinear dynamic systems whose prediction is considered a challenging task.

© Springer Nature Switzerland AG 2020
A. P. Pandian et al. (Eds.): ICICCS 2019, AISC 1039, pp. 9–19, 2020.
https://doi.org/10.1007/978-3-030-30465-2_2

African stock markets are not unique to such evolutionary dynamism. The case for Africa is worsened by the opaqueness of African markets which are characterized by low levels of liquidity, high volatility, low growth potential and inefficiency. Although other emerging and frontier markets have achieved to net investors some of their best return in recent years, no tangible gain has been realised from Africa's 27 markets. Despite the afore-mentioned, there still remains a massive appetite for African stocks especially by institutional investors.

Thus, a key question that needs address is; can an investor achieve substantial return and profit from such markets? It is thus the contribution of this paper to address this challenge by developing a prediction model that can help investors predict market trends in such small markets with increased certainty and achieve a good return on investment. The key contribution of this paper is testing the usefulness of GAMs in stock price prediction especially for African stock markets. In addition, the other contribution of this paper is in stock market pattern discovery through GAMs for portfolio construction, diversification and optimisation. Pattern prediction for African stock markets is essential to investors and traders investment decision making processes. Investments will be done from a data centric perspective, hence the ability to increase active trading strategies on African bourses.

The remainder of this paper is organised as follows; Sect. 2 discusses related work on GAM applications. Section 3 presents the methodology used for stock market direction and stock price predictions using GAM. Section 4 discusses the experimental results and Sect. 5 concludes the paper.

2   Related Work

Statistical approaches have stood the test of time in prediction initiatives. Owing to their flexibility in model specifications [6], GAMs have being applied to various domains such as spatial science modelling [6, 7], ecology [8], Gaussian process modelling [9], environmental modelling [10], classification and regression modelling [11], energy modelling [12] and bankruptcy prediction [13]. However, application of GAMs to big data domains such as financial forecasting of stock market indices has been scanty and is therefore the focus of this research paper.

As a blend of key features of generalized linear models and additive models, GAMs are nonparametric generalizations of generative linear models in which linear predictors are replaced by additive predictors. The key assumption on which GAMs rely on is that the simulators response can approximately be decomposed as a sum of univariate functions [9]. GAMs use a smooth function that permits identifying nonlinear relationships between the response and predictor variables and still retain much of the intelligibility of linear models [8, 11].

A general consensus amongst academic researchers is that GAMs are widely used in data analysis to explore the nonlinear effects of the covariant on the response variable. They are not only reasonably flexible for prediction but also have the advantage of avoiding the curse of dimensionality [14]. GAMs are statistical approaches for nonparametric or semi parametric modelling and have demonstrated their ability to capture nonlinear relationships between explanatory variables and response variables [15]. Thus the strength of GAMs is not only in providing predicted values but also an understanding of the influence of the co-variables used on the response variable [16]. The GAM model adopted for this research is in [15] and is a sum game of smooth functions as shown in the following formula;

$$g(E(y_i)) = \beta_0 + f_1(x_{i1}) + \ldots + f_p(x_{ip}) + \varepsilon_i \tag{1}$$

where $y_i\tilde{}$ is some exponential family distribution, and $i = 1, \ldots, N$, g is a link function (identical, logarithmic or inverse), y is a response variable, $x_1, \ldots, x_p$ are independent variables, β_0 is an intercept, $f_1, \ldots, f_p$ are unknown smooth functions and is an i.i.d. random error. Therefore, the model can be written in a linear way like this:

$$g(E(y)) = \beta X + \varepsilon \tag{2}$$

GAMs are known for exhibiting higher prediction performance when properly learned from a sufficiently large amount of data [17]. They are also known for detailed underlying data pattern discovery [13]. It is for this reason that GAMs are considered for stock price prediction in this research since stock markets are characterized by high noise, non-linearity, dynamic and deterministically chaotic data, evolutionarity, non-randomness, non-stationarity, volatility and voluminous data transactions [18–26].

3   Methodology

Data and Data Sources: For the purposes of this research, daily stock market index data from 10 African stock markets namely Botswana, Egypt, Kenya, Mauritius, Morocco, Nigeria, South Africa, Tunisia, Zambia and Zimbabwe was considered. The Bloomberg database was used to collect the daily stock index data in the form of open, high, low, volume and closing prices for the research period starting from 2012 to 2018. The input variables are open, high, low and volume whilst closing price is the output or target variable.

Statistical Models: GAM can map nonlinear data and predict with reasonable accuracy. GAM is a statistical approach for nonparametric or semi parametric modelling and it has demonstrated its ability to capture nonlinear relationships between explanatory variables and response variables [15]. It is for this reason that this statistical model is compared to deep neural networks to determine if it could predict data nonlinearity with much success. Herewith are the steps which were undertaken in building the GAM predictive model (Table 1).

Table 1. Six Steps in designing a GAM forecasting model

Step 1: Importation of libraries
Step 2: Data Uploading and formatting according to GAM prophet format
Step 3: Modeling holidays
Step 4: Model fitting
Step 5: Validating the model
Step 6: Backfitting the Algorithm and Evaluation

Python 3.6 was used to model this GAM pattern discovery model for the stock markets. Subsequent to data uploading, formatting and seasonal decomposition, the GAM model is trained to take into consideration holidays. The impact of a particular holiday on the time series is often similar year after year, so it is important to incorporate it into the forecast. The model is fit through the stan package. Adopting the model in [27], the financial time series is decomposed into three components namely trend, seasonality, and holidays and expressed as follows;

$$y(t) = g(t) + s(t) + h(t) + \varepsilon t \qquad (3)$$

where g(t) is the trend function modeling non-periodic changes in the value of the time series, s(t) represents periodic changes (e.g., weekly and yearly seasonality), and h(t) represents the effects of holidays which occur on potentially irregular schedules over one or more days. The error term t represents any idiosyncratic changes which are not accommodated by the model. The forecasting problem is framed as a curve-fitting exercise, which is inherently different from time series models that explicitly account for the temporal dependence structure in the data. An important benefit of the decomposable model is that it allows us to look at each component of the forecast separately. The horizon or forecasting period for this model was pegged at 365 days. The model predicts stock prices for the next year. A confidence interval of 95% is embedded in the model; therefore, GAM results will be assessed on 5% significance level. Automatic check points are also included in the model to show major events that defined the trend.

In validating the model, a procedure called simulated historical forecasts (SHF) is used to generate historical forecast errors to fit this model. SHFs are used to produce k forecasts at various cut off points in the history. This SHF procedure is an adaption from the rolling origin forecast evaluation procedures. SHF slices data into time segments across the forecasting horizon instead of sampling data points across time. Each segment is used to predict values in the subsequent year. For instance, in this research, HSF was segmented into twelve (12) time periods over the forecasting horizon. The model is then back fitted to assess model fitness. Mean prediction error plots are then produced and model is evaluated on mean absolute deviation (MAD), mean squared error (MSE), root mean square error (RMSE), mean absolute percentage error (MAPE) and Accuracy.

4   Prediction Results and Discussion

Table 2 below shows that GAM was able to predict stock price indices for the 10 African stock markets. The lowest prediction accuracy was for Egypt with 97.55% whilst the highest was for the Kenyan bourse with 100% prediction accuracy. The GAM prediction results outperformed deep neural network results for an African stock market prediction initiative in [28]. It can be noted that GAMs proved to be effective for financial stock price prediction for the 10 African stock markets under consideration. Though deep neural networks exhibited satisfactory results for the African markets, GAM superseded the deep neural networks. It can therefore be concluded that for African stock markets, use of nonlinear statistical prediction models yields better prediction accuracy.

In addition to superior prediction capabilities, GAM models also predicted stock market directional better than the deep neural networks. Owing to the fact that GAM decomposes the time series into trend, seasonality and holidays, investors can make us of the results in Table 3 to aid investment decision making and portfolio construction, diversification and optimisation.

Table 2. Summary of prediction accuracy results

Algorithm	GAM	RNN	BRNN	LSTM	BLSTM GRU		BGRU
Stock market	%	%	%	%	%	%	%
Botswana	99.76	95.28	95.76	96.05	91.22	95.95	95.59
Egypt	97.55	90.05	90.31	92.95	91.50	86.50	91.42
Kenya	100.00	71.76	81.00	67.55	68.36	74.51	84.71
Mauritius	99.21	97.00	97.01	96.57	97.35	95.43	97.50
Morocco	99.50	97.67	97.53	97.70	97.35	97.69	97.74
Nigeria	99.32	97.42	96.72	98.96	98.88	98.50	98.65
South Africa	99.58	97.36	97.94	97.74	97.66	97.45	97.90
Tunisia	99.88	82.53	84.18	97.95	96.25	94.75	95.71
Zambia	99.59	76.45	76.32	76.80	75.95	76.27	76.29
Zimbabwe	99.52	97.96	97.72	97.66	96.94	95.62	97.46

Figures 1, 2, 3, 4 and 5 depict the time-series predictions for the African stock markets showing the 2019 forecasts.

A common feature in GAM is that whenever there are key events in the history of the country, the model is able to detect that. For example in Botswana, the period preceding 2014 general elections shows a decline in the market. The market self adjusts after elections and becomes bullish again in 2015. The Botswana Stock Exchange is forecasted to be bearish post 2017 period. Investors can therefore take market positions given the expected downward trend post 2017. The invaluable nature of automatic

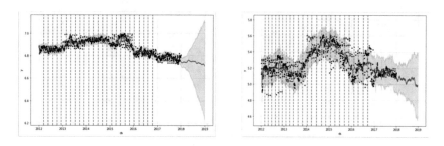

Fig. 1. Graphical representation of GAM results for Botswana and Egypt stock exchange

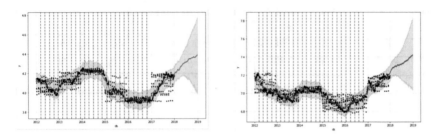

Fig. 2. Graphical representation of GAM results for Mauritius and Morocco stock exchange

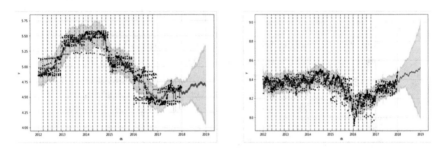

Fig. 3. Graphical representation of GAM results for Nigeria and Johannesburg stock exchange

change point detection in forecasting initiatives is in concurrence to previous studies in [29–31] who in their studies also take on POCID as a key metric.

It can also be noted in Fig. 1 that major shocks hit the Egypt stock market in the period 2012–2014. The market self adjusted from 2014 onwards but with a further slump in the period 2016 right into the forecast period. A notable trigger of such market upsets was political related events which drive up the equity risk, volatilities and correlations of stock returns [32]. The 2013 coup and the 2012 and 2014 Presidential elections had an impact on the stock market movements. As exhibited in Fig. 2, there is a notable positive impact on stock markets during election periods.

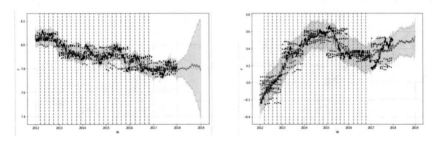

Fig. 4. Graphical representation of GAM results for Tunisian and Nairobi STOCK exchange

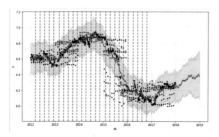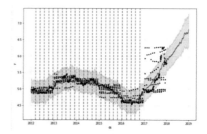

Fig. 5. Graphical representation of GAM results for Lusaka and Zimbabwe stock exchange

This is in agreement to studies to ascertain impact of political events to stock market performance in Egypt [33, 34]. A negative impact was realized during the 2011 uprisings for pro-democracy in Egypt on the 25th of January 2011 as indicated in [33]. The effect lasted till 2012 when the market was low for a considerable time period. Author in [35] also concurs to these findings in a study conducted on the Egyptian markets after the 2011 revolutionary upsets in the country. The conclusion from this study is that political instability does play an important task in carrying out the stock market function. After the 2014 up-market movements, the market started to become bearish in 2016 and post 2017.

Due to an economic recession in Nigeria in 2016 as a result of structural bottle-necks and a fall in global oil prices, the local bourse reacted negatively to it. Figure 3 reveals a fairly stable market over time in South Africa. However, the market exhibits a decline from 2015 into 2016. It then self corrects to have an upward trend post 2016 with a sharp rise or positive market direction. The economic recession in South Africa in 2016 caused the markets to slump down aggressively. The possibility of spill overs from Nigeria and South Africa as Africas two economies could have induced the low performance in all the other African markets in 2016.

A key observation of the decomposed time series into seasonalities brought about valuable insights for investment decision making.

Prediction Summary on GAM: Table 3 is a summary of the different country indices when they are lowest and highest. This is done on a weekly and monthly basis. Investment and trading practitioners can use this information to place their market orders. There is a strong resemblance of results on a weekly basis than on a monthly basis.

Botswana Stock Exchange is at its lowest levels in January and the market picks generally over the entire trading year with the highest peak in April. The results are in contrast to the January effect where market prices are believed to be high and provide the best returns in January. Weekly trends are revealing that the market starts on its lowest on Monday and Wednesday with its peak on Thursday. Such weekly trends are beneficial to market traders who can take positions to go long on a Monday and Wednesday and go short on a Thursday. The afore-mentioned results are in agreement to the day of the week studies in [36, 37] which account for the lowest returns on a Monday.

Table 3. Summary of Weekly and monthly highs and lows

Country	Weekly	Monthly
	Lowest - highest	Lowest - highest
Botswana	Monday - Thursday	January - April
Egypt	Friday - Tuesday	January - October
Kenya	Monday - Wednesday	December - April
Mauritius	Wednesday - Tuesday	November - May
Morocco	Wednesday - Tuesday	October - February
Nigeria	Monday - Wednesday	February - October
South Africa	Wednesday - Tuesday	January - May
Tunisia	Thursday - Friday	January - August
Zambia	Thursday - Tuesday	February - August
Zimbabwe	Thursday - Friday	August - October

With regards to the Stock Exchange of Egypt, a downward trend is expected post 2017. Weekly trends reveal that the stock market is at its highest and best return platform on a Tuesday and it is lowest on a Friday. The low returns on a Friday are consistent with findings in [38] who also found out that least returns are either on a Thursday or Friday and highest returns on a Tuesday or Wednesday. The month of October is the best time to go short and January is the best time to go long. Results are pointing to October as the month of best returns in Egypt. This contrasts evidence from prior research on the month effect for Egypt were in [39] it was found out that Egypt exhibits the month of the year effect in January where the best returns for the bourse are realized.

Overall, in consideration of weekly results, Kenya and Nigeria show similar results where the markets are lowest and highest on a Monday and Wednesday respectively. Botswana posts similar results except that the highest is on a Thursday. Three countries stock markets namely Mauritius, Morocco and South Africa have their lowest and highest on Wednesday and Tuesday. Tunisia and Zimbabwe have their lowest and highest days on a Thursday and Friday. Zambia and Egypt have their highest on a Tuesday and lowest on Thursday and Friday respectively. An investor in Africa can make use of this information to construct investment portfolios that generate high returns. Thus an investor can through clustering, maximise on the correlations for a basket of portfolios. They may consider the Mauritius, Morocco and South Africa cluster or Kenya, Nigeria and Botswana cluster, the Tunisia and Zimbabwe cluster or the Zambia and Egypt cluster on a weekly basis in order to maximise returns. This works well for investors with a high risk profile. If the investor appetite for risk is low and there is need to diversify, they can pick stocks from stock markets clusters that are uncorrelated.

In addition, a key observation from the graphical and seasonal components of forecast time series results is that all ten stock markets experienced a slump in 2016. This could have been a result of economic crisis in two of Africas major markets, namely South Africa and Nigeria which could have spilled over to other countries. This suggests that if there is a regional crisis from Africas biggest stock markets, the shock is transmitted to other stock markets.

5   Conclusion

African markets are predictable. The direction of the market is highly predictable through GAM modelling and prediction accuracy levels achieved by GAM are highly remarkable compared to deep neural network forecasts of African stock markets.

Co-movements in market trends where discovered in African stock markets. Weekly forecasts reveal co-movements between Kenya, Nigeria and Botswana as they exhibit similar weekly trends. Mauritius, Morocco and South Africa post similar results whilst Tunisia and Zimbabwe also show similar trends on a weekly basis. Zambia and Egypt are also correlated on a weekly basis. In consideration of monthly trends, African markets are at their lowest mostly in January and February and at their highest peak in October. Therefore, weekly least returns for the African stock markets in this study are mostly on a Thursday and greatest returns on a Tuesday. GAM is a great tool for modelling stock market close price and market direction predictions. Future studies can focus on possible returns that can be realised from investing in frontier and emerging stock markets in Africa to aid investment decision making for African stock market investors. An African stock market cointegration study and returns generation possibility is a study that can sequel this current study.

References

1. Hsu, C.: A hybrid procedure with feature selection for resolving stock futures price forecasting problems. Neural Comput. Appl. **22**, 651–671 (2013)
2. Bagheri, A., Peyhani, H.M., Akbari, M.: Financial forecasting using anfis networks with quantum-behaved particle swarm optimisation. Expert Syst. Appl. **41**, 6235–6250 (2014)
3. Leung, M.T., Daouk, H., Chen, A.S.: Forecasting stock indices: a comparison of classification and level estimation models. Int. J. Forecast. **16**(2), 173–190 (2000)
4. Abbasi, N.M., Aghaei, M.A., Fard, M.M.: An integrated system based on fuzzy genetic algorithm and neural networks for stock price forecasting: case study of price index of Tehran stock exchange. Int. J. Qual. Reliabil. Manage. **31**(3), 281–292 (2014)
5. Anish, C.M., Majhi, B.: Hybrid nonlinear adaptive scheme for stock market prediction using feedback flann and factor analysis. J. Korean Stat. Soc. **45**, 64–76 (2015)
6. Fang, X., Chan, K.S.: Additive models with spatio-temporal data. Environ. Ecol. Stat. **22**, 61–86 (2015)
7. Schratz, P., Muenchow, J., Iturritxa, E., Ritcher, J., Brenning, A.: Performance evaluation and hyperparameter tuning of statistical and machine learning models using spatia data. arXiv. 1803.11266v1A (2018)
8. Jeong, C., Min, J.H., Kim, M.S.: A tuning method for the architecture of neural network models incorporating gam and ga as applied to bankruptcy prediction. Expert Syst. Appl. **39**, 3650–3658 (2012)
9. Durrande, N., Ginsbourger, D., Roustant, O.: Additive kernels for gaussian process modeling, pp. 1–17. ArXiv.1103:4023v1 (2010)
10. Chen, B., Liu, H., Huang, B.: Environmental controlling mechanisms on bacterial abundance in the South China sea inferred from generalized additive models (GAMs). J. Sea Res. **72**, 69–76 (2012)
11. Lou, Y., Caruana, R., Gehrke, J.: Intelligible models for classification and regression, pp. 150–158. ACM. 978-1-4503-1462-6/12/08 (2012)

12. Ploennings, J., Chen, B., Schumann, A., Brady, N.: Exploiting generalized additive models for diagnosing abnormal energy use in buildings. ACM. 978-1-4503-2431-1/13/11(2013)

13. Chulwoo, J., Jae, H.M., Myung, S.K.: A tuning method for the architecture of neural network models incorporating GAM and GA as applied to bankruptcy prediction. Expert Syst. Appl. **39**(3), 3650–3658 (2012)

14. Hastie, T., Tibshirani, R., Friedman, J.: The Elements of Statistical Learning. Springer, New York (2001)

15. Hastie, T., Tibshirani, R.: Generalised Additive Models. Stat. Sci. **1**, 297–318 (1986)

16. Moro, A., Garbinato, B., Chavez-Demoulin, V.: Discovering demographic data of users from the evolution of their spatio-temporal entropy. arXiv: 1803.04240v1 (2018)

17. Matsushima, S.: Statistical learnability of generalized additive models based on total variation regularization, pp. 1–15. Arxiv. 1802.03001v2 (2018)

18. Kazem, A., Sharifi, E., Hussain, F.K., Saberi, M., Hussain, O.K.: Support vector regression with chaos based firefly algorithm for stock market price forecasting. Appl. Soft Comput. **13**, 947–958 (2013)

19. Lu, C.-J.: Hybridising nonlinear independent component analysis and support vector regression with particle swarm optimisation for stock index forecasting. Neural Comput. Appl. **23**, 2417–2427 (2013)

20. Hsu, C.: A hybrid procedure for stock price prediction by integrating self-organizing map and genetic programming. Expert Syst. Appl. **38**, 14026–14036 (2011)

21. Wong, C., Versace, M.: CARTMAP: a neural network method for automated feature selection in financial time series forecasting. Neural Comput. Appl. **21**, 969–972 (2012)

22. Araujo, R.: A morphological perceptron with gradient-based learning for Brazilian stock market forecasting. Neural Netw. **28**, 61–68 (2012)

23. Mohapatra, P., Raj, A.: Indian stock market prediction using differential neural network model. Int. J. Electron. Commun. Comput. Technol. (IJECCT). **2**(4), 159–166 (2012)

24. Araujo, R.: A quantum inspired evolutionary hybrid intelligent approach for stock market prediction. Int. J. Intell. Comput. Cybern. **3**(1), 24–54 (2010)

25. Mostafa, A., Atiya, A.F.: Introduction to financial forecasting. Appl. Intell. **6**, 205–213 (1996)

26. Hall, J.W.: Adaptive selection of U.S stocks with neural nets in trading on the edge: neural, genetic and fuzzy systems for chaotic financial market. Wiley, New York (1994)

27. Konstantinos, V., Marianne, C.: A Bayesian journey to a better research workflow. In: Proceedings of the 17th Python in Science Conference, pp. 137–144 (2018)

28. Murekachiro, D., Mokoteli, T., Vadapalli, H.: Predicting emerging and frontier stock markets using deep neural networks. In: Bi, Y., Bhatia, R., Kapoor, S. (eds.) Intelligent Systems and Applications. IntelliSys 2019. Advances in Intelligent Systems and Computing, vol. 1037. Springer, Cham (2020). https://doi.org/10.1007/978-3-030-29516-5_68

29. Oliveira, F.A., Nobre, E.N., Zarate, L.E.: Applying artificial neural networks to prediction of stock price and improvement of the directional prediction index-case study of PETR4, Petrobas, Brazil. Expert Syst. Appl. **40**, 7596–7606 (2013)

30. Ma, Z., Dai, Q., Liu, N.: several novel evaluation measures for rank based ensemble pruning with applications to time series prediction. Expert Syst. Appl. **42**, 280–292 (2015)

31. Nayak, S.C., Misra, B.B., Behera, H.S.: Artificial chemical reaction optimization of neural networks for efficient prediction of stock market indices. Ain Shams Eng. J. 1–20 (2015)

32. Pastor, L., Veronesi, P.: Political uncertainty and risk premia. J. Financ. Econ. **110**(3), 520–545 (2013)

33. Ahmed, W.: The impact of political regime changes on stock prices: the case of Egypt. Int. J. Emerg. Mark. **12**(3), 508–531 (2017)

34. Nezerwe, Y.: Presidential election and stock returns in Egypt. Rev. Bus. Financ. Stud. **4**(2), 63–68 (2013)
35. Abdelbaki, H.H.: The impact of Arab spring on stock market performance. Br. J. Econ. Manage. Trade **3**(3), 169–185 (2013)
36. Cross, F.: The behaviour of stock prices on Fridays and Mondays. Financ. Anal. J. **29**(6), 6769 (1973)
37. Gibsons, M.R., Hess, P.: Day of the week effects and asset returns. J. Bus. **54**(4), 579–596 (1981)
38. Bayar, A., Kan, O.B.: Day of the week effects: recent evidence from nineteen stock markets. Cent. Bank Rev. **2**, 77–90 (2002)
39. Alagidede, P.: Month of the year and pre-holiday effects in African stock markets. SAJEMS NS **16**(1), 64–74 (2012)

Optimization of Flow Shop Scheduling Through a Hybrid Genetic Algorithm for Manufacturing Companies

Amelec Viloria[1]([⊠]), David Martínez Sierra[2], Sonia Ethel Duran[3],
Etelberto Pallares Rambal[4], Hugo Hernández-Palma[4],
Jairo Martínez Ventura[5], Alberto Roncallo Pichon[6],
and Leidy José Jinete Torres[7]

[1] Universidad de la Costa, St. 58 #66, Barranquilla, Atlántico, Colombia
aviloria7@cuc.edu.co
[2] Universidad Simón Bolívar Barranquilla, Barranquilla, Colombia
dmartinez@unisimonbolivar.edu.co
[3] Fundación Universitaria Unicolombo Internacional, Cartagena, Colombia
sduran@unicolombo.edu.co
[4] Universidad del Atlántico, Puerto Colombia, Colombia
etelbertopallares@uniatlantico.edu.co,
hugohernandezp@mail.uniatlantico.edu.co
[5] Corporación Universitaria Latinoamericana, Barranquilla, Colombia
academico@ul.edu.co
[6] Corporación Universitaria Minuto de Dios – UNIMINUTO, Bogotá, Colombia
alberto.roncallo@uniminuto.edu
[7] Universidad Libre Seccional Barranquilla, Barranquilla, Colombia
josejinete@hotmail.com

Abstract. A task scheduling problem is a process of assigning tasks to a limited set of resources available in a time interval, where certain criteria are optimized. In this way, the sequencing of tasks is directly associated with the executability and optimality of a preset plan and can be found in a wide range of applications, such as: programming flight dispatch at airports, programming production lines in a factory, programming of surgeries in a hospital, repair of equipment or machinery in a workshop, among others. The objective of this study is to analyze the effect of the inclusion of several restrictions that negatively influence the production programming in a real manufacturing environment. For this purpose, an efficient Genetic Algorithm combined with a Local Search of Variable Neighborhood for problems of n tasks and m machines is introduced, minimizing the time of total completion of the tasks. The computational experiments carried out on a set of problem instances with different sizes of complexity show that the proposed hybrid metaheuristics achieves high quality solutions compared to the reported optimal cases.

Keywords: Hybrid Genetic Algorithm · Scheduling · Flow Shop · Variable Neighborhood Search

© Springer Nature Switzerland AG 2020
A. P. Pandian et al. (Eds.): ICICCS 2019, AISC 1039, pp. 20–29, 2020.
https://doi.org/10.1007/978-3-030-30465-2_3

1 Introduction

On many situations, the structured production chain for the sequencing of processes requires that each task to carry out must go through all the productive stages in the same order [1, 2]. This problem is generally known as Regular Flow Variant or Flow Shop Scheduling Problem (FSSP), which, like many others in this field, is difficult to solve and is technically classified as a solution in a non-polynomial time (NP-Hard) [3, 4]. In general, since the 1950s, scientists in the area of Operations Research (OR) have relied on mathematical optimization methods for solving scheduling problems, contributing with a wide variety of methods that include from linear programming to different advanced techniques of Artificial Intelligence (AI) [5, 6]. Most of the methods described above guarantee, in many cases, optimal solutions that are achieved in a reasonable period of time. However, they generally exclude situations that arise in industrial practice. Unfortunately, the problems of scheduling in the real world, when presenting a set of constraints typical of these environments, turn into much more challenging problems, making difficult to apply these methods [7, 8]. Modeling is a topic to consider when solving scheduling problems in a real environment. The types of manufacturing processes with their own particularities make it difficult to build generally applicable models. Some restrictions are difficult to represent mathematically. In addition, finding the right constraints to model reality depends on extensive knowledge of the domain, which may be unavailable to researchers and is only usable for a particular type of production process. Among these kinds of restrictions, the following ones are mentioned [9, 10]:

- Release dates of the machines: Before starting the production process in manufacturing companies, the machines are initially configured according to the type of task they can perform.
- Sequence-dependent configuration times: The operations of reference change in the machines constantly vary. They take longer when the incoming task is very different from the outgoing one, impacting the total performance of the operations.
- Transportation times: The tasks need a time interval to be transported from one machine to another, making companies revise the policy of generating good production schedules. In turn, it reinforces the need to automate problem modeling by applying algorithms that can be adapted to the characteristics of a real environment. All the above illustrates the need to search for other strategies, as well as new and efficient algorithms to solve the problem. For this reason, this paper presents a proposal for a Hybrid Genetic Algorithm which is combined with a Local Search of Variable Neighborhood for the solution of the FSSP under the effect of a set of restrictions that frequently occur in real manufacturing environments. The proposed algorithm is tested with several instances of the problem showing that it achieves excellent solutions compared to the optimal ones.

2   Theoretical Review

2.1   Definition of the Scheduling Problem

Within the scheduling theory, a large number of problems can be distinguished, in which there is a set of N tasks that have to be processed on a set of M resources or physical machines following a flow pattern or technological route [11–13]. Frequently, these processes must be executed for all tasks in the same order, implying that they follow the same flow pattern, as happens in various real manufacturing environments. This problem is referred to as the Regular Flow Variant or FSSP. The restrictions identified and to which the FSSP is subject are the following [14, 15]:

- Only one machine tool of each type per stage is available.
- Technological restrictions are well defined and are previously known, and are inviolable.
- It is not allowed that two operations of the same task are processed simultaneously.
- No task can be processed more than once on the same machine.
- Each task is processed until it is finished, once an operation is started, it is interrupted only when completed.
- No machine can process more than one task at a time.
- The release dates of the machines are taken into account.
- Configuration times that depend on the sequence are known in advance.
- The time of transportation of the task between stages is considered.

The objective is to find a sequence of tasks by stages under the restriction that the processing of each task has to be continuous with respect to the objective of minimizing the makespan or *Cmax* as it is also known. Thus [16–18]:

If we have a $r(j)$ as the time of release of the machine j, $t(i, j, l)$ as the time of transportation of the task i from the machine j to the machine l, $p(i, j)$ as the time of task processing i on the machine j, $s(i, k, j)$ as the time of configuration dependent on the sequence between task i and task on the machine j, and a permutation of tasks ($J1, J2,$ ..., Jn), then the total completion time denoted by *Cmax* is calculated as follows:

$$C(J_1, 1) = r(1) + p(J_1, 1) \tag{1}$$

$$C(J_i, 1) = s(J_{i-1}, J_i, 1) + C(J_{i-1}, 1) + p(J_i, 1) \ for \ i = 2, \ldots, n \tag{2}$$

$$C(J_1, j) = max\{r(j), C(J_1, j-1) + t(i, j-1, j) + p(J_1, j)\} \ for \ j = 2, \ldots, m \tag{3}$$

$$C(J_i, j) = max\{C(J_{i-1}, j) + s(J_{i-1}, J_i, j), \ C(J_i, j-1) + t(i, j-1, j) + p(J_i, j)\}$$
$$for \ i = 2, \ldots, n; \ for \ j = 2, \ldots, m \tag{4}$$

$$C_{max} = C(J_n, m) \tag{5}$$

Under these conditions, the total processing time corresponds to the task completion time of the last task in the last machine. In other words, it is the time necessary to complete all the tasks [19].

3   Materials and Methods

The solution of the FSSP under the effect of real restrictions of the production using the Genetic Algorithms (GA) begins with the design of the solution representation of the given problem. The representation of the possible solutions within the search space of a problem defines the structure of the chromosome that will be manipulated by the algorithm. There are different types of representations, the choice of use will always depend on the characteristics of the problem to be solved [20].

In the case of the problem focused in this study, a representation was selected similar to that used by [21], according to which the chromosome represents the natural sequence in which the tasks are processed. Figure 1 shows the chromosome that represents a solution for the FSSP where 8 tasks are processed. The sequence of numbers represents the task number and the order in which they will be processed. The aptitude of each chromosome will be given by the calculation of C_max from the task sequence represented in it.

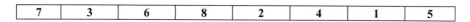

| 7 | 3 | 6 | 8 | 2 | 4 | 1 | 5 |

Fig. 1. Representation of order

The workflow of the proposed Hybrid Genetic Algorithm (AGH$_{VNS}$) is presented through a block diagram in Fig. 2 [22].

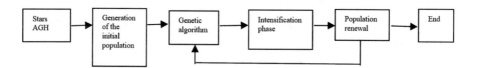

Fig. 2. Workflow of AGH$_{VNS}$

When the initial population (*pobini*) is generated, the objective is to acquire a diversified population. In the proposed AGH$_{VNS}$, the initial population is generated randomly and the size is variable depending on the number of tasks (n) and machines (m) where $pobini = n * m$. In step 2, a basic GA is used to improve the population. In our implementation, the selection of parents is based on the classic selection scheme by Ranking where $pobini/2$ pairs of parents are selected. For the selection of each pair of parents, the chromosomes are matched in couples from greater to lower fitness [23]. After this, the crossing of each mating is done. In this case, the Two Crossing Point method is applied. It is necessary to mention that when applying this method directly, illegal chromosomes can be generated (in our case it means that a task can be on a single chromosome twice) [24]. This problem is corrected using the methodology studied in [25] which is based on the substitution of the tasks that are repeated by those that are not in the mentioned chromosome. After the combination, the mutation process would be carried out, which consists of selecting two points of the chromosome randomly and changing these task positions. Figure 3 shows this process [26].

Genes to mutate

Descendant	7	3	6	8	2	4	1	5

Descendant mutate	7	4	6	8	2	3	1	5

Fig. 3. Mutation per Value for the FSSP

[27] emphasize the importance of the balance between the search of the GA and the local search to find high quality FSSP solutions in acceptable computational times. In step 3 of the algorithm, a Local Variable Neighborhood Search is applied to explore the neighborhood of a subset of the population complementing the GA. In each generation, the *pobini/2* individuals obtained in step 2 are selected. This intensification operates on each of the selected individuals by applying a neighborhood operator which is chosen randomly. The value and detail of each operator is described as follows [28, 29]:

1. **Exchange Operator:** It works similarly to the mutation operator.
2. **Transfer or insertion operator:** A task is selected randomly and moved to another position of the randomly selected chromosome.
3. **Operator 2-Op:** Two consecutive tasks are selected randomly, then a position is randomly selected inside the chromosome and these are moved to that position.
4. No operation is performed on the chromosome.

The main idea behind this intensification is to diversify the population and achieve a balance between exploration and exploitation.

Finally, in step 4 of the algorithm, the population is renewed from the new individuals generated. Specifically, the *pobini/2* less fitted individuals of the current population are replaced by the solutions obtained in the intensification phase.

4   Analysis and Results

In order to analyze the performance of the proposed algorithm, 10 instances are used. If the search space for the FSSP is $n!$, these instances were created with small dimensions in order to perform an exhaustive search in this space to determine the optimal solution and compare these values to those obtained by the AGH$_{VNS}$.

In the same way, these results are compared with the variant of the GA proposed in this study excluding the intensification phase. In this way, the effect of this phase on the solution of the problem is analyzed. The dimension of the instances is: $5 \times 3, 5 \times 4, 5 \times 5, 7 \times 6, 7 \times 7, 8 \times 8, 9 \times 4, 9 \times 9, 10 \times 8$ and 10×10. Random numbers were generated to create the processing times, sequence-dependent configuration times, release dates, and transportation times. Each instance was executed 10 times. To determine the quality of the solutions, the Relative Mean Error (RME) is defined as [30]:

$$ERM = \sum_{I=1}^{10} \left[\frac{MK_i - UB_i}{UB_i} * 100 \right] / 10 \tag{6}$$

Where MK is the solution obtained by the algorithm and is the optimum obtained by the exhaustive search. AGH_{VNS} was implemented in Java and all executions were performed on a Pentium IV PC at 2.4 GHz with 1 GB of RAM.

Calibration of Parameters. A large number of executions revealed that using the number of AGH_{VNS} generations as a stop condition was not adequate to control its convergence. In many cases, the algorithm reached a local optimum before completing the number of generations. As a result, a maximum processing time was introduced depending on the size of the problem to be solved. It was arbitrarily defined as $tmax = n * m/5$ s, which in computational terms is a practical time and, in all cases, allows the algorithm to finish in a shorter time than defining the number of generations as stop condition. The values of the other main parameters of the AGH_{VNS} are summarized in Table 1. These values are taken from [31] and [32], in which the authors conduct a study of their behavior in the solution of the classic FSSP.

Table 1. Initial parameters of the proposed GA.

Parameter	Value
Crossing factor	0.62
Mutation factor	0.045
Selection method	*Ranking*
Crossing method	Two crossing points
Mutation method	Value mutation
Type of model	Stationary model

To illustrate the problem that is solved in this research, an instance example is introduced. It describes a problem with 5 tasks that must be processed on 5 machines (5 × 5). Table 2 shows the processing time of the task on the machines.

Table 2. Processing times

Task/machine	Processing times				
	J0	J1	J2	J3	J4
M0	10	6	11	8	11
M1	15	9	14	10	14
M2	12	11	9	10	6
M3	8	4	8	9	12
M4	6	6	8	6	3

Table 2 is interpreted as the time it takes a task to be processed in each of the machines. For example, task J0 needs 10, 15, 8, 12 and 6 units of time to be processed in each of the 5 machines.

As mentioned above, given the lack of instances of the problem addressed in this research, instances of different sizes of complexity were randomly generated. Specifically, 10 instances (nxm) of $5 \times 3, 5 \times 4, 5 \times 5, 7 \times 6, 7 \times 7, 8 \times 8, 9 \times 4, 9 \times 9, 10 \times 8$ and 10×10 were created, where n represents the number of tasks and m the number of machines respectively. The proposed algorithm was executed 10 times [31] for each of them and the average value of makespan ($Cmax(prom)$) is calculated. In addition, in order to analyze the performance of the proposed algorithm, the intensification process was disabled obtaining a basic GA, which, like the AGHVNS, was executed 10 times for each of the instances. These results were taken into account for the comparative study. Also, given that these instances were intentionally generated with a small level of complexity (given by the number of tasks and machines), the optimal value of $Cmax$ was determined for each of them by performing an exhaustive search. The RME was calculated from these values and those obtained by the two variants of GAs. Table 3 summarizes the results obtained. This also shows the lowest value of makespan ($Cmax(min)$) obtained in the executions for both variants. The last row shows the average of RME.

Table 3. Summary of the results obtained.

Instance	Optimum	AGHVNS		RME (%)	AG		RME (%)
		$Cmax(min)$	$Cmax(prom)$		$Cmax(min)$	$Cmax(prom)$	
5×3	103	103	103.00	0.0000	103	103.00	0.0000
5×4	117	117	117.00	0.0000	117	117.90	0.0077
5×5	140	140	140.00	0.0000	144	144.00	0.0286
7×6	207	207	207.00	0.0000	208	208.50	0.0072
7×7	236	236	236.00	0.0000	236	236.80	0.0034
8×8	256	257	257.00	0.0039	262	262.90	0.0270
9×4	185	185	185.80	0.0043	193	194.30	0.0503
9×9	292	292	293.80	0.0062	300	302.00	0.0342
10×8	289	290	290.40	0.0048	296	297.10	0.0245
10×10	338	339	339.90	0.0056	347	347.00	0.0266
Average:				**0.0025**			**0.0209**

Based on the results from Table 3, it can be seen that the proposed algorithm is capable of obtaining good results and its applicability to scheduling problems is shown. Restrictions that occur frequently in real manufacturing environments are taken into account. The results correspond to what is stated in the specialized literature if considering the gap that exists between production programming problems that are commonly solved by specialists in the subject and those that are presented in practice.

For the 10 proposed instances, except for the instance of 8×8, 10×8, and 10×10, the AGHVNS reached the optimal value of makespan. For the first five instances, it achieved the optimum value in all its executions. The average RME was just 0.0025%.

In addition, the GA without the intensification phase just reached the optimal value in the first two instances although for the rest the obtained values of makespan were close to the optimum. The average RME was 0.0209%. The results obtained in Table 3 show the importance of achieving a balance between exploration and exploitation from the intensification phase. It allowed the proposed algorithm in the search process in that space to achieve better solutions than the GA variant without the local search process. This statement is evidenced when establishing a comparison between the average of RME between the AGHVNS and the basic GA. The RME of the AGHVNS, in all cases, was lower than that obtained by the basic GA, and this one was well below 1%.

5   Conclusions

This paper presented the application of a hybrid method for the solution of the FSSP under the effect of restrictions that occur in real production environments. The proposed algorithm is constituted by three main components: a generation of the initial population; a GA and an intensification process through a local search named VNS. The AGHVNS was evaluated through a set of ten instances of problems and the results obtained were compared to the optimal values and those obtained by the basic variant of the GA demonstrating its efficiency and effectiveness in terms of the quality of the solutions. It is important to mention that the proposed metaheuristics constitutes an interesting alternative to solve complex optimization problems. It should also be noted that the proposed algorithm is simple and easy to implement.

References

1. Rossit, D.A., Tohmé, F., Frutos, M.: The non-permutation flow-shop scheduling problem: a literature review. Omega **77**, 143–153 (2017)
2. Wilson, J.M.: Alternative formulations of a flow-shop scheduling problem. J. Oper. Res. Soc. **40**(4), 395–399 (1989)
3. Neufeld, J.S., Gupta, J.N.D., Buscher, U.: A comprehensive review of flowshop group scheduling literature (2016)
4. Ronconi, D.P., Birgin, E.G.: Mixed-integer programming models for flowshop scheduling problems minimizing the total earliness and tardiness, pp. 91–105. Springer, New York (2012)
5. Reza Hejazi, S., Saghafian, S.: Flowshop-scheduling problems with makespan criterion: a review. Int. J. Prod. Res. **43**(14), 2895–2929 (2005)
6. Phanden, R.K., JainJan, A.: Assessment of makespan performance for flexible process plans in job shop scheduling. IFAC-PapersOnLine **48**(3), 1948–1953 (2015)
7. Semančo, P., Modrák, V.: A comparison of constructive heuristics with the objective of minimizing makespan in the flow-shop scheduling problem. Acta Polytech. Hungarica **9**(5), 177–190 (2012)
8. Palmer, D.S.: Sequencing jobs through a multi-stage process in the minimum total time—a quick method of obtaining a near optimum. J. Oper. Res. Soc. **16**(1), 101–107 (1965)
9. Gupta, J.N.D.: A heuristic algorithm for the flowshop scheduling problem. Rev. Fr. d'Autom. Inform. Rech. Oper. **10**(2), 63–73 (1976)

10. Gupta, J.N.D.: A functional heuristic algorithm for the flowshop scheduling problem. J. Oper. Res. Soc. **22**(1), 39–47 (1971)
11. Nawaz, M., Enscore, E.E., Ham, I.: A heuristic algorithm for the m-machine, n-job flowshop sequencing problem. Omega **11**(1), 91–95 (1983)
12. Campbell, H.G., Dudek, R.A., Smith, M.L.: A heuristic algorithm for the N job, M machine sequencing problem* f, no. 10 (1970)
13. Alharkan, I.M.: Algorithms for sequencing and scheduling
14. Pugazhenthi, R., Anthony Xavior, M., Somasundharam, E.: Minimizing makespan of a permutation flowshop by, pp. 110–112 (2014)
15. Prodromidis, A., Chan, P.K., Stolfo, S.J.: Meta learning in distributed data mining systems: issues and approaches. In: Kargupta, H., Chan, P. (eds.) Book on Advances in Distributed and Parallel Knowledge Discovery. AAAI/MIT Press (2000)
16. Parthasarathy, S., Zaki, M.J., Ogihara, M.: Parallel data mining for association rules on shared-memory systems. Knowl. Inform. Syst. Int. J. **3**(1), 1–29 (2001)
17. Grossman, R.L., Bailey, S.M., Sivakumar, H., Turinsky, A.L.: Papyrus: a system for data mining over local and wide area clusters and super-clusters. In: Proceedings of ACM/IEEE Conference on Supercomputing, Article 63, pp. 1–14 (1999)
18. Chattratichat, J., Darlington, J., Guo, Y., Hedvall, S., Kohler, M., Syed, J.: An architecture for distributed enterprise data mining. In: Proceedings of 7th International Conference on High Performance Computing and Networking, Netherlands, 12–14 April, pp. 573–582 (1999)
19. Wang, L., Tao, J., Ranjan, R., Marten, H., Streit, A., Chen, J., Chen, D.: G-Hadoop: MapReduce across distributed data centers for data-intensive computing. Future Gener. Comput. Syst. **29**(3), 739–750 (2013)
20. Butenhof, D.R.: Programming with POSIX Threads. Addison-Wesley Longman Publishing Company, USA (1997)
21. Bhaduri, K., Wolf, R., Giannella, C., Kargupta, H.: Distributed decision-tree induction in peer-to-peer systems. Stat. Anal. Data Min. **1**(2), 85–103 (2008)
22. Rafailidis, D., Kefalas, P., Manolopoulos, Y.: Preference dynamics with multimodal user-item interactions in social media recommendation. Expert Syst. Appl. **74**(1), 11–18 (2017)
23. Izquierdo, N.V., Lezama, O.B.P., Dorta, R.G., Viloria, A., Deras, I., Hernández-Fernández, L.: Fuzzy Logic applied to the performance evaluation. honduran coffee sector case. In: Tan, Y., Shi, Y., Tang, Q. (eds.) Advances in Swarm Intelligence, ICSI 2018. Lecture Notes in Computer Science, vol. 10942 (1), pp. 1–12. Springer, Cham (2018)
24. Lezama, O.B.P., Izquierdo, N.V., Fernández, D.P., Dorta, R.L.G., Viloria, A., Marín, L.R.: Models of multivariate regression for labor accidents in different production sectors: comparative study. In: International Conference on Data Mining and Big Data, vol. 10942 (1), pp. 43–52. Springer, Cham (2018)
25. Suárez, J.A., Beatón, P.A., Escalona, R.F., Montero, O.P.: Energy, environment and development in Cuba. Renew. Sustain. Energy Rev. **16**(5), 2724–2731 (2012)
26. Sala, S., Ciuffo, B., Nijkamp, P.: A systemic framework for sustainability assessment. Ecol. Econ. **119**(1), 314–325 (2015)
27. Singh, R.K., Murty, H.R., Gupta, S.K., Dikshit, A.K.: An overview of sustainability assessment methodologies. Ecol. Ind. **9**(2), 189–212 (2009)
28. Varela, N., Fernandez, D., Pineda, O., Viloria, A.: Selection of the best regression model to explain the variables that influence labor accident case electrical company. J. Eng. Appl. Sci. **12**(1), 2956–2962 (2017)
29. Yao, Z., Zheng, X., Liu, C., Lin, S., Zuo, Q., Butterbach-Bahl, K.: Improving rice production sustainability by reducing water demand and greenhouse gas emissions with biodegradable films. Sci. Rep. **7**(1), 1–12 (2017)

30. Suárez, D.F.P., Román, R.M.S.: Consumo de água em arroz irrigado por inundação em sistema de multiplas entradas. IRRIGA **1**(1), 78–95 (2016)
31. Amelec, V.: Increased efficiency in a company of development of technological solutions in the areas commercial and of consultancy. Adv. Sci. Lett. **21**(5), 1406–1408 (2015)
32. Amelec, V.: Validation of strategies to reduce exhausted shelf products in a pharmaceutical chain. Adv. Sci. Lett. **21**(5), 1403–1405 (2015)

Rational Approximation of Fractional-Order System with Multiple Fractional Powered Terms - A Comparative Study

Jaydeep Swarnakar$^{(\boxtimes)}$ and Wandarisa Sungoh

Department of Electronics and Communication Engineering, School of Technology, North-Eastern Hill University, Shillong 793022, Meghalaya, India
jaydeepswarnakar@gmail.com

Abstract. Rational approximation of the infinite dimensional fractional-order system (FOS) is necessary for their practical implementations. In this paper, two well known continuous-time approximation methods, namely, Charef approximation method and Oustaloup approximation method have been individually applied on a fractional-order transfer function having multiple fractional powered terms to obtain two different integer order approximants of the same FOS. The frequency response and the time response resulted from both the approximation methods have been compared subsequently using simulation results.

Keywords: Fractional-Order System (FOS) · Charef approximation · Oustaloup approximation

1 Introduction

The initiation of the fractional calculus emanated from the communication between two eminent mathematicians Leibniz and L'Hospital over three hundred years back [1]. Since then, the subject has become popular more and more among the research community due to its inherent ability to express the natural phenomena more accurately as compared to the integer order calculus. In recent era, the manifestations of the fractional-order systems (FOS) have been widespread to model several dynamical processes related to science and engineering. For example, fractional-order controllers like CRONE controller, $PI^{\lambda}D^{\mu}$ controller etc. have substantiated their superior performance over classical PID controller through incorporating extra tuning parameters [2]. The infinite dimensional FOSs are having infinite memory and therefore, it is indispensable to convert them within a finite dimension for practical applications. This calls for the rational approximation of the FOS. The idea of finding rational approximation of FOS aims to obtain the suitable integer order model that demonstrates analogous frequency response and time response characteristics like original FOS. Many methods have been developed for the continuous-time approximation of the fractional operator in s-domain as pointed out in [3, 4]. Out of them, two recognized methods namely Charef method [5–9] and Oustaloup method [10–14] have been chosen to apply separately on a class of FOS containing multiple fractional powered terms. Both the approximation methods are then compared based on the frequency

© Springer Nature Switzerland AG 2020
A. P. Pandian et al. (Eds.): ICICCS 2019, AISC 1039, pp. 30–37, 2020.
https://doi.org/10.1007/978-3-030-30465-2_4

response and time response behaviour. The paper is organized in four sections. Section 1 gives the introduction. Section 2 elaborates the mathematical background of Charef method and Oustaloup method for approximating fractional-order operator. In Sect. 3, simulation results are presented by taking benchmark example from the literature. The frequency characteristics, error analysis and step responses obtained from two methods are compared with respect to the original FOS. Section 4 draws the final conclusion of the entire work.

2   Integer-Order Approximation of FOS

Let, the transfer function of the FOS with multiple fractional powered terms is represented as below:

$$T(s) = \frac{A}{\alpha_n s^{m_c} + \alpha_{n-1} s^{m_{c-1}} + \alpha_{n-2} s^{m_{c-2}} + \ldots + \alpha_1 s^{m_1} + \alpha_0} \tag{1}$$

where $s^{m_i} (i = 1, 2, 3\ldots c)$ are the different fractional powered terms, $\alpha_k (k = 0, 1, 2\ldots\ldots n)$ are the corresponding real coefficients and A represents the d.c gain. $T(s)$ is approximated using Charef method and Oustaloup method to obtain $T_{CA}(s)$ and $T_{OA}(s)$ respectively. To obtain the overall approximation of $T(s)$, each fractional powered term is to be approximated individually. The mathematical detail of approximating each fractional powered term using both the methods are given in the succeeding subsections.

2.1   Charef Approximation Method

The transfer function of the fractional-order operator s^m is represented as below:

$$G(s) = s^m \quad (0 < m < 1) \tag{2}$$

As mentioned in [7], Eq. (2) can be modelled by the Eqs. (3) and (4).

$$s^m \approx G'(s) = \left(\omega_{lo}\sqrt{10^{e/10m} - 1}\right)^m \left(1 + \frac{s}{\omega_{lo}\sqrt{10^{e/10m} - 1}}\right)^m \tag{3}$$

$$s^m \approx G_{CH}(s) = \left(\omega_{lo}\sqrt{10^{e/10m} - 1}\right)^m \frac{\prod\limits_{i=0}^{L}\left(1 + \frac{s}{z_i}\right)}{\prod\limits_{i=0}^{L}\left(1 + \frac{s}{p_i}\right)} \tag{4}$$

$$L = \text{integer}\left[\frac{\log(100\omega_{hi}/z_0)}{\log(\beta\chi)}\right] + 1 \tag{5}$$

The parameters needed in Eqs. (4) and (5) are listed as below:

$$\beta = 10^{n/10(1-m)}, \chi = 10^{n/10m}, z_0 = \left(\omega_{lo}\sqrt{10^{e/10m}-1}\right) \times 10^{n/20m}, p_0 = \beta z_0$$

$$p_i = p_0(\beta\chi)^i \ (i = 0,\ 1,\ 2\ldots\ldots L) \text{ and } z_i = z_0(\beta\chi)^i \ (i = 0,\ 1,\ 2\ldots\ldots L).$$

The approximation of s^m is obtained within the frequency range $(\omega_{lo}, \omega_{hi})$, e represents the maximum error between the fractional-order slopes as depicted in Eqs. (2) and (3), n is the prescribed error in dB between the fractional-order models represented by Eqs. (3) and (4) respectively.

2.2   Oustaloup Approximation Method

The Lth order approximation of s^m $(0<m<1)$ within a frequency band $(\omega_{lo}, \omega_{hi})$ obtained using Oustaloup approximation method is given as below [11–13]:

$$s^m \approx G_{OUS}(s) = \omega_{hi}^m \prod_{j=1}^{L} \frac{s+\omega'_j}{s+\omega_j} \tag{6}$$

where,

$$\omega_{uo} = \sqrt{\omega_{hi}/\omega_{lo}}, \ \omega'_j = \omega_{lo}\omega_{uo}^{(2j-1-m)/L} \text{ and } \omega_j = \omega_{lo}\omega_{uo}^{(2j-1+m)/L}$$

3   Simulations and Results

Let us consider the following FOS given as below [15]:

$$T(s) = \frac{A}{\alpha_2 s^{m_2} + \alpha_1 s^{m_1} + \alpha_0} \tag{7}$$

Taking $A = 5$, $\alpha_2 = 1$, $\alpha_1 = 1.3$, $\alpha_0 = 1.25$, $m_2 = 2.3$ and $m_1 = 0.9$,   following fractional-order transfer function is resulted as shown below:

$$T(s) = \frac{5}{s^{2.3} + 1.3s^{0.9} + 1.25} \tag{8}$$

Now, $s^{2.3} = s^2 \times s^{0.3}$. Thererefore, Eq. (8) is having two fractional powered terms $s^{0.3}$ and $s^{0.9}$ which are approximated using Charef method and Oustaloup method as per the methodology described in the previous section to obtain the two different approximated versions of $T(s)$, i.e. $T_{CA}(s)$ and $T_{OA}(s)$ respectively. Considering, $\omega_{lo} = .001$ rad/s, $\omega_{hi} = 1000$ rad/s, $e = 10^{-5}$ and $n = 3$ dB, we obtain $\beta = 2.6827$, $\chi = 10$, $z_0 = 8.761 \times 10^{-6}$, $p_0 = 2.35 \times 10^{-5}$ and $L = 8$ while approximating $s^{0.3}$ $(m = 0.3)$ using Charef method as deliberated from Eqs. (2) to (5). Under the similar condition, we obtain $\beta = 1000$, $\chi = 2.1544$, $z_0 = 2.348 \times 10^{-6}$, $p_0 = 2.348 \times 10^{-3}$ and $L = 4$ while approximating $s^{0.9}$ ($m = 0.9$) using Charef method. Using Eq. (4), the Charef approximation of $s^{0.3}$ and $s^{0.9}$ have finally resulted ninth order and fifth order transfer functions for $L = 8$ and $L = 4$ respectively as shown in Table 1.

Table 1. Rational approximation of $s^{0.3}$ & $s^{0.9}$ obtained using Charef method

Fractional powered terms	Charef approximation
$s^{0.3}$	$\dfrac{\begin{aligned}&1.263 \times 10^{8} s^{9} + 3.804 \times 10^{14} s^{8} + 2.705 \times 10^{19} s^{7} + 8.835 \times 10^{22} s^{6} + 1.076 \times 10^{25} s^{5} \\ &+ 4.88 \times 10^{25} s^{4} + 8.255 \times 10^{24} s^{3} + 5.205 \times 10^{22} s^{4} + 1.222 \times 10^{19} s + 1.03 \times 10^{14}\end{aligned}}{\begin{aligned}&8.157 \times 10^{5} s^{9} + 5.342 \times 10^{12} s^{8} + 1.257 \times 10^{18} s^{7} + 1.102 \times 10^{22} s^{6} + 3.598 \times 10^{24} s^{5} \\ &+ 4.379 \times 10^{25} s^{4} + 1.987 \times 10^{25} s^{3} + 3.361 \times 10^{23} s^{2} + 2.116 \times 10^{20} s + 4.789 \times 10^{15}\end{aligned}}$
$s^{0.9}$	$\dfrac{\begin{aligned}&9.336 \times 10^{14} s^{5} + 4.725 \times 10^{22} s^{4} + 1.109 \times 10^{27} s^{3} \\ &+ 1.209 \times 10^{28} s^{2} + 6.114 \times 10^{25} s + 1.435 \times 10^{20}\end{aligned}}{\begin{aligned}&1.537 \times 10^{5} s^{5} + 7.776 \times 10^{15} s^{4} + 1.826 \times 10^{23} s^{3} \\ &+ 1.99 \times 10^{27} s^{2} + 1.006 \times 10^{28} s + 2.361 \times 10^{25}\end{aligned}}$

Using Table 1 and Eq. (8), the Charef approximation of $T(s)$ is obtained as below:

$$T_{CA}(s) \approx \frac{\begin{aligned}&6.267 \times 10^{11} s^{14} + 3.172 \times 10^{22} s^{13} + 9.523 \times 10^{29} s^{12} + 4.934 \times 10^{36} s^{11} + 1.201 \times 10^{42} s^{10} \\ &+ 2.256 \times 10^{46} s^{9} + 1.129 \times 10^{50} s^{8} + 3.638 \times 10^{52} s^{7} + 6.167 \times 10^{53} s^{6} + 2.402 \times 10^{54} s^{5} \\ &+ 1.008 \times 10^{54} s^{4} + 1.926 \times 10^{52} s^{3} + 5.033 \times 10^{49} s^{2} + 2.523 \times 10^{46} s + 5.654 \times 10^{41}\end{aligned}}{\begin{aligned}&1.941 \times 10^{13} s^{16} + 9.823 \times 10^{23} s^{15} + 2.546 \times 10^{31} s^{14} + 5.676 \times 10^{37} s^{13} + 5.553 \times 10^{42} s^{12} \\ &+ 6.996 \times 10^{46} s^{11} + 1.78 \times 10^{50} s^{10} + 2.231 \times 10^{52} s^{9} + 2.107 \times 10^{53} s^{8} + 6.366 \times 10^{53} s^{7} \\ &+ 9.556 \times 10^{53} s^{6} + 9.174 \times 10^{53} s^{5} + 2.59 \times 10^{53} s^{4} + 4.845 \times 10^{51} s^{3} + 1.26 \times 10^{46} s^{2} \\ &+ 6.308 \times 10^{45} s + 1.413 \times 10^{41}\end{aligned}} \quad (9)$$

Again, considering $\omega_{lo} = .001$ rad/s and $\omega_{hi} = 1000$ rad/s, the Oustaloup approximation of $s^{0.3}$ ($m = 0.3$) and $s^{0.9}$ ($m = 0.9$) are obtained using Eq. (6) by maintaining the parity in order of approximation as obtained from Charef method. Ninth order and fifth order approximation are derived accordingly for the fractional power terms $s^{0.3}$ and $s^{0.9}$ using Oustaloup method as shown in Table 2.

Table 2. Rational approximation of $s^{0.3}$ and $s^{0.9}$ obtained using Oustaloup method

Fractional powered terms	Oustaloup approximation
$s^{0.3}$	$\dfrac{\begin{aligned}&7.943 s^{9} + 3733 s^{8} + 3.109 \times 10^{5} s^{7} + 5.375 \times 10^{6} s^{6} + 1.986 \times 10^{7} s^{5} \\ &+ 1.577 \times 10^{7} s^{4} + 2.694 \times 10^{6} s^{3} + 9.833 \times 10^{4} s^{2} + 744.8 s + 1\end{aligned}}{\begin{aligned}&s^{9} + 744.8 s^{8} + 9.833 \times 10^{4} s^{7} + 2.694 \times 10^{6} s^{6} + 1.577 \times 10^{7} s^{5} \\ &+ 1.986 \times 10^{7} s^{4} + 5.375 \times 10^{6} s^{3} + 3.109 \times 10^{5} s^{2} + 3733 s + 7.943\end{aligned}}$
$s^{0.9}$	$\dfrac{501.2 s^{5} + 3.875 \times 10^{4} s^{4} + 1.778 \times 10^{5} s^{3} + 5.129 \times 10^{4} s^{2} + 929.6 s + 1}{s^{5} + 929.6 s^{4} + 5.129 \times 10^{4} s^{3} + 1.778 \times 10^{5} s^{2} + 3.875 \times 10^{4} s + 501.2}$

Using Table 2 and Eq. (8), the Oustaloup approximation of $T(s)$ is obtained as below:

$$T_{OA}(s) = \frac{\begin{array}{l} 5s^{14} + 8372s^{13} + 4.21 \times 10^6 s^{12} + 6.624 \times 10^8 s^{11} + 3.848 \times 10^{10} s^{10} + 8.518 \times 10^{11} s^9 \\ + 6.552 \times 10^{12} s^8 + 1.967 \times 10^{13} s^7 + 2.21 \times 10^{13} s^6 + 8.746 \times 10^{12} s^5 + 1.369 \times 10^{12} s^4 \\ + 7.704 \times 10^{10} s^3 + 1.51 \times 10^9 s^2 + 1.089 \times 10^7 s + 1.991 \times 10^4 \end{array}}{\begin{array}{l} 7.943 s^{16} + 1.112 \times 10^4 s^{15} + 4.189 \times 10^6 s^{14} + 4.878 \times 10^8 s^{13} + 2.173 \times 10^{10} s^{12} \\ + 3.566 \times 10^{11} s^{11} + 2.179 \times 10^{12} s^{10} + 6.201 \times 10^{12} s^9 + 1.018 \times 10^{13} s^8 + 1.194 \times 10^{13} s^7 \\ + 8.256 \times 10^{12} s^6 + 2.646 \times 10^{12} s^5 + 3.704 \times 10^{11} s^4 + 1.989 \times 10^{10} s^3 + 3.828 \times 10^8 s^2 \\ + 2.738 \times 10^6 s + 4987 \end{array}} \quad (10)$$

Thus, Eqs. (9) and (10) represent the two seperate approximated transfer functions $T_{CA}(s)$ and $T_{OA}(s)$ of the original fractional-order transfer function $T(s)$. The frequency responses of $T(s)$, $T_{CA}(s)$ and $T_{OA}(s)$ have been plotted together in Fig. 1. It is obvious that the Oustaloup method has produced commendable magnitude approximation and phase approximation in the entire frequency band, whereas, the Charef approximation seems to be consistent in the mid frequency and high frequency range. In the low frequency part, Charef approximation deviates from the frequency response of the original FOS and thus becomes exposed to magnitude error and phase error. The magnitude error and phase error resulted from two methods have been plotted in Fig. 2. The maximum absolute magnitude error and phase error for both the approximation methods have been listed in Table 3 to compare both the methods. The magnitude error and phase error are less in Oustaloup approximation as compared to the Charef approximation over the whole frequency range. Similarly, the step responses of $T(s)$, $T_{CA}(s)$ and $T_{OA}(s)$ have been plotted together in Fig. 3. Again, Oustaloup approximation performs better as compared to Charef approximation as the similarity between the step responses of $T(s)$ and $T_{OA}(s)$ is more as compared to the similarity found between the step responses of $T(s)$ and $T_{CA}(s)$. Therefore, Oustaloup method delivers the better approximation of the original FOS with multiple fractional powered terms as compared to the Charef approximation.

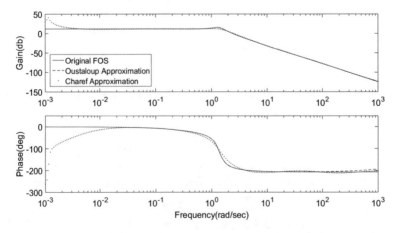

Fig. 1. Comparison of frequency responses using two approximation methods

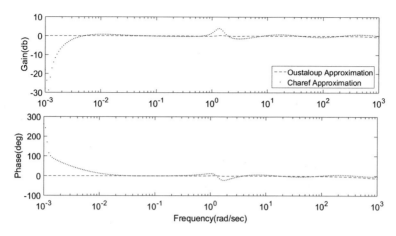

Fig. 2. Error graphs obtained using two approximation methods

Table 3. Maximum errors obtained in the frequency range .001 rad/s to 1000 rad/s employing both the approximation methods

Method	Maximum magnitude error in dB (absolute value)	Maximum phase error in degree (absolute value)
Oustaloup	0.763	13.57
Charef	28.76	262

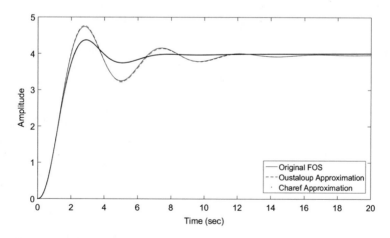

Fig. 3. Comparison of step response curves using two approximation methods

4 Conclusions

In this paper, two rational approximation methods have been compared while approximating a class of FOS containing multiple fractional powered terms. The work has been done in two stages. In the first stage, the fractional powered terms are approximated individually using Charef method and Oustaloup method. In the later stage, the individual integer order approximants generated from the respective methods are combined to find out the rational approximation of the overall FOS. Simulation results have been analysed and it has been established that the Oustaloup method has produced better approximation of the original FOS as compared to Charef method both in the frequency domain and also in the time domain.

References

1. Das, S.: Functional Fractional Calculus for System Identification and Controls. Springer, Heidelberg (2008)
2. Chen, Y.Q., Petras, I., Xue, D.: Fractional-order control—a tutorial. In: Proceeding of the American Control Conference (ACC), pp. 1397–1411, IEEE, St. Louis (2009)
3. Vinagre, B.M., Podlubny, I., Hernandes, A., Feliu, V.: Some approximations of fractional-order operators used in control theory and applications. Fract. Calc. Appl. Anal. 3(3), 231–248 (2000)
4. Shrivastava, N., Varshney, P.: A new improved technique for frequency band implementation of fractional-order functions. Int. J. Math. Models Methods Appl. Sci. 12, 185–193 (2018)
5. Charef, A., Sun, H.H., Tsao, Y.Y., Onaral, B.: Fractal system as represented by singularity function. IEEE Trans. Autom. Control 37(9), 1465–1470 (1992)
6. Ladaci, S., Charef, A.: On fractional adaptive control. Nonlinear Dyn. 43(4), 365–378 (2006)
7. Charef, A.: Analogue realisation of fractional-order integrator, differentiator and fractional $PI^\lambda D^\mu$ controller. In: IEE Proceeding-Control Theory Applications 153(6), 714–720 (2006)
8. Mansauria, R., Bettayeb, M., Djennoune, S.: Approximation of high order integer systems by fractional-order reduced-parameters models. Math. Comput. Model. 51(1–2), 53–62 (2010)
9. Meg, L., Xue, D.: An Approximation algorithm of fractional-order pole models based on an optimization process. In: Proceedings of 2010 IEEE/ASME International Conference on Mechatronics and Embedded Systems and Applications, Shenyang, pp. 486–491 (2010)
10. Oustaloup, A., Levron, F., Mathieu, B., Nanot, F.M.: Frequency-band complex non-integer differentiator: characterization and synthesis. IEEE Trans. Circ. Syst. I: Fundam. Theory Appl. 47(1), 25–39 (2000)
11. Monje, C.A., Chen, Y.Q., Vinagre, B.M., Xue, D., Feliu, V.: Fractional-Order Systems and Controls. Springer, London (2010)
12. Baranowski, J., Bauer, W., Zagorowska, M., Dziwinski, T., Piatek, P.: Time-domain oustaloup approximation. In: 20th Internatonal Conference on Methods and Models in Automation and Robotics (MMAR), pp. 116–120. IEEE (2015)
13. Swarnakar, J., Sarkar, P., Singh, L.J.: Realization of fractional-order operator in complex domains—a comparative study. In: Bera, R., Sarkar, S., Chakraborty, S. (eds.) Lecture Notes in Electrical Engineering, vol. 462, pp. 711–718. Springer, Singapore (2018)

14. Swarnakar, J., Sarkar, P., Singh, L.J.: Rational approximation methods for a class of fractional-order SISO system in delta domain. In: Bera, R., Sarkar, S. K., Singh, O.P., Saikia, H. (eds.) Lecture Notes in Electrical Engineering, vol. 537, pp. 395–402. Springer, Singapore (2019)
15. Senol, B., Yeroglu, C.: Filter approximation and model reduction comparison for fractional-order systems. In: International Conference on Fractional Differentiation and its Applications (ICFDA), pp. 1–6. IEEE (2014)

Analysis and Prediction of Stock Market Using Twitter Sentiment and DNN

T. P. Sahana[✉] and J. Anuradha

Vellore Institute of Technology, Vellore, India
tpsahana27@gmail.com, januradha@vit.ac.in

Abstract. The analysis of market is one of the important tasks for data analysts. The stock market is very volatile and many models are developed to predict the market by training the model on historical data. But sentimental analysis approach to predict the market is not common. Here we are using REST APIs to capture the twitter data and sentiment analysis using Google's Natural Language Cloud API along with machine learning techniques like linear regression, Random forest and DNN are used to predict the bear and bullish curves of the stock data. It was found that positive response in the tweets is observed to result in a rise in market value and negative tweets results in fall of the market value. Neural-networks with Feed-forward technique are found to predict prices of stocks with less error in comparison with other used prediction techniques.

Keywords: Google's Natural Language Cloud API · Twitter sentiment · DNN · Stock prediction

1 Introduction

Stock market and value prediction and analysis is necessary for investors and traders to decide whether to buy or sell the shares. Studying the past market and present market will help to make better decisions and more profits to the traders. Fundamental analysis and technical analysis are the two kinds of stock market analysis. Fundamental analysis refers to data and reports that are internal to the company. Technical analysis is the prediction of future value based on past values. Sometimes the market is also event driven i.e. based on some event the market value changes.

Currently the stock market value depends on many unpredictable factors and has a random pattern and traditional methods have failed to provide less than 50% accurate results [4]. The stock value fluctuates based on news, any new releases in the products of the company and any major events in the company [5].

Social media has provided a platform to express emotions for the public and has a major impact on the overall public opinion. One of the most popular application for micro-blogging is Twitter. Everyday millions of people post 100s of millions of tweets a day. There are many instances where information from twitter is used for some predictions. Asur and others [1] have predicted the collections of movie before release using public opinion from twitter. Flu disease outbreak was predicted based on twitter data [2], in [3] the public mood analyzed from twitter was in sync with DJIA value.

© Springer Nature Switzerland AG 2020
A. P. Pandian et al. (Eds.): ICICCS 2019, AISC 1039, pp. 38–45, 2020.
https://doi.org/10.1007/978-3-030-30465-2_5

The Google's Cloud Natural Language API uses powerful models of machine learning to predict the structure and underlying meaning of any given text. Blogs, articles in news channels and even text document's features and information about people or any events or about places can be extracted using this API. Customer sentiment on a product or reaction of public regarding an event from social media or a blog or a messaging app can be captured using the Google's API. Google's Cloud Natural Language API can also be used to determine the sentiment of the tweets. El Alaoui et al. [6] have used Google's Cloud Natural Language API to verify the sentiment scoring from their analyzer.

Machine learning is one of the hottest topics of discussion. Classification and algorithms for predictions are used in every field and stock prediction is also one among them. Every prediction should be first checked whether it is a linear model hence linear regression is first used and Random forest is one more efficient algorithm that is used in the prediction of stocks.

2 Literature Survey

Mizumoto et al. [7] proposed sentimental analysis using semi-supervised machine learning techniques for analysis of stock prediction. Stock market news is used to construct a polarity dictionary and sentiment is analyzed. Zhang [8] was able to find that there is a correlation between stock values and twitter sentiment and correlation depends on the choice of words. Bollen [3] determined that the mood of people obtained from twitter is correlated with DIJA value. Prediction was done using Fuzzy neural networks. Maini et al. [9] proposed Random forest and Support vector machine models for predicting whether there is a rise or fall in the next day's stock value. They used a hybrid data set consisting from Reddit news channel and Guardian's news API for predicting the sentiment and DIJA stock values. Pagolu et al. [10] built their own sentiment analyzer to predict the sentiment of twitter data and used SVM model to predict the rise and fall of stock price.

Tiwari et al. [11] propose several models for stock prediction including ARIMA, Linear Regression and Neural networks. They used historical data of stock and found that neural network performs the best prediction. Google's Cloud Natural Language API is used in many applications. El Alaoui et al. [12] have used Google's Cloud Natural Language API to verify the sentiment scoring from their analyzer. The approach used to obtain a sentiment score from the API and train our model to predict the stock prices.

3 Methodology

3.1 Data Collection and Data Pre-processing

Data from twitter is collected using web scraping technique. Four companies Facebook, Apple, Google and Tesla are used for the analysis in this paper. Keywords like #FB, #Facebook, #DeleteFaceBook, etc. were used to scrape tweets from twitter [14] for Facebook and similarly for other companies we used keywords like #Apple, #Iphone, #Google, #GooglePlay, #GoogleAndroid, #GooglePixel, #Tesla, #ElonMusk, #Tesla-Model3 etc. These tweets are used to get the sentiment of the companies. The pre-processing is done on these tweets by removing the user names, numbers, emoticons, URLs and links. The test is converted to lower case and each word is tokenized and right and left spaces are stripped and the words are all joined again and made as clean tweets. The rise and fall of Stock values is taken from Yahoo! Finance [13]. When there is a weekend or holiday the market is closed and value from previous day is assumed for these days. So that the values are continuous and the curve is smooth.

3.2 System Architecture

Fig. 1. Proposed overall system architecture

The stock prediction system depicted in Fig. 1. has four main components.

  i. Data collection (tweets and stock values)
 ii. Data pre-processing
iii. Sentiment analysis
 iv. Stock prediction.

3.3 Analysis of Sentiment

The Google's Natural Language Cloud API works on machine learning algorithms and models that are powerful and helps us to reveal the inner meaning or structure of the text. The API analyzes sentiment and determines whole attitude of the given text or any document and it gives scores and magnitude values. Natural Language Cloud API can be used when we have language module. Language module can be imported from the library of google-cloud-language.

  First it is necessary to acquire credentials from google developers and then we can communicate with the services of Natural Language Cloud API. The requests and responses of the API are encapsulated by the wrapper Client Language Service library. A score between −1 to +1 is assigned to each tweets using this API. A value between

−1 to −0.25 depicts negative tweet, −0.25 to +0.25 is considered neutral and +0.25 to +1 is considered positive tweet.

Table 1. shows tweets with their sentiment scores obtained after using Google's Natural Language Cloud API. There are many sentiment analyzers available, but the Google Cloud Natural Language API is found to be good and is one of the most used and popular one among other sentiment analyzer.

Table 1. Tweets with their sentiment scores

	Date	Text	Score	SScore
0	2018-01-01	Love usa uk fb the laptop lifestyle is attract…	0.0	0.8
1	2018-01-01	People who hurry do not love well drmm fb love	0.0	0.0
2	2018-01-01	Good you re smart i get infuriated by the excu…	0.0	−0.8
3	2018-01-01	Love usa uk fb the laptop lifestyle is attract…	0.0	0.8
4	2018-01-01	Instagram recommends photos in your feed now u…	0.0	−0.9

3.4 Analysis Between Prices of Stock and Twitter Sentiment

Facebook, Apple, Google and Tesla company stock data is used to analyze the developed prediction system. If there is a rise in the stock value of current day than previous day 1 is marked else 0 is marked. The three classes negative, neutral and positive are used as features for training and the bull and bears of stock market are predicted. A window size of three days is used, selected after trial and error. Three attributes are fed with 70% portion to train and 30% portion for testing.

The stock price is predicted using three prediction algorithms namely Linear regression, Random Forest Regressor and DNN. Linear regression is used to check if the model is linear. It is said that any prediction model should be tried and tested with linear model first and upon results other model can be used. Random forest model with 20 estimators or trees are used for training the model. A neural network with 4 hidden layers is constructed with ReLU as activation function and ADAM optimizer is used for adjusting weights iteratively instead of traditional stochastic gradient descent. TensorFlow library is used extensively to build the neural network.

4 Results

It was found that the model of Linear Regression predicted stock values with the coefficients for negative count, neutral count and positive counts for Facebook, Apple, Google and Tesla as shown in Table 2.

In Fig. 2 the graphs of the model linear regression which shows a comparison of actual stock values v/s predicted stock values are present.

Table 2. Coefficients and variance score of companys using linear regression

Company	neg(Coefficient)	neu(Coefficient)	Pos(Coefficient)	Variance(Score)
FaceBook	0.69083513	0.26113717	0.03555986	0.013549613
Apple	0.80585225	0.1200181	-0.18278813	0.031531156
Google	2.69487268	1.04477421	0.08601251	0.019603821
Tesla	−0.15248737	0.68161707	0.14441053	0.006883488

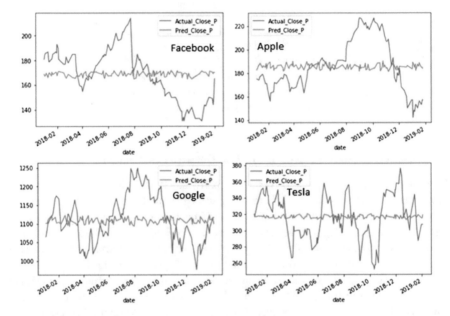

Fig. 2. Actual stock value v/s predicted stock close value for linear regression model (Facebook, Apple, Google and Tesla)

In Figs. 3 and 4. the graphs of the model Randomforest regressor and DNN which shows a comparison of actual stock values v/s predicted stock values.

The mean of absolute error, MSE, RMSE and MAPE (Mean of Absolute Percentage Error) for the used models of prediction i.e. Linear regression, RandomForest Regressor and DNN are listed in the Table 3.

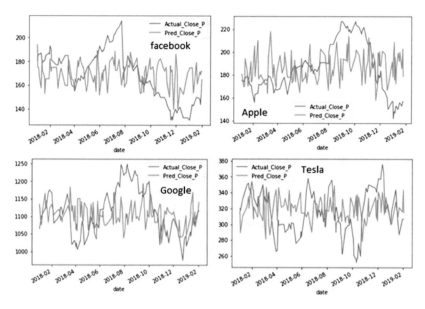

Fig. 3. Actual stock value v/s predicted stock close value RandomForest regressor (Facebook, Apple, Google and Tesla)

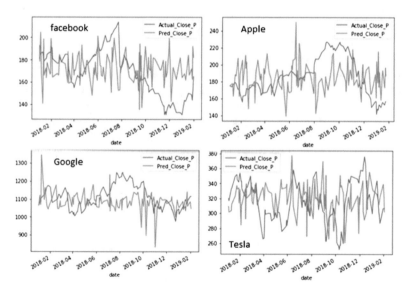

Fig. 4. Actual stock value v/s predicted stock close value for DNN with epochs = 500 (Facebook, Apple, Google and Tesla)

Table 3. Comparison of error rates for different models for different company stock values

Company	Model	MAE	MSE	RMSE	MAPE
FaceBook	Linear regression	18.70529467	522.812194	22.86508679	11.52444
FaceBook	Random forest	16.52015278	410.4864557	20.26046534	10.02677
FaceBook	DNN	16.47196065	383.8084335	19.59102941	10.01534
Apple	Linear regression	22.20091273	764.5163475	27.64988874	12.087
Apple	Random forest	19.69819444	604.6225029	24.58907283	10.852
Apple	DNN	18.17984038	506.7808701	22.51179402	9.913932
Google	Linear regression	70.06368306	7983.428369	89.35003284	6.201243
Google	Random forest	53.95108333	4639.5921	68.11455131	4.814049
Google	DNN	52.44517918	4175.50642	64.61815859	4.699897
Tesla	Linear regression	29.60085882	1316.492979	36.28350836	9.664281
Tesla	Random forest	26.15069444	1016.807878	31.88742507	8.555957
Tesla	DNN	23.62170512	801.4702363	28.31024967	7.659892

5   Conclusion

It is found that there is a strong positive correlation between the sentiment of the public and the stock market value fluctuations. The tweets were classified as negative, neutral, and positive. It can be seen that with positive tweet responses there was bullish curves and negative sentiments resulted in bearish curves. It can be seen from results that DNN has lower percentage of mean error when compared to linear regression and Random forest algorithms.

The twitter data is used to determine the sentiment of the people and it should be noted that all stock traders and financial analysts may not use twitter to give their opinions. Many news feeds like The Wall Street Journal, Bar-chart, Seeking Alpha, Etc. can be used in determining the sentiment. The number of tweets used to train the model can be increased and the analysis is done only on one year's data and this can be extended to several years. Google's Cloud Natural language libraries are used to assist sentiment prediction.

References

1. Asur, S., Huberman, B.A.: Predicting the Future with social media. In: Proceedings of the ACM International Conference on Web Intelligence, pp. 492–499 (2010)
2. Aramaki, E., Maskawa, S., Morita, M.: Twitter catches the flu: detecting influenza epidemics using Twitter. In: Proceedings of the Conference on Empirical Methods in Natural Language Processing. Association for Computational Linguistics (2011)
3. Bollen, J., Mao, H., Zeng, X.: Twitter mood predicts the stock market. J. Comput. Sci. **2**(1), 1–8 (2011)
4. Qian, B., Rasheed, K.: Stock market prediction with multiple classifiers. Appl. Intell. **26**(1), 25–33 (2007). https://doi.org/10.1007/s10489-006-0001-7

 5. Fama, E.F.: The behavior of stock-market prices. J. Bus. **38**(1), 34105 (1965). https://doi.org/10.2307/2350752
 6. El Alaoui, I., Gahi, Y., Messoussi, R., et al.: J. Big Data **5**, 12 (2018). https://doi.org/10.1186/s40537-018-0120-0
 7. Mizumoto, K., Yanagimoto, H., Yoshioka, M.: Sentiment analysis of stock market news with semi-supervised learning. In: 2012 IEEE/ACIS 11th International Conference on Computer and Information Science, Shanghai, pp. 325–328 (2012). https://doi.org/10.1109/icis.2012.97
 8. Jansen, B., Zhang, M., Sobel, K., Chowdury, A.: Twitter power: tweets as electronic word of mouth. J. Am. Soc. Inform. Sci. Technol. **60**, 2169–2188 (2009)
 9. Maini, S.S., Govinda, K.: Stock market prediction using data mining techniques. In: 2017 International Conference on Intelligent Sustainable Systems (ICISS), Palladam, pp. 654–661 (2017). https://doi.org/10.1109/iss1.2017.8389253
10. Pagolu, V.S., Reddy, K.N., Panda, G., Majhi, B.: Sentiment analysis of Twitter data for predicting stock market movements. In: 2016 International Conference on Signal Processing, Communication, Power and Embedded System (SCOPES), Paralakhemundi, pp. 1345–1350 (2016). https://doi.org/10.1109/scopes.2016.7955659
11. Tiwari, S., Bharadwaj, A., Gupta, S.: Stock price prediction using data analytics. In: 2017 International Conference on Advances in Computing, Communication and Control (ICAC3), Mumbai, pp. 1–5 (2017). https://doi.org/10.1109/icac3.2017.8318783
12. El Alaoui, I., Gahi, Y., Messoussi, R., Chaabi, Y., Todoskoff, A., Kobi, A.: A novel adaptable approach for sentiment analysis on big social data. J. Big Data **5** (2018). https://doi.org/10.1186/s40537-018-0120-0
13. Home page of Yahoo Finance. https://in.finance.yahoo.com/. Accessed 30 Mar 2019
14. Home page of Twitter. https://developer.twitter.com/. Accessed 30 Mar 2019

Multiband Antenna with Indoor Applications Using Soft Computation Method

K. Kalaiarasan[1(✉)], A. Kavitha[2], and J. N. Swaminathan[3]

[1] Vel Tech Rangarajan Dr. Sagunthala R&D Institute of Science
and Technology, Chennai, Tamilnadu, India
kalaiarasan@veltech.edu.in
[2] Vel Tech Multi Tech Dr. Rangarajan Dr. Sakunthala Engineering College,
Chennai, Tamilnadu, India
Kavivenkat99@gmail.com
[3] Godavari Institute of Engineering and Technology, Rajahmundry,
Andhra Pradesh, India
jnswaminathan@giet.ac.in

Abstract. To improve the state-of-art data transferring in wireless indoor devices, it is necessary to have an integrated multiple-input multiple-output (MIMO). This paper proposes a fully defined MIMO antenna system for 4G, 5G and wireless LAN networks (WLAN) to obtain high data rate transfer. The integrated MIMO antenna system is able to transfer signals for the frequency of 5 GHz for 5G standards, 13 GHz for future 5G mm-wave band and 2.4 GHz for Wireless LAN network. The design of the antenna system is aimed to achieve more than 5dBi of directivity with an efficiency greater than 60%.

Keywords: Multiple-Input Multiple-Output · Multiband antenna

1 Introduction

The speedy evolution of unbounded communication brings outstanding changes in setting new wireless standard for communication system. Multiple-input multiple-output (MIMO) system is one of the significant technologies that can enhance data throughput in 5G network. The antenna design covers the frequency range of WLAN, 4G and 5G networks. This system of multiple antennas requires MIMO characterization and as such the evolution of multiple antenna is very important in various metrics such as frequency of operation, impedance matching and gain of an antenna. Few antenna designs were reported very recently that works for 5G LTE applications.

2 Related Works

A pair of reactive loaded monopoles is present in the MIMO antenna system design in which the mm-wave array has a planar 2 by 4 slot antennas proposed by Hussain et al. The proposed system works in the frequency range of 1870 to 2530 MHz for 4G, and it is used for 5G in the wave band of 28 GHz [1]. The antenna proposed by Dhar et al.,

A. P. Pandian et al. (Eds.): ICICCS 2019, AISC 1039, pp. 46–52, 2020.
https://doi.org/10.1007/978-3-030-30465-2_6

was designed on a substrate of $50 \times 90 \times 0.76$ mm^3. To improve the overall system performance the amplifier and the antenna are codesigned and integrated, instead of designing them separately with $50 - \Omega$ interface. The integrated MIMO antenna system radiates with a small gain of 14.17 dBi and ECC less than 0.33 [2]. The array configuration of UWB MIMO system for four element and eight elements was proposed by Sipal et al. [3].

The four element has the dimensions of $38 \times 38 \times 0.762$ mm^3 and the eight has dimensions of $38 \times 90 \times 0.762$ mm^3. These prototypes produce effective impedance bandwidth between 3 to 15 GHz [3–5]. Li et al. proposed a small sized coupled printed antenna covers the in frequency of 698–960 MHz and 1710–2690 MHz for 8 - band LTE/GSM/UMTS [6, 7].

Tarot et al. proposed the Dual - band coplanar waveguide - fed smiling monopole antenna for 5G LTE and WIFI applications, covers the frequency range of 2.3–3.0 GHz and 4.7–5.9 GHz [8]. A very pliable slot meander microstrip patch antenna for 4G portable and handheld devices was proposed by Ibrahim et al., covers the frequency range of 1.68–3.88 GHz, 0.5–0.75 GHz, 1.1–2.7 GHz, 3.3–3.9 GHz [9]. The low-profile monocone antenna proposed by Sun et al., has a patch which is cup shaped and it is shorted to ground with metallic wires. This antenna was fed by probe-ring-slot structure. The compact structure is formed by integration of two antennas, which is common to the antenna radiators. This broadband patch antenna yields an impedance bandwidth of 44% from 1.8 to 2.9 GHz, covering the bands UMTS/LTE2300/2500 [10]. Roslan et al. investigated Long-Term Evolution (LTE) applications for 2.6-GHz rectangular dielectric resonator antenna (RDRA). The feed mechanisms used in RDRA are coaxial probe and coplanar waveguide (CPW) [11] and [12]. The impedance bandwidth is measured for ports 1 and 2 and it is found to be 47% (2.09–3.38 GHz) and 25% (2.40–3.09 GHz) respectively. For the mobile phone applications, the proposed antenna well suits. Second, the proposed antenna has simpler structure and wider operating band. So, it is most suitable and promising for 4G, future 5G and WLAN indoor applications [11].

3 Antenna Design

The integrated antenna design resonates at the frequency of 2.4 GHz for WLAN, 5 GHz for 4G and 13 GHz for 5G network. The designed antenna follows the dimension of 4.45 mm $\times$ 3.4 mm $\times$ 1.678 mm using FR_4 as a substrate material. The antenna design was simulated using ADS (Advanced Design System) software. The three antennas are designed and simulated separately and then finally integrated on a single substrate. Hence there will be different input signals for each of the antennae. Hence it is MIMO. The structure of the 4G antenna and the 5G antenna is having inverted – E shape with slot at the center. The shape of WLAN is inverted – Y. The proposed antenna dimensions are manipulated using the formulae given in Eqs. 1 and 2.

$$\frac{Antenna\ Length}{width} = \frac{C}{2f_r\sqrt{\varepsilon_r}} \tag{1}$$

where, C is Speed of light
f_r is resonating frequency

$$Antenna\ Height = \frac{0.822L}{2} \tag{2}$$

where, L is length of antenna and all parameters are in mm (Table 1).

Table 1. Designed specifications of proposed antenna dimensions for 4G, future 5G and WLAN for indoor applications.

Network	Design frequency (GHz)	Length/width	Height (mm)
4G	5	13.98	5.74
Future 5G	13	5.82	2.39
WLAN	2.4	29.20	29.20

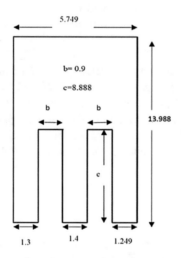

Fig. 1. Design of 4G antenna

The design dimensions of 4G, future 5G and WLAN antennae are shown pictorially. Figure 1 shows the design dimensions of 4G antenna. It is an inverted E – shaped antenna.

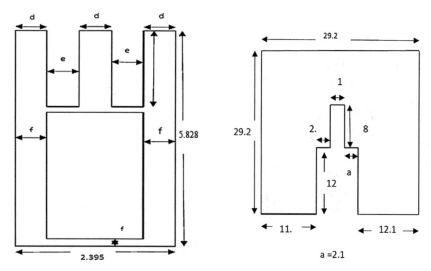

Fig. 2. Design of future 5G antenna   **Fig. 3.** Design of WLAN antenna

Figure 2 shows the design dimensions of future 5G antenna. It is an inverted E – shaped antenna with a slot at the centre. Figure 3 shows the design dimensions of WLAN antenna. It is an inverted Y – shaped antenna.

4   Paramater Analysis

The WLAN antenna was designed with the dimensions of 29.14 mm × 11.9 mm × 1.67 mm, the dimension of 4G antenna is 13.987 mm × 5.74 mm × 1.67 mm and the dimension of future 5G antenna is 5.828 mm × 2.395 mm × 1.67 mm. FR_4 the substrate material used with the thickness of 1.678 mm and the permittivity of the material is 4.6. Copper is used as a patch conductor. The three antennas are designed and simulated separately and then it integrated into a single substrate.

Figure 4 shows the fabrication of proposed MIMO antenna for 4G, future 5G and WLAN indoor application. The main parameter for the antenna to radiate is return loss. Return loss for an antenna should be above −10 dB. From the simulated output it is observed that the return loss is −19.138 dB for 2.37 GHz, −12.237 dB for 4.95 GHz and −31.583 dB for 12.57 GHz. Figures 5 and 6 shows the simulated S-Parameter output for Integrated Antenna and its marker values at different frequencies using ADS. Figure 7 shows the current distribution of Integrated antenna for 4G, future 5G and WLAN indoor application.

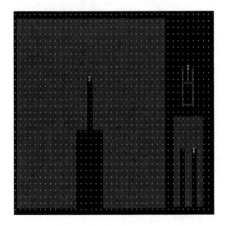

Fig. 4. Integrated antenna design

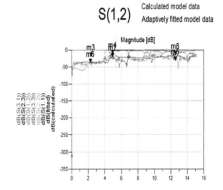

Fig. 5. Simulated S-parameter output for the integrated antenna

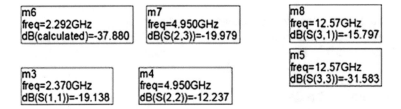

m6	m7	m8
freq=2.292GHz	freq=4.950GHz	freq=12.57GHz
dB(calculated)=-37.880	dB(S(2,3))=-19.979	dB(S(3,1))=-15.797

		m5
		freq=12.57GHz
		dB(S(3,3))=-31.583

m3	m4
freq=2.370GHz	freq=4.950GHz
dB(S(1,1))=-19.138	dB(S(2,2))=-12.237

Fig. 6. Simulated S-parameter output for the integrated antenna with marker values

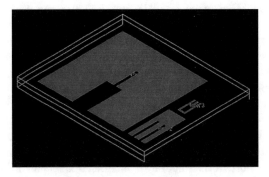

Fig. 7. Current distribution of integrated antenna

5 Result and Discussion

From the simulated results WLAN exhibits omni-directional radiation pattern with the directivity of 6.26675 dBi and gain of about 2.4906 dBi. The 4G antenna exhibits bidirectional radiation pattern with the directivity of 8.05749 dBi and gain of about 2.29573 dBi. The 5G antenna exhibits omni directional radiation pattern with the directivity of 9.27589 dBi and gain of about 3.4264 dBi (Table 2).

Table 2. Results for integrated antenna design for 4G, future 5G and WLAN indoor application

Network	Obtained frequency (GHz)	Return Loss (dB)	Gain (dBi)	Directivity (dBi)	Efficiency (%)
4G	4.95	−12.24	2.30	8.06	66.54
Future 5G	12.57	−31.58	3.43	9.28	66.01
WLAN	2.37	−19.14	2.50	6.27	61.91

6 Conclusion

The design of multiband antenna for 4G, future 5G and WLAN are designed for indoor applications and simulated using ADS software. The designed antenna has a directivity more than 5 dBi and efficiency more than 60%. The gain of antenna can be improved by using array of antenna.

References

1. Hussain, R., Alreshaid, A.T., Podilchak, S.K., Sharawi, M.S.: A compact 4G MIMO antenna integrated with a 5G array for current and future mobile handsets. IET Microwaves Antennas Propag. 11(2), 271–279 (2017)
2. Dhar, S.K., Sharawi, M.S., Hammi, O., Ghannouchi, F.M.: An active integrated ultra-wideband MIMO antenna. IEEE Trans. Antennas Propag. 64(4), 1573–1578 (2016)
3. Sipal, D., Abegaonkar, M.P., Koul, S.K.: Easily extendable compact planar UWB MIMO antenna array. IEEE Antennas Wirel. Propag. Lett. 16, 2328–2331 (2017)
4. Kavitha, A., et al.: Design of flexible textile antenna using FR4, jeans cotton and teflon substrates. Microsyst. Technol. 25(4), 1311–1320 (2019)
5. Amsaveni, A., et.al.: Design and performance analysis of low SAR hexagonal slot antenna using cotton substrate. Microsyst. Technol. 25(6), 2273–2278 (2018)
6. Li, J.L.-W., Wu, Y., Ban, Y.-L., Chen, J.-H.: Small-size printed coupled-fed antenna for eight-band LTE/GSM/UMTS wireless wide area network operation in an internal mobile handset. IET Microwaves Antennas Propag. 7(6), 399–407 (2013)
7. Swaminathan, J.N., et.al.: Performance analysis of LMS filter for SSPA linearization in different modulation conditions. In: Elsevier ICECIT-2012, pp. 49–52 (2012)
8. Tarot, A.C., Las-Heras, F., Mantash, M., de Cos, M.E.: Dual-band coplanar waveguide-fed smiling monopole antenna for WiFi and 4G long-term evolution applications. IET Microwaves Antennas Propag. 7(9), 777–782 (2013)

9. Ibrahim, N., Elamin, M., Rahman, T.A., Abdulrahman, A.Y.: New adjustable slot meander patch antenna for 4G handheld devices. IEEE Antennas Wirel. Propag. Lett. **12**, 1077–1080 (2013)
10. Sun, L., Huang, W., Sun, B., Sun, Q., Fan, J.: Two-port pattern diversity antenna for 3G and 4G MIMO indoor applications. IEEE Antennas Wirel. Propag. Lett. **13**, 1573–1576 (2014)
11. Roslan, S.F., Kamarudin, M.R., Khalily, M., Jamaluddin, M.H.: A coplanar waveguide rectangular dielectric resonator antenna (RDRA) for 4G applications. J. Teknol. **73**(1), 97–100 (2015)
12. Secmen, M.: Multiband and wideband antennas for mobile communication systems. In: Recent Developments in Mobile Communications—A Multidisciplinary Approach, pp. 143–163. InTech, New York (2011)

Integration of Statistical Techniques to Evaluate the Fatigue of Operators on the Productivity of a Company

Alexander Parody[1], Amelec Viloria[2(✉)], Marleidis Hernandez[3],
Arlis Niño[3], and Jorge Cervera[1]

[1] Facultad de Ingeniería, Universidad Libre Seccional Barranquilla,
Barranquilla, Colombia
{alexandere.parodym, jorgee.cerverac}@unilibre.edu.co
[2] Facultad de Ingeniería, Universidad de la Costa, Barranquilla, Colombia
aviloria7@cuc.edu.co
[3] Facultad de Ingeniería, Universidad Autónoma del Caribe,
Barranquilla, Colombia
sofiamhr@hotmail.com, arlismaria@yahoo.com

Abstract. The present study seeks to determine the influence of fatigue according to the Yoshitake's physical, mental, and general classification on the percentage of performance in workers of an assembly line of energy accumulators in a plant located on the north coast of Colombia. The study shows that the type of fatigue affects the performance of workers and this effect varies depending on: the work shift (morning, afternoon, or evening), the position (operator, assistant, or packager), and the experience or the age of the worker. All relationships were established from multiple linear statistical models, applying a 95% of confidence in the models for the three studied shifts, presenting values of R squared above 76% showing that the models are highly explanatory and reliable.

Keywords: Multiple linear regression · Physical fatigue · Mental fatigue · General fatigue · Productivity

1 Introduction

Currently, due to the increase in technology, there are many companies with automated production lines. However, some manufacturing sectors still maintain workers for assembling the most critical parts of their products to avoid damages and errors in the processes. These manual works are repetitive and monotonous and are usually done in work environments that can generate discomfort in the workers due to the inherent conditions of the product manufacture. The discomfort feeling is generated due to the personnel physical and mental efforts that lead to the appearance of fatigue, understood as a decrease of resistance and capacity in the work [1], and resulting in low productivity of the company, stops in the production line, problems in the assembly line, among other issues, National Institute of Safety and Hygiene at Work [2].

© Springer Nature Switzerland AG 2020
A. P. Pandian et al. (Eds.): ICICCS 2019, AISC 1039, pp. 53–62, 2020.
https://doi.org/10.1007/978-3-030-30465-2_7

At the same time, the automobile industry is experiencing growth in its production levels and, with the increase in the acquisition of vehicles, other industries are having an economic awakening, as in the case of factories of energy accumulators. Accumulators can be defined as "a series of elements that generate electricity through reversible chemical reactions, that is, they transform the chemical energy into electrical energy" [3]. In this industry, the products can be classified in two types, non-rechargeable batteries whose elements produce irreversible reactions, thus when the reaction has finished (electron flow) the production of energy finishes. On the other hand, rechargeable batteries contain elements whose chemical reactions (movement of electrons) are reversible, so they can be charged and discharged during their use.

The employees of the energy accumulator manufacturing companies are susceptible to fatigue and other issues inherent to the motivation and psychology of the employees since the work environment affects the way they work, requiring physical and mental demand and generating different levels of fatigue. The quality of the work and productivity is directly affected by the following environmental conditions: high temperatures, noise, exposure to chemical agents, repetitive and monotonous work, intense days, etc. Therefore, several authors have conducted studies measuring the impact of fatigue on production [4].

Based on the above, this paper focuses on assessing fatigue in workers of a production plant through the use of the Yoshitake's survey, analyzing the presence of fatigue in three main areas: mental, physical, and proprioceptive, for determining the relationship that exists between these three spheres and productivity.

This research allows to identify the key factors to assess the fatigue present in workers and their relationship with productivity in the work; thus contributing with the identification of strategies of particular improvements associated with a work in order to safeguard the physical health of the employee and increase productivity and profits.

Currently, there are many studies related to fatigue, but its link with productivity has so far been very low. That is the main contribution of this research aiming to provide guidelines to prevent and control the presence of fatigue as part of the psychosocial risk and its influence on productivity from a statistical perspective, avoiding losses to the company and improving work-man balance.

2 Methodology

The scope of this research is focused on a descriptive-inferential study. Firstly, it is descriptive because it pretends to visualize and summarize data originated from the phenomena studied, in this case, the productivity and fatigue, whose values or data can be numerically or graphically summarized. So, the objective is to analyze the incidence of fatigue on productivity, identifying the factors that influence fatigue in workers during their work. Secondly, it is inferential because it generates the models and predictions associated with the phenomena considering the randomness of the observations and the facts in their natural environment, that is, in the production area of the company. For this purpose, the data collection was carried out in a transversal way since it was done in a single time for describing the aforementioned variables and analyzing their incidence at a given time.

A sample of 29 workers was taken from the energy accumulator manufacturing process, and the survey was applied daily. The sample was divided into three groups of workers that make up the three shifts stated in the organization, both at the beginning and at the end of the work shift. Then, the Yoshitake's fatigue test was applied to determine the fatigue levels of the group of workers. This test was built in [5], and is used in studies of fatigue at the beginning and the end of the working day, and was validated since it offers similar results to objective tests.

For measuring the operator's performance, the finished product counting from the process (packaging) was taken as a main source of information, and the activities of the studied population was recorded in the data collecting sheet as a structured tool of the company for obtaining relevant and precise data about the performance reached by the workers in the area. This information was taken at the end of each shift, recording the total production of the workers in the data collection sheet, considering all issues that may affect the production process such as stops, adjustments, changes of mold, among others.

The statistical analysis focuses on the generation of a multiple linear regression model for each of the work shifts, to evaluate if age, experience, position, physical, mental, or general fatigue are related to the worker's productivity percentage in the accumulator manufacturing process [6].

3   Results

A generalized linear regression model was applied to determine which of the independent variables studied were related to the percentage of performance of the worker in the process, with confidence level of 95% (level of significance of 5%). The variables studied were: position, age, years of experience, entry general fatigue (general fatigue 1), exit general fatigue (general fatigue 2), entry mental fatigue (mental fatigue 1), exit mental fatigue (mental fatigue 2), entry physical fatigue (physical fatigue 1), exit physical fatigue (physical fatigue 2). Considering that the work shift may have some special effect on fatigue behavior, the analysis was divided by work shift, where shift 1 is that of the morning, shift 2 is afternoon, and shift 3 is the turn of the night.

3.1   Results Shift 1

The P-value in the ANOVA table for percentage of performance is less 0.05 (0.0023), it establishes that there is a statistically significant relationship between the performance and the predictor variables with a confidence level of 95.0%. Interactions between each of the variables and their respective P-value can be observed (see Table 1). The linear model generated explains 83.82% of the variability obtained in the percentage of performance during the analyzed shift (shift 1), which shows that these variables with their respective interactions are sufficient to explain the behavior of % of performance, concluding that the variables that show the greatest influence are: the position, the years of age, and the entry general fatigue (general fatigue 1).

Table 1. Sum of squares type III shift 1.

Source	Sum of squares	GL	Middle square	Reason-F	P-value
Position	1573.93	3	524.644	8.3	0.0077
F General 1	1131.67	1	1131.67	17.89	0.0029
Position*Years	3054.47	3	1018.16	16.1	0.0009
Position*Experience	1479.41	3	493.136	7.8	0.0093
Position*F General 1	2285.87	3	761.957	12.05	0.0025
Experience*F General 1	1071.43	1	1071.43	16.94	0.0034
Experience*F General 2	956.304	1	956.304	15.12	0.0046
Position*Years*Experience	1536.62	3	512.207	8.1	0.0083
F General 1 * F General 2	843.005	1	843.005	13.33	0.0065
Shrinkage	505.964	8	63.2454		
Total (corrected)	10554.7	27			

Table 2 shows the variables or associations of variables that most negatively affect production: position of packager with an estimated coefficient value of −13136.8 with confidence limits between −19596.2 and −6677.34, that is, with a confidence of 95%. This variable has the most negative impact on productivity, followed by the interaction position (operator)*years with estimated value of −441,932 with confidence limits between −656,381 and −227,483. It is observed that these values also have a negative sign, indicating that they affect the fulfillment of the production goal. Finally, the position of assistant/packager with −338,289 with confidence limits between −523,783 and −152,795.

Respecting to the fatigue measurements, it is evident that the general fatigue that the worker presents at the beginning of the shift is the one that most negatively affects the worker's performance with a coefficient of −64.4697. Therefore, the entry general fatigue with which the worker starts his work shift has a stronger negative influence compared to mental fatigue and physical fatigue [7–13].

3.2   Results Shift 2

The P-value in the ANOVA table for percentage of performance is less than 0.05 (0.0005), it is established that there is a statistically significant relationship between the performance and the predictor variables with a confidence level of 95.0%. In the same way, the interactions between each of the variables and their respective P-value are observed (see Table 3). The generated linear model explains 76.45% of the variability obtained in the percentage of performance during the analyzed shift (shift 2), which shows that these variables with their respective interactions are sufficient to explain the behavior of the percentage of performance, concluding that the variables that show the greatest influence are: the position, years of age, years of experience, physical fatigue, mental fatigue, and entry general fatigue (general fatigue 1).

Table 2. Confidence limits of 95.0% for the estimated coefficients to estimate the percentage of performance of shift 1. Source: Own elaboration by the author

Parameter	Estimate	Lower limit	Upper limit
Constant	272.282	139.527	405.037
Position (operator)	13793	7095.99	20489.9
Position (assistant)	−338.289	−523.783	−152.795
Position (shaper)	−13136.8	−19596.2	−6677.34
F General 1	−64.4697	−99.6154	−29.324
Position (operator)*Years	−441.932	−656.381	−227.483
Position (assistant)*Years	3.10274	0.0359911	6.16949
Position (packager)*Years	433.891	221.287	646.494
Position (operator)*Experience	−99.6393	−148.381	−50.8977
Position (assistant)*Experience	22.8721	8.10901	37.6351
Position (packager)*Experience	73.6015	33.1399	114.063
Position (operator)*F General 1	−174.618	−258.181	−91.0552
Position (assistant)*F General 1	45.7184	20.3127	71.1242
Position (packager)*F General 1	92.1866	47.2848	137.088
Experience*F General 1	0.46465	0.204323	0.724977
Experience*F General 2	−0.29489	−0.46977	−0.120011
Position (operator)*Years*Experience	3.44624	1.7622	5.13027
Position (assistant)*Years*Experience	−0.426943	−0.688211	−0.165675
Position (packager)*Years*Experience	−2.93669	−4.42854	−1.44485
F General 1*F General 2	1.51499	0.558081	2.4719

Table 4 shows the variables or associations of variables that most negatively affect production: the position of operator with mental fatigue 1 (entry) since it presents an estimated value of −30.8748, that is, with a 95% of confidence, this individual variable is the one that most negatively affects the production, followed by the operator's position in interaction with general fatigue 2, with −11.4168. The variable with the higher incidence is the mental fatigue 2 (entry), with an estimated value of −10,567, which indicates that the mental fatigue of the worker will contribute to the decrease in the % of performance.

Table 3. Sum of squares type III shift 2.

Source	Sum of squares	Gl	Mean square	Reason-F	P-value
Position	1208.04	3	402.678	16.34	0.0001
F Mental 1	437.852	1	437.852	17.77	0.0009
F Mental 2	708.762	1	708.762	28.76	0.0001
Position*Years	978.21	3	326.07	13.23	0.0002
Position*F Mental 1	992.713	3	330.904	13.43	0.0002
Position*F General 2	468.604	3	156.201	6.34	0.0061
Position*F Physical 2	308.824	3	102.941	4.18	0.0262
Experience*F Mental 1	637.266	1	637.266	25.86	0.0002
Experience*F Physical 2	290.171	1	290.171	11.78	0.0041
Position*Years*Experience	706.24	3	235.413	9.55	0.0011
F General 1*F General 2	182.373	1	182.373	7.4	0.0166
F Mental 1*F Mental 2	238.304	1	238.304	9.67	0.0077
Shrinkage	344.978	14	24.6413		
Total (corrected)	3977.08	38			

In shift 2, the variables that mostly affect production are the interactions between the position (operator) with mental fatigue 1 (entry) and position (operator) with general fatigue 2 (exit). Likewise, the individual variable that has more impact on the performance is the mental fatigue 2 (exit), so the fatigue at the beginning of the working day are generating a decrease in the performance of the group of workers in the delta process [7]. It is important to remember that the performance is the total produced by the three employee positions (operator, assistant, and packager) and, if any of them is affected from home, it will unquestionably harm the group performance.

3.3   Results Shift 3

The P-value in the ANOVA table for percentage of performance is lower 0.05 (0.0106), and establishes that there is a statistically significant relationship between the performance and the predictor variables with a confidence level of 95.0%. The interactions between each of the variables and their respective P-value can be observed [8]. The generated linear model explains 86.39% of the variability obtained in the percentage of performance during the analyzed shift (shift 3), which shows that these variables with their respective interactions are sufficient to explain the behavior of the percentage of performance, concluding that the variables that show the greatest influence are: the position, years of age, physical fatigue, mental fatigue, and entry general fatigue (general fatigue 1), see Table 5.

Table 4. Confidence limits of 95.0% for the estimated coefficients.

Parameter	Estimate	Lower limit	Upper limit
Constant	65.6963	53.3917	78.0009
Position (operator)	105.071	58.7142	151.428
Position (assistant)	−7.52218	−57.3304	42.286
Position (packager)	−25.9048	−87.9038	36.0942
F Mental 1	22.3447	10.9756	33.7138
F Mental 2	−10.567	−14.7929	−6.3411
Position (operator)*Years	−1.39092	−2.50212	−0.279718
Position (assistant)*Years	0.231221	−0.727844	1.19029
Position (packager)*Years	−0.88463	−2.62419	0.854939
Position (operator)*F Mental 1	−30.8748	−41.4573	−20.2923
Position (assistant)*F Mental 1	1.93407	−15.3832	19.2514
Position (packager)*F Mental 1	26.9343	11.4137	42.4549
Position (operator)*F General 2	−11.4168	−17.3301	−5.50352
Position (assistant)*F General 2	7.59846	1.71506	13.4819
Position (packager)*F General 2	2.75518	0.109684	5.40067
Position (operator)*F Physical 2	8.59729	3.09173	14.1029
Position (assistant)*F Physical 2	−3.74161	−15.4021	7.91893
Position (packager)*F Mental 2	−4.12239	−12.6633	4.41851
Experience*F Mental 1	−0.59240	−0.842252	−0.342557
Experience*F Physical 2	0.097723	0.0366449	0.158802
Position (operator)*Years*Experience	−0.0068	−0.0130367	−0.000575758
Position (assistant)*Years*Experience	−0.00745	−0.012713	−0.00219861
Position (packager)*Years*Experience	0.006362	−0.00146819	0.0141938
F General 1*F General 2	0.33843	0.0716178	0.605242
F Mental 1*F Mental 2	3.74202	1.1612	6.32284

Table 6 shows the variables or associations of variables that most negatively affect the production: position (packager) given that they have an estimated value of −755,858, that is, with a 95% of confidence, concluding that this individual variable is the one that most negatively affects production. In relation to interactions, the highest incidence is presented by the position of assistant/packager*Years of age, with an estimate of −16,645. Finally, the physical fatigue 1 (entry) with an estimated value of −7,1129.

After performing the multivariate statistical analysis for shift 3, with the percentage of performance as a dependent variable, it can be concluded that the independent variables that stand out in shift 3 are: position, physical fatigue 1 (entry) and the interaction position*years. Since this shift works at night, the different types of fatigue influenced in a more decisive way. However, Table 6 shows that the variable with the most relevance is the position of packager (−755,858) followed by the interaction position (assistant) for years (−16,645). Finally, the physical fatigue 1 (−7,1129). The wide difference between the estimate of the first variable and the interaction can be observed, indicating that the fatigue presents a much lower value than expected.

Table 5. Sum of squares type III shift 3.

Source	Sum of squares	Gl	Mean square	Reason-F	P-value
Position	1397.68	3	465.893	34.42	0.0009
Experience	583.453	1	583.453	43.11	0.0012
F Physical 1	426.91	1	426.91	31.54	0.0025
F Physical 2	604.008	1	604.008	44.63	0.0011
F General 1	140.892	1	140.892	10.41	0.0233
F General 2	115.909	1	115.909	8.56	0.0328
F Mental 1	541.837	1	541.837	40.03	0.0015
Position*Experience	979.522	3	326.507	24.12	0.0021
Position*Years	1405.65	3	468.55	34.62	0.0009
Shrinkage	67.6711	5	13.5342		
Total (corrected)	1989.24	20			

Table 6. Confidence limits of 95.0% for the estimated coefficients to estimate the percentage of performance of shift 3.

Parameter	Estimate	Lower limit	Upper limit
Constant	−23.4622	−65.7448	18.8204
Position (operator)	63.8525	14.8722	112.833
Position (assistant)	471.159	277.188	665.131
Position (packager)	−755.858	−1034.12	−477.598
Experience	0.859012	0.522698	1.19533
F Physical 1	−7.1129	−10.3685	−3.85733
F Physical 2	7.6606	4.71285	10.6084
F General 1	2.28123	0.463725	4.09873
F General 2	−3.01526	−5.66385	−0.366664
F Mental 1	18.9548	11.254	26.6555
Position (operator)*Experience	−0.66706	−0.988991	−0.345128
Position (assistant)*Experience	2.72411	1.65419	3.79402
Position (packager)*Experience	−1.54834	−2.05647	−1.04022
Position (operator)*Years	1.16111	0.264456	2.05776
Position (assistant)*Years	−16.645	−23.1891	−10.101
Position (packager)*Years	19.3645	12.4097	26.3193

4   Conclusions

From all the aforementioned issues, it can be concluded that the fatigue present in the workers of a manufacturing company negatively affects the productivity [9] when generated by the working conditions inherent to the activity developed and the daily life of the operator outside of the work place. Therefore, the personal and labor well-being of any worker within an organization is a significant factor for the achievement of the goals. So, if high levels of productivity, quality, and competitiveness are desirable in the company, it is important to provide safety, health, and well-being to the workers both in the workplace and their personal life.

Acknowledgment. To the Universidad de la Costa and the Universidad Libre Seccional Barranquilla, especially the GIDE research group, for making this research possible.

References

1. Grandjean, É.: Ergonomía: aspectos físicos y psicológicos, Fatiga General. Enciclopedia de salud y seguridad en el trabajo. Chantal Dufresne, Madrid, BA (1998)
2. Horan, K.A., Singh, R.S., Moeller, M.T., Matthews, R.A., Barratt, C.L., Jex, S.M., O'Brien, W.H.: The relationship between physical work hazards and employee withdrawal: the moderating role of safety compliance. Stress Health **35**(1), 81–88 (2019)
3. Parody, A., Viloria, A., Lis, J.P., Malagón, L.E., Calí, E.G., Hernández, P.H.: Application of an experimental design of D-optimum mixing based on restrictions for the optimization of the pre-painted steel line of a steel producer and marketing company. In: Tan, Y., Shi, Y., Tang, Q. (eds.) Data Mining and Big Data. DMBD 2018. Lecture Notes in Computer Science, vol. 10943 (1), pp. 1–12. Springer, Cham (2018)
4. Kelley, B.P.: Fabricación de acumuladores de plomo. Enciclopedia de salud y seguridad en el trabajo. Chantal Dufresne, Madrid, BA (1998)
5. Viloria, A., Bucci, B., Luna, M., Lis-Gutiérrez, J.P., Parody, A., Solano, B.D.E., Borrero, L. A.L.: Determination of dimensionality of the psychosocial risk assessment of internal, individual, double presence and external factors in work environments. In: Tan, Y., Shi, Y., Tang, Q. (eds.) Data Mining and Big Data. DMBD 2018. Lecture Notes in Computer Science, vol. 10942 (1), pp. 149–158. Springer, Cham (2018)
6. Laurig, W., Vedder, J.: Ergonomia: Introducción. Enciclopedia de salud y seguridad en el trabajo. Chantal Dufresne, Madrid, BA (1998)
7. Yoshitake, H.: Three characteristic patterns of subjective fatigue symptoms. Ergonomics **21** (3), 201–233 (1978)
8. Izquierdo, N.V., Lezama, O.B.P., Dorta, R.G., Viloria, A., Deras, I., Hernández-Fernández, L.: Fuzzy logic applied to the performance evaluation. honduran coffee sector case. In: Tan, Y., Shi, Y., Tang, Q. (eds.) Advances in Swarm Intelligence. ICSI 2018. Lecture Notes in Computer Science, vol. 10942 (1), pp. 1–12. Springer, Cham (2018)
9. Parody, A., Viloria, A., Gaitán, M., Malagón, L.E., Henry, M.A., Palma, H.H., Barrios, L.M.C.: Application of a central design composed of surface of response for the determination of the flatness in the steel sheets of a colombian steel. In: International Conference on Data Mining and Big Data, vol. 10943 (1), pp. 739–747. Springer, Cham (2018)
10. Bucci, N., Luna, M., Viloria, A., García, J. H., Parody, A., Varela, N., López, L.A.B.: Factor analysis of the psychosocial risk assessment instrument. In: International Conference on Data Mining and Big Data, vol. 10943 (1), pp. 149–158. Springer, Cham (2018)
11. Mok, M.M., Liu, C.K., Lam, M.F., Kwan, L.P., Chan, G.C., Ma, M.K., Chan, T.M.: A longitudinal study on the prevalence and risk factors for depression and anxiety, quality of life, and clinical outcomes in incident peritoneal dialysis patients. Perit. Dial. Int. **39**(1), 74–82 (2019)

12. Bhalaji, R.K.A., Bathrinath, S., Ponnambalam, S.G., Saravanasankar, S.: A fuzzy decision-making trial and evaluation laboratory approach to analyse risk factors related to environmental health and safety aspects in the healthcare industry. Sādhanā **44**(3), 48–55 (2019)
13. Witt, K., Milner, A., Spittal, M.J., Hetrick, S., Robinson, J., Pirkis, J., Carter, G.: Population attributable risk of factors associated with the repetition of self-harm behaviour in young people presenting to clinical services: a systematic review and meta-analysis. Eur. Child Adolesc. Psychiatry **28**(1), 5–18 (2019)

Modelling and Simulation for Damping of Power System Network Oscillations by Meticulous Tuning of PSS Parameters Employed in DAVR of Practical Alternators

A. Nalini[1(✉)], E. Sheeba Percis[1], R. Shankar[2], J. Jayarajan[3],
and J. Jenish[3]

[1] Dr. MGR Educational and Research Institute, Maduravoyal, Chennai, India
nalinitosiva@gmail.com
[2] TNEB, Chennai, India
[3] Department of EEE, Dr. MGR Educational and Research Institute,
Maduravoyal, Chennai, India

Abstract. In the modern Power Network, the Generators connected are prone to small and large disturbances. The AC generator adjusts itself to the small disturbances and maintains its synchronism. The Power System Stabilizer is a special feature provided in the Digital Automatic Voltage Regulator (DAVR) of a Generator Excitation Control system, which plays a vital role in enhancing the damping of low frequency oscillations. The Central Electricity Authority mandates technical standards for connectivity of the grid that every generating unit above 100 MW shall have PSS. The Indian Electricity Grid Code (IEGC) also mandates that PSS in DAVR of Generating owner shall carryout tuning of PSS from time to time and Central Transmission Utility (CTU) and Regional Power Committee (RPC) will be carry out checking the PSS and further tuning, wherever considered necessary. Tuning of PSS is mandatory in accordance with IEGC & CEA amendment regulation 2013. This paper deals with optimum tuning of PSS for enhancement of small signal stability and confirms it by Modeling and simulation by using MATLAB Simulink.

Keywords: PSS · Oscillations · DAVR · Alternator

1 Introduction

Power System Stability is a system property that enables the system to persist in pristine condition under standard condition and to restore its equilibrium after being interrupted. Stability is influenced by the rotor angle dynamics and the interaction between the power angle [1]. Stability is mainly categorized into stability of the rotor angle and voltage.

1.1 Rotor Angle Stability

The ability of the interconnected synchronous machines to maintain synchronism is Rotor angle stability. This involves the electromechanical oscillations. As the rotor

© Springer Nature Switzerland AG 2020
A. P. Pandian et al. (Eds.): ICICCS 2019, AISC 1039, pp. 63–69, 2020.
https://doi.org/10.1007/978-3-030-30465-2_8

oscillates, the power output of the synchronous machines varies [2]. As the rotor is driven by the turbine, an alternating voltage is induced in the three phase armature windings of the stator due to the rotating magnetic field of the field winding. The frequency and the current of the connected load are influenced by the speed of the rotor, so the frequency is synchronized with the rotor mechanical speed. As more number of machines are interconnected in the Power System, the machines must have the same frequency and maintain synchronism.

1.2   Voltage Stability

The ability of the power network to maintains its voltage level under normal operating states and also when subjected to a disturbance. This occurs when there is a change in the system load which leads to voltage sag [3, 4]. The main issue is the inability to compensate the demand for the reactive power. The bus voltage increases as the injection of reactive power of the same bus increases. The system is stable if V-Q sensitivity is positive and unstable if it is negative.

The Rotor Stability comprising of Transient stability, Small signal stability, voltage stability and voltage lag. Fast isolation of faulty electrical network from the healthy one and simultaneous acting of automatic voltage regulators with no dead band and high degree of response improves the steady state stability of the power network. The PSS provide control over the generator excitation system to enhancing the small signal stability of the power system in most cost effective way.

2   Generator Excitation Control System

The excitation system provides direct current to the field winding of the synchronous machine and also provides control and protective functions. Generally the excitation system automatically adjusts the field current of the Generator to maintain the terminal voltage [5]. The exciter uses the auxillary stabilising signal and the terminal voltage error signal to control the field voltage to damp the oscillations known as Power System Stabilizer.

2.1   Principle Operation of an Excitation System

The exciter provides dc power for the Synchronous generator field winding. The Regulator amplifies the exciter control input, the terminal voltage transducer and the load compensator senses the voltage and load compensation is provided [6, 7]. The Power System Stabilizer provides the regulator with additional input signals such as velocity deviation, accelerating energy and frequency deviations, etc. to dampen the oscillations. The limiters and protective circuits ensure the exciter's capacity limit. The performance of the excitation control depends on the behavior of the excitation system, generator and Power Network [8]. Since the system is non-linear, the dynamic performance is classified into large signal performance & small signal performance.

3 System Control and Protection Function

The Generator excitation system is implemented by DAVR and provides DC to the rotor winding of the AC Generator. It provides control and protection function for the voltage there by reactive power flow [9, 10]. Also to confirm that the generator capability limits not exceeded to the desired value. The DAVR regulates the machine terminal voltage to its desired level.

3.1 Functional Parameter of Practical DAVR

There are various parameters of DAVR that controls/limits various parameters of excitation system within the capability limit of the synchronous machine namely, Minimum field current Limiter (I_{min}), Capacitive stator current limiter (I_{gcd}), Attenuation factor (KIG_c), Load Angle Limiter (D_{ref}).

3.2 Role of Power System Stabilizer

PSS is a device that enhances power system stability by damping out low frequency power oscillations at the synchronous machines by controlling the excitation. As a benefit of the action, several PSS at several synchronous machines, the stability of the whole power network enhances [11, 12].

The low frequency oscillations that can be damped by the PSS are divided into the following modes.

- Local mode-1.0 to 2.0 Hz (one generator swings against rest of the system)
- Inter area mode-Around 0.3 Hz (Two Area's swing against each other)
- Torsional Mode-10–40 Hz (Generator Geometry).

The load angle of the AC Generator is formed by difference in magnetic field between rotor and stator. Under steady state condition the load angle is continuous and remains constant and has a value ensuring balance between mechanical driving torque and electrical o/p torque that provides synchronizing torque and there by maintain machines are in synchronism state.

If load angle disturbs the electrical torque changes and brings the load angle back to equilibrium position of machine.

Due to inertia of the machine this can take place periodically in the form of more or less well damped oscillations. To damp the oscillations damping torque is required. The damping torque has to be proportional to difference between actual and synchronous speed i.e., slip. The damper windings provide this torque.

The PSS is a special feature provided in the Digital Automatic Voltage Regulator (DAVR) of a Generator Excitation control system [11]. The stabilizer must produce a component of electrical torque in phase with the deviations in rotor speed to provide damping.

3.3   Results of PSS Tuning

PSS tuning was performed at the DAVR of 600 MW unit at different load with ±2% and ±4% step response test and the behavior of DAVR at the −4% when PSS off and PSS on is given below as shown in Fig. 1.

(a) Response of DAVR system when PSS OFF by giving a step input of −4% to Uref.
(b) Response of DAVR system when PSS ON by giving a step input of −4% to Uref.

4   Modelling and Simulation of DAVR

4.1   Modelling of DAVR and PSS

The structure of PSS is shown in Fig. 2. The block diagram of DAVR/Exciter block is shown in Fig. 3. The representation of exciter with Gex(s) represent the transfer function of AVR and exciter as shown in Fig. 4.

4.2   Simulation Results Using MATLAB

Simulation performed by using the various parameters of the Alternator, DAVR and PSS and the functional parameters is fine-tuned to till the desired results achieved. The simulation performed with PSS and without PSS for different change in reference voltage. The simulation model is shown in Fig. 5 and the expected results are shown in Fig. 6.

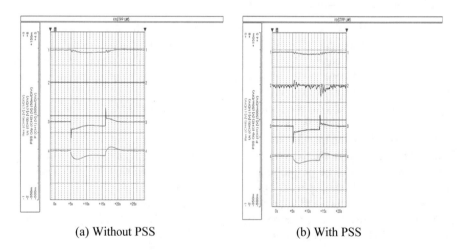

(a) Without PSS   (b) With PSS

Fig. 1. Response of DAVR with and without PSS

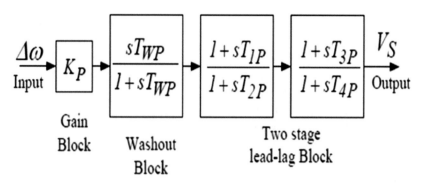

Fig. 2. Structure of PSS

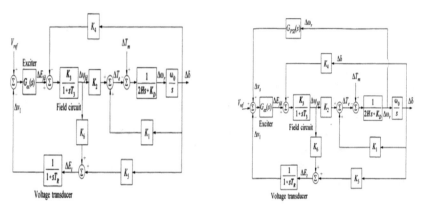

Fig. 3. Control diagram exciter and DAVR.

Fig. 4. Control diagram of AVR and PSS.

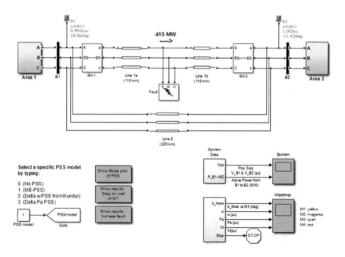

Fig. 5. Simulink model

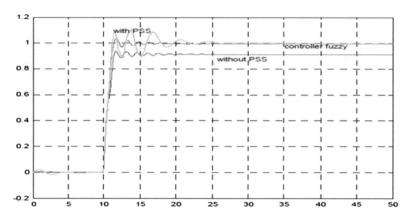

Fig. 6. Output of PSS

5   Conclusion

The study results shows that the roll of PSS is very critical for enhancing the power network stability during the Small signal disturbances such as momentary change in Demand and supply. The optimum selection of PSS parameters is providing better performance of DAVR against low frequency network oscillations. The practical test results as well as Simulink simulation results are indicates that the voltage over shoots are settling to the steady state value much quicker when compared to the PSS is in off condition there by best response despite shooting and settling time.

References

1. M/s BHEL/India: The manufacturer Manual of DAVR, AC Generator parameters and Site test procedure and results
2. Panda, S., Padhy, N.P., Patel, R.N.: Modelling, simulation and optimal tuning of TCSC controller. Int. J. Simul. Modell. **6**(1), 37–48 (2007)
3. Jena, M.K., Panigrahi, B.K., Samantaray, S.R.: A new approach to power system disturbance assessment using wide area post disturbance records. IEEE Trans. Ind. Inform. **14**(3), 1253–1261 (2018)
4. Cai, L., Thornhill, N.F., Kuenzel, S., Pal, B.C.: Wide area monitoring of power systems using principal component analysis and k-nearest neighbor analysis. IEEE Trans. Power Syst. **33**(5), 4913–4923 (2018)
5. Kim, J., Choi, K., Kim, Y., Kim, W., Do, K., Choi, J.: Delay monitoring system with multiple generic monitors for wide voltage range operation. IEEE Trans. Very Large Scale Integr. (VLSI) Syst. **26**(1), 37–49 (2018)
6. Nalini, A., Manivannan, S., Sheeba Percis, E.: Intelligent identification for wide area monitoring in power system. ARPN J. Eng. Appl. Sci. **10**(20), 9401–9407 (2015)
7. Sheeba Percis, E., Arunachalam, P., Nalini, A., Rajam, S.: Modeling and intelligent control of hybrid microgrid ina wide area system. Int. J. Pure Appl. Math. **120**(6), 11437–11446 (2018)

8. Nalini, A., Sheeba Percis, E., Sivanesan, S., Rama, S.T., Bhuvaneswari, S.: Wide area monitoring system for an electrical grid. Elsevier Sci. Direct Energy Proc. **160**, 381–388 (2019)
9. Sheeba Percis, E., Manivanna, S., Nalini, A.: Electric vehicle as an energy storage for grid connected solar power system. Int. J. Power Electron. Drive Syst. **6**(3), 567–575 (2015)
10. Kumar, A.: Power system stabilizers design for multimachine power systems using local measurement. IEEE Trans. Power Syst. **31**(3), 2163–2171 (2016)
11. Esmaeil, G., Kamwa, I.: Local and wide area PMU based decentralised dynamic state estimation in multimachine power systems. IEEE Trans. Power Syst. **31**(1), 547–562 (2016)
12. Bian, X.Y., Geng, Y., Lo, K.L., Fu, Y., Zhou, Q.B.: Coordination of PSSs and SVC damping controller to improve probabilistic small-signal stability of power system with wind farm integration. IEEE Trans. Power Syst. **31**(3), 2371–2382 (2016)

Band Limited Realization of Fractional-Order Proportional Integral Controller for a Class of Finite Dimensional System

Jaydeep Swarnakar[(✉)] and Bhanita Adhikary

Department of Electronics and Communication Engineering,
School of Technology, North-Eastern Hill University,
Shillong 793022, Meghalaya, India
jaydeepswarnakar@gmail.com

Abstract. Fractional-order controller (FOC) is an important area of study nowadays in control science. Due to their infinite dimensional nature, these controllers are approximated within a finite band for practical uses. In this paper, a fractional-order proportional integral controller (FOPI) has been realized within a limited band for a class of finite dimensional plant model. Firstly, the FOC has been tuned for the defined plant model to achieve a set of frequency domain objectives and then, the Oustaloup approximation has been employed to realize it within a limited band. The impact of the control design scheme is tested by connecting both the original FOC and the approximated FOC seperately with the plant model and subsequently comparing the open loop frequency responses with respect to the proposed design objectives through simulation studies. The robustness of the overall controlled system is also studied deviating the plant gain by a significant amount.

Keywords: Fractional-order controller (FOC) · Oustaloup approximation · Robustness

1 Introduction

The natural phenomena that surround us are dynamic in nature. In order to describe such phenomena, non-integer order calculus could be an excellent tool as it can model the system dynamics more adequately as compared to the traditional integer order calculus. In the past few years, the application of fractional-order systems have been perceived in many areas of science, engineering and applied mathematics [1, 2]. Unsurprisingly, the field of modern control theory is not an exception to explore such systems. In control science, the concept of fractional calculus was first introduced by Manabe [3]. For recent work, we cite [4]. Back in 1942, Ziegler and Nichols first presented the design and tuning of PID controllers. Since then, it has been most commonly used algorithm for industrial applications [5]. As an extension of conventional PID controllers, Podlubny first proposed a $PI^\lambda D^\mu$ controller with two extra tuning parameters μ and λ, which delivered robust performance as compared to its classical counterpart [6]. Fractional-order controllers have been applied in numerous

A. P. Pandian et al. (Eds.): ICICCS 2019, AISC 1039, pp. 70–77, 2020.
https://doi.org/10.1007/978-3-030-30465-2_9

applications like active reduction of vertical tail buffeting, controlling of a power electronic bulk converter, fractional sliding mode control etc. [7, 8]. Different tuning methods are proposed from [9–12] based on some frequency domain design specifications. In this work, an analytical design approach is adopted [10, 12] to design a fractional-order proportional integral controller (FOPI) for a class of finite dimensional plant model. Unlike integer order system, fractional-order system is basically infinite dimensional. That is why, implementation requires proper approximation within an appropriate range of frequencies. Therefore, the key emphasis of this paper has been given towards the finite dimensional realization of the FOPI controllers within a limited band for satisfying the desired objectives of the overall controlled system. To this aim, the well-known Oustaloup approximation has been chosen [13, 14]. The entire work has been subdivided into four sections. Section 1 gives the introduction. The design methodology of FOPI controller is explained in Sect. 2. Simulation and results are given in Sect. 3. Section 4 concludes the whole work.

2 Design Methodology of FOC

A typical first order plant is considered having the transfer function given below:

$$P(s) = \frac{H}{\tau s + 1} \tag{1}$$

This plant has to satisfy the following three design specifications:

- Phase Margin = φ_m degree
- Robustness through exhibiting flat phase characteristics around gain cross over frequency
- Gain cross over frequency = ω_c rad/sec

A fractional-order proportional integral (FOPI) controller is chosen to satisfy the above objectives. The transfer function of the controller is given as

$$C(s) = k_p \left(1 + \frac{k_i}{s^\eta} \right) \tag{2}$$

The open loop transfer function (OLTF) is given as follows:

$$G_{OL}(s) = P(s)C(s) = \frac{H}{\tau s + 1} k_p \left(1 + \frac{k_i}{s^\eta} \right) \tag{3}$$

Design specifications are expressed as below:

$$\angle G_{OL}(j\omega_c) = -\tan^{-1} \left(\frac{\frac{k_i}{\omega_c^\eta} \sin\left(\eta \frac{\pi}{2}\right)}{1 + \frac{k_i}{\omega_c^\eta} \cos\left(\eta \frac{\pi}{2}\right)} \right) - \tan^{-1}(\tau \omega_c) = -\pi + \varphi_m \tag{4}$$

$$\left(\frac{d\angle(C(j\omega)P(j\omega))}{d\omega}\right)_{\omega=\omega_c} = \frac{d}{d\omega}\left(-\tan^{-1}\left(\frac{\frac{k_i}{\omega^\eta}\sin\left(\eta\frac{\pi}{2}\right)}{1+\frac{k_i}{\omega^\eta}\cos\left(\eta\frac{\pi}{2}\right)}\right)-\tan^{-1}(\tau\omega)\right)_{\omega=\omega_c} = 0 \tag{5}$$

$$|G_{OL}(j\omega_c)| = \left|\frac{H.k_p.\sqrt{1+\frac{(k_i)^2}{(\omega_c)^{2\eta}}+2\frac{k_i}{\omega_c^\eta}\cos\left(\eta\frac{\pi}{2}\right)}}{\sqrt{1+(\tau\omega_c)^2}}\right| = 1 \tag{6}$$

In order to design the fractional-order controller (FOC), it is needed to find out the unknown values of the controller parameters namely k_p, k_i and η from Eqs. (4), (5) and (6). Reframing Eqs. (4) and (5), following equations are obtained:

$$k_i = \frac{-\tan(\tan^{-1}(\tau\omega_c)+\varphi_m)}{\omega_c^{-\eta}\sin(\eta\pi/2)+\omega_c^{-\eta}\cos(\eta\pi/2).\tan(\tan^{-1}(\tau\omega_c)+\varphi_m)} \tag{7}$$

$$k_i^2\left(\frac{\tau\omega_c^{-2\eta}}{1+(\tau\omega_c)^2}\right)+k_i\left(2\omega_c^{-\eta}\frac{\tau}{1+(\tau\omega_c)^2}\cos\left(\eta\frac{\pi}{2}\right)-\eta\omega_c^{-\eta-1}\sin\left(\eta\frac{\pi}{2}\right)\right)$$
$$+\frac{\tau}{1+(\tau\omega_c)^2}=0 \tag{8}$$

Using Eqs. (7) and (8), two graphs are plotted taking k_i as a function of η where the intersection of the two graphs determines the desired values of k_i and η. The remaining parameter k_p is determined from Eq. (6) as shown below:

$$k_p = \frac{\sqrt{1+(\tau\omega_c)^2}}{H\sqrt{1+\frac{k_i^2}{\omega_c^{2\eta}}+2\frac{k_i}{\omega_c^\eta}\cos\left(\eta\frac{\pi}{2}\right)}} \tag{9}$$

So, the FOC is finally obtained by finding k_p, k_i and η. The FOC thus obtained is connected to the first order plant and subsequently the open loop frequency response is plotted to verify the required design objectives. Once the design criterions are met, then the FOC is converted to its integer order equivalent performing Oustaloup approximation of the fractional power term. Then, the finite dimensional FOC is again connected to the first order plant and the integer order OLTF is computed which is denoted as $G'_{OL}(s)$. The frequency response of $G'_{OL}(s)$ and $G_{OL}(s)$ are plotted simultaneously to investigate the similarity between them. The greater similarity between the frequency response of $G'_{OL}(s)$ and $G_{OL}(s)$ will ensure the fact that the proposed design criterions are satisfied by the finite dimensional FOC too. Employing Oustaloup approximation, the Nth order approximation of s^η $(0<\eta<1)$ within a frequency band (ω_b,ω_h) is given as below:

$$s^\eta \approx K \prod_{i=1}^{N} \frac{s + \omega_i'}{s + \omega_i} \tag{10}$$

where,

$$\omega_i' = \omega_b \omega_u^{(2i-1-\eta)/N}, \quad \omega_i = \omega_b \omega_u^{(2i-1+\eta)/N}, \quad K = \omega_h^\eta \text{ and } \omega_u = \sqrt{\omega_h/\omega_b}.$$

3 Simulations and Results

Taking $H = 1$ and $\tau = 0.3$, the plant transfer function is obtained as below:

$$P(s) = \frac{1}{0.3s + 1} \tag{11}$$

The design objectives are as follows:

- $\varphi_m = 80° = 0.44\pi$
- Robustness through exhibiting flat phase characteristics around gain cross over frequency
- $\omega_c = 8$ rad/sec.

By substituting $\tau = 0.3$, $\varphi_m = 80° = 0.44\pi$ and $\omega_c = 8$ rad/sec in Eqs. (7) and (8), we obtain the equations as given below:

$$k_i = \frac{0.64}{8^{-\eta} \sin(\eta\pi/2) - 0.64 \times 8^{-\eta} \cos(\eta\pi/2)} \tag{12}$$

$$\begin{aligned}0.04438 \times 8^{-2\eta} k_i^2 + 0.125\eta \times 8^{-\eta} \sin(\eta\pi/2) \times k_i \\ -0.08876 \times 8^{-\eta} \times \cos(\eta\pi/2) \times k_i + 0.04438 = 0\end{aligned} \tag{13}$$

Both the Eqs. (12) and (13) are plotted as k_i versus η and subsequently we obtain $k_i = 4.6464$ and $\eta = 0.8871$ from the intersection of the two graphs as shown in Fig. 1. Now, in Eq. (9), putting $H = 1$, $\omega_c = 8$ rad/sec, $\varphi_m = 80° = 0.44\pi$, $\tau = 0.3$, $k_i = 4.6464$ and $\eta = 0.8871$ we obtain $k_p = 1.9387$.

Therefore, the transfer function of the FOPI controller is obtained as follows:

$$C(s) = 1.9387 \left(1 + \frac{4.6464}{s^{0.8871}}\right) \tag{14}$$

The OLTF is obtained as follows:

$$G_{OL}(s) = P(s)C(s) = \left(\frac{1}{0.3s+1}\right) \times 1.9387 \left(1 + \frac{4.6464}{s^{0.8871}}\right) \tag{15}$$

The frequency response of $G_{OL}(s)$ is shown in Fig. 2. From the Fig. 2, we obtain gain cross over frequency $\omega_c = 8$ rad/sec, phase margin $\varphi_m = (180° - 100°) = 80°$ and a flat phase around gain cross over frequency. Therefore, all the requirements have been satisfied by employing the FOC to control the plant.

Now, the FOC has one fractional power term $s^{0.8871}$ as shown in Eq. (14). This term has been approximated to a third order transfer function within a frequency range .01 rad/sec to 100 rad/sec using Oustaloup approximation as mentioned in Eq. (10). Taking $\eta = 0.8871$, $\omega_b = .01$, $\omega_h = 100$ and $N = 3$, the fractional power term $s^{0.8871}$ is approximated as below:

$$s^{0.8871} \approx \frac{59.46s^3 + 344.1s^2 + 88.17s + 1}{s^3 + 88.17s^2 + 344.1s + 59.46} \tag{16}$$

Using Eq. (16), the integer order equivalent of $G_{OL}(s)$ i.e. $G'_{OL}(s)$ is obtained as below:

$$G'_{OL}(s) = \frac{124.3s^3 + 1461s^2 + 3271s + 537.6}{17.84s^4 + 162.7s^3 + 370.6s^2 + 88.47s + 1} \tag{17}$$

The frequency response of $G_{OL}(s)$ and $G'_{OL}(s)$ together have been plotted in Fig. 3. The frequency response of $G'_{OL}(s)$ is almost similar to the frequency response of $G_{OL}(s)$. Therefore, all the design requirements are also met by employing finite dimensional controller obtained after rational approximation of FOC. To examine the robustness of the overall system, the actual plant gain $H = 1$ has been altered upto 20%, i.e. by making $H = 1.2$ and $H = 0.8$ respectively. The frequency responses obtained for $H = 1$, $H = 0.8$ and $H = 1.2$ are shown in Fig. 4. It is noticable that the phase margin still remains constant in each cases which indicate the robustness of the overall system even when the plant gain changes.

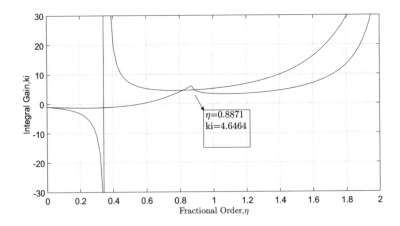

Fig. 1. Plot of integral gain k_i versus η

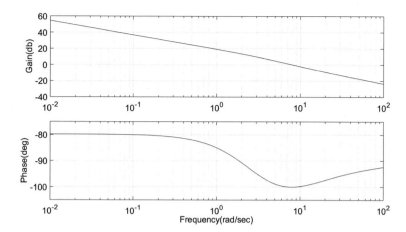

Fig. 2. Frequency response of $G_{OL}(s)$ with FOC

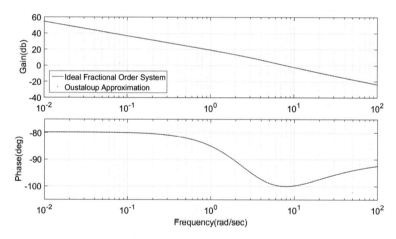

Fig. 3. Frequency response comparison between $G_{OL}(s)$ and $G'_{OL}(s)$ employing ideal FOC and band limited FOC respectively with the same plant model

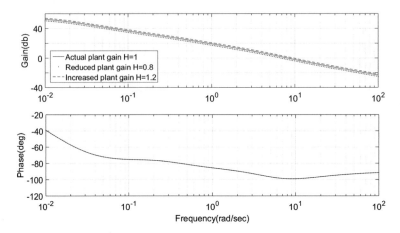

Fig. 4. Frequency responses of $G'_{OL}(s)$ with band limited FOC at different plant gains

4   Conclusions

In this paper, an analytical design approach has been recasted for realizing a FOPI controller within a finite band for a typical first order plant model. Oustaloup approximation has been employed to represent the FOC within a finite band. It is apparent that the overall controlled system has achieved all the proposed objectives by employing both the FOC and its band limited counterpart. The robustness of the entire system has also been verified deviating the plant gain by a significant amount.

References

1. Monje, C.A., Vinagre, B.M., Feliu, V., Chen, Y.Q.: Tuning and auto-tuning of fractional order controllers for industry applications. J. Control Eng. Pract. **16**(7), 798–812 (2008)
2. Freeborn, T.J.: A survey of fractional order circuit models for biology and biomedicine. IEEE J. Emerg. Sel. Top. Circuits Syst. **3**(3), 416–424 (2013)
3. Manabe, S.: The non-integer integral and its application to control systems. J. Inst. Electr. Eng. Japan **80**(860), 589–597 (1960)
4. Barbosa, R.S., Tenreiro Machado, J.A., Ferreira, I.M.: A fractional calculus perspective of PID tuning. In: ASME 2003 International Design Engineering Technical Conferences and Computers and Information in Engineering Conference, pp. 651–659 (2003)
5. Ziegler, J.G., Nichols, N.B.: Optimum settings for automatic controllers. J. Trans. ASME **64** (11), 759–768 (1942)
6. Podlubny, I.: Fractional order systems and $PI^\lambda D^\mu$ controllers. IEEE Trans. Autom. Control **44**(1), 208–214 (1999)
7. Enchez, Y.S.: Fractional-PID Control for Active Reduction of Vertical Tail Buffeting. Master's thesis, Saint Louis University, St. Louis, USA (1999)
8. Calderon, A.J., Vinagre, B.M., Feliu, V.: Fractional sliding mode control of a DC-DC buck converter with application to DC motor drives. In: Proceedings of the 11th International Conference on Advanced Robotics, Coimbra, Portugal, pp. 252–257 (2003)

9. Monje, C.A., Vinagre, B.M., Chen, Y.Q., Feliu, V., Lanusse, P., Sabatier, J.: Proposals for fractional $PI^{\lambda}D^{\mu}$ tuning. In: Proceedings of the First IFAC Workshop on Fractional Differentiation and its Applications, vol. 38, Bordeaux, France, pp. 369–381 (2004)
10. Wang, C.Y., Luo, Y., Chen, Y.Q.: An analytical design of fractional order proportional integral and [Proportional Integral] controllers for robust velocity servo. In: 4th Conference on Industrial Electronics and Applications (ICIEA), pp. 3448–3453. IEEE (2009)
11. Wu, H., Su, W., Liu, Z.: PID controllers: design and tuning methods. In: 9th IEEE Conference on Industrial Electronics and Applications, pp. 808–813. IEEE (2014)
12. Luo, Y., Chen, Y.Q.: Fractional order [Proportional Derivative] controller for robust motion control: tuning procedure and validation. In: American Control Conference, pp. 1412–1417. IEEE (2009)
13. Baranowski, J., Bauer, W., Zagorowska, M., Dziwinski, T., Piatek, P.: Time-domain oustaloup approximation. In: 20th International Conference on Methods and Models in Automation and Robotics (MMAR), pp. 116–120. IEEE (2015)
14. Swarnakar, J., Sarkar, P., Singh, L.J.: Rational approximation methods for a class of fractional-order SISO System in delta domain. In: Bera, R., Sarkar, S.K., Singh, O.P., Saikia, H. (eds.) Lecture Notes in Electrical Engineering, vol. 537, pp. 395–402. Springer, Singapore (2019)

Improving the Effectiveness of Energy Savings Measures at Companies by Means of a New Baseline Adjustment Strategy

Andrés David Rodríguez Toscano[1], Julio César Mojica Herazo[2],
Rafael Humberto Rojas Millan[2], Aurora Patricia Piñeres Castillo[2],
Moisés Hinojosa Rivera[3], and Jesús Silva[4(✉)]

[1] Energy Department, Universidad de la Costa (CUC), Barranquilla, Colombia
arodrigu83@cuc.edu.co
[2] Industrial, Agro-Industrial and Operations Engineering Department,
Universidad de la Costa (CUC), Barranquilla, Colombia
{jmojica5, rrojas}@cuc.edu.com, apineres2@cuc.edu.co
[3] Faculty of Mechanical and Electrical Engineering, Universidad Autónoma de
Nuevo León (UANL), Monterrey, Mexico
moises.hinojosar@uanl.mx
[4] Universidad Peruana de Ciencias Aplicadas, Lima, Peru
jesussilvaUPC@gmail.com

Abstract. This paper discusses a strategy for establishing an energy consumption baseline for the effects of defining and applying new strategies to improve the effectiveness of energy savings measures. Through this analysis, the energy baseline is adjusted to the dynamics of a typical operation, reducing uncertainty about operating data when it is not possible to determine that a given energy consumption level is typical. The strategy enables focusing efforts on the points in the operation with greatest impact on energy efficiency as a function of frequency of operation.

Keywords: Energy indicators · Energy baseline · Pareto

1 Introduction

Energy consumption continues to increase worldwide [1, 2]. In order to reduce energy consumption, standards and policies have been developed that enable the definition of scenarios that are suitable for improving energy efficiency in different economic sectors [3, 4]. The policies and standards are effectively implemented through strategies based on an energy baseline that characterizes the dynamics of the operation and energy consumption in the case under study [5].

In the specialized literature, energy efficiency is measured based on energy performance indicators (EPIs) against an energy baseline (EnB) [6]. The latter should represent the energy consumption dynamics and variability of the subject being studied [3]. In some cases, the strategies needed to improve energy efficiency are simple and of low cost [6]; however, others may involve switching technologies and substantial investments. Such investments must be carefully assessed using technical-economic

© Springer Nature Switzerland AG 2020
A. P. Pandian et al. (Eds.): ICICCS 2019, AISC 1039, pp. 78–85, 2020.
https://doi.org/10.1007/978-3-030-30465-2_10

criteria. At many companies, energy costs are associated with operating costs, and it is possible that the interaction of frequency of operation and energy performance indicators will improve the feasibility of energy efficiency projects.

Several studies display consumption dynamics in which operating points are concentrated in one area of the energy baseline [1, 7]; however, this characteristic is not associated with a typical frequency of operation as an input to better assess the savings measures to be implemented. The objective of the study is to develop a new strategy that will improve the effectiveness of energy savings measures by selecting the data with highest frequency of operation that represent the typical energy consumption behavior. This enables eliminating uncertainty regarding energy efficiency, while avoiding unnecessary efforts in non-typical operating areas with low probability of use.

2   Methodology

The specialized literature shows that at many companies electricity consumption is normally concentrated in certain points of their production, activity or variable [1, 7], i.e., the points of operation are often concentrated in a range of an activity, production level or a variable that has significant influence on the variability of electricity consumption. In such areas, the dynamics of electric energy consumption associated to production or the significant energy variable is expressed more clearly because the other points of the operation (atypical) may be affected by other dynamics associated with other factors that are often unknown and do not display a normal pattern of electricity consumption in a company or machine. For this reason, the proposal is to use a Pareto distribution diagram to define the baseline, in order to identify such behavior and implement the baseline based on points of operation that display the typical behavior of the subject of the study, as well as to adjust and focus savings measures in such points. This does not eliminate the use of data filtering methods; in fact, data filtering is implemented in the usual manner in this study. However, use of this tool enables obtaining a more realistic and consistent baseline. It should be highlighted that this strategy should only be implemented at companies where consumption is largely concentrated in a typical region of the operation and where it does not significantly affect the correlation of the baseline after normal data filtering.

Lastly, the results are compared using the cumulative sum (CUSUM) graph, using the usual methodology to establish the electric energy baseline by applying the strategy of selecting the operating points based on the Pareto distribution.

2.1   Case Study

In many countries, hotels are among the main drivers of the service economy and some of the largest energy consumers [5, 8, 9]. Use of energy at hotels is closely associated with each hotel's specific operating characteristics [9].

In Colombia, the services sector accounts for close to 6% of energy consumption and has been the fastest-growing sector in recent years [12–16]. Additionally, forecasts indicate that overall energy consumption in the country will grow quickly in the next few years [12, 13]. In Barranquilla, according to UPME, companies of this type consume more energy per square meter than in other cities of the country [10, 11, 15]. Companies of this type display seasonal operating trends due to the characteristics of the business. For this reason, the selected case study is a hotel in the city of Barranquilla, which has 98 rooms, a parking lot, a swimming pool, 5 air handling units for air conditioning, 98 mini-split air conditioners, 110 mini-refrigerators, 5 refrigerators, 5 electric motors and one LED lighting system.

Electric energy consumption performance at the hotel was measured using the energy performance indicators defined in the standard ISO 50006 [3, 4] and the standard ISO 50001 [4] was used as guide for the implementation of energy planning. Measurements were taken using a network analyzer and energy sensors installed by electric energy provider on the hotel property's boundary and a thermographic camera. The measurements were taken from October 2015 to October 2018.

2.2   Calculation of the Energy Baseline (EnB or $E_{baseline}$)

The electric energy baseline describes the dynamics of electric energy consumption as a function of the variables that significantly affect its variability. Normally at hotels, the EnB is established by means of a lineal correlation between hotel room daily occupancy (HDO) during the month and monthly electricity consumption.

$$EnB = m \bullet HDO + E_0 \tag{1}$$

m = Electric energy consumption associated with the HDO
E_0 = Electric energy consumption not associated to the HDO.

2.3   Calculation of the Energy Consumption Index (CI)

The energy consumption index expresses the relationship between electricity consumption and HDO. It is established for the purpose of increasing energy efficiency through HDO planning.

$$CI = m + \frac{E_0}{HDO} \tag{2}$$

2.4   Calculation of the Cumulative Sum Graph (CUSUM)

The cumulative sum indicator shows the variation between electric energy consumption against the EnB in a given period, to show the actual amount of energy saved based on current HDO.

$$CUSUM = \left((E_{real} - E_{baseline})_i + (E_{real} - E_{baseline})_{i-1} \right) \tag{3}$$

When the result is negative, it means that energy was saved, and when it is positive it means that energy was wasted.

EnB is the energy baseline at hotels.

3   Results

The results are shown below:

3.1   Determining the EnB and Selection of Operating Points that Characterize Consumption Dynamics at the Hotel

Figure 1 displays the EnB with no data filtering. It highlights that most operating points (indicate percentage) are concentrated in the red shared area.

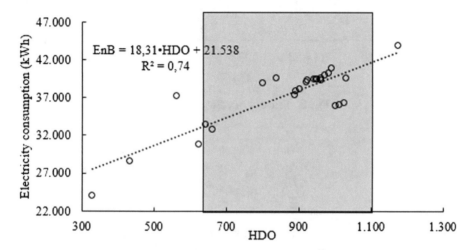

Fig. 1. Behavior of the baseline with no data filtering (Source: authors)

In order to identify the operating points with greater precision, a Pareto distribution graph was developed, as shown in Fig. 2, which indicates that 80% of operating points are concentrated in the range 500–1099 HDO. The 500–699 range only displays 2 of the 4 operating points.

Figure 3 displays EnB after having filtered out the atypical data and having selected the operating points that reflect the hotel's characteristic consumption dynamics as a function of HDO. The baseline shows that the energy associated with HDO is 18,304, in other words, each HDO consumes 18,304 kWh. The energy not associated with HDO is 22.181 kWh.

Figure 4 displays the behavior of CI in terms of the operating points and indicates that it worsens as HDO increases compared to the theoretical CI.

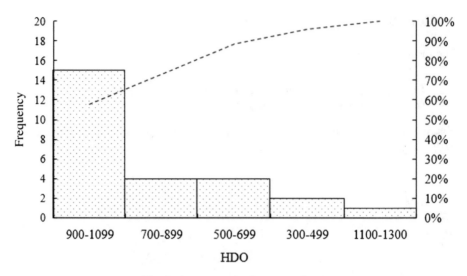

Fig. 2. Pareto graph (Source: authors)

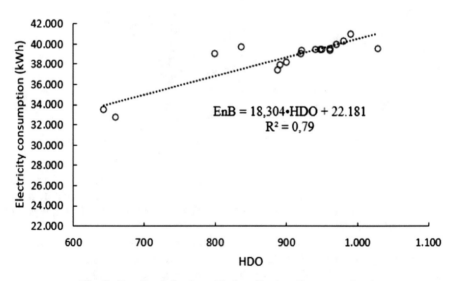

$$EnB = 18{,}304 \cdot HDO + 22.181$$
$$R^2 = 0{,}79$$

Fig. 3. Baseline behavior with data filtering (Source: authors)

3.2   Action Plan to Increase Energy Efficiency at the Hotel

With the aim of increasing energy efficiency, measurements were taken at the times when actual CI was below the expected level, and at the points of greatest energy consumption for those HDO. This enabled determining that the largest energy consumers were the air conditioning and refrigeration systems (based on a census of load and a Pareto distribution). It was found that whenever HDO was between 900 and 950, the employees would switch on the air handling units on every floor, regardless of

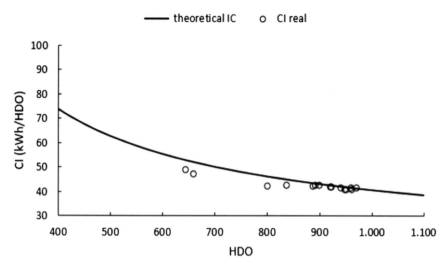

Fig. 4. CI behavior compared to operating points (Source: authors)

whether it was required or not. This was because rooms were randomly assigned to guests, and when occupancy rates were high the employees found it difficult to remember which floors had been switched on or off. Consequently, when occupancy was high, all units were turned on. It was decided to assign the staff on duty on each floor the duty of switching the units on or off as required.

Additionally, the temperature control set points of the mini-split units were increased to 2 °C higher than the minimum set point of the standard configuration. Also, presence sensors were installed in the rooms in order to shut down electric equipment whenever the guests leave the room. Automatic door closing timers were adjusted and the form and priority for filling the 5 largest refrigeration units were redistributed. The latter enabled the systems to complete the refrigeration cycle, because the thermal load of certain items that were placed in other refrigerators were too high for the HDO level and the compressors of these systems were practically working all day long. Lastly, a team was assigned to continuously monitor and control equipment use in accordance with standard demand at the hotel.

3.3 Comparison of the CUSUM Control Graph with and Without Selection of the Operating Control Points that Describe the Hotel's Dynamics

Figure 5 shows CUSUM after implementation the selection of the operating points that characterize electricity consumption dynamics as a function of HDO and after having implemented the energy efficiency measures over a period of 6 months, compared to the CUSUM without the strategy of selecting the operating points. It displays significant electricity savings with the strategy by not focusing on the operating points that are not representative of normal electricity consumption dynamics as a function of HDO.

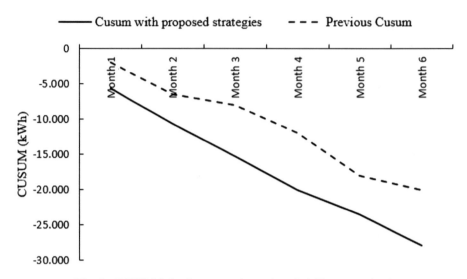

Fig. 5. CUSUM behavior over a 6-month period (Source: authors)

4 Conclusions

By implementing and focusing energy planning efforts on the most relevant operation points enables improving energy performance for more time and with better results. It also facilitates the identification of variables or causes that affect energy consumption and reduces uncertainty regarding its effects. In this case study it was found that in only 6 months the hotel was able to save approximately 28.000 kWh in electricity thanks to the selection of operating points, and without this strategy it would have saved 20.000 kWh over the same time period. This strategy improved energy management and produced savings in electricity consumption of up to 28,5%, compared to the standard procedure.

The study demonstrates that it is crucial for companies in the hotel industry to perform energy planning. It also indicates that in defining energy efficiency actions it is advisable to first focus on properly associating HDO to energy consumption, to use ON/OFF control systems based on presence parameters, to use equipment in accordance with the production dynamics that characterize electricity consumption and the hotel's HDO, to adjust the temperature set points in refrigeration and air conditioning equipment, because these are the largest energy consumers and are most used at the operating points with high operating frequency.

Lastly, this strategy should be further studied in different operational dynamics with different energy baselines in order to assess its effects in different sectors, because homogeneous operating conditions may exist that may have an influence on results. Additionally, it should be highlighted that performing a load census offers the potential of identifying with greater precision the effect of the systems that most consume energy in the company's most representative operating conditions. It should also be noted that the strategy implemented in a baseline with few points may not be sufficiently reliable because of the minimum requirements in terms of sample data, because the frequency of operation is low.

References

1. Lai, J.H.: Energy use and maintenance costs of upmarket hotels. Int. J. Hospitality Manage., 10 (2016)
2. Junghun, L., Seunghwan, Y., Jonghun, K., Doosam, S., Hakgeun, J.: Improvements to the customer baseline load (CBL) using standard. Energy **144**, 11 (2018)
3. ISO, 4 12 2014. https://www.sis.se/api/document/preview/918420/
4. ISO: ISO 50001. Requirements with guidance for Use. Energy Management Sytems (2011)
5. Cabello Eras, J.J., Sousa Santos, V., Sagastome Gutirrez, A., Guerra Palencia, M.A., Haeseldonckx, D., Vandecasteele, C.: Tools to improve forecasting and control of th electricity consumption in hotels. J. Cleaner Prod. **137**, 803–812 (2016)
6. Grimaldo Guerrero, J.W., Rodríguez Toscano, A.D., Vidal Pacheco, L., Osorio Tovar, J.: Analysis of the energetic and productive effects derived by the installation of a conveyor belt in the metal-mechanic industry. Int. J. Energy Econ. Policy **8**(6), 196–201 (2018)
7. Sagastume Gutiérrez, A., Cabello Eras, J.J., Sousa Santos, V., Hernández Herrera, H., Luc, H., Vandecasteele, C.: Electricity management in the production of lead-acid batteries: the industrial case. J. Cleaner Prod. **198**, 37 (2018)
8. Xu, P., Chan, E.H.W., Visscher, H.K., Zhang, X., Wu, Z.: Sustainable building energy efficiency retrofit for hotel buildings using EPC mechanism analytic Network Process (ANP) approach. J. Cleaner Prod. **107**, 378–388 (2015)
9. Hotel Energy Solutions: Analysis on Energy Use by European Hotels: Online Survey and Desk Research. Hotel Energy Solutions project publications (2011)
10. UPME: UPME (2006). http://www.upme.gov.co/sigic/documentosf/vol_2_diagnóstico_fnce.pdf
11. Unidad de planación Minero Energetica (UPME): Guía para el consumo consciente, racional y eficiente de la energía. Industria hotelera, comercial e institucional. Zonas cálidas tropicales de Colombia. San Andrés, Providencia y Santa Catalina, Amazonas y Chocó Charlies Impresores Ltda. (2014). http://www.si3ea.gov.co/LinkClick.aspx?fileticket= wru7z0gVd%2FI%3D&tabid=123&mid=449&language=en-US, ISBN: 978-958-8363-22-6
12. Unidad de Planeación Minero Energética (UPME): Proyección de Demanda de Energía en Colombia (2010)
13. Govierno de Australia: Energy Efficiency Opportunities in the Hotel Industrial Sector. Department of Industry, Turism and resources (2002)
14. Deng, S.: Energy and water uses and their performance explanatory indicators in hotels in Hong Kong. Energy Build **35**, 775–784 (2003)
15. UPME (2006). http://www.upme.gov.co/upme12/2007/upme13/caracterizacion_energetica_sectores.pdf
16. Amelec, V.: Validation of strategies to reduce exhausted shelf products in a pharmaceutical chain. Adv. Sci. Lett. **21**(5), 1403–1405 (2015)

Object Detection for Autonomous Vehicle Using TensorFlow

Sadanand Howal, Aishwarya Jadhav, Chandrakirti Arthshi[✉],
Sapana Nalavade, and Sonam Shinde

RIT, Islampur, India
sadanand.howal@ritindia.edu,
aishwaryadjadhav2014@gmail.com,
chandrakirtiarthshi@gmail.com,
sapananalawade121@gmail.com, sonam.97shinde@gmail.com

Abstract. The area of computer vision is emerging continually with the increasing interaction and development to provide a comfortable interaction between human and machines. One of the key aspects in the process of computer vision is object detection. Either objects can be identified partially or close to the original objects. The accuracy in detecting the objects can be improved by using state-of-the-art deep learning models like faster-Regional Convoluted Neural Network (faster-RCNN), You Only Look Once model (YOLO), Single Shot Detector (SSD) etc. Traditional algorithms can't recognize objects as efficiently due to its limitations. Whereas the deep learning models require large amount of data for training the dataset, which has more resource and labour intensive in nature. The selection of algorithm determines its precision in object detection as well as its reliability. The recognition and classification of object begins with preparing dataset followed by splitting the dataset into training dataset and test dataset. The task of training the dataset can be assisted by both traditional as well as modern deep neural networks. The loss per step or epoch is calculated on the training dataset to signify the efficiency and accuracy of the model. In this model, the loss per step is 2.73. We have achieved a maximum accuracy of about 85.18% after training the dataset used.

Keywords: Object detection · Deep neural network · Image segmentation · Computer vision · Tensorflow object detection API · Self-driving vehicle

1 Introduction

Object detection is the blooming research area in the field of computer vision. The ability to identify and recognize objects either in single or more than one image frame can gain extreme importance in various ways as while driving the vehicle, the driver cannot identify objects properly due to the dearth of attention, reflection of light, anonymous objects etc. which may lead to fatal accidents. In order to overcome such perceptible problems, autonomous vehicles and Advanced Driver Assistance System (ADAS) took the generous task of object detection and classification. The task of computer vision is performed in the following steps:

© Springer Nature Switzerland AG 2020
A. P. Pandian et al. (Eds.): ICICCS 2019, AISC 1039, pp. 86–93, 2020.
https://doi.org/10.1007/978-3-030-30465-2_11

1. Classification of object in image
2. Localization of object in image
3. Object detection
4. Segmentation of image.

The application of object detection can be found in advanced robotics, defense systems, surveillance systems, space research, face recognition and many more. The idea of self-driving vehicles has been evolving with the advancement in techniques related to the task of identifying and extracting features from the objects. Object detection for self-driven vehicles is a non-trivial task in order to navigate on the road.

The evolution of deep neural networks have changed the aspect of computer vision over the traditional methods. Conventional machine learning and computer vision models plays a prevalent role in the process of object classification, however the industry now heavily relies on the deep learning based classifiers. The emergence of graphical processing units (GPU) has led to more efficient and convenience in achieving the task of object classification through deep neural network models. These models try to learn important features corresponding to each class that are inspired from the biological structure of neurons in humans. Google's Tensorflow is one such machine learning framework which works on dataflow programming among a range of tasks. Nodes in TensorFlow represent mathematical operations and the graph edges represent multidimensional arrays called as Tensors. Tensorflow object detection API is capable in detecting objects in an image with good accuracy it is also able to detect objects in live streaming video with a good degree of precision in which speed of frames is about 20–30 frames per second.

We propose the use of Tensorflow object detection API for our dataset to train and test the dataset in order to detect objects successfully for an autonomous vehicle.

2 Related Literature

The deep learning models for object detection are capable of identifying the objects but the accuracy of detection for each model varies. In [1] the authors have practically performed object detection using SSD model, which was very fast in producing the results but the accuracy was very less. Whereas in case of Faster-RCNN the accuracy of the detected object was high as compared to SSD but the time required to produce the results was also more as compared to SSD.

In [2] the researchers have used faster-RCNN for feature extraction and object detection on the dataset prepared by them for docking space crafts at space station. They concluded that although some visible features that has low difference from the surroundings or any other region of the station were never detected. Thus selecting appropriate model is of significant importance as per requirement and resources available.

In [13] the authors have supported the expressivity of Deep Neural Network (DNN) for object detector. But the results came at some computational cost at training time i.e., one has to train a network per object type and mask type.

Thus it was critically very essential to select a model which has good accuracy as well as fast result production in order to meet the requirement of object detection in self driving vehicles.

TensorFlow overcomes the problem of both accuracy and speed in a live video stream which is equivalent to problem of object detection for a driverless vehicle. Moreover, object detection API in TensorFlow along with the MobileNet neural network allows us to create dataset with low computation cost which can be exported easily for the task of object detection.

3   Methodology

Image classification can achieve various coups but the major problem is that the model can identify only one class per image. Whereas in case of object detection model more than one class per image can be classified along with the position of the object in the image with a bounding box around the object (Fig. 1).

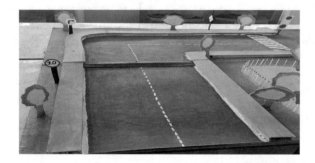

Fig. 1. The test-bench created for the demo vehicle

We began by collecting our dataset of the test bench that we have created for the navigation of the demo test car. The test car was mounted with a camera setup for capturing the images of objects on the test bench. The rest process is followed in the steps as follows:

1. Preparing dataset: The camera setup mounted on the test car captures the images of objects on the test bench. Ideally 250–300 test images are to be captured of each object on the test-bench. The collected images were to be split into two sub datasets i.e., train and test. Of the total images captured 10% were used as testing and remaining images were used for training the dataset.

2. Creating bounding box: For creating the bounding box around the test images, the image's height, width and each class with parameters like x_{min}, x_{max}, y_{min}, y_{max} are required. The bounding box captures exactly the class of the object in the image. This follows the task of creating labels for the test images. Labels are created by using 'labelImg' tool. The labels are stored into individual xml label for each image which further need to be converted into csv file for training (Figs. 2 and 3).

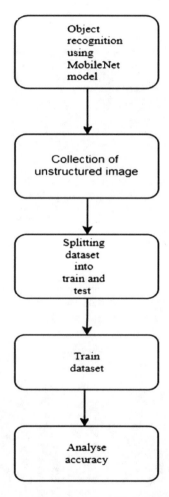

Fig. 2. Proposed workflow for object detection model

3. Converting csv file into Tensorflow Record (TFRecord): For each training and testing dataset, a csv file is obtained which is further converted into TFRecord. The TFRecord is a format for storing the sequential structured data into binary strings (Fig. 4).

4. Selecting a model: SSD model along with MobileNet neural network is selected as it provides moderate efficiency and the rate of result production is faster. The MobileNet is a light weight neural network as it consumes low processing power (Fig. 5).

Fig. 3. Object from the testbench with label and accuracy percentage

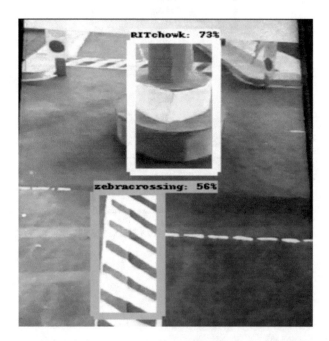

Fig. 4. Two classes of two different objects captured during run-time execution

5. Retraining the model with data: A file (pbtxt) containing records of all the classes with their attributes is created and stored in the training directory. The configuration file for the selected model is executed such that the training of dataset starts showing the losses and checkpoints at step-wise (Fig. 6).
6. Generating Loss graph: The proper working of the module can be estimated when the loss per step is under 3. The lower loss per step implies to greater accuracy. In our model the loss per step is 2.73. The loss per step decreases on increasing the number of steps thus ultimately increasing the number of images in the dataset.

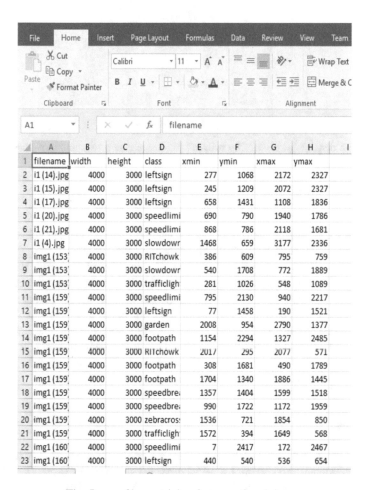

	A	B	C	D	E	F	G	H	I
1	filename	width	height	class	xmin	ymin	xmax	ymax	
2	i1 (14).jpg	4000	3000	leftsign	277	1068	2172	2327	
3	i1 (15).jpg	4000	3000	leftsign	245	1209	2072	2327	
4	i1 (17).jpg	4000	3000	leftsign	658	1431	1108	1836	
5	i1 (20).jpg	4000	3000	speedlimi	690	790	1940	1786	
6	i1 (21).jpg	4000	3000	speedlimi	868	786	2118	1681	
7	i1 (4).jpg	4000	3000	slowdowr	1468	659	3177	2336	
8	img1 (153	4000	3000	RITchowk	386	609	795	759	
9	img1 (153	4000	3000	slowdowr	540	1708	772	1889	
10	img1 (153	4000	3000	trafficligh	281	1026	548	1089	
11	img1 (159	4000	3000	speedlimi	795	2130	940	2217	
12	img1 (159	4000	3000	leftsign	77	1458	190	1521	
13	img1 (159	4000	3000	garden	2008	954	2790	1377	
14	img1 (159	4000	3000	footpath	1154	2294	1327	2485	
15	img1 (159	4000	3000	RITchowk	2017	295	2077	571	
16	img1 (159	4000	3000	footpath	308	1681	490	1789	
17	img1 (159	4000	3000	footpath	1704	1340	1886	1445	
18	img1 (159	4000	3000	speedbrei	1357	1404	1599	1518	
19	img1 (159	4000	3000	speedbrei	990	1722	1172	1959	
20	img1 (159	4000	3000	zebracros:	1536	721	1854	850	
21	img1 (159	4000	3000	trafficligh	1572	394	1649	568	
22	img1 (160	4000	3000	speedlimi	7	2417	172	2467	
23	img1 (160	4000	3000	leftsign	440	540	536	654	

Fig. 5. csv file containing features of each image

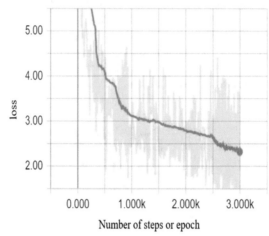

Fig. 6. Graph displaying relationship between the loss per step vs number of steps (drawn on TensorBoard)

4 Conclusion

In this work we leveraged the task of object detection for self-driving vehicle by using TensorFlow API followed by MobileNet neural network. The efficiency in detection for objects is about 85.18%, which is above average, but the rate of result production is very fast. The loss per step or epoch is 2.73 (under 3) that supervises the reliability of the model. As for now we have tested the model on the dataset prepared from the testbench. The model works fine in identifying object in an image but for multiple objects in an image the bounding box shifts from one object to another inconsistently. Such inconsistency can be overcome by increasing the computation cost as well as dataset. We are planning to augment the model on the actual electric vehicle for performing object recognition and classification.

References

1. Shetty, J., Jogi, P.S., Pandian, D., et al. (eds.): Study on Different Region Based Object Detection Models Applied to Live Video Stream and Images Using Deep Learning. Proceedings of the International Conference on ISMAC in Computational Vision and Bio-Engineering 2018 (ISMAC-CVB). Lecture Notes in Computational Vision (2018). https://doi.org/10.1007/978-3-030-00665-5_6
2. Fomin, I., Gromoshinskii, D., Bakhshiev, A., Kryzhanovsky, B., et al. (eds.): Advances in Neural Computation, Machine Learning, and Cognitive Research. SCI, vol. 736. https://doi.org/10.1007/978-3-319-66604-4_12

3. Lin, C., Li, L., Luo, W., Kelvin, C.P., Wang, J.G.: Transfer learning based traffic sign recognition using inception-v3 model. https://doi.org/10.3311/PPtr.11480
4. Saha, S., Tairin, S., Khaled, M.A.B., Saha, S., et al.: An efficient traffic sign recognition approach using a novel deep neural network selection architecture. In: Proceedings of IEMIS 2018, vol. 3. https://doi.org/10.1007/978-981-13-1501-5_74
5. Talukdar, J., Gupta, S., Rajpura, P.S., Hegde, R.S.: Transfer learning for object detection using state-of-the-art deep neural networks. In: 5th International Conference on Signal Processing and Integrated Networks (SPIN). 978-1-5386-3045-7/18
6. Abadi, M., et al.: TensorFlow: large-scale machine learning on heterogeneous distributed systems. arXiv:1603.04467, pp. 1–19 (2016)
7. Sapp, A.S., Ng, A.Y.: A fast data collection and augmentation procedure for object recognition. In: Proceedings of the AAAI, Chicago, IL, USA, pp. 1402–1408 (2008)
8. Hoffman, C., Thiagarajan, D.: Continuity report: Revisiting grocery recognition using tensorflow, to be published
9. Harzallah, H., Jurie, F., Schmid, C.: Combining efficient object localization and image classification. In: Proceedings of the IEEE 12th International Conference Computer Vision, pp. 237–244, September/October 2009
10. Cheang, E.K., Cheang, T.K., Tay, Y.H.: Using convolutional neural networks to count palm trees in satellite images (2017). https://arxiv.org/abs/1701.06462
11. Shetty, S., Karpathy, A., Toderici, G., Leung, T., Sukthankar, R., Fei-Fei, L.: Large-scale video classification with convolutional neural networks. In: Proceedings of the IEEE Conference Computer Vision and Pattern Recognition, pp. 1725–1732, June 2014
12. Hinton, G.E., Srivastava, N., Krizhevsky, A., Sutskever, I., Salakhutdinov, R.R.: Improving neural networks by preventing co-adaptation of feature detectors (2012). https://arxiv.org/abs/1207.0580
13. Szegedy, C., Toshev, A., Erhan, D.: Deep neural networks for object detection. In: Advances in Neural Information Processing Systems, pp. 2553–2561 (2013)
14. Avramović, A., Ševo, I.: Convolutional neural network based automatic object detection on aerial images. IEEE Geosci. Remote Sens. Lett. 13(5), 740–744 (2016)
15. Ahmad, T., Ahmad, T., Ilstrup, D., Emami, E., Bebis, G.: Symbolic road marking recognition using convolutional neural networks. In: 2017 IEEE Intelligent Vehicles Symposium (IV), pp. 1428–1433 (2017)
16. Bruno, D.R., Osorio, F.S.: Image classification system based on deep learning applied to the recognition of traffic signs for intelligent robotic vehicle navigation purposes. In: 2017 Latin American Robotics Symposium (LARS) and 2017 Brazilian Symposium on Robotics (SBR) (2017). https://doi.org/10.1109/sbr-lars-r.2017.8215287
17. Szegedy, C., Vanhoucke, V., Ioffe, S., Shlens, J., Wojna, Z.: Rethinking the inception architecture for computer vision. In: 2016 IEEE Conference on Computer Vision and Pattern Recognition (CVPR), pp. 2818–2826 (2016)
18. Zaklouta, F., Stanciulescu, B.: Random forests CAOR traffic sign classification using K-d trees and Random forests traffic sign classification using K-d trees and random forests. In: International Joint Conference on Neural Networks (IJCNN), August 2011
19. Sermanet, P., Lecun, Y.: Multi-scale CNNs sermanet traffic sign recognition with multi-scale convolutional networks traffic sign recognition with multi-scale convolutional networks. In: International Joint Conference on Neural Networks (IJCNN), August 2011
20. Sudarshan, D.P., Raj, S.: Object recognition in images using convolutional neural network. In: 2nd International Conference on Inventive Systems and Control (ICISC) (2018). https://doi.org/10.1109/icisc.2018.8398893

Threat Modeling for Breaking of CAPTCHA System

Divya Suvarna[(⊠)] and Sujata Pathak[(⊠)]

Department of Information Technology,
K. J. Somaiya College of Engineering, Mumbai, India
{divya.suvarna, sujatapathak}@somaiya.edu

Abstract. The online websites are accessed by millions of people and the information present on it holds value. To secure them from attacker, one such mechanism is "Completely Automated Public Turing Test to keep the Computers and Humans Apart". They are used to ensure that internet user's activity is performed by humans only and not the bots. CAPTCHAs are solved by people every day to prevent Denial of Service attack and online spam attack. But unfortunately, it is now possible to break them by using Machine Learning. This paper presents, the Vulnerabilities related to Text-based CAPTCHA System, compromised system using Machine Learning and proposed Algorithm. A Threat Modeling was performed on the website using a Text-based CAPTCHA System in order to discover various Attack Vectors with the help of a Tool and performs detailed analysis on affected areas. Lastly, a solution is provided to the website service provider to overcome the exsisting system flaws and also to make them even more strong and secure.

Keywords: CAPTCHA · Machine Learning Algorithm · Cyber Attacks · Threat Modeling · Word Cloud

1 Introduction

Many websites, online services, e-banking services, online tolling and digitize books use the completely Automated Public Turing tests to identify whether the user is Computer or Human. For Example, Facebook restrict the creation of fake profiles to spam legitimate users, Google use re-CAPTCHA to improve its services by blocking access to automated spammers, and Yahoo mail improves by blocking bots from spamming the Mail.

Today we have a different form of CAPTCHAs such as Image, Audio, Video, 3D, Text based CAPTCHA. And threats against the text based CAPTCHA is increasing. As it is understood annoying but yet bulletproof. But this understanding is challenged today. Various research studies state that there is no standard methodology for designing and evaluating it. With the use of existing technology hackers can easily break the existing security mechanism. By finding the vulnerability in the existing system it is possible to disrupt the availability of the services. Prakash et al. in this paper, authors have studied different types of CAPTCHA and a survey has been done

© Springer Nature Switzerland AG 2020
A. P. Pandian et al. (Eds.): ICICCS 2019, AISC 1039, pp. 94–104, 2020.
https://doi.org/10.1007/978-3-030-30465-2_12

in order to list down the drawbacks of different types of CAPTCHAs [11]. The idea behind this system is good in theory but the design of it is weak.

One of the creators of CAPTCHA is Luis Von Ahn, he created something that could challenge bots online, a brainpower that is capable to read distorted text, gimpy text, 3D image, image with lines and text and many more. This brainpower can only be solved by Humans. But unfortunately the technology has become more advanced that bot use ML, which is capable of learning online activities and behaving somewhat like humans. For Example: GATE Exam Online registration form, internet banking and social media. In this study, a Threat Modeling is done on the text based CAPTCHA System to address the threat related to it. Threat Modeling involves identifying all attack cases related to the system workflow and arranging them in the form of attack trees. ML is considered as the Subset of Artificial Intelligence (AI), an application of AI that allows machine (system) to automatically learn from their experiences. Earlier Humans use to learn from past experiences and machine use to follow instructions given by the human to perform task, but now they can train machine very easily which is called ML Algorithm. There are various ML capable tools available such as Tesseract, GOCR, SimpleOCR, and MicrosoftOneNote. One such tool used in this project is Tesseract.

Tesseract is an Optical Character Recognition Engine for Character accuracy. Tesseract OCR is executed from Command Line Interface. Tesseract uses some Machine Learning concept and that is Long Short-Term Memory (LSTM). LSTM is a small unit of Recurrent Neural Network (RNN). These LSTM network composed of a cell, input gate, output gate and forget gate. The cell remembers the arbitrary value and the other three gates contribute to the incoming and outgoing flow of information.

In this paper, Sect. 1 gives a background of security system that website uses. The motive behind selecting the topic and concern related to it. In Sect. 2, Literature Survey on the research topic is elaborated in detail. In Sect. 3, Methodology explains the breaking of Text-based CAPTCHA by using Tesseract OCR and created algorithm. In Sect. 4, results are provided after threat modeling of weak system and importance of it to the website owner. Lastly, conclusion is provided with all the possible outcomes that are achieved followed by the references.

2 Literature Survey

In the paper [1], author Chandavale et al. 2009 reviewed the importance of CAPTCHA to prevent bots attack. It is of two types: (i) Visual and (ii) Audio CAPTCHA. The implementation work is divided into different phases such as pre-processing, segmentation, feature extraction and character recognition. Following the system architecture, authors could be able to get a success of at least 80% image.

Moradi et al. [2] addresses the most crucial issues related to CAPTCHA system such as security, usability and accessibility are becoming more controversial. In this paper, efforts were made to review various state-of-the-art CAPTCHA techniques, so that an alternative solution is provided in a tabular view. Based on the classification of various CAPTCHA, problems associated with it was mentioned and then it is compared. The evaluation aims to bring new alternatives in the near future to defeat the incompleteness in the security mechanism.

Lee et al. 2011 [3] main focus was to study the conducted experiment of various age group and distortion types on the CAPTCHA task performance (Text based), objective visual fatigue and subjective workload. There were 24 participants in two age groups. The result of non-parametric data Analysis was done were one type with no distortion and five types with common distortion techniques were considered. They found out that the two age groups differ in response time, error rate and NASA-TLX score. There was a significant effect of distortion type on error rate. The study has vital indications for the design of a CAPTCHA system. Before bringing to the public it is very important to test and verify its usability. The paper addresses the user to solve it in three general steps: recognition, rehearsal and motor response. Also the authors expressed certain issues related to gender difference, learning curve and interaction between native language and character set used in it are also important.

Internet is a free service and with that computer legitimate users are under serious tension. Due to online threats hackers have become more versatile with their nefarious behaviour. Attacks like Spoofing, Man in the Middle attack have become more common. To protect online services, one such method is called CAPTCHA. In this paper Umadevi et al. [4] aimed to present a Model using neural network called CAPTCHA Recognition Using Neural Network (CRNN). Following is the work done by the authors: a system to secure from the bot, a method to extract text from any image type, pre-processing of image and each step is explained in detailed. They used the method called "gradient method" from which the features are extracted. Results are provides in the form of graph which shows simulation convergence and validation check.

Google's invention re-CAPTCHA after CAPTCHA is considered to be the improvised security mechanism. But unfortunately according to author Jonathan Jung re-CAPTCHA is considered to be unethical under several different frameworks and is also in violation of American law(s). These theoretical arguments can be overcome by making few changes to make it legal and ethical. With minimal CAPTCHA and re-CAPTCHA method is shown in the paper in which the current re-CAPTCHA has two set of work to be completed: CAPTCHA to determine whether solver is human and the re-CAPTCHA portion is for productivity [5].

Internet is the backbone of Web Application, and to make it secure Google CAPTCHA is quite effective. Visual CAPTCHAs are commonly but we also have Audio CAPTCHAs for visually impaired users. Most of them can be broken by using ML technique. In this paper, the authors Darnstadt et al. [6] have presented an attack to break Audio CAPTCHAs by using active and semi-supervised learning method.

Sharma et al. [7] mentioned the need for developing new CAPTCHAs in order to safely secure against bots, e-mail accounts fraud, online polls fraud and worm contained in e-mails. Being a survey paper authors focuses on three important Optical Character Recognition CAPTCHAs. This survey paper provides various Applications and Attacks on Text Based CAPTCHA. The authors also explain the working of an OCR system with a Diagram.

Nasser Mohammed Al-Fanna discusses seven existing safeguards to protect CAPTCHA against cyber-attacks. Attackers now-a-days have found a way to attack them with the help of a bot. But these bot are meant for breaking one type of

CAPTCHA. Author's main intension was the build a Defence in depth layer. In this paper [8], CAPTCHAs safeguards comparison is mentioned in a table format that is important to fight against the bot and should be implemented.

Every Online Web Application includes security features in order to keep their companies and user data safe from the hackers. In this paper, author Sharma et al. [9] says that OCR is not able to read CAPTCHA and still it is possible to attack CAPTCHA image. OCR text, image, audio, video all are vulnerable to attacks by computer. Authors in their research mentioned that they have identified vulnerabilities and categorized them in three domains: breaching client-site trust, manipulating server-side implementation and attacking the CAPTCHA image. In the end authors state that one must try to increase the security of CAPTCHA system in order to prevent attacks.

In this paper, Jain et al. [10] explains that if we elevate the state of CAPTCHA security by adding noise and distortion it will be very difficult for the humans to read it. Because there is a limit to which humans can read distorted text due to which results higher error rate. These highly distorted texts can be segmented with OCR technique.

In the paper [12], Yan et al. 2007 explained how Visual CAPTCHA is widely used across the internet. They documented the breaking of it and provided at Captchaservice. org. The system was not harmed by attacks conducted using a high-quality OCR program and hence proved to be secured. But it was noticed the attack was performed on fatal design flaw and developed a simple attack with 100 success rate. This was possible due to design error which was discovered in each scheme.

3   Methodology

In Sect. 3 of Methodology, the paper presents method, ML OCR, and algorithm to break the system. The flow of the methodology starts from Manual breaking of CAPTCHA using Machine Learning Algorithm (OCR), and Algorithm for breaking of CAPTCHA System.

3.1   Manual Breaking of Text-Based CAPTCHA Using Machine Learning Algorithm OCR

Web CAPTCHA is intended to secure your html page where clients send data to others clients or databases. With this CAPTCHA no one but human can finish the security code and just a human will have the capacity to utilize your page. In this experiment, we considered text based CAPTCHA so that we understand the Vulnerable System and flaws in it. The step followed as: First the CAPTCHAs are saved in multiple folder based on their types, which then act as an input to the logic code. Next the logic in the code explains that the code with ask for choices: On selecting the desired option the code will take the input from the source folder and with the help of Machine Learning Algorithm the image CAPTCHA is broken down into piece (decoded).

Once the CAPTCHA is decoded it will send the output to the destination text file. The Machine Learning Algorithm used in the experiment is "Tesseract-OCR". This package contains associate degree OCR engine – libtesseract and a command program – tesseract. Tesseract 4 adds a brand new neural web (LSTM) primarily based OCR engine that is concentrated on line recognition. Tesseract has Unicode (UTF-8) bolster, and can perceive in excess of 100 dialects "out of the crate".

Tesseract bolsters different yield positions: plain-content, hocr (html), pdf, tsv, undetectable content just pdf. ImageMagick is a free and open-source programming suite for showing, changing over, and altering raster picture and vector picture documents. It can peruse and compose more than 200 picture record positions. Some features related to ImageMagick are: Format conversion, transform resize image, add shapes or text to an image and many more. Once the image is processed the output in obtained in the given folder path and file name. The CAPTCHA input image size is of around 4 to 6 KB.

The CAPTCHA Decoding process is explained in Five Steps. Step 1 is to open and execute the main program from an IDE, Step 2 is display of choices which is to be entered by the user, Step 3 is the algorithm decoding the CAPTCHA, Step 4 is Segmentation performed by the given Logic by (Tesseract) and Lastly Step 5 gives the output specified to the destination path.

STEP 1: Open the Logic (Code) in Eclipse IDE for CAPTCHA breaking. In Fig. 1, the Pre-processing logic in the code is being executed. By using IDE one can run the code.

Fig. 1. Code snippet for pre-processing logic

STEP 2: Run the Program and Choose any Option
In Fig. 2, choices provided by the code are (1) Captcha with Noise and Lines (2) Captcha in black ball (3) Captcha with noise (4) Captcha with lines (5) Captcha with background and exit mode. One can choose any choices and press enter to see the result.

Fig. 2. Enter choice for CAPTCHA

STEP 3: Take the Input File and Pre Process it in Machine Learning Algorithm
In Fig. 3, by using the code logic and OCR the text based CAPTCHA selected by the user can be broken.

Fig. 3. Succesfully text based CAPTCHA broken

STEP 4: Segmentation performed by the given Logic. Segmentation is performed itself by the Machine Learning Algorithm (Tesseract-OCR)
STEP 5: Output is provided in the given destination file (Solved CAPTCHA). In Fig. 4, the result file will be generated by the code itself and will be saved in the given specified path.

Fig. 4. Desired output

3.2   Algorithm for Breaking Text Based CAPTCHA

The below given algorithm is the logic behind breaking the Text-based CAPTCHA. The Steps involved are given below.

Step 1: Initialize the code (Start the program).
Step 2: Initialize the image path.
Step 3: Take the input CAPTCHA image by the user.
Step 4: Users are allowed to choose any one options from the choices available in the program for different types of CAPTCHAs.
Step 5: If Case 1 is selected then condition 1 is executed or Case 2 is selected then condition 2 is executed or Case 3 is selected then condition 3 is executed or Case 4 is selected then condition 4 is executed or Case 5 is selected then condition 5 is executed and Case can be chosen in random order.

Case 1: APPLY IMAGE PROCESSING BY THIS QUERY. Input_image –colorspace gray –threshold 50% output input
Case 2: APPLY IMAGE PROCESSING BY THIS QUERY. Input_image –fuzz 50% -fill black –opaque black –bordercolor white –border 2 –fill black –draw "color 010 floodfill" –alpha off –negate output_image
Case 3: APPLY IMAGE PROCESSING BY THIS QUERY. Input_image –gaussian – blur o –threshold 25% -point 1 output_image
Case 4: Input_image –negate –morphology erode octagon: 1 -negate -threshold 70% output_image
Case 5: Input_image –threshold 50% output_image

Step 6: if Invalid Choice is selected then the program will End (Stop).
Step 7: Once the Condition is done with image processing the output is given to the OCR (Tesseract).
Step 8: Text format is created.
Step 9: Stop

4   Results

Threat Modeling for a Web Application Using Text-based CAPTCHA System: In this chapter, results for Threat Modeling performed for a Web Application using Text-based CAPTCHA System and followed by Attack tree Model.

4.1   Threat Modeling

Threat Model is a deep observed method that is used to identify threats from more strategic perspective. It helps in discovering unknown threats and vulnerabilities. Steps involved in creating Threat Model are Target system specification followed by determining and ranking of threats and finally determining countermeasure(s) and mitigations techniques. One can use STRIDE Model or an Attack Tree Model. For our Threat Modeling, we adopted Attack Tree method in order to study the new threats associated with CAPTCHA system on a Dummy Web application. The tool used in this paper for threat modeling is "AD Tool" (Attack Defense Tree). An AD tree tool has a root node representing Goal which defines the processes that might lead to attacks target system. The sub-goals are represented in branch from in the tree. The leaf nodes of the tree are the actions to be implemented by the malicious user. In the AD tool, ovals/circle in red color represents attack node and rectangle in green color represents defense node. In this paper, Threat Modeling is performed on the CAPTCHA System used by a dummy web application. In our discussions we mainly focused on attacking part of the tree.

4.2   Discussion on Web Application for Using Text-Based CAPTCHA System

In this paper, a Web Application with in-built Text based CAPTCHA System is created so that we can perform breaking of the system. The attack is achievable by simply taking control over the Victim's PC. The Web Application is based on Java and being hosted on the Tomcat Server. The Web Application has two pages such as the first page consists of three component Login domain- username and password and CAPTCHA System. The second page contains a simple dashboard.

  The main motive behind this is to show that a hacker can easily target the weak mechanism and can try to attack Victim's Web or try to take control over their computer activities. With this it gives hacker an advantage to stay one step ahead of the legitimate user in committing Internet crime. The flow of the attack is easy to understand, a hacker try guessing the user credentials and tries for login attempt but hacker can only be sure after successful filling of CAPTCHA. If the provided input is correct, the hacker will be allowed to enter the Victim's account and if it is incorrect, then hacker is not allowed to access the account. Hence it is important that all the details are correctly filled in by the hacker. The take away from this attack is though it is Google's well known security feature it is vulnerable at some point which can be exploited. Therefore the web owner should follow "Threat Modeling" in order to secure their webpages.

4.3   Attack Tree

An Attack Tree is provided for the Text based CAPTCHA breaking System. This is done in order to discover existing vulnerability and threat on the system. It gives an elaborative and descriptive overview of the scenario. By creation an Attack tree one can easily prevent threat and try to get countermeasures. The Attack tree is provided for Four Attack scenario, such as (1) Dos attack on Victim's Operating System (2) Dos attack on Dummy Server (3) Enumeration Attack on Username credentials (4) AD Tree for Weak CAPTCHA system.

1. **Denial of Service attack on Victim's Operating System-PING (Unavailability)**

    Aim: To Target the Victim's OS by flooding with PING requests (Fig. 5).

Fig. 5. Denial of service attack on Victim's OS

Scenario steps for Exploitation: The target system for DoS attack is the Victim's Operating System. The attacker will first try to locate the Victim's location and IP address. This is possible with Foot-printing or Social Engineering attack. Once IP address is obtained attacker can now PING the Victim to check whether the system is responsive or alive. If the PING reply comes, that means system is awake and ready for attack and if it does not reply then it is not possible to perform ping request. Once PING request is possible, attacker can send huge PING request and slow down the operating system. Due to which user will experience performance issue. Denial of Service attack Tools: Nemesy and PING of Death.

2. **Denial of Service attack on Dummy Server-PING (Unavailability)**

    Aim: To Target the Victim's Web Server by flooding with PING requests (Fig. 6).

Scenario Steps for Exploitation: This time the target system for DoS attack is the Victim's Web Server. The attacker will first try to gather resources related to Victim's location and IP address and Services running. This is possible with Foot-printing or

Fig. 6. Denial of service attack on webserver

Social Engineering attack. Once IP address of Server is obtained attacker can now PING the Server to check whether the system is responsive, alive or dead. If the PING reply comes, that means system is awake and ready for attack and if it does not reply then it is not possible to perform ping request. Once PING request is possible, attacker can send huge PING request and slow down the Server until all server's memory is been consumed and the server crashes. Due to which user will experience Service unavailability issues.

3. **Enumeration Attack on Username credentials (User Authentication Threat)**

   Aim: To Exploit target Victim's Credentials (Fig. 7).

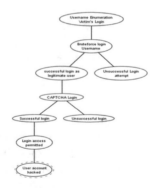

Fig. 7. Username enumeration attack

Scenario Steps for Exploitation: First step is to Footprint Victim and then take over Victim's Webpage by session Hijacking. The attacker will check for various attempts such as default username and password or weak username and password. Next challenge is to crack CAPTCHA system, once that is done attacker can get access to the Victim's account. Otherwise Web Application will block the user from entering.

4. **AD Tree for Weak CAPTCHA system**

   Aim: To provide Countermeasure for weak CAPTCHA system (Fig. 8).

Fig. 8. Alternative solution for weak CAPTCHA system.

Scenario steps for Exploitation: Initially attacker will get access of Victim's browser. Next step is to enter the credentials rightly in the login options. Also the attacker needs to break the CAPTCHA. Once all this activity is performed one can say that Account is been hacked. Though CAPTCHAs are used by many website, but the assurance level of security is very low. Possible attacks on Weak CAPTCHAs Websites are Man in the middle attack, User impersonation, Cookie hijacking, and DoS/DDoS attacks Therefore we will introduce Word Cloud in our Future to make it strong.

5   Conclusion

A great deal of work has been done in Improving CAPTCHA's ease of use and Security, one such model is the utilization of re-CAPTCHA but since it has not proved to be a smart idea, hackers have an eye on such weak system. The development of late appearances made it difficult to anticipate computerized bots and the different perilous spammers against CAPTCHA assaults. This paper reviews the use of CAPTCHAs in web applications, adverse effects of Cyber Attacks for Legitimate computer users, AD Tree Tool representation for attacks on Weak CAPTCHA. Experienced hackers are trying to invade the Security Mechanism by which they can disrupt computer user Experiences. Though we cannot build a full proof Security Controls, there will be certain Vulnerabilities present which can be exploited. Vulnerabilities such as abuse of functionality, and authentication bypass. The results of breaking of text based CAPTCHA was achieved successfully. On realizing such threat, a threat modeling was performed on web application using such weak system. This could help every Web owner to take preventive measures from attack vectors. Solutions like dynamic CAPTCHA system, more robust system, and use of multi factor CAPTCHA model can be used by the Website owner. Therefore this paper provides research, concerns, and alternative to Text based CAPTCHA System.

References

1. Chandavale, A.A., Sapkal, A.M., Jalnekar, R.M.: Algorithm to break visual CAPTCHA. In: 2009 2nd International Conference on IEEE Emerging Trends in Engineering and Technology (ICETET) (2009)
2. Moradi, M., Keyvanpour, M.: CAPTCHA and its alternatives: a review. Secur. Commun. Networks **8**(12), 2135–2156 (2015)
3. Lee, Y.-L., Hsu, C.-H.: Usability study of text-based CAPTCHAs. Displays **32**(2), 81–86 (2011)
4. Zhang, T., Zheng, H., Zhang, L.: Verification CAPTCHA based on deep learning. In: 2018 37th Chinese Control Conference (CCC), Wuhan, pp. 9056–9060 (2018)
5. Lung, J.: Ethical and legal considerations of reCAPTCHA. In: 2012 Tenth Annual International Conference on Privacy, Security and Trust (PST). IEEE (2012)
6. Darnstädt, M., Meutzner, H., Kolossa, D.: Reducing the cost of breaking audio CAPTCHAs by active and semi-supervised learning. In: 2014 13th International Conference on Machine Learning and Applications (ICMLA). IEEE (2014)
7. Sharma, S., Seth, N.: Survey of text CAPTCHA techniques and attacks. Int. J. Eng. Trends Technol. **22**(6) (2015)
8. Al-Fannah, N.M.: Making defeating CAPTCHAs harder for bots. arXiv preprint arXiv:1704.02803 (2017)
9. Sharma, P., Tyagi, N., Singhal, D.: CAPTCHAs: vulnerability to attacks. Int. J. Emerg. Trends Technol. Comput. Sci. **2**(2) (2013)
10. Azad, S., Jain, K.: CAPTCHA: attacks and weaknesses against OCR technology. Global J. Comput. Sci. Technol. **13**, 14–18 (2013)
11. Singh, V.P., Pal, P.: Survey of different types of CAPTCHA. Int. J. Comput. Sci. Inf. Technol. **5**(2), 2242–2245 (2014)
12. Yan, J., El Ahmad, A.S.: Breaking visual CAPTCHAs with naive pattern recognition algorithms. In: Twenty-Third Annual Computer Security Applications Conference, ACSAC 2007. IEEE (2007)

Design and Randomness Evaluation of Mixed-Signal TRNG for Low Power Applications

E. Kavinkartik, S. Suseendiran, and N. Mohankumar[(✉)]

Department of Electronics and Communication Engineering, Amrita School of Engineering, Coimbatore, Amrita Vishwa Vidyapeetham, Coimbatore, India
kavince28@gmail.com, suseendiran.s@gmail.com, n_mohankumar@cb.amrita.edu

Abstract. A system design for security and privacy plays a major role in maintaining a firm, running a business, storing the database, adhering to fair play in entertainment sector. This requires generation of secret codes, passwords, security pin, which must be unpredictable and secure. To address these purposes, a method has been proposed to generate random numbers based on different criteria and one of the ways is through exploiting the available randomness exhibited by a physical source in any form. In this paper, we designed a True Random Number Generator Circuit which consumes very less power and utilizes the noise from the physical world to generate a random sequence. Basically, the circuit design comprises of a ring oscillator and a simple sample and hold block, which utilizes minimum power in generating the random output. Field Programmable Analog Array is used for hardware realization and implementation of the circuit. Ten different sequences having thousand bits each is tested for randomness using the NIST test suite. The positive results from these standard tests prove that the circuit generates true random number sequences.

Keywords: True Random Number Generator (TRNG) · Sample and hold · FPAA · Design for security · Hardware security

1 Introduction

True random numbers are extensively used in areas like cryptography, numerical simulations, lottery, randomized algorithms, etc. proving their extreme importance in the security domain. They have also found a place in probabilistic, hazard games and in numerous applications of our daily day life such as, online payments, internet trade, e-mail access, cyber security, ATM, e-bookings, team viewer, online software in which all has an activity of generating one time password (OTP) which will be generated randomly with varying time.

Disturbing the entropy source during the randomness extraction will make the process data predictable. There are mainly two types of random number generators available. The first type captures the random signal from the real world to create its sequence which is nothing but a true random number generator (TRNG). In this type it is hard to predict the next sequence because the physical entropy does not go par with any mathematical models instead, they are just some random noise voltages [1]. The

© Springer Nature Switzerland AG 2020
A. P. Pandian et al. (Eds.): ICICCS 2019, AISC 1039, pp. 105–113, 2020.
https://doi.org/10.1007/978-3-030-30465-2_13

second type generates random numbers by algorithms with unpredictable outputs (assume that initial conditions were unknown). This technique of generating random numbers were called as pseudo-random number generators (PRNG).

Oscillators are the final category of true random number generators. Use of basic hardware makes more convenient for the installation. Inverter gates are used to construct oscillator. The inverter gates are taken in odd numbers for the construction of the oscillator [2]. Since the current oscillates in a sine wave pattern, the output is undefined with respect to time. The manufacturing of this gates are never perfect and causes a random deviation from sine wave. This deviation is called as jitter which is used as an entropy source in simple random number generator. Oscillator is of easy use and very cheap to install when compared to other devices.

The paper is organised as follows. A literature survey of concepts that we have studied from different papers is given in Sect. 2. Section 3 consists of a brief explanation of design and implementation of the proposed model. Section 4 consists of system results and inference. Section 5 provides the concluding remarks.

2   Literature Survey

Random numbers are widely used for conditional key generators and also as a password source also. The security of the applications depends mainly on the randomness of the source [3]. Classical TRNG utilizes random physical fluctuations but at present the most frequently used is the embedded TRNGs. Embedded TRNG is a jitter noise of any clock pulses generated through MOS devices. However, the proper utilization of the randomness in extracting the random number through proper devices is important. Devices like FPGA or any other microprocessors are used based on the desired output [4, 5].

For classical TRNG chaotic circuit is an effective alternative. Instead of using commonly used sources of randomness they use a analog deterministic circuits which shows chaos like thermal noise or any white Gaussian noises [2]. Chaotic systems are determined by a "sensible dependence on initial conditions" that is a large change observed in the state is caused by small perturbation. In the initial state there will be a slightest uncertainty which is unavoidable in all the implementation of analog which after sometime causes a very large uncertainty. Because of this uncertainty the system can be predicted only for some time.

The ring oscillator is used to produce randomness. There has been work previously done by generating control signal using modulo-2 division [1]. The odd number of ring oscillator is used because if the input was '1' then the output at the third will be '0' and again the output is given as the input. Now the input will be '0' and the output will be '1'. Like this ring oscillator generates alternative outputs so that alternatively the random signal input will be send to the generator circuit which increases the chance of more randomness in the output. Here the generator switch which allows the input was controlled by the oscillator. The output of the oscillator is clock signal which automatically operates the switch [4]. This oscillator also helps to control the bits generated at the output per second that is bit rate at the output by controlling the clock signal. The frequency of clock signal is the main factor to operate the TRNG at high throughput or low power mode. In design and implementation section we will discuss more about this oscillator and the generator circuit.

For this circuit implementation we use Field Programmable Analog Array (FPAA). It is a reconfigurable platform for analog circuits and also very cost efficient [6]. The circuit design is analyzed by the FPAA chip AN231E04. It is called switched capacitor FPAA chip. We can physical signals directly or we can use nonlinear function block to generate random signal and a sample hold block are required for each entropy core and all these can be built using the resources available in the Computational Analog Block (CAB). If not with only single entropy we can also use dual entropy which might give better random results. The survey helps us to realize a dual chaos system design using Bernoulli maps as entropy source [7].

The Field programmable Arrays (FPAA) device recently became popular as an interesting alternative to FPGA for analog implementations. For specific applications FPAA is used and not FPGA [8]. A 3.3V DC was given to the chip and driven by the master clock of 16 MHz frequency which is required to facilitate the CAB component clocks. For some applications [9], FPGA's are indicated, for others, clearly FPAA's are more suitable because of dynamically reconfigurable properties and the system is completely designed using analog circuit with predefined functional block. Good analysis of our system is to be done to wisely choose the necessary tools for implementation.

The main advantages of TRNG over PRNG are TRNG can be used in high security applications [10]. The entropy source selected is directly proportional to the randomness in the output. True random number generators are more fragile when compared to pseudo random number generator. This is because of the passive and active attacks. A passive attack is much harder to detect but the active attack leaves a footprint after altering the system.

However true random number generators suffer from active attack but the pseudo random number generator suffers from passive attack also in which random sequence are easily predictable. So it's wise to use true random number generators. So after a deep study we have preferred TRNG over PRNG.

3 Design and Implementation

The proposed technique is to provide random numbers with high randomness with less power and area utilization. The random numbers should be generated with a high randomness in it to avoid predictions.

The circuit designed is a mixed signal circuit which comprises a simple sample and hold block, ring oscillator, ADC converter. The input to the design is a physical noise signal or simply called as "entropy", which is actually sampled and held with respect to the output of the ring oscillator as the control signal. Digital binary bits are generated as the final output and the randomness is verified using standard NIST tests. Figure 1 represents the circuit block diagram.

The main challenge is to generate random numbers with high degree of randomness. So input to the functional block is an important factor in obtaining desired results.

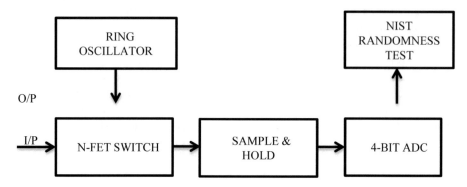

Fig. 1. Block diagram.

In our proposed technique we are using noise from the chaotic physical entropy source from the external environment which has randomness in it [11]. This random analog signal should be converted into binary by analog to digital convertor (ADC) for each sample.

In our proposed plan ring oscillator plays an important role in adding randomness of the source output signal. The randomness of the noise source is increased by the ring oscillator circuit. A 3 stage ring oscillator circuit is used in the design. The important phenomenon called jitter can be observed in ring oscillator design with different W/L ratio which provides a support in random sampling of the input signal [12]. The ring oscillator circuit is shown in Fig. 2.

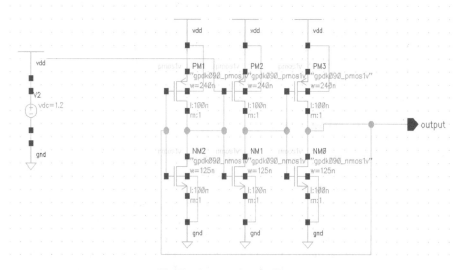

Fig. 2. 3 stage ring oscillator

The output of the oscillator is given as the input gate voltage to the nFET which bridges the buffer block and the input noise signal. When the output from the oscillator is above the threshold value of the control nFET used then the source signal is allowed through the FET into the hold circuit which is nothing but a sample of values have been taken and vice versa. Thus the oscillator output decides when the random input signal has to be sampled and held.

When the input to gate terminal is high, the nFET turns on thus capacitor connected to it charges. The main task of capacitance is to store or hold the source random analog signal for a long time till it gets processed. Thus the input random noise signal is sampled. When the input to the nFET gate is low the circuit behaves as an open circuit as the threshold voltage is not sufficient.

This stage makes the capacitor isolate from receiving input from the source and it holds the voltages till the last input value. The random noise signal input is thus held by the capacitor in this stage. The capacitor then starts discharging slowly due to the leakage currents. Capacitor is immediately connected to the buffer circuit which is nothing but an operational amplifier with unity gain which is designed in cadence and the same is displayed in Fig. 3. This circuit is also called as voltage follower circuit. Signal to be is applied at input of unity feedback amplifier (Buffer). Input impedance of unity feedback amplifier is very high so input signal to be held and discharged is not loaded. While sampling output of the unity feedback amplifier is same as the input signal to the unity feedback amplifier.

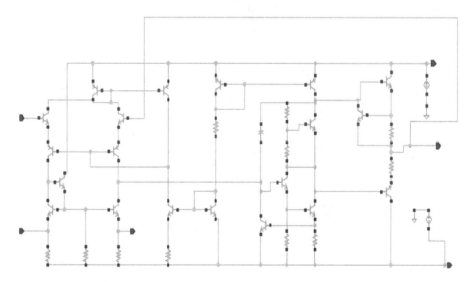

Fig. 3. Operational amplifier.

Unity feedback amplifier is used because of its unity gain and high input impedance. It transfers the input signal directly to the next stage without any change in input signal. The entire circuit is shown in Fig. 4. The entire circuit is designed and simulated in cadence and the power analysis is done and the results are discussed in the upcoming section.

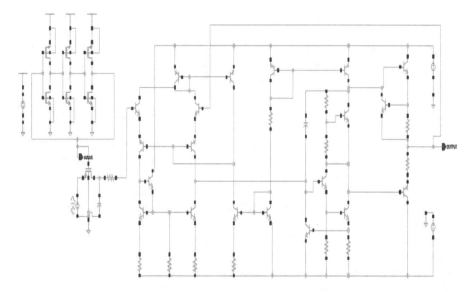

Fig. 4. True random number generator circuit

4 Result and Analysis

AnadigmDesigner2 tool is used to perform the implementation in FPAA and the necessary functional blocks used to represent our circuit design is shown below in Fig. 5.

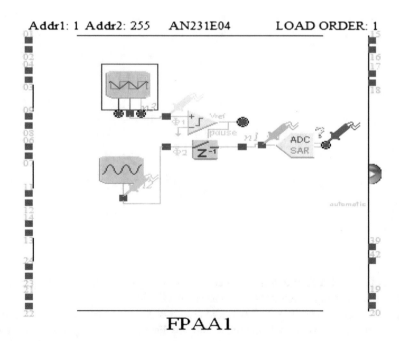

Fig. 5. TRNG circuit in FPAA

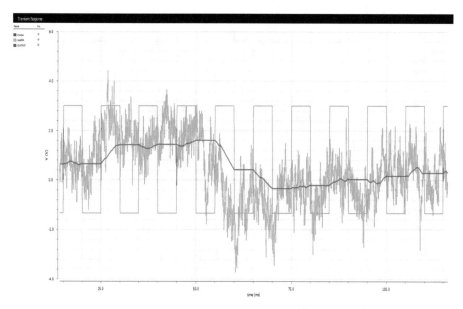

Fig. 6. (GREEN) Input Noise Entropy, (RED) Output generated from the ring oscillator, (VIOLET) Generated Random Analog Signal.

The final output obtained from the TRNG is shown in Fig. 6. Violet colour represents the final output of the generator which is then converted to bits using 4- bit ADC converter. The converted binary bits are then given as input to the NIST test.

Power analysis is one of the most important criteria to prove the efficiency of the circuit. Static power, Dynamic power and Average power of our TRNG circuit is calculated using Cadence Virtuoso and the same is represented in the Table 1.

Table 1. Power analysis of the circuit design.

Power profile	Power (W)
Static power	0.054×10^{-3}
Total average power	0.2881
Dynamic power	0.288046

Static power is evaluated using all DC input to the TRNG and then the dynamic power is calculated by subtracting the static power from the total average power.

Dynamic power is almost same as average power because static power is very low and for any circuit the dynamic power is close to average since the power is utilized more during switching.

Table 2. NIST test results of the random sequence.

Tests	Seq1	Seq2	Seq3	Seq4	Seq5	Seq6	Seq7	Seq8	Seq9	Seq10
Frequency (Monobit) test	0.486	0.184	0.375	0.751	0.486	0.113	0.311	0.658	0.612	0.612
Frequency test within a block	0.509	0.227	0.515	0.608	0.913	0.597	0.329	0.520	0.526	0.548
Runs test	0.715	0.050	0.501	0.797	0.763	0.579	0.393	0.042	0.855	0.307
Overlapping template matching test	1	1	1	1	1	1	1	1	1	1
Non-Overlapping template matching test	0.406	0.406	0.406	0.406	0.406	0.406	0.406	0.406	0.406	0.406
Cumulative sum test	1	1	1	1	0.924	1	1	1	1	1
Spectral Test	0.411	0.411	0.837	0.837	0.681	0.837	0.304	0.304	0.538	0.304

NIST test results have been projected in Table 2. A set of statistical tests for checking randomness is performed. It is useful to detect the deviations and the randomness of the input or the final binary sequence that is to be checked for randomness. There are various statistical tests that can be used on the sequence to evaluate the sequence and verify whether it is a random sequence. Randomness is a probabilistic property; the properties of a random sequence can be characterized and described in terms of probability. Since there are many tests that prove the randomness, there are no fixed rules stating the number of tests to be performed. During testing of the sequence, the statistical testing must be interpreted with some care and caution to avoid incorrect conclusions about a specific generator. For this we performed seven different tests to conclude the analysis of the sequences. For a sequence to be random it should follow 1% decision rule. The result of the Matlab code we used gives the probability value or the P-value. If P-value $\geq$ 0.01, then the test concludes that the sequence is random. If P-value < 0.01, then the sequence is nonrandom and we can say the sequence failed the test. Here 10 sequences of 1000 bits each are used to analyze the randomness of the sequence.

5   Conclusion

The area and power analysis proves very less average power is consumed, which is an advantage over other complex circuits but it might compromise on the performance to some extent. This low power utilization paves way for its apt usage into many low power applications like low power gaming applications where the system needs good randomness as well as low power usage. The randomness mainly depends on the input noise entropy and the output of the ring oscillator which can be controlled by adjusting the W/L ratio of nFET or pFET used in the design of the CMOS ring oscillator circuit thus providing us with random signals due to the presence of jitter in the design as we realized earlier. Thus jitter and the chaos in source signals help in generating random bits. The improvement can be brought about knowing the requirement of our application needs. Complex circuits are used generally for generating very high random numbers for maximum security which also increases the power and area utilization.

References

1. Aksshaya, B., Madhura, L.V.G., Nivethashri, S., Vishnuvarthini, T., Mohankumar, N.: Design and analysis of analog TRNG using sample and hold circuit. Int. J. Eng. Technol. **7** (3.8), 69–73 (2018)
2. Çicek, İ., Pusane, A.E., Dundar, G.: A novel design method for discrete time chaos based true random number generators. Integration VLSI J. **47**(1), 38–47 (2014)
3. Mohan, A.K., Devi, N., Sethumadhavan, M., Santhya, R.: A selective generation of hybrid random numbers via android smart phones. Int. J. Pure Appl. Math. **118**(8), 311–317 (2018)
4. Tuncer, T., Avaroğlu, E., Türk, M., Özer, A.B.: Implementation of non-periodic sampling true random number generator on FPGA. J. Microelectron. Electron. Compon. Materials **44** (4), 296–302 (2014)
5. Drutarovsky, M., Galajda, P.: Chaos-based true random number generator embedded in a reconfigurable hardware. J. Electr. Eng. **57**(4), 218–225 (2006)
6. Deese, J., Jimenez, C.N.: Utilization of field "Programmable analog arrays (FPAA) to emulate power system dynamics". In: Proceedings of the ISCAS 2009 (2009)
7. Cicek, A.P., Dundar, G.: A new dual entropy core true random number generator. Analog Integr. Circuits Sig. Process. **81**(1), 61–70 (2014)
8. Boyacı, O., Tantuğ, A.C.: A Random Number Generation Method Based on Discrete Time Chaotic Maps. IEEE (2017)
9. Yang, B., Rožic, V., Mentens, M.G.N., Verbauwhede, I.: ES-TRNG: a high-throughput low-area true random number generator based on edge sampling, international association for cryptologic research. IACR Trans. Cryptographic Hardware Embed. Syst. **2018**(3), 267–292 (2018)
10. Lee, K., Lee, S., Seo, C., Yim, K.: TRNG (True Random Number Generator) method using visible spectrum for secure communication on 5G network, IEEE, Special section on recent advances on radio access and security methods in 5G networks, vol. 6, pp. 12838–12847 (2018)
11. Addabbo, T., Alioto, M., Fort, A., Rocchi, S., Vignoli, V.: A technique to design high entropy chaos-based true random bit generators. In: Proceedings - IEEE International Symposium on Circuits and Systems, ISCAS (2006)
12. Cicek, I., Dundar, G.: A chaos based integrated jitter booster circuit for true random number generators. In: European Conference on Circuit Theory and Design (ECCTD), pp. 1–4 (2013)

Early Discovery of Disaster Events from Sensor Data Using Fog Computing

Kethavath Srinivas$^{(\boxtimes)}$ and Mohit Dua

Department of Computer Engineering, NIT Kurukshetra, Kurukshetra, India
seenucbit@gmail.com, er.mohitdua@nitkkr.ac.in

Abstract. Disaster event such as hurricane, blizzard, and winter storm always demand an early response. The lesser the time it takes to respond, the more damage can be prevented. In a disaster event, predicting the happening and alerting the concerned authorities should be done with a minimal latency. Today's existing technologies highly rely on information disposal to a far away control station. Hence, we aim at achieving an almost zero latency in Natural Disaster discovery. In this paper, the early discovery of disaster events are achieved with the help of Fog Computing infrastructure. Here, we have proposed a machine learning based prediction with Weather sensors. Machine Learning as a tool offers quick and highly reliable predictive models. These models once trained, it will make use of the basic computational operations. Hence they are perfectly suitable for various emergency situations. With Fog computing, the latency in data upload has been minimized. Also, for prediction purpose we have used data from over 5 years by using a Weather API. Multiple machine learning models were trained on this data, and the best model in terms of computation time has been deployed for evaluation. Our evaluation metrics show an impressive 96% accuracy of the deployed model and the response time remains as less as milliseconds.

Keywords: Disaster · Internet of Things (IoT) · Fog Computing · Support Vector Machine (SVM)

1 Introduction

In the event of a disaster, delivering a quick response to the desired damage remains as a crucial factor. Given the advancements in Internet of Things a fine level of sensing and tracking of environment is made possible. If this information can be analyzed immediately after sensing, it can eventually make the disaster recovery faster. IoT architectures use the cloud as brain for various operations. Cloud platforms are present in huge datacenters located at remote locations. Data transfer from IoT device to a remote location takes a considerable amount of time in which the event of disaster is not desirable.

Fog Computing brings the computation to be in proximity with IoT devices, which essentially makes the data transmission times zero [1]. In an architecture with Fog Computing helps to aid operations of IoT devices to deliver quick responses. With the disaster problem in mind we aim to adopt Fog architecture to solve the emerging

© Springer Nature Switzerland AG 2020
A. P. Pandian et al. (Eds.): ICICCS 2019, AISC 1039, pp. 114–121, 2020.
https://doi.org/10.1007/978-3-030-30465-2_14

challenges. Machine Learning (ML) provides tools to classify and identify disasters, which has given enough labelled data. Machine learning is a subset of Artificial Intelligence (AI), which aims to make the computer systems to learn automatically without being programmed explicitly [2]. The ML mainly focuses on the computer program's development which can access the data and learn themselves.

Table 1. Fog computing V/s cloud computing.

Requirements	Fog computing	Cloud computing
Latency	Low	High
Delay Jitter	Very low	High
Server node's location	At the edge of local network	Within the internet
Awareness of location	Yes	No
Security	Defined	Undefined

The machine learning methods are mainly classified as Supervised Learning, Unsupervised Learning, Semi-supervised Learning, and Reinforcement Learning. In this paper we are mainly concerning with Support Vector Machine which is presented in supervised learning. In supervised learning, the machine is provided with labeled data for training i.e. the machine is trained using past input data with their desired output. The machine learns new rules until it maps inputs and outputs with better accuracy. Supervised machine learning includes regression and classification algorithms. When outputs are having a certain range of numerical values the regression algorithms are used and when outputs are restricted to a limited set of values classification algorithms are used. In this paper we are dealing with classification algorithms to predict the occurrence of disasters.

The rest of the paper is structured as follows: In Sect. 2, the related works on forest fire is presented. While in Sect. 3, our objectives and goals are mentioned. Our proposed framework and implementation are presented in Sects. 4 and 5 respectively. In Sect. 6, the results are discussed. Finally, in Sect. 7 the conclusions and future plans are exposed.

2   Related Work

The researchers have been discovered few alerting systems for discovering the natural disasters early, likewise Early Warning Systems (EWS) [3, 4] which plays a crucial role in minimizing the effects of natural disasters. EWSs gathers the data from sensors, performs various analysis of these gathered data such as simulation based predictions and results are passes to users. Authors made use of arial data (Images) to counter disasters [5]. Instead of using satellite captured images, this model is using the arial images which are being captured by the Unmanned Arial Vehicles (UAVs) as this takes very less time as compared to satellite. Crowdsourcing is used to annotate the required features of captured images and these human annotated features had used for training the supervised learning machine to learn and to recognize such features in unseen new images. Social Networks and Crowd sourcing were used in disaster management improvement [6]. However, this model is unreliable. When disaster occurs, the injured people can save their lives by contacting emergency medical treatment centers using CroudHelp application which is installed in victim's mobile or also can accessed by website, this approach takes very large to process to get a real time response. Naa aziz et al. reviewed different country's disaster management systems using wireless sensors networks [7, 8] which are closely related to our work but machine learning algorithms is present in those proposed models. Along with these authors [9] used natural language processing sentiment based analysis to identify disasters, but this method becomes more complex.

2.1 Machine Learning in Predicting Disasters

Support Vector Machines (SVM) are non-probabilistic classifiers [10]. They are particularly useful because of the infinite dimensional space used for classification. Given that our data has features coming from multiple sensors, we employ SVMs in this context. Each feature is considered as an individual dimension by the SVM.

2.2 Fog Computing Models

Fog is a prominent architecture that provides computation, communication, control and storage closer to end terminal nodes though the internet [11]. Fog computing is not only brings cloud services to edge of the network but also reduces the communication overhead of the cloud. The Table 1 shows better comparisons of fog and clod computing. In [12] the author presented the smart phone based emergency alerting system using fog computing. This Emergency Help Alert Mobile Cloud (E-HAMC) provides fast way of sending emergency alerts to relevant department by using the services of fog for offloading as well as pre-processing purposes. Though the model is quick but not predicts future occurrence of the disasters.

3 Objectives and Goals

It takes a lot of time to move data to a cloud and to analyze it. At an event of emergency like a natural calamity, we need immediate detection and alerting the concerned authorities. So, deploying the said application at a Fog node will enable us to address the problem. Our Objectives and Goals through the publication are as follows:

- To develop an application that discovers disaster events with sensor data.
- An application capable of discovery in minimal possible amount of time.
- Efficient and deploy-able on limited to resource constrained devices.

4   Proposed Framework

The architecture of our proposed solution is as follows:

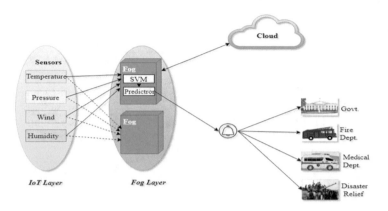

Fig. 1. Architecture of the proposed approach

4.1   IoT Layer

IoT layer consists of terminal devices or sensors, which are capable of communicating over the internet. We have in consideration sensors from different categories like temperature, wind-speed, pressure, humidity. All these sensors will communicate to the Fog layer via secure channels.

4.2   Fog Layer

In our architecture this is the disaster prevention layer. On this layer, incoming data is fed to the deployed Machine Learning model. If the model predicts a disaster immediate alert will be sent to all concerned authorities and the data is forwarded to Cloud. Otherwise the data is just forwarded to the Cloud for storage and analysis.

4.3   Disaster Alert System

This may include Govt. agencies and other disaster management authorities. Essentially communication end points of these agencies are part of this block.

4.4   Cloud

Cloud stores the data which is coming from the terminal nodes or sensors via fog nodes. And also analyses the stored data for further investigations of discovering the disasters on the demand of disaster management authorities. The Fig. 1 shows the architecture in detail.

Table 2. List of Natural Disasters in the presented period.

Date from	Date to	Disaster type	Country
16/09/2017	02/10/2017	Hurricane Maria	Florida and Puerto Rico
30/08/2017	13/09/2017	Hurricane Irma	Florida, South Carolina, Georgia, and Puerto Rico
17/08/2017	02/09/2017	Hurricane Harvey	Texas, Louisiana, and Alabama
28/09/2016	10/10/2016	Hurricane Matthew	Florida, Georgia, The Carolinas
19/01/2016	29/01/2016	January 2016 United States blizzard	Southeast through Mid Atlantic to Northeast
13/11/2014	26/11/2014	November 2014 North American winter storm	Buffalo, New York, and Great Lakes region
07/02/2013	18/02/2013	February 2013 nor' easter	Eastern US
22/10/2012	02/11/2012	Hurricane Sandy	Eastern US

5   Implementation

To understand and come up with solutions for disaster events we have taken the Historic Hourly Weather Data from public domain Kaggle [13] and analysis.

5.1   Data-Set Description

Using the Weather API the data-set has a collection of hourly data spanning over 5 years. From 10-Oct-2012 to 30-Nov-2017. 36 cities from the United States and Canada were chosen for collection. Given the amount of data with ranges on dates, we have identified 3 major Hurricanes in Canada and x number of disasters in the United States as shown in the Table 2.

Also, we have plotted the data for a subject town 'Vancouver' over the features (Figs. 2 and 3).

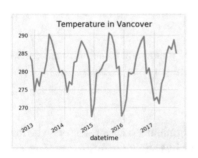

Fig. 2. Temperature variation

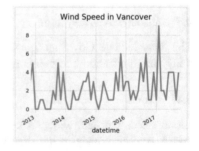

Fig. 3. Wind speed variation

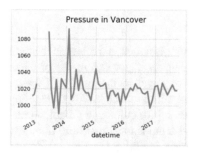

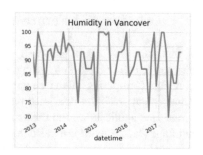

Fig. 4. Pressure variation   **Fig. 5.** Humidity variation

From the plots it is clear that the variations are actually unique and they can be used as features. However, we can see that in pressure a considerable amount of data was missing. We sampled this data from a geographic neighbor for out usage (Fig. 4 and 5).

5.2 Methodology

We have labelled a given day as one of the three tags: normal, pre-disaster, disaster. The values are generated on a separate column to make our dataset for training the ML model. The values ranging in the disaster zone as marked in Table 2 are given a value 1 and the remaining a value 0. The columns humidity, temperature, pressure, wind-speed were considered as input features. Features were normalized before training the SVM. We also performed a training task with out normalization, results are presented in the corresponding section.

Date as a feature was left out for SVM training as SVMs are used with numerical data. However, date contains valuable information. It is from intuition that a Snow Blizzard is likely to happen in November compared to May. This information is not lost, it is being captured in the features such as temperature and humidity. Hence, the omission of date is logical.

6 Results

Once our model was trained we have tested it with the held out data-set. Data in held out set was 0.33, 0.5 and 0.1 was used to generate results. The best and most realistic results was at a 0.33 percentage. With 0.33 held out, model shows an accuracy of 96%. A linear model gave around the same accuracy. 0.5 ratio of held out set gave an accuracy of 88% and the 0.1 held out was over-fitting the problem. Given the nature of application, over-fitting of data is not an option. As not every year will see the same rainfall and not every extreme temperature will cause a disaster. A more optimistic model would be the model with 0.33 held out.

7   Conclusion and Future Work

In this paper we have proposed, implemented and discussed a real time Natural Disaster detection method by using various Climate sensors. Approaching the problem with Fog Computing at the center, it becomes more novel in the field of Disaster Management. We have discussed the related techniques, importance of Fog Computing in real time applications and the use of Machine Learning for quick and reliable model building. Dataset with critical climate attributes like temperature, pressure, humidity and wind speed were considered. Results showed a successful estimation of 96% of natural disasters. Also, analysis of response time with Fog in center was presented. In a nutshell, we have achieved our objectives:

- Developed an application that uses sensor data to discover disasters.
- The amount of time taken was almost constant.
- Our model runs well on the resource constrained devices.

7.1   Future Work

We have considered 'Vancouver' as the target city. This makes our model useful in geographies similar to Vancouver and with climate patterns similar to it. However, the model is to be trained to each of the city/geography separately. This can be streamlined by piping in the new sensor readings directly to the model, and updating the model whenever weights are obsolete. A more generalized model can be built in the near future.

References

1. Yannuzzi, M., et al.: Key ingredients in an IoT recipe: fog computing, cloud computing, and more fog computing. In: IEEE 19th International Workshop on Computer Aided Modeling and Design of Communication Links and Networks (CAMAD), pp. 325–329 (2014)
2. Kotsiantis, S.B., Zaharakis, I., Pintelas, P.: Supervised machine learning: a review of classification techniques. In: Emerging Artificial Intelligence Applications in Computer Engineering, vol. 160, pp. 3–24 (2007)
3. Balis, B., et al.: The urbanflood common information space for early warning systems. Procedia Comput. Sci. 4, 96–105 (2011)
4. Bahrepour, M., et al.: Use of wireless sensor networks for distributed event detection in disaster management applications. Int. J. Space-Based Situated Comput. 2(1) (2012)
5. Ofli, F., et al.: Combining human computing and machine learning to make sense of big (aerial) data for disaster response. Big Data 4(1), 47–59 (2016)
6. Besaleva, L.I., Weaver, A.C.: Applications of social networks and crowdsourcing for disaster management improvement. In: International Conference on Social Computing. pp. 213–219. IEEE (2013)
7. Ab Aziz, N.A., Ab Aziz, K.: Managing disaster with wireless sensor networks. In: 13th International Conference on Advanced Communication Technology (ICACT2011), pp. 202–207 (2011)
8. Al-Amin Hoque, M., et al.: Tropical cyclone disaster management using remote sensing and spatial analysis: a review. Int. J. Disaster Risk Reduction 22, 345–354 (2017)

9. Beigi, G., et al.: An overview of sentiment analysis in social media and its applications in disaster relief. In: Sentiment Analysis and Ontology Engineering, pp. 313–340 (2016)
10. Suykens, J., Vandewalle, J.: Least squares support vector machine classifiers. Neural Process. Lett. **9**(3), 293–300 (1999)
11. Chiang, M., Zhang, T.: Fog and IoT: an overview of research opportunities. IEEE Internet Things J. **3**(6), 854–864 (2016)
12. Aazam, M., Huh, E.-N.: E-HAMC: leveraging fog computing for emergency alert service. In: 2015 IEEE International Conference on Pervasive Computing and Communication Workshops (PerCom Workshops), pp. 518–523 (2015)
13. Selfish Gene. Historical Hourly Weather Data 2012–2017, December 2017. https://www.kaggle.com/selfishgene/historical-hourly-weather-data#pressure.csv

Inequality in Writing Competence at Higher Education in Colombia: With Linear Hierarchical Models

Elcira Solano Benavides[1], Fernando Cabarcas Charris[1], and Amelec Viloria[2(✉)]

[1] Universidad del Atlántico, Barranquilla, Colombia
elcirasolano@hotmail.com,
fernandocabarcas@mail.uniatlantico.edu.co
[2] Universidad de la Costa, Barranquilla, Colombia
aviloria7@cuc.edu.co

Abstract. The objective of this study is to analyze the factors that determine the inequality in writing competence at higher education in Colombia. A linear hierarchical econometric model was estimated for obtaining the factors that determine the quality in higher education. Quality was used as a proxy through the results in the Saber Pro test in writing competence for the year 2014. The main results were that there is inequality in quality between the groups of students (90%), universities (5.8%), departments (2.3%), and regions (1.9%), where the first group presents the biggest gap. The factors that determine the difference in writing competence are: personal (being a woman, unmarried, and the value of the tuition); family (occupation of the mother and father, having benefit in health, having dependents, being granted a scholarship, having a greater number of academic credits, among others; and regional (dwelling in regions with high levels of poverty).

Keywords: Quality of education · Human capital · Writing competence

1 Introduction

From the human capital theory, studies have been conducted to determine if it is worth the effort in time and investment in formal education. Two elements are considered in order to assess the relevance of education from the neoclassical school: the first one from an individual perspective related to wages or rates of return on education, and the other one from the approach of inequality in the sense that education can overcome the socio-economic gaps between groups of individuals. This work focuses on the second element, that is to say, academic performance as a proxy for productivity of students in society.

One of the first works in this line was the Coleman report, which establishes the factors that affect academic performance differences between student groups, institutions, regions, and countries. So, this study aims to describe the higher education and academic performance measurements that assess the educational system quality. Studies by Celis, Jimenez, and Jaramillo 0, Escobar and Orduz 0, Rodriguez, Ariza,

A. P. Pandian et al. (Eds.): ICICCS 2019, AISC 1039, pp. 122–132, 2020.
https://doi.org/10.1007/978-3-030-30465-2_15

and Ramos 0, and Valens 0, consider the State tests in specific competencies of each academic program in Colombia as a quality variable, however, the results cannot be compared among them[1].

In this study, generic competence in writing is considered as the quality variable and results can be compared between academic programs. Because the writing competence test is the same for all university students. In addition, the contribution of this research is to include factors of departments and regions of Colombia that affect the quality of education. Among the reasons to justify this research is the fact that no other studies consider quality as a proxy for the generic writing competencies. The reason for including this variable is that the test is the same for all students. In addition, there are few studies where econometric methodologies are used in generic competencies to develop educational policies. So, the objective of this research is to analyze the factors that determine the inequality in writing competence in Colombian higher education.

2   Theoretical References

From the microeconomics, classic authors addressed the relationship between education and productivity. One of the first authors to analyze this relationship was Smith 0 in his book referring to the stock of human capital and deducing that training in job skills generates productivity. The authors Petty 0 and Nicholson 0 were the pioneers in measuring the stocks of human capital in an economy. However, Becker 0 and Schultz 0 are the ones who formally begin the school of human capital. At the beginning, the expectations of investment in education was reflected in educational policies, promoting changes in educational systems and universal coverage in formal education for population groups that had no access because of low incomes.

The Coleman report is one of the most referenced studies that relate the academic performance with socio-economic profiles in different groups of students in the United States. The methodological proposals in empirical investigations about the factors that determine students' academic performance are framed by: (a) the family background, (b) the social and cultural environment; (c) the favorable effect on academic performance that the institution is able to offer the student and may alter the unequal distribution of this factor, which would be key to educational policy. However, Solano and Navarro 0 suggest not underestimating the effect of educational quality in economic growth for teachers, salaries, expenditure on educational institutions, bibliographic resources, etc.

3   Method

3.1   Linear Hierarchical Model

The econometric methodology of linear hierarchical models is used in the studies of social phenomena from the application of nested structures. The pioneers of these

---

[1] Each test is different for the specific competence, according to each profession.

models were Goldstein, Healy, and Rasbash 0 who rigorously developed efficient statistical estimates of the regression coefficients. In addition, they provide tools to explain and explore that the differences in test scores between students in universities may be due to individual and family factors, policies of educational institutions where they study, and regional differences by practices of public policies in each department or region where each student lives 0, 0, 0, 0.

4 Results of Linear Hierarchical Model

4.1 Database of Linear Hierarchical Model

For the level Student, information used is provided by the Icfes from its information system on the Internet of the Saber Pro Tests 2014 of Icfes. Considering the information of 150,382 students from 81 universities in Colombia: 75,085 students from 49 private universities and 75,297 students from 32 public universities and the variables that contain the information shown in Table 1. For the level University, the information of all universities in Colombia that report to SNIES[2] is used. The financial information of the State universities is obtained from the CHIP system[3] that is found on the website of the Public Accounting office in Colombia, the variables that contain the information is shown in Table 2. For the level Department, the information was obtained from the website of the Dane; the variables used are shown in Table 3.

Table 1. Variables level students that the student belongs for the linear hierarchical model

Level students			
Personal information	Academic institutions	Socio-economic	Family
Genre	Credits	Socioeconomic status	Housing construction
Marital Status	Course Preparation	Access to Cell Phone	Access to electrical appliances
Age	Academic Bachelor	Salary range	Father's Occupation
Scholarship	Normal Bachelor	Dependent People	Mother's Occupation
Disability	Technical Bachelor	Subsidized Health	Lives with parents
Head of family	Academic Program	Access to Internet	Father's Education
Place of residence	Terms completed	Tv	Mother's Education
		College payment	Access to tv
		Works	Access to car
		Family Income	

---

[2] The National Information System of Higher Education.

[3] Consolidator of Finance and Public Information.

Table 2. Variables level University that belong to the student for the linear hierarchical model

Level University					
Nature of the university	Financial resources	Academic	Infrastructure	Teachers	Students
Public	Undergraduate Tuition Value	No. of postgraduate offered	Mts2 built	No. of full-time teachers	No. of students in National Mobility
Private	Owned by the University	No. undergraduate programs offered		No. of teachers in mobility	No. of students in Abroad Mobility
	Postgraduate Tuition value	Institutional Accreditation			
	Taxes	No. of research groups			
	Sales of services	No. of Indexed Journals			
	Nation's Contribution	No. of articles published			
	Department Contribution				
	Investment				

Table 3. Variables level Department that the student belongs for the linear hierarchical model

Level Department	
Financial	Social
Revenue department	UBN (Unmet Basic Needs)
	Misery
	GDP
	Constant GDP

4.2   Linear Hierarchical Model

Null Model: The null model or empty model is a model that does not have explanatory variables and it is estimated to know whether there are significant differences in the test of writing competence between groups of students, universities, departments, and regions in Colombia.

If the null model shows there are significant differences in the writing competence exam, it is necessary to estimate a model with explanatory variables. The null model is shown in the following equation.

$$writing_{ijkl} = \beta_{0ijkl} + e_{0ijkl} + u_{0jkl} + v_{0kl} + f_{0l} \tag{1}$$

Where the suffixes are:

i = 1, 2, 3, units of first level (students who took the Saber Pro tests).
j = 1, 2, 3, units of second level (group of students who belong to the University j in Colombia).
k = 1, 2, 3, are the units of third level (group of students who live in the Department k in Colombia).
l = 1, 2, 3, are the units of fourth level (group of students who belong to the Region l).
$writing_{ijkl}$, is the highest score in writing obtained by the student i, of the university j, department k, and of the region l.
β_{0ijkl}, represents the overall mean of student performance by university, department, and region,
e_{0ijkl}, is the residual variance of the first level assuming a normal distribution,
u_{0jkl}, is the deviation of the mean among universities (level 2) j with respect to its actual value, assuming a normal distribution.
v_{0kl}, is the residual variance of the third level assuming a normal distribution,
f_{0l}, is the deviation of the region l, assuming a normal distribution.

4.3   Estimating the Null Model in Four Levels

Table 4. Null model for writing in four levels

Fixed Part		
Average performance in writing	β_0	10,053 (0,003):
Random Part		
Level: Student: Variance in writing	$\sigma^2_{e_0}$	0.940 (0,027):
Level: University: Variance in writing	$\sigma^2_{u_0}$	0.060 (0,028):
Level: Department: Variance in writing	$\sigma^2_{v_0}$	0.024 (0,008):
Level: Region: Variance in writing	$\sigma^2_{f_0}$	0.020 (0,003):
-2*loglikelihood		428,379.925
Number of parameters		5
N		148,706

Note: standard error in parentheses

The fixed part of the model represents the overall mean in writing competence in Colombia, which was 10.05. This value represents the constant and is significant in the null model, see Table 4. The random part of the model shows how variances of the waste in writing competence have resulted statistically significant in all four levels. In this sense, there is inequality in the outcome of the writing competence test among students, universities, departments, and regions in Colombia. It is important to note that the variance, not explained in the results in written communication in the four levels, is a reason to continue estimating the model extended with explanatory variables. See the following equations that show the variance significance test.

$$\frac{\beta_{0ijkl}}{sd} = 3.351 > 1{,}96 \tag{2}$$

In the level Student, the variance is much greater than the standard error that is to say that there is inequality in the results of writing competence between students, see equation.

$$\frac{\sigma^2_{e_{0ijkl}}}{S_{\sigma^2_{e_{0ijkl}}}} = 34{,}81481 > 1{,}96 \tag{3}$$

At the level University, the variance is greater than the standard error, showing great differences in writing competence between universities:

$$\frac{\sigma^2_{u_{0jkl}}}{S_{\sigma^2_{u_{0jkl}}}} = 2{,}14286 > 1{,}96 \tag{4}$$

In the level Department, the variance is greater than the standard error. In the same way, there is inequality in the results of writing competence between departments in Colombia:

$$\frac{\sigma^2_{v_{0kl}}}{S_{\sigma^2_{v_{0kl}}}} = 3 > 1{,}96 \tag{5}$$

The level Region presents enough variance without explaining, that is to say there are differences in the tests between the regions in Colombia.

$$\frac{\sigma^2_{f_{0l}}}{S_{\sigma^2_{f_{0l}}}} = 6{,}66667 > 1{,}96 \tag{6}$$

Additionally, the proportion of variance explained by the null model indicates that the difference in writing competence between students is the level that has greater inequality in the test, with a 90.04%, while the difference between universities is 5.75%, between departments 2.3%, and between regions 1.92%. See the following equations:

Sean:

$$e_{0ijkl} = \sigma^2 \; ; u_{0jkl} = \tau_\pi; \quad v_{0kl} = \tau_\beta \quad f_{0l} = \tau_\alpha \tag{7}$$

$$\text{Proportion of variance of level 1:} \; \frac{\sigma^2}{\left(\sigma^2 + \tau_\pi + \tau_\beta + \tau_\alpha\right)} = 0{,}9004$$

$$\text{Proportion of variance of level 2: } \tau_\pi / \left(\sigma^2 + \tau_\pi + \tau_\beta + \tau_\alpha\right) = 0,0575 \tag{8}$$

$$\text{Proportion of variance of level 3: } \tau_\beta / \left(\sigma^2 + \tau_\pi + \tau_\beta + \tau_\alpha\right) = 0,0230 \tag{9}$$

$$\text{Proportion of variance of level 4: } \tau_\alpha / \left(\sigma^2 + \tau_\pi + \tau_\beta + \tau_\alpha\right) = 0,0192 \tag{10}$$

4.4 Multilevel Model with Explanatory Variables and Nested Structure

The result of the null model demonstrates the need for the estimation of linear hierarchical model with explanatory variables in each of the levels. The model with explanatory variables aims to find the factors at each level that explain the inequality in the results in writing competence of the State test. See Table 5.

4.5 Results of the Model for Fixed Part

Level Student. The students' personal factors that explain the inequality in writing competence are: sex, being a woman gets 0.174 points more than men, and students who are single obtained 0.06 points higher than those who are not. The corresponding to the variable Family; those who dwell in the usual home have 0,035 points less than those living in temporary homes for reasons of study or other. The significant factors in parent's occupation are: students whose father's occupation is entrepreneur reduce their score in 0.027 points compared to those parents who have another occupation, and if the occupation of the mother is housewife the students' scores increase in 0.016 points. The socio-economic factors related to family income that were significant are: for each additional person the student is in charge of, performance on the test decreases in 0.009 points. The students of higher stratum and those living in rural areas get 0.052 and 0.152 points less respectively; those who work obtain 0.036 points more than those who do not work. Those students who do not have sisben[4] obtain 0.026 points more than those students with sisben level one; those who belong to the level three obtained an average of 0.023 points more than those who are at level 1; and Level 2 obtained 0.021 more points than level one. The tuition paid by students was significant, i.e. students who pay less tuition fees get less points in the test; those who paid tuition fees between \$500.000 and \$1.000.000 during the year prior to the test reduced their performance in 0.049 points, and those who paid between \$1.000.000 and \$3.000.000 for tuition during the year 2013 reduced their performance in 0.041 points, i.e. the lower the tuition payment, the lower the result in the competence.

Referring to the component Family Resources, having a computer indicates 0.071 points more in writing than those without this equipment. The number of bedrooms at home has a positive differential of 0.005 points per additional bedroom in the house. About the factors corresponding to the institution in secondary students, graduates from

---

[4] Is thel Beneficiary Selection System for Social Programs (SISBEN - Sistema de Selección de Beneficiarios para Programas Sociales).

an academic and technical institution obtain 0.116 and 0.076 points less than those who graduated from normal schools. About the institutional framework of the university in the test in reading competence, for each additional point obtained by students in reading, the writing score increases 0.145 points, followed by citizenship competencies with 0.117 points, Competence in English with 0.057 points, and competence in quantitative reasoning with 0.038 points per additional unit of score; scholarship recipients during the year prior to the submission of the test obtain 0.091 points more than students without scholarships. The percentage of credits approved by students has a positive effect on the performance of students, where those who have completed between 81 and 90% of the credits in their study programs have 0.029 points more than those who have approved less than 80% of credits, and those who have completed more than 90% of the credit increase their score on 0.021 points with respect to those who have completed less than 80% of credits. On average, obtaining another preparation different from preicfes courses has a positive effect on students' scores in 0.049 points.

Level: University. Academic programs that positively impact the writing test are: humanities and social sciences which obtain 0.458 and 0.461 additional points respectively. If the students come from a public university, their performance increases in 0.059 points with respect to students attending private universities, while those who study at institutionally accredited universities obtain 0.112 points more than those who are studying in not accredited universities.

Level: Department. The largest proportion of students by department with unmet basic needs has a negative effect on students' scores. In each 1% additional of students in UBN, the score in writing is reduced by 0.002 points.

Level: Region. A level of statistical significance with the Andean region is observed. Students who live in this region have 0.157 points more than students in other regions.

Table 5. Full writing model for fixed part

Constant	β_0	6.023 (0.039):	Another preparation	β_{23}	0.049 (0.013):
Reading	β_1	0.145 (0.003):	No. of bedrooms at home	β_{24}	0.005 (0.002):
Quantitative Reasoning	β_2	0.038 (0.003):	No. of people in charge	β_{26}	−0.009 (0.003):
English	β_3	0.057 (0.002):	Father Entrepreneur	β_{27}	−0.027 (0.007):
Citizenship Competencies	β_4	0.117 (0.003):	Mother works at home	β_{28}	0.016 (0.005):
Gender (female)	β_5	0.174 (0.005):	Management	β_{29}	0.155 (0.009):
Single	β_6	0.060 (0.008):	Social Sciences	β_{30}	0.461 (0.017):
Scholarship	β_7	0.091 (0.008):	Communications	β_{31}	0.206 (0.015):
Academic	β_8	−0.116 (0.013):	Accounting	β_{32}	0.131 (0.011):

(continued)

Table 5. (*continued*)

Technician	β_9	−0.076 (0.014):	Laws	β_{33}	0.396 (0.010):
Current Home	β_{10}	−0.035 (0.006):	Economics	β_{34}	0.339 (0.019):
High	β_{11}	−0.052 (0.010):	Education	β_{35}	0.191 (0.009):
Rural Areas	β_{12}	−0.152 (0.059):	Nursing	β_{36}	0.051 (0.017):
Works	β_{13}	0.036 (0.005):	Humanities	β_{37}	0.458 (0.031):
Sisben level 2	β_{14}	0.021 (0.008):	Engineering	β_{38}	0.078 (0.009):
Sisben level 3	β_{15}	0.023 (0.011):	Medicine	β_{39}	0.136 (0.016):
No Sisben	β_{16}	0.026 (0.007):	Psychology	β_{40}	0.250 (0.013):
Computer	β_{17}	0.071 (0.008):	Public universities	β_{41}	0.059 (0.006):
Between $500 K and $1 M	β_{18}	−0.049 (0.007):	Institutional Accreditation	β_{42}	0.112 (0.006):
Between $1 M and $3 M	β_{19}	−0.041 (0.006):	UBN	β_{43}	−0.002 (0,000):
More than 90% credits approved	β_{20}	0.021 (0.006):	Andean Region	β_{44}	0.157 (0.006):
Between 81 and 90% credits approved	β_{21}	0.029 (0.006):			

Note: standard error in parentheses

4.6   Results of the Model for Fixed Part

The random component corresponds to the estimators of the part that is not explained by the model obtaining information about the components "variance and covariance". According to the results, universities have inequality in the written test considering the performance of their students in generic competences as writing, reading, quantitative reasoning, English, and citizenship competencies. There are differences in the outcome in writing between universities considering if the student belongs to the programs of humanities and social sciences. When they are positive, it indicates that the differences are higher in the high scores than in low ones. There is covariance between reading and social sciences. When positive, there is evidence that these differences are greater between the highest scores and less significant in the lowest scores.

The covariance between the general components of the test were significant. Their negative signs indicate that the joint action of these variables represent differences in the performance of students from different universities in the lowest scores. While the covariance between English/sex indicates the differences in performance in English of the students with feminine gender in the lowest scores according to the university. The performance of students in quantitative reasoning differs between universities with respect to gender. These inequalities are to a greater extent in those who obtained high scores in the test. See Table 6.

Table 6. Full model writing for random part

Level: Student			Covariance English and Reas.	σ_{μ_7}	−0.011 (0.002):
Variance in writing	σ_e	0.793 (0.023)	Covariance English and Reas. Quantit.	σ_{μ_8}	−0.077 (0.020):
Level: University			Covariance English and Reas. City.	σ_{μ_9}	−0.162 (0.021):
Variance in writing	σ_{μ_0}	2.362 (0.265)	Covariance English and Reas. Quantit.	$\sigma_{\mu_{10}}$	−0.008 (0.002):
Variance in Reading	σ_{μ_1}	0.008 (0.002)	Covariance English and Reas. Quantit.	$\sigma_{\mu_{11}}$	−0.003 (0.001):
Variance in quantit. reasoning.	σ_{μ_2}	0.017 (0.002)	Covariance reading and social sciences	$\sigma_{\mu_{12}}$	0.008 (0.001):
Variance in comp. Citizenship	σ_{μ_3}	0.004 (0.001)			
Variance in English	σ_{μ_4}	0.017 (0.002)	**-2*loglikelihood**		**397.632,162**
Variance in Humanities	σ_{μ_5}	0.358 (0.047)	**Number of parameters**		**60**
Variance in social sciences	σ_{μ_6}	0.211 (0.023)	**N**		**148.557**

Note: standard error in parentheses

5   Conclusions and Discussion of Results

Using the null model, it is concluded that the groups of students, universities, departments, and regions in Colombia present statistically significant differences in quality. However, the group of students present the greatest inequality in the results with a 90%, followed by 5.8% between universities, 2.3% between departments, and 1.92% between regions.

Using the extended model, the factors that explain these differences in quality were obtained between the groups or levels (students, universities, departments, and regions). In the level Students, the factors that explain the differences in writing competence and that are positive are: being female, unmarried, working, having mother working at home, not belonging to the Sisben, having more rooms in the house, being granted a scholarship, having more credits approved to the presentation of the test, taking a preparation course preicfes for the test. The factors that negatively affect the quality are: dwell in a permanent home, having a father entrepreneur, having an additional person in charge, belonging to higher strata, paying less tuition, and being academic bachelor. Likewise, for each additional point obtained in the examination by students in the competencies of reading, citizenship, English, quantitative reasoning, the result in written competence increases.

In the group Universities, the factors that positively affect the quality are belonging to the program of humanity or social sciences, studying in public university, and belonging to high quality accredited universities. In the studies mentioned above, explanatory variables relating to university levels, departments, and regions were not used. In the level Departments, the factor that negatively affects the quality is to study in departments with Unmet Basic Needs (UBN). In the level Regions, the factor that positively affects quality is to study in universities geographically located in the Andean Region.

References

1. Coleman, J.S., Campbell, E.Q., Hobson, C.J., McPartland, J., Mood, A.M., Weinfeld, F.D., York, R.: Equality of Educational Opportunity. Department of Health, Education & Welfare, Washington (1966)
2. Celis, M., Jiménez, O., Jaramillo, J.: Cuál es la Brecha de la Calidad Educativa en Colombia en la Educación Media y en la Superior? Estudios sobre la calidad de la educación en Colombia, pp. 67–98 (2012)
3. Escobar, S., Orduz, M.: Determinantes de la Calidad en la Educación Superior en Colombia. Facultad de Ciencias Económicas y Administrativas, Pontificia Universidad Javeriana, Bogotá (2013)
4. Rodríguez, A., Ariza, M., Ramos Ruíz, J.: Calidad Institucional y Rendimiento Académico: El Caso de las Universidades del Caribe Colombiano. Perfiles educativos **143**, 10–29 (2014)
5. Valens, M.: Calidad de la Educación Superior en Colombia: un Análisis Multinivel con Base en el ECAES de Economía 2004. Sociedad y Economía. **13**, 132–154 (2007)
6. Petty, W.: Political arithmetic. In: Petty, S.W. (ed.) The Economic Writings. Cambridge University Press (1899). vol. 1, C.H. Hull (1690)
7. Smith, A.: Investigación sobre la Naturaleza y Causas de la Riqueza de las Naciones. FCE, México (1997)
8. Nicholson, J.S.: The Living Capital of the United Kingdom. Econ. J. **1**, 95–107 (1891)
9. Becker, G.S.: Human capital: a theoretical and empirical analysis. With special reference to education. In: 3rd The National Bureau of Economic Research, New York, pp. 169–172 (1993)
10. Schultz, T.W.: Investment in Human Capital. Am. Econ. Rev. **51**, 1–17 (1961)
11. Solano, E., Navarro, G.: Remuneración Salarial Eficiente en la Educación Superior en Colombia. Economía **33**, 43–63 (2012)
12. Goldstein, I., Healy, M.J., Rasbash, J.: Multilevel time series models with applications to repeated measures data. Stat. Med. **13**(16), 1643–1655 (1994)
13. Vasquez, C., Torres, M., Viloria, A.: Public policies in science and technology in Latin American countries with universities in the top 100 of web ranking. J. Eng. Appl. Sci. **12** (11), 2963–2965 (2017)
14. Torres-Samuel, M., Vásquez, C., Viloria, A., Lis-Gutiérrez, J.P., Borrero, T.C., Varela, N.: Web visibility profiles of Top100 Latin American Universities. In: Tan, Y., Shi, Y., Tang, Q. (eds.) Data Mining and Big Data. DMBD 2018. LNCS, vol. 10943, pp. 1–12. Springer, Cham (2018)
15. Viloria, A., Lis-Gutiérrez, J.P., Gaitán-Angulo, M., Godoy, A.R.M., Moreno, G.C., Kamatkar, S.J.: Methodology for the design of a student pattern recognition tool to facilitate the teaching – learning process through knowledge data discovery (Big Data). In: Tan, Y., Shi, Y., Tang, Q. (eds.) Data Mining and Big Data, DMBD 2018. LNCS, vol. 10943, pp. 1–12. Springer, Cham (2018)
16. Vásquez, C., Torres-Samuel, M., Viloria, A., Lis-Gutiérrez, J.P., Crissien Borrero, T., Varela, N., Cabrera, D.: Cluster of the Latin American Universities Top100 according to webometrics 2017. In: Tan, Y., Shi, Y., Tang, Q. (eds.) Data Mining and Big Data. DMBD 2018. LNCS, vol 10943, pp. 1–12. Springer, Cham (2018)

3D Reconstruction of an Indoor Environment Using SLAM with Modified SURF and A-KAZE Feature Extraction Algorithm

S. Srividhya[1(✉)], S. Prakash[2], and K. Elangovan[3]

[1] Department of ISE, BNM Institute of Technology, Bangalore, India
s.srividhyaa@gmail.com
[2] East Point College of Engineering and Technology, Bangalore, India
prakash.hospet@gmail.com
[3] Er.Perumal Manimekalai College of Engineering, Hosur, India
drelangovank@gmail.com

Abstract. 3D reconstruction of an environment has applications in various fields like city planning, gaming, robotic mapping, virtual environment, motion capture, augmented reality, etc. If spatial data is constructed it can be used for various applications like object tracking, building a path, etc. conventional approaches of accumulating spatial data like physical measurement with visual scrutiny and total survey are manual intensive and protracted. It also provides inaccurate measurements due to lack of ability, knowledge and experience of a labour. An alternative approach to this could be senor technology which can generate 3D point cloud data of an environment with accurate information about presence of object. On the other hand, it's expensive and implementation of the sensor requires experienced operators. To overcome this challenge Kinect sensors has drawn the attention because of its less cost, ease of use, and higher accessibility with the development of computer vision-based techniques. In this paper, the proposed algorithm is used to extract more features in less time and use random sample consensus (RANSAC) to filter correspondences.

Keywords: Point cloud · 3D reconstruction · A-KAZE · SURF · RANSAC · SLAM

1 Introduction

Computer vision based techniques can be used for navigation in an indoor environment using an autonomous Micro Aerial Vehicle (MAV). Perhaps the estimation of the position of a robot in the chosen environment inflicts several confronts for different circumstances. The GPS signals can be used in indoor or cluttered surroundings of urban region. But there are chances for inaccurate measurements in GPS position. Such errors are called as user equivalent range error. In [1] these errors are categorized as deviation of values and random noise. In [2] it is proved that the total user equivalent range error is roughly 4.0 m (σ) which is a random noise. It makes GPS based navigation system undependable for various applications when accurate actions are required [3]. Therefore sensor information's can be used in such scenarios with

© Springer Nature Switzerland AG 2020
A. P. Pandian et al. (Eds.): ICICCS 2019, AISC 1039, pp. 133–143, 2020.
https://doi.org/10.1007/978-3-030-30465-2_16

accurate information. All these issues have commenced the use of cameras in MAVs. It is for vision based navigation in an unknown environment where the GPS sensor is undependable. In this research work we have used Kinect sensor as camera mounted on an MAV to capture the depth image and use the same for creating the map of the environment. In Visual SLAM methods, a MAV can manoeuvre in an unknown surrounding using the sensor to simultaneously build a map of its environment and identify the position of itself in the map [4, 5]. SLAM with path planning techniques allow robots to securely navigate and discover an unknown surroundings autonomously. The robot uses these sensor readings to create a map and to localize itself on that map. The map can be a sparse or dense map. Efficient loop closure, re-localization, kidnapped robot problems can solved using visual SLAM.

Loop closure is utilized to identify when a robot has come back to a zone that has been mapped earlier. In turn it is used to correct the error that has occurred during the last visit of the area. The loop closure occurs if a landmark or set of landmarks are recognized more than a threshold value. According to the new observation of the robot's position, the previous observations are updated.

The interpreted data's are stored in map which is the centre part of SLAM. The map is the illustration of an environment which can be of metric and topological form. In robotics area, metric maps are common which describes the world coordinates using 2D or 3D point cloud data. Point clouds are set of coordinates in metric space representing the occurrence of obstacles based on the sensor values. Point clouds are also used for surface reconstruction, object mesh generation, etc. these point clouds are used for robot navigation in an unknown environment.

In 3D reconstruction techniques the retrieval algorithms with localized key points are used widely. The key point is a landmark that is observed in an environment. It is significantly detected on regions with variation in edges and the descriptors are invariant to changes in orientation of images. In market there are many 3D reconstruction software packages are available to extract the features from images.

SIFT [8] and SURF [9] is the commonly used algorithm feature detectors and descriptors. SURF detects the maximum number of features where as SIFT detects only less number of features. The recent algorithms like KAZE and A-KAZE also provides good features when compared to SIFT with better computational efficiency. In this paper the proposed method uses a modified SURF and A-KAZE for feature extraction which in turn provides better results in terms of number of features and computational efficiency [10] (Fig. 1).

2   System Architecture

2.1   Feature Extraction Using SURF

SURF is a local feature detector and descriptor proposed by Herbert Bay et al. at an European Conference on Computer Vision during 2008 [11]. It can perform object recognition, or registration or classification or 3D reconstruction. It is faster than SIFT and identifies more features than SIFT and KAZE. The interest points are detected by using an integer approximation of the determinant of Hessian blob detector.

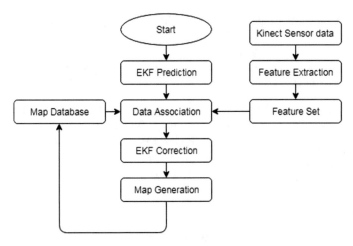

Fig. 1. System architecture

Precomputed integral image undergoes three integer operations to identify the interest points. The calculated feature descriptor will be the sum of the wavelet response around the interest point. The descriptor can be used to identify an object, people, to make 3D scenes, to track an object, etc.

The steps involved in SURF algorithm are:

- Integral image
- Key point detection
- Orientation assignment
- Feature descriptor generation

The algorithm strength lies in identifying the patch direction before extracting its feature vector. In a loop structure environment with respect to horizontal and vertical Haar wavelets, based on Gaussian weighted responses the interest point direction is identified. The interest points are distorted into polar coordinates and divided based on the resolution of an angle which is predefined. The distorted vectors that belong to the same partition are collected and the angle of the highest vector will be chosen as the patch direction. The patch is revolved based on the direction and a real vector descriptor is evaluated by performing the Haar wavelet on the discrete sample.

2.2 Feature Extraction Using A-KAZE

Alcantarilla et al. [12] proposed the Accelerated-KAZE (A-KAZE) during 2013. The features are extracted by means of non linear scale spaces in an image. In order to speeds up the non linear scale space computation, the algorithm was developed from KAZE [13] by embedding mathematical Fast Explicit Diffusion (FED) in order to increase the computational speed. It consists of three steps:

- The sequence time is specified by a non linear scale space construction using FED.
- Feature detection detects the feature points in an image that are the local maxima of the determinant of the H matrix.
- Finally the detected feature points are represented using Modified-Local Difference Binary (M-LDB).

In the step one, A-KAZE builds pyramidal structure of nonlinear scale space. The scale space is divided into a series of octaves (O) and sub-levels (S). Discrete octave (o) represents the set of octaves and a sub level (s) represents the set of sub levels and they are mapped to their corresponding scale (Σ) as shown in Eq. (1) and M represents the sum of filtered images as shown in Eq. (1)

$$\sum i(o,s) = 2^{(o+s)/S}, o \in (0,\ldots O-1), s \in (0,\ldots S-1), i \in (0,\ldots M) \quad (1)$$

The values of discrete scale levels represented in pixel units are distorted into Σi time units. The nonlinear diffusion filtering in time term is defined as shown in Eq. (2)

$$Ti = \frac{1}{2}\sum i^2, i = 0\,to\,M \quad (2)$$

In the step two, the local maxima feature points are detected from the filtered images (Fi), by calculating the determinant of the H matrix in the nonlinear scale space. H matrix of the images are calculated by a normalized scale factor as shown in Eq. (3),

$$F_H^i = \sum_{i,normal}^2 \left(F_{xx}^i F_{yy}^i - F_{xy}^i F_{xy}^i\right) \quad (3)$$

In the step three, the M-LDB descriptor is used to calculate the binary strings for image patches using slope and intensity values from non linear scale space [13]. The average intensity is calculated by dividing the image into various sizes of max grid cells. It is highly sensitive to the rotational changes as it requires huge resources. Therefore, A-KAZE uses scale dependent sampling for attaining the scale and rotation invariance, by identifying the orientation information. In this step it constructs the descriptor by using xy derivatives from the identified features.

In this research work, a nearest neighbour approach is used for proficient feature matching. It is the ratio of the distance to the nearest and the next nearest neighbours. Once this is done RANSAC method is used to detect the laser readings that lie closer to the best fit line.

2.3   Proposed Method

The proposed method uses modified SURF and A-KAZE algorithm in order to identify maximum features in less computation time. It uses the combinational key point detection of SURF and A-KAZE algorithm to have maximum features with maximal match. The steps involved in the algorithm are:

(i)   Feature detection: The features are extracted by means of non linear scale spaces in an image.
(ii) The mapping of the discrete octave and the sub level has to be done as shown in Eq. (1).
(iii) The nonlinear diffusion filtering in time term is defined as shown in Eq. (2).
(iv) H matrix of the images are calculated by a normalized scale factor as shown in Eq. (3).
(v)   By using the horizontal and vertical derivative of the identified feature the descriptor is developed.

3 Data Association in VSLAM

The data association phase is used to map the newly extracted features with the existing features and use them to localize the robot and to estimate the landmark. If the robot moves to a new position the odometry value is transformed and through odometry update phase it is updated in the kalman filter as a repetitive process. The heart of the SLAM process is Kalman filter, which is conscientious for informing where the robot thinks is based on the identified features. The data association is based on visual place recognition techniques that are classified into following categories [14]:

3.1 Cooperative SLAM

CoSLAM is a vision-based simultaneous localization and mapping (SLAM) in a dynamic environments with multiple cameras. These cameras shift independently and can be ascended on different stages. A global map of the environment is built by means 3D positions of stationary background points and paths of the moving forefront key points. The localization and mapping process includes inter-camera pose estimation and mapping to deal with dynamic objects. To improve the robustness of the system, the ambiguity in position of each map point has to be maintained. For collision detection and path planning it is important to reconstruct the 3D trajectories of the moving objects.

3.2 Loop Closure Detection

A graph is constructed with nodes that represent the location of a key point in an image and links represent the path to the reference point. Loops in the graph are the previously visited place. When the robot moves it observes for a closing loop for every perspective of an image [15]. For every new image if the loop closure is not detected it is added to the graph as new node. The loop closure detection involves following steps:

1. A fresh perception picture is obtained.
2. If the match between the current image and previous image is more than a threshold, it returns to step 1 else it proceeds to step 3.
3. The current image and all the reference images loop closure is endeavoured.

4. If the loop closure is identified, a link is added among the previously visited node in the graph and the node connected with the matched reference image.
5. If the loop closure is not identified, a new allusion of an image is obtained and it is connected as new node in the graph.
6. Repeat the process from the step 1.

3.3   Kidnapped SLAM

The "Kidnapped robot problem" is intimately connected to multiple session mapping. In the Kidnapped robot problem, the goal is to calculate the robot's position with respect to a prior map given no *a* priori information about the robot's position. In multisession SLAM, in combination with this global localization problem the robot should start mapping straight away and upon localization the map from the current session should be included into the global map from previous sessions.

3.4   EKF Prediction and Correction

The position of the robot from the odometry data and landmark predictions is evaluated approximately using the Extended Kalman Filter. Using state estimation the EKF can be described. The SLAM process can be represented in three steps when the land mark extraction and the data association are in position [16]:

1. Using the odometry data update the current state estimation.
2. From re-observed landmarks update the estimated state.
3. The current state is added with new landmarks.

The first step is very simple where the controls of the robot are added to the previously estimated state. E.g. with rotation theta the robot is at point x, y and dx, dy are the control points and d theta is the rotation change. After the step one the robot will be in new state x + dx, y + dy with rotation theta + dtheta. Re-observed landmarks will be considered in step two. It is possible to identify the position of the landmark by using the estimation of the current position. The difference between that is called innovation. Therefore the innovation based on robot view is the difference between the estimation of the robot position and actual robot position. In step two the ambiguity of each observed landmark is updated to replicate current changes. For example, less uncertainty in current landmark position. Re-observation of the landmark from this location with less ambiguity will enhance the landmark certainty i.e., the discrepancy of the landmark with respect to the present position of the robot. In step three the robot map of the environment is added with new landmarks. This can be performed by updating the information about present position and accumulating the relation between the new landmark and the old landmarks information.

4 Map Generation

The SLAM software running on the base station was capable of reducing the physical memory usage of the SLAM map stored in memory to well below the performance threshold of 500 MB for real-time operation. Using the Point Clouds Library (PCL), the entire world map is quantized into a grid, each element (cell) of which was set to be a cubic centimeter in volume. In order to reduce redundancy, the visualization software requires that only one point ever occupy a given grid cell (cubic centimeter region) at one time. If a point already occupies a specific grid cell, additional points that try to provide data for that cell are discarded immediately. The size of each cell in this grid is called the leaf size. A leaf size of a cubic centimeter was found, through experimentation, to provide a good tradeoff between visual quality and performance. The size of an unfiltered point cloud in memory is 640×480 (the resolution of the Kinect's cameras), or 307, 200, points times the size of each point (4 bytes), resulting in a size of 1,228.8 KB for each scan from the Kinect.

By using the export to PCL functionality it is possible to load the reconstructed world into PCL and use all functions included. Among other things PCL has a fast triangulation of point cloud datasets which can generate a 3D mesh from unordered point clouds. The result of the reconstruction may look something like what is presented in Fig. 2. The matching of re-projected points to old points is also displayed in Fig. 3. Both these images are available at all times during the reconstruction process.

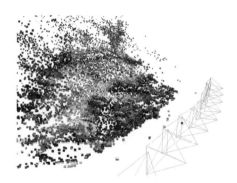

Fig. 2. Perspective view of a typical indoor environment

Fig. 3. Feature extraction using proposed method

The objects scanned are monitor, processor and a book placed on a flat surface. The scanned objects are textured in order to provide the features that may be detected. The camera performs a motion similar to a running track. It starts out with a straight stretch and then performs a half circle motion. After this it performs another straight stretch and another turn. In the end the camera pose should return to the start position. The aim of this test is to show how well an orthogonal symmetric object is reproduced. Is it still orthogonal from every view? Is the length equal to the width? Apart from this the trajectory of the cameras are also checked. The measures used are the standard deviation of the height component in the camera position compared to ground truth, as well as the ratio between the length and width of the track. The reconstruction resulted in data that may be viewed in Figs. 4, 5 and 6.

The trajectories associated with this reconstruction gave the following numerical results. The standard deviation of the height component of the camera position expressed as a percentage of the width of the track is 0.0059709. The relation between the length and the width of the track is 1.2732, which is to be compared to ground truth of 1.2535. This gives 1.57% difference between the reconstructed path relation and the true path relation.

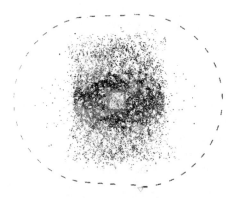

Fig. 4. Top view of reconstructed data

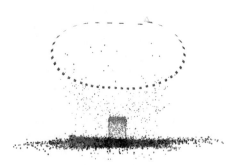

Fig. 5. Side view of reconstructed data

Fig. 6. Side view of reconstructed data

Notice that there are a lot of points present. In order to more clearly visualize the points related to the actual object the points below a certain level are removed from

Figs. 7 and 8. The red dashed lines plotted in the graphs represent an orthogonal cube positioned in the center of reconstructed box. This represents the ground truth and hence it is a true image of monitor with all sides of equal length. These images illustrate orthogonally as well as how scaling of depth versus width or height behaves when reconstructing.

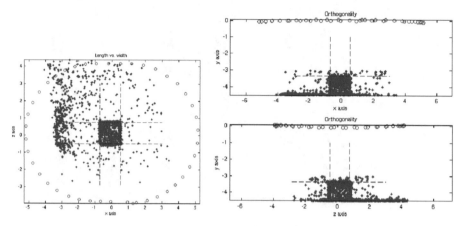

Fig. 7. Top view of synthetic data   **Fig. 8.** Side view of synthetic data

It is clear to see, especially in Figs. 4 and 7, that there exists a small drift in pose for the last frames compared to the first. The effect of this is also noticeable as a thicker density of points on the side of the monitor facing these frames. Likely the same points have been triangulated twice polluting the map. Overall orthogonally and scaling seems to hold up well considering that no loop closing or global bundle adjustment is performed. An overview including the visualization of the edges holding the key frames together can also be seen. A line in this graph means that the two frames connected share at least one 3D point in the reconstruction [17].

These results demonstrate a dramatic improvement over the state-of-the-art methods, as demonstrated by Kinect SLAM work. The Kinect mapping techniques were largely infeasible for areas larger than a single room, this improvement makes mapping on a floor- or building-wide scale a feasible problem for Kinect SLAM (Fig. 9).

Fig. 9. Reconstruction of the test environment

5 Conclusion

This research presented a 3D reconstruction method of an indoor environment using modified SURF and A-KAZE feature extraction algorithm to perceive exact corresponding key points and to reconstruct accurate 3D point clouds of the environment automatically. As the chosen environment mostly consists of texture-less objects with planar surfaces, traditional feature extraction algorithms encountered challenges and loop closure is taken care while scanning the environment. To solve this problem, the proposed feature extraction algorithm was employed in this research. To measure the performance of the proposed algorithm as feature detector and descriptor, analysis of detecting and matching results has been conducted. The results presented that the proposed algorithm is reliable and efficient in detecting resultant features from high-resolution images which were obtained in different view of angles. This research contributes to establish a comprehensive 3D reconstruction method for any chosen environment.

References

1. Parkinson, B.: Global Positionig System: Theory and Applications. American Institute of Aeronautics and Astronautics, Reston (1996)
2. Zogg, J.: Essentials of satellite navigation, Tech. rep. u-blox (2009)
3. Munguia, R., Urzua, S., Bolea, Y., Grau, A.: Vision-based slam system for unmanned aerial vehicles. Sensors **16**(3), 372 (2016)

4. Artieda, J., Sebastian, J.M., Campoy, P., Correa, J., Mondragon, I.F., Martinez, C., Olivares, M.: Visual 3-D SLAM from UAVs. J. Intell. Robot. Syst. **55**, 299–321 (2009)
5. Weiss, S., Scaramuzza, D., Siegwart, R.: Monocular-SLAM based navigation for autonomous micro helicopters in GPS-denied environments. J. Field Rob. **28**(6), 854–874 (2011)
6. Lowe, D G.: Object recognition from local scale-invariant features. In: Proceedings ICCV, vol.2, pp. 1150–1157 (1999)
7. Bay, H., Tuytelaars, T., Van Gool, L.: SURF: speeded up robust features. In: Proceedings European Conference Computer Vision (2006)
8. Alcantarilla, P.F., Bartoli, A., Davison, A.J.: KAZE features. In: Proceedings European Conference Computer Vision (2012)
9. Alcantarilla, P.F., Nuevo, J., Bartoli, A.: Fast explicit diffusion for accelerated features in nonlinear scale spaces. In: Proceedings British Machine Vision Conference (2013)
10. Bay, H., et al.: Speeded-up robust features (SURF). Comput. Vis. Image Underst. **110**(3), 346–359 (2008)
11. Alcantarilla, P.F., Nuevo, J., Bartoli, A.: Fast explicit diffusion for accelerated features in nonlinear scale spaces. In: Proceedings of the British Machine Vision (2013)
12. Alcantarilla, P.F., Bartoli, A., Davison, A.J.: KAZE features. In Proceedings of European Conference on Computer Vision, October 2012, Berlin, Germany, pp. 214–227 (2012)
13. Yang, X., Cheng, K.T.: LDB: an ultra-fast feature for scalable augmented reality on mobile devices. In: Proceedings of Mixed and Augmented Reality (ISMAR), pp. 49–57 (2012)
14. SriVidhya, S., Akki, C.B., Prakash, S.: A survey on data association methods in VSLAM. Int. J. Eng. Trends Technol. (IJETT), V30(2), 83–88 (2015). ISSN: 2231-5381
15. Wolf, D., Sukhatme, G.: Mobile robot simultaneous localization and mapping in dynamic environments. Autonomous Robots **19**(1), 53–65 (2005)
16. Wang, C., Thorpe, C., Thrun, S., Hebert, M., Durrant Whyte, H.: Simultaneous localization, mapping and moving object tracking. Int. Robot. Res. **26**(9), 889 (2007)
17. Angeli, A., Doncieux, S., Meyer, J.A., Filliat, D.: Incremental vision based topological slam. In: Proceedings of the IEEE/RSJ International Conference on Intelligent Robots and Systems, 22–26 September, pp. 1031–1036 (2008)

Optimal Participation of Hybrid Renewable Energy Sources in Standalone Microgrid

Manas Ranjan Nayak[1], Manoj Saha[1], and Chinmay Kumar Nayak[2(✉)]

[1] Department of Electrical Engineering, Biju Patnaik University of Technology,
Rourkela 769015, India
manasnk72@gmail.com, manojsaha666@gmail.com
[2] Department of Electrical Engineering, Indira Gandhi Institute of Technology,
Sarang, Dhenkanal 759146, Odisha, India
chinmaynayak009@gmail.com

Abstract. Being a tropical country, India has huge potential of sustainable energy tanks like solar, wind and micro-hydro energy to deliver electricity to the remote areas in a stand-alone mode. The nonlinear nature of renewable energy sources poses severe issues to the regulation of outcome capacity in order to meet the load demand. To mitigate these reliability issues, Diesel generators (DGs) along with devices capable of storing electrical energy may be utilized to compensate the fluctuations in renewable energy supply. This paper elaborates the analysis of optimized hybrid sources (combination of solar powered energy system, wind energy system, micro-hydro power, DG and battery) to enable the remote areas without the provision of grid power supply to enjoy the benefits of electricity. In this paper "Interior point" algorithm is employed in optimization solver "fmincon" as it is able to resolve complex programming problems with huge amount of data with speed by considering all the constraints.

Keywords: Hybrid system · Optimal operation control · Cost minimization · Renewable energy · Optimization algorithm

1 Introduction

Renewable energy sources, battery storage systems and DGs supplement each other while supplying power to the load demand from both technical and economic point of views. This combination reliably supplies the load demand by optimal utilization of all these energy sources [1]. Use of storage systems increases the cost of the system hopelessly. So by using the hybrid system, it reduces the size of the energy storage system and the capital cost of the entire system to a greater extent [2].

Ample development of solar photovoltaic (PV) systems and wind generators facilitate the use of renewable sources for rural and remote operations. Geographical areas with enough water resources have the advantage of harnessing micro-hydro electric energy to cater the load demand at a lower cost of energy [3]. In micro-hydro plants, the kinetic energy (KE) of the flowing water rather than the KE of the falling water in case of conventional hydro power plants is utilized to drive the generators. Additionally, DGs have the ability to supply power on demand with a very little initial

© Springer Nature Switzerland AG 2020
A. P. Pandian et al. (Eds.): ICICCS 2019, AISC 1039, pp. 144–153, 2020.
https://doi.org/10.1007/978-3-030-30465-2_17

cost. But they lack the economic viability in the perspective of higher operation and maintenance costs while serving partial loads [4]. A non-algorithmic method is utilized to optimize the contribution of power to the particular hybrid system.

Researchers of the contemporary era have found artificial intelligent (AI) techniques to be quite useful to optimize hybrid renewable energy systems consisting of different sources like PV system, wind energy system, Micro-hydro power, DG and battery in order to establish optimum participation of the sources in stand-alone microgrid. This paper investigates the optimal contribution of various renewable energy sources and as well as conventional energy sources like battery and DG. Proposed optimization solver, i.e. "fmincon" problem solver in MATLAB [5] and Interior point algorithm (IPA) techniques are utilized to optimize each renewable energy sources and standard energy sources.

This paper investigates the optimal contribution of various renewable energy sources and as well as conventional energy sources like electric battery and Diesel generator (DG). Proposed optimization solver i.e. "fmincon" problem solver in MATLAB [5] and Interior point algorithm (IPA) techniques are utilized to optimization of each renewable energy sources and standard energy sources.

In Sect. 2 of this paper, the system architecture is presented. The problem under investigation is formulated and different operational constraints are discussed in Sect. 3. In Sect. 4, "fmincon" solver and IPA are detailed. The simulation results are illustrated and analyzed in the Sect. 5 followed by the conclusion presented in the final section of the paper.

2 System Modelling

The proposed model consists of a combination of micro hydro power, PV and WT along with battery and DG. Here the battery is being charged by the generated renewable energy, whereas the switching of DG only to supply the load in the case of energy deficiency. While operating all the components care is taken to ensure the supply-demand matching.

2.1 PV System Modelling

The dc power output of a PV array (P_{pv-dc}) is the product of solar irradiance (E_{ir}), the PV array area ($A_{pv} = 33\,m^2$), the efficiency of the cell at STC ($\eta_{pv} = 15\%$), efficiency of maximum power point tracker ($\eta_{pt} = 100\%$), derating factor for manufacture tolerance ($f_{man} = 97\%$), derating factor due to dirt ($f_{dirt} = 95\%$) and switching and cable loss between the array and inverter ($\eta_{pv-inv} = 97\%$) [6, 7]. Mathematically, P_{pv-dc} can be written as

$$P_{pv-dc} = A_{pv} \times \eta_{pv} \times \eta_{pt} \times f_{man} \times f_{dirt} \times f_{cell} \times \eta_{pv-inv} \times E_{ir}(t) \times 10^{-3} \quad (1)$$

The temperature induced derating factor is a function of ambient temperature $(T_{amb} = 14.05 \,°C)$ and solar irradiance. It can be expressed as

$$f_{cell} = [1 - T_{co} \times (T_{amb} + \left(\frac{NOC - 20}{800} \times E_{ir}(t)\right) - T_{ref} \tag{2}$$

where T_{co} (=0.005) is the temperature coefficient, T_{ref} (=25 °C) is the reference temperature, NOC is the nominal operating cell temperature.

The PV power output in alternating form $P_{pv-ac}(t)$ can be mathematically represented as

$$P_{pv-ac}(t) = P_{pv-dc}(t) \times \eta_{inv} \times \eta_{inv-sb} \tag{3}$$

where η_{inv} (=95%) is the inverter efficiency, η_{inv-sb} (=99%) is the factor responsible for reduction of power output due to cable and switching loss between the inverter and primary switch board.

2.2   Modelling of the Wind Turbine

The power generated by a wind turbine essentially is a function of the velocity of wind. Mathematically,

$$Pw = \begin{cases} 0 & V < V_{cut,in}, V > V_{cut,out} \\ V^3 \left(\frac{P_{rated}}{V_{rated}^3 - V_{cut,in}^3}\right) - P_{rated}\left(\frac{V_{cut,in}^3}{V_{rated}^3 - V_{cut,in}^3}\right), & V_{cut,in} \le V < V_{rated} \ge 0 \\ P_{rated} & V_{rated} \le V \le V_{cut,out} \end{cases} \tag{4}$$

where P_{rated} is the rated power of the WT (=6 kW), V_{rated} is the wind speed at which the WT generates rated power (=14 m/s), $V_{cut,in}$ is the cut-in speed (=2 m/s) i.e. at this velocity the WT starts generating electrical power and $V_{cut,out}$ is the cut-out speed beyond which the WT does not generate any power (=20 m/s).

Wind velocity and GHI data of one day (24 h) were collected from the Indian meteorological department (IMD) [9] for the location of Bhubaneswar, Odisha.

2.3   Modelling of the Battery

The power output of a battery (P_{Bat}) is calculated as follows

$$P_{Bat} = P_{Bat,max/U} * X_B * \eta_B \tag{5}$$

where $P_{Bat,max/U}$ is the rated power of the battery, X_B is the battery contribution factor [0, 1] and η_B is the numbers of battery strings in parallel (Figs. 1 and 2).

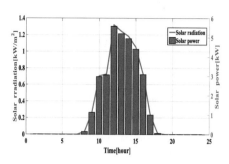 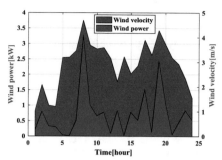

Fig. 1. Solar radiation and solar power P_{pv}

Fig. 2. Wind velocity and generated wind power

The battery bank consists of parallel strings of series-connected batteries. The ratio of DC bus voltage and the terminal voltage of each battery determines how many batteries to be connected in series. The numbers of parallel strings of batteries depend on the current requirement at the DC bus and η_B of each battery. In discrete-time domain the battery state of charge (SOC) can be mathematically represented as given below [10–12].

$$SOC_{(J)} = SOC_{(0)} - t_s \frac{\eta_{Bat}}{E_{nom}} \sum_{i=1}^{J} P_{Bat}(i) \qquad (6)$$

where η_{Bat} is the bi-directional power flow efficiency of the battery, t_s is the sampling time and E_{nom} is the rated energy storing capacity of the battery bank. The battery bank power output in one day is shown in Fig. 3.

2.4 Modeling of Micro-hydro Power Generation

Micro-hydro generators have the ability to produce power under significantly lower head and lower flow rate conditions [14]. The particular water release is estimated on such basis as topography and here only one-day rainfall data taken from nearby rainwater gauge station. The resultant power of micro-hydro generator P_{Hydro} is given by

$$P_{Hydro} = \rho \times g \times H \times Q(t) \qquad (7)$$

Where ρ is the water density (=1025 kg/m^3), g is the acceleration due to gravity (=9.81 m/sec^2), Q is the quantity of water discharged through the turbine (in our case Q is considered to be 387.1 Lit/sec), H is the effective water head (=4 m).

2.5 Modeling of Load

The one hour basis i.e. hourly load data has been taken at the location of Bhubaneswar, Odisha, for 2018, and the load variation is illustrated in Fig. 4.

2.6   Modeling of DG

The cost function of diesel generator is expressed as a quadratic equation:

$$C_{dgi,t} = a_i \left(P_{dg_i}(t)\right)^2 + b_i \left(P_{dg_i}(t)\right) + c_i \tag{8}$$

where a_i, b_i, c_i are the DG's cost function coefficients. The cost associated with the diesel generator is very high [15]. So, in this work, the DGs are included in the generation only for back-up. DG's cost function coefficients are $a_i = 0.0074$, $b_i = 0.233$, $c_i = 0.4200$. $P_{dg}(t)$ is the DG output power at any time t.

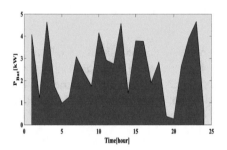

Fig. 3. Battery power P_{Bat} in 24 h   **Fig. 4.** Hourly load demand P_{load} in a day

3   Problem Formulation

From the perspective of a shorter duration of one day, the operational costs of the sources except DG's fuel consumption cost are insignificant. So the fuel cost is taken into account for economic consideration.

3.1   Objective Function

This paper addresses the purpose to minimize the daily fuel consumption cost from the DG. The long term cost of such a system generally consists of maintenance, fuel, lubricants, components replacement costs. As the optimization window is of only 24 h in this paper, costs other than fuel cost of the DG are excluded off the scope of analysis. This can be expressed as:

$$\min \int \left(aP_{dg}^2 + bP_{dg} + c\right) dt \tag{9}$$

3.2   Constraints

The equality constraint in this case is as follows:

$$P_{Hydro}(t) + P_{PV}(t) + P_W(t) + P_{DG}(t) + P_{Bat} = P_{Load}(t) \tag{10}$$

The inequality constraints are as detailed below:

$$0 \leq P_{Hydro} \leq P_{Hydro}^{max} \tag{11}$$

$$0 \leq P_{PV} \leq P_{PV}^{max} \tag{12}$$

$$0 \leq P_W \leq P_W^{max} \tag{13}$$

$$0 \leq P_{dg} \leq P_{dg}^{max} \tag{14}$$

$$-P_{Bat}^{max} \leq P_{Bat} \leq P_{Bat}^{max} \tag{15}$$

where P_{Hydro}^{max} is the maximum hydro power output, P_{PV}^{max} is the maximum PV power output, P_W^{max} is the maximum WT power output, P_{dg}^{max} is the maximum DG power output and P_{Bat}^{max} is the maximum battery power output.

4   Proposed Optimization Solver and Algorithm

The majority of conventional optimization techniques are based on finding out the local minima. But the operational behavior of the proposed hybrid system is extremely non-linear in nature [13] which instills a number of local minima in the solution search space. So, conventional optimization strategies may not be effective in discovering a global minimum in the search space. The optimization problem, discussed above, can be solved by "fmincon" in MATLAB.

The power outputs from different sources are assigned to the functions of the variable "x". The set-up has been complied with in the following way.

$$P_{Hydro} = P_1 = x(1:N) = [x_1, x_2] \tag{16}$$

$$P_{PV} = P_2 = x(N+1:2N) = [x_3, x_4] \tag{17}$$

$$P_W = P_3 = x(2N+1:3N) = [x_5, x_6] \tag{18}$$

$$P_{dg} = P_4 = x(3N+1:4N) = [x_7, x_8] \tag{19}$$

$$P_{Bat} = P_5 = x(4N+1:5N) = [x_9, x_{10}] \tag{20}$$

5   Simulation Results and Analysis

The purpose of the present simulation is to reduce the everyday operational cost of the developed hybrid system with different renewable energy sources subjected to variable loading. An hourly based load profile and information sources are utilized to evaluate the everyday operational cost savings for simulated hybrid system.

Figures 5, 6, 7 and 8 demonstrate how the load demand "P_L" is met effectively as the optimum electrical power flows from the renewable energy sources for one whole day. It is observed that the micro hydro turbine contributes significantly in comparison to the other participating renewable sources and consequently it has major impact on reduction of DG fuel consumption and battery charging/discharging procedure.

Figure 5 illustrates the optimized load demand during a whole day on hourly basis. During the early and late hours of a day, the absence of P_{PV} and P_W compel the micro-hydro turbine to cater to the load demand. It is observed that load demand during morning and night drastically reduces.

From Fig. 6, it is understood that contributions from other sources inhibit the PV arrays to generate at their maximum capacity, which in turn, reduces the costs related to PV power generation. In the evening from 17:00 h, the load demand slowly rises and attains its peak in between 19:00 h and 20:00 h. It starts reducing significantly after 21:00 h. As a result, from 17:00 h to 18:00 h, PV array supplies the load along with a little share from the DG and battery. After the sun sets, the absence of solar irradiation restricts the PV array to cater to the load demand even if the demand is rising. So, the power supply from the battery bank as well as the DG comes into picture to satisfy the demand.

In Fig. 7, optimal P_{Hydro} is presented. The optimal power output coincides with the maximum power generation capacity and it does not affect the economy adversely as micro-hydro units are cost effective.

From Fig. 8, it is observed that the WT does not produce significant amount of power from 8:00 h to 17:00 h and after 21:00 h in the evening. So during these hours, wind power is predominantly used to charge the battery. The remaining wind power, if available, supplies the load. During the remaining hours the wind power aids to other available renewable power to meet the load demand.

Figure 9, sheds light on variation of p-

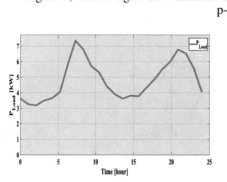

Fig. 5. Optimized load demand for 24 h

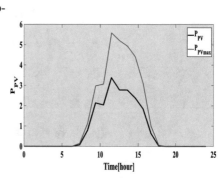

Fig. 6. Optimized P_{PV} in 24 h

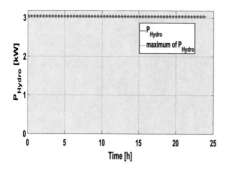

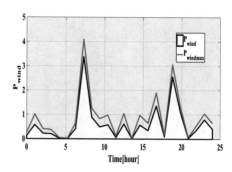

Fig. 7. Optimized P_{Hydro} for 24 h

Fig. 8. Optimized P_W for 24 h

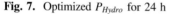

o-

wer flowing into or out of the battery bank. During the early hours up to 8:00 h the battery discharges to supply the load as there is none or very little PV and wind power. As the PV power becomes enough to cater the load demand, surplus renewable power charges the battery bank. It starts discharges at 18:00 h, charges for the next hour and continues discharging further when the renewable power is deficient. In this figure, negative and positive values of P_{Bat} denotes the process of charging and discharging respectively. In no case P_{Bat} exceeds its maximum capacity. As illustrated in the Fig. 10, SOC of the battery bank lies well within its allowable minimum (20%) and maximum (95%) limit which ensures the health of the battery bank and its life span. After the 24th h, the SOC continues to reduce and remains constant on reaching its minimum value until and unless it charges further.

DG plays a vital role here to stabilize the energy required. It is switched off as quickly as there is adequate electrical power from the other energy sources. Working hours and output power of the DG are largely dependent on the load demand, battery SOC and the generated renewable power during the studied period of time. It is observed from Fig. 11 that the DG is never employed to re-charge the battery bank. It only is switched on to supplement the deficiency of load demand.

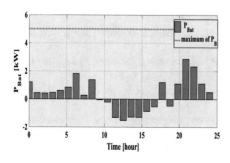

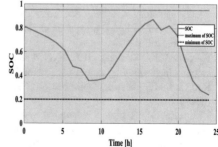

Fig. 9. Optimized P_{Bat} in 24 h

Fig. 10. Optimized SOC in a day

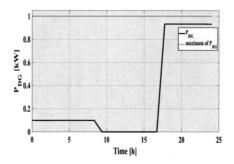

Fig. 11. Optimized P_{DG} in 24 h

6 Conclusion

The highly unpredictable nature of renewable energy sources and load demand eclipse the idea of implementing renewable energy sources to supply loads situated at locations which are out of the reach of conventional electricity grids. Using battery banks and DGs to compensate the non-linearity is a viable option, but very high costs associated with DG's fuel consumption shadows its economic feasibility. This paper derives an approach to mitigate the issue of unreliable power supply at remote locations keeping an eye on economic profitability. The hourly water discharge, solar insolation, wind speed and load demand information have been utilized as input data for simulation purposes. The proposed model intents upon lessen the use of DG and to shift the load demand to hybrid energy sources. This optimization problem is solved effectively within the purview of operational constraints using "fmincon" solver with Interior – point algorithm. The optimization tool proves to be fast, reliable and capable of solving complex computational problems.

References

1. Tazvinga, H., Xia, X., Zhang, J.: Minimum cost solution of photovoltaic–diesel–battery hybrid power systems for remote consumers. Sol. Energy **96**, 292–299 (2013)
2. Supriya, C.S., Siddarthan, M.: Optimization and sizing of a grid-connected hybrid PV-wind energy system. Int. J. Eng. Sci. Tech. **3**(5), 4297–4323 (2011)
3. Paish, O.: Small hydro power: technology and current status. Ren. Sust. Energ. Rev. **6**(6), 537–556 (2002)
4. Kusakana, K., Vermaak, H.J.: Hybrid renewable power systems for mobile telephony base stations in developing countries. Renewable Energy **51**, 419–425 (2013)
5. Rao, S.S.: Engineering Optimization Theory and Practice. John Wiley & Sons, Hoboken (2009)
6. Kusakana, K.: Optimal scheduled power flow for distributed photovoltaic/wind/diesel generators with battery storage system. IET Renew. Power Gener. **9**(8), 916–924 (2015)
7. Accredited, Clean Energy Council. Grid-connected Solar PV Systems (No Battery Storage) Design Guidelines for Accredited Installers (2013)

8. Diaf, S., Notton, G., Belhamel, M., Haddadi, M., Louche, A.: Design and techno-economical optimization for hybrid PV/wind system under various meteorological conditions. Appl. Energy **85**(10), 968–987 (2008)
9. Indian Meteorological Department, Bhubaneswar, India. www.imd.gov.in
10. Kasturi, K., Nayak, M.R.: Multi-objective assessment of wind/BES siting and sizing in electric distribution network. In: Pattnaik, P., Rautaray, S., Das, H., Nayak, J. (eds.) Progress in Computing, Analytics and Networking, pp. 407–418. Springer, Singapore (2018)
11. Nayak, C.K., Nayak, M.R.: Technoeconomic analysis of a grid-connected PV and battery energy storage system considering time of use pricing. Turk. J. Electr. Eng. Comput. Sci. **26**(1), 318–329 (2018)
12. Nagaraj, R.: Renewable energy based small hybrid power system for desalination applications in remote locations. In: IEEE 5th India International Conference on Power Electronics (IICPE). IEEE (2012)
13. Buss, M., Glocker, M., Hardt, M., Von Stryk, O., Bulirsch, R., Schmidt, G.: Nonlinear hybrid dynamical systems: modeling, optimal control, and applications. Modelling, Analysis, and Design of Hybrid Systems, pp. 311–335. Springer, Berlin, Heidelberg (2002)
14. Güney, M.S., Kaygusuz, K.: Hydrokinetic energy conversion systems: a technology status review. Renew. Sustain. Energy Rev. **14**(9), 2996–3004 (2010)
15. Mahmoud, M.M., Ibrik, I.H.: Techno-economic feasibility of energy supply of remote villages in Palestine by PV-systems, diesel generators and electric grid. Renew. Sustain. Energy Rev. **10**(2), 128–138 (2006)

Design and Implementation of LNA for Biomedical Applications

Malti Bansal[(⊠)] and Gaurav Srivastava

Department of Electronics and Communication Engineering, Delhi
Technological University, Delhi 110042, India
maltibansal@gmail.com

Abstract. LNA (low noise amplifier) remains as the key element of RF
transceiver system. These are used for amplifying a weak, low power signal
without degrading its signal to noise ratio. LNA used in radio communication
system and ISM band application is capable of handling wide dynamic range of
input signal without getting overloaded and undergoing saturation. In this paper,
use of LNA in biomedical applications includes neural, biosensors, ECG and
EEG application and comparison of different LNA topologies in terms of chip
area, power consumption and method of implementation has been done.

Keywords: LNA · Neural · Biosensors · ECG · EEG

1 Introduction

LNA stands for Low Noise Amplifier. In a RF system, LNA is used for amplifying a
signal of very low strength received from antenna. These are considered as the significant part of transceiver components of a RF system. LNA is the first component of
receiver that intercepts the signals and modulates the signal to noise ratio (SNR). These
are low power devices that have a current consumption ranging from 10 to 100 mA.
LNA in biomedical domain [1] is usually implemented via cascade inductor with
degenerate topology for amplifying a weak signal and provides a low noise figure and
high gain. In [2], fully differential RF front end amplifier is designed on 0.13 μm
CMOS technology for obtaining low power and direct conversion of short range radio
signal at low cost. The RF signal is down-converted to a baseband signal that contains
dc offsets and flicker noise, which tends to corrupt the SNR. Therefore common gate
LNA (CG-LNA) is used instead of common source LNA (CS-LNA). Eliminating on
chip inductor and providing a good input matching, it improves stability, linearity and
robustness against PVT variation in CS topology. LNA of small size and low power
consumption are preferred in biomedical applications. Therefore in [3], LNA using g_m
boost technique is implemented on 130 nm IBM technology and processed, providing
a good matching of 600 Ω and current consumption up to 500 μA. LNA with CG
topology is preferred as it consumes less power and has less sensitivity (Fig. 1).

© Springer Nature Switzerland AG 2020
A. P. Pandian et al. (Eds.): ICICCS 2019, AISC 1039, pp. 154–167, 2020.
https://doi.org/10.1007/978-3-030-30465-2_18

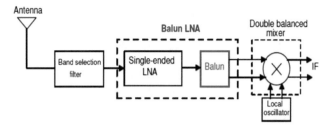

Fig. 1. BALUN LNA [3]

In [38], RF front end for biomedical applications is implemented using ultra low power and using 0.18 μm technology. It uses single ended compatible current-reuse LNA (CRLNA). The LNA is capable of achieving a power gain of 20 dB with IIP3 of −8.1 dBm and noise figure of 2.8 dB. The RF front end consists of coupling CRLNA with quadrature folded mixer, with local oscillator buffer, achieving a conversion gain of 28.7 dB and NF of 5.5 dB, consuming less power of 500 μW, with chip area of 0.7 mm².

(i) **LNA in Neural Applications**

In biomedical applications, neural signals are dealt with the learning and comprehensive study of brain function and its healing, curative and prosthetic application. Neural signals are interfaced in large number of channels for the study of neurons specifically for brain microcircuits. The entire system consists of dealing with important parameters like noise performance, CMRR and power consumption. The entire system consists of action potentials carrying information to the neural cells. These are formed by the process of polarisation and depolarisation occurring across neural cell membrane; or these are caused due to the fast movements of ions across cell membrane. In animal cells, the ion's like Na^+/K^+ are powered by ATP molecules forming basic building blocks. Whenever channel consisting of ions is activated electrically or biochemically through mechanisms like depolarisation, voltage sensitivity is caused resulting in the opening of ions membrane containing Na^+ to open large number of positive charges across cell. However, the balance is obtained by the inflow of K^+ ions across the cell membrane. For getting sneak peak for action potential, an amplifying method is required. The neural amplifier should reject the dc offsets of the resulting signal amplitudes. The amplifier should have high SNR and proper CMRR and PSRR. The majority of neural amplifiers successfully extract input referred noise between 3–7 μVrms. Without the controlling mechanism, this is entirely dominated by noise. The action potentials (Aps) across the cell membrane are found to be at 100 mV. The main component of noise is mainly from muscle contraction, affecting neural recording and intensifying brain activity. Environmental noise includes the noise from radio and electrical signals which degrade the performance drastically. Thermal and flicker noise are caused by intrinsic property of the semiconductor. In [4], the topology like Operational Transconductance Amplifier (OTA) which is common for executing LNA, is executed using two stage op-amps and current mirror OTA's. The folded cascaded combinations are shown in Fig. 3. It is expected that the topology should yield zero leakage current when compared to different amplifiers. Chae et al. [5] implemented OTA with self biased topology, attaining gain of 40.5 dB and Noise Efficiency Factor (NEF) of 4.5 and consuming power up to 12.5 μW. However current feedback instrumentation amplifier possesses a high impedance and an offset in mV range. Chopping at high frequency modulates noise, resulting in white band noise. This chopping at a high frequency is effectively reduced by using a notch filter in passband.

Capacitive Feedback Topology

The capacitive feedback topology is shown in Fig. 4. It is optimized for very low noise power operation through a dedicated circuit design topology. The architecture achieved a gain of 39.5 dB, power consumption of 80 μW and NEF of 4. Majidzadeh et al. [6] achieved a modified and optimized capacitor feedback topology having a power consumption of 7.92 μw and input referred noise of 3.5 μV$_{rms}$ (Fig. 2).

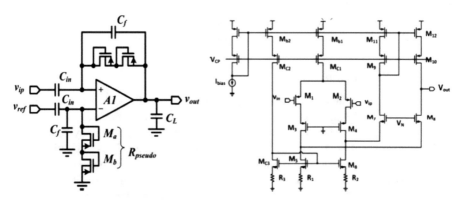

Fig. 2. Folded cascaded active feedback topology [5]

Fig. 3. Capacitive feedback topology [6]

In this topology, a segment of a signal was filtered out, which was used as a feedback for suppressing a signal component. This is one of the efficient ways for extracting the low frequency component of a signal, which makes it a very stable system. Wei et al. employed topology for active feedback, attaining 6 μW power consumption and NEF of 3.1, having a bandwidth of 8.9 kHz and a midband gain of 46 dB on a chip area of 0.022 mm^2. In [7], as shown in Fig. 5, a MOS bipolar pseudo resistor (implemented using a cross coupled configuration) was used for implementing AC coupling and to reject the large DC offsets caused due to contact potential. Every transistor in MOS diode is connected such that a parasitic source-bulk diode is obtained in anti-parallel. If the voltage across the device is small, then neither of the diode conducts, since the effective resistance is very large (>10 GΩ). For minimizing the input referred noise, two strategies are adopted; first amplifier is reduced to a single branch which is operating at full current. The reference current so obtained is found to be 10 times the amplifier bias current, which does not contribute to the total power consumption. The second strategy [8] is to drive the gate of MP$_1$ and MN$_1$. Since the input is AC coupled, the gates of transistor MP_1 and MN_1 are avoided of DC coupling by operating it within a range of frequency. This results in doubling of the transconductance of common source amplifier while keeping output noise constant and scaling the input noise by a factor of 2. The aspect ratio of MP_1 and MN_1 are chosen for maximizing the g_m/I_d ratio and these are operated in weak inversion region. The reference current is adjusted to maximize the bias current at 770 nA and obtaining the highest gain of 44 dB and a bandwidth of 1.9 kHz. Thus, this topology provides the method for detection of action potentials required for neural recording (Table 1).

Table 1. LNA in neural applications

S. No.	[9]	[6]	[10]	[11]	[7]	[12]
Gain (dB)	40–56	39.4	60	76.2	39.2	15
Bandwidth (kHz)	1–10	7.2–10	<9.1–10	–	7.2	2.4–2.5 (GHz)
Noise (rms) (µV)	2.2	3.5	4.8	2.89	2.2	–
Power consumption	68	7.92	50	15.174	80	7 mW
CMRR	N/A	70.1	N/A	>120 dB	83	
NEF	6.62	3.35	11.42	2.35	2.9	

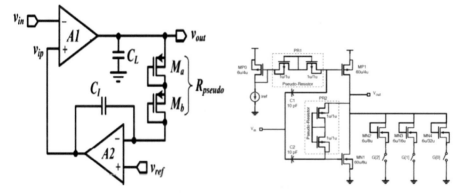

Fig. 4. Active feedback LNA topology   **Fig. 5.** Bipolar pseudo resistor [7]

In [11], for low power application, the LNA is executed with a wireless medical telemetry requiring a portable device capable of operating at a low supply voltage, provided from a battery and a energy source. A non coherent RF front end receiver and super-regenerative RF receiver are crucial design parameters of this module. An RF envelope detector implemented using ON-OFF keying requires a specific local oscillator executed using phase locked loop. This causes increase in the power consumption. However, for the demodulation at the RF receiver, the input signal is mixed with the signal from mixer by the process of down and upconversion of frequency. The signal is converted back to the baseband signal. Example of this is gilbert cell implemented with a CMOS simulation process, having high conversion gain and port to port isolation.

LNA are implemented using cascaded topology and since LNA implemented using current reuse technique consumes more power and because its efficiency is low at low power, these are implemented using folded topology. However for low power consumption and low voltage, RF front end amplifier are implemented using two stage common source topology. In [12], CMOS OOK is implemented at 2.4 GHz requiring low voltage. LNA topology is implemented using current reuse method providing high gain and noise figure. After post simulation at 10 ns, digits 0 and 1 are received successfully recovering OOK data's at 100 Mbits/s. At low frequency, the noise factor is high due to presence of two noises: thermal and flicker noise. In [11], EKV model using PMOS transistors was executed for reducing flicker noise. It had large gate area and was operated in weak inversion mode. Such techniques consumed 2.89 µVrms at the supply current of 4.8 µA. And power consumed was 15–17 µW. This can be used for neuroscience research, as the system is capable of measuring weak neural signals ranging from (0.5–5 mV) and for clinical and neuro prosthetic applications having frequency ranging from 100 Hz to 7 kHz. LNA designed consisted of single ended output in which M_1 and M_2 served as a PMOS I/P transistors and M_3 and M_4 as PMOS cascade stage transistors. NMOS transistors M_5 and M_6 made current mirror load. M_7 and M_8 acted as current mirror circuits. For minimising flicker noise, PMOS acted as an input for which noise power density was given by

$$E_n^2 = \frac{8kT\gamma}{g_{m1}} + 8kT\gamma\left(\frac{g_{m5}}{g_{m1}}\right)^2 + \frac{2k_p^{1/f}}{C_{ox}W_1L_1f} +$$
$$\frac{2k_n^{1/f}}{C_{ox}W_5L_5f}\left(\frac{g_{m5}}{g_{m1}}\right)^2 \qquad g_{m1} \gg g_{m5},$$

And since,

$$\left(\frac{W}{L}\right)_1 \gg \left(\frac{W}{L}\right)_5$$

$$E_n^2 = \frac{8kT\gamma}{g_{m1}} + \frac{2K_p^{1/f}}{C_{ox}W_1L_1f}$$

The circuit was designed on cadence virtuso 6.1.3 and simulated on spectra simulator at a temperature of 36.9 °C and operating at 100 Hz frequency and having a duty cycle of 10. The noise efficiency factor was obtained by

$$NEF = Vrms\sqrt{\frac{2I_{tot}}{\pi U_T 4kT.BW}}$$

In the above formula, V_{rms} represents an input noise voltage and BW represents amplifiers bandwidth.

(ii) LNA in Biosensors

The term 'Biosensor' is short for "biological sensor". The device consists of transducer and biological elements consisting of enzymes, antibody or nucleic acids. The bioelements interact with the analyte being tested and biological response is converted to electrical signal by transducers. The obtained biological response of biosensors is examined by biocatalytic membrane, achieving the conversions of reactant to the product. Immobilised enzymes provide a large number of advantages making them applicable for biological systems. These can be reused ensuring catalytic activity for a series of analyses. Thus, biosensors are widely used in the field of healthcare, medical diagnosis, etc. In [13], CMOS on chip sensor was used for measuring the dielectric constant of an organic chemical. The dielectric constant was measured by observing changes in the tank capacitance producing LC oscillation in VCO, due to the exposure to the liquid. Frequency change causes change in the voltage measured, using on chip digitised analog to digital converter. The calibration of sensor was used for the measurements of dielectrics. The measurements were made for frequency ranging from 7–9 GHz. The advancements in wireless communication and network technologies have led to the significant advancement in wearable and implantable sensors creating an impact on e-health and telemetry systems. In [14], m-health comprised of mobile computing, medical sensors and communication technologies. For diagonsis and monitoring of diseases like diabetes, a large number of biosensors are worn and implanted. The interconnections are made inside a human body and the entire system is called BASN (body area sensor network) performing two operations: the transmitting of processed data and; through a secured channel, transmitting of data takes place between human body and biosensors placed at the same node. The BASN differentiates between the individuals based on their psycological and behavioural characterstics. Introduction of the ubiquitous (present everywhere) technology in healthy living regimen and medical field is called as ubiquitous healthcare [U-health] [15, 16] which means the patient can collect and accept the information of the diagnosis therapy and prognosis management with information and communication technology. The patient is provided with biomedical digital assistant, monitoring patient health at any moment through wireless and wired mode of telecommunication. The biomedical signal is sent through a biomedical digital assistant through a remote server, providing doctor with the relevant information about patient. These devices are portable and wearable to patient. They are expected to have light weight and should be operable for 24 h a day. The biosensors used are ECG and PPG sensors. The biomedical devices are implemented using two methods: cellular phones and PDAs. However, wired sensors implemented using PDA have shorter time, therefore they should be implemented wirelessly, which improves the processing and storage function of the sensors (Fig. 6).

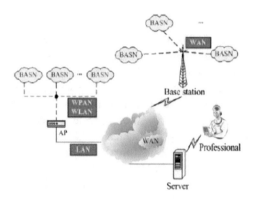

Fig. 6. BASN health model [15]

In [17], WBSN in health care and monitoring system was implemented consisting of wireless sensors forming a network. The data was collected through these sensors and adjacent sensors were decided by the header containing the address of adjacent node. The WBSN consisted of control path and wireless communication nodes isolated from each other. Control path added controllability and scheduling of the system at the transceiver (Table 2).

Table 2. LNA in biosensors

S. No.	[17]	[18]	[19]
Power consumption	1440 µW	4 mW	2.34 mW
Chip area	–	(91.4 * 123) µm^2	350 * 20 µm^2
Implementation	TSMC (0.18 µm)	0.65 µm CMOS	0.35 µm (CMOS)
Current (µA)	800	52 fA	1–10 µA
Supply (V)	1.8	–	3.3

(iii) **LNA in ECG**

Biomedical signals provide information about the physiological activities of organisms, produced by gene and protein sequences. The well-known bio signals are ECG, EEG, EMG and MMG. ECG is the acquisition of electrical activity observed from the heart and captured by the electrodes attached to the skin. These are very weak signals and because of their weak nature, noise gets added, for which signal processing is required. The ECG or electrocardiography signals are produced due to the potential difference between the two electrodes and since these are recorded for long time, the compression of data is required both at transmitting and receiver ends. The contraction of the muscles and cardiac pattern ECG signals are coupled with noise. Therefore, filtering is not easy. To solve this problem, the three algorithms applied are stress testing, ambulatory monitoring and intensive care monitoring. After signal processing of received ECG signals, its morphological characteristics and specific heart rhythms are rectified. In [20], low power and compact wearable sensors were designed. The signals were wirelessly transmitted and monitored using ECG filtering and QRS detection methods. This topology was implemented using successive approximations register (SAR) ADC using 2 SPI interfaces and on chip SRAM having central control unit of 8 Kbits. The ECG signals were sampled at a rate of 256 Hz using QRS complexes and the ECG signals were then continuously written on asynchronous FIFO and were forwarded to CPU, thus consuming less power. Wearable ECG monitoring systems use electrode that requires the skin preparation in advance, so these are not suitable for high level of activity due to high level of spikes of noise that appears in the data. Therefore in [22], as shown in Fig. 8, a new wearable and ultra low power wireless sensor node called eco was formed. This reduced the size and weight of the system and provided reliable operation. In [23], the conditioned biosignal was received by the encoder from a low noise amplifier and compression of N input samples gave M output samples and these were digitised and transmitted using ADC. The RF front elements were made to operate at Nyquist rate, having a conversion factor of N/M which was greater than 1, and these signals were downconverted to output y(t) (Fig. 7).

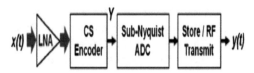

Fig. 7. LNA in ECG measurement system

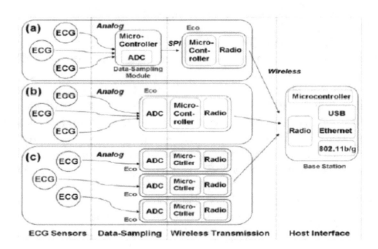

Fig. 8. Architecture of ECG monitoring system [23]

Table 3. LNA in ECG

S. No.	[21]	[22]	[24]	[25]	[26]
Implementation	On body	On cloth	0.18 μm CMOS (TSMC)	0.5 μm CMOS	0.18 μm CMOS
Power consumption	9.6 μW	30 μW	–	30 μW	110 μW
Technology	QRS and SAC diagnosis	Capacitive sensing	ARM displayer	Adaptive sampling with DSP	Current F/B topology
Area	5.74 mm^2	–	–	–	–

(iv) **LNA in EEG**

The EEG signal which stands for electroencephalogram is a signal produced due to the electrical activity of the brain and recorded from the scalp. The recorded waveform consists of cortical, electrical activity. These are very weak signal and are measured in microvolts. The main frequency of the human EEG wave consists of Delta (frequency of 3 Hz or below), Theta (frequency of 3.5 Hz and 7.5 Hz), Alpha (frequency between 7.5 and 13 Hz) and Beta (frequency > 14 Hz). Since these are weak signal, the voltage fluctuation measured at the electrodes is very small; the recorded data is then digitised and sent to amplifier. The amplified data can then be displayed as a sequence of voltage values. This is one of the fastest imaging techniques having high sampling rate. In [26], brain computer interfacing as a technique was used for setting up a communication path between the brain and outside environment. The authors described the method of signal acquisition and signal pre-processing. Signal acquisition using non invasion methods like electroencephalography (EEG), fMRI, NIRS and MES is considered a good practice. EEG signals are non-stationary and have low spatial resolution EEG signals are prone to artefacts caused due to muscle activities, heartbeat, blinking of eyes or eye movements, etc. Signal pre-processing is defined as a process of signal enhancement. The contamination of a signal received is caused due to artefacts and noise. Various signal processing methods include Common average deferencing (CAR), ICA (independent component analysis), etc. In [27], EEG signals were extracted through brain computer interface (BCI) techniques during mental activity. In [28], for neurological diagnosis, lossless compression of EEG signals played a crucial role. The authors presented a predictor and adaptive error modelling technique. In [29], EEG was examined using HHT (Hibertz-Huang Transform) process. For this, a de-noising process was applied for smoothening of EEG signals and was based on EMD (empirical mode decomposition) and Monte Carlo process. In [30], EEG analysis was carried out at different frequency. Experimental analysis was carried out to obtain different EEG data (Table 3).

(v) **LNA in Hearing Aids**

Hearing aids comprise of electroacoustic devices. They are typically worn in or behind the ears and they modulate sound for hearing impaired. The microphone transforms sound to electrical signals which on further amplification is converted back to the sound by receiver. The entire process is powered by a source of a battery (Fig. 9).

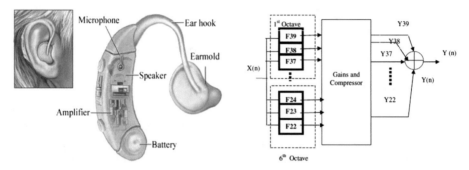

Fig. 9. Hearing aid device   **Fig. 10.** Parallel band structure [31]

In [31], the digital signal processing technique eased the filter bank designing for various applications. Filter bank algorithms were subsequently used in noise reduction, speech encoding and auditory compensation methods. The entire system design depended upon the type of filter used. In [32], binaural hearing aids were designed such that they can communicate wirelessly to both multimedia devices and smart phones. A wireless multimedia device was used along with a customised Bluetooth device with adapter. This was provided with fitting parameters controlling process which can control volume of the hearing aids using graphical user interface. Thus, system designed was efficient enough without any controlling mechanism. The wireless communication control panel was designed using FPGA kits implemented using a DSP processor. This system was capable of attaining a data rate or baud rate of 4.6 bits/sec and device was having a height of 12 mm, length 46 mm and width 15 mm. In [33], an energy efficient analog front end circuit was designed and executed for hearing aid applications. Adaptive-SNR (ASNR) and combined gain control technique was implemented having low power consumption. The ASNR technique operated by changing the value of SNR, according to the input amplitudes, for optimising SNR; and reducing power consumption. The method was capable of achieving 86 dB SNR at a supply voltage of 0.9 V. An automatic gain control was required for preamplifying hearing ability and to avoid unnecessary loud sounds (Fig. 10).

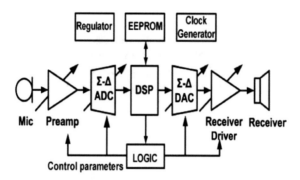

Fig. 11. Block diagram of hearing-aid system [33]

(vi) LNA in Medical Implantable Devices

An implant is a medical device built for replacing a misplaced biological structure or supporting an injured biological structure. Some implantable devices are bioactive which are subcutaneous and are present in the form of implantable pills or drug eluting-stents. Implantable devices are residing within human body and are used for monitoring, diagnostic or therapeutic purposes because of their unique responsivity and rational module. These are facilitated with minimum damage caused when host tissue or biomolecule are removed. Their application areas include orthopaedic, dental, cardiovascular, etc (Fig. 11).

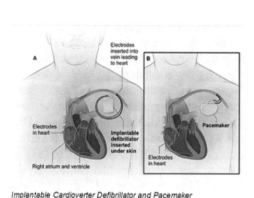

Fig. 12. Medical implantable device

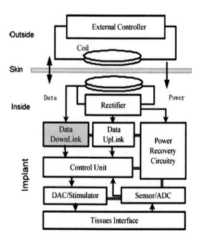

Fig. 13. Inductive link to transmit power/data [37]

In [34], biomedical implantable devices (IMDs) were designed with traditional goals such as safety and utility; without compromising the patient health. It consisted of drug delivery system, neurostimulators, pacemakers, etc. These are helpful in determination of diseases like diabetes, cardiac arrhythmia, etc. IMD ubiquitous continues to safeguard 25 million US citizens and gaining their confidence for life critical functions. In today's scenario, IMD's provide with the features of delivering telemetry for remote monitoring over a long distance and providing high wireless bandwidth and forming a communication channel with other IMDs. In [35], implantable medical devices were electronic devices implanted within a human body for therapeutic monitoring and authentication. Because of their different nature, IMD's are not fully secured or effective. However, these are only accessible by IMD programmer and medical personnel. However, the attacks like compilation of fake commands and stealing of the health related data ensures that IMD is not fully secured. Till now, we don't have distinct security standards. The authors presented basic architecture with access controlling mechanism with logics using different controlling models. The biomedical IMDs can be designed with wireless inductive link capable of consuming less power and communication of data through wireless modes. In [36], BPSK modulator was used for designing COSTAS loop topology dedicated to medical devices. The analysis revealed data rate or baud rate of 1.12 Mbps; and power less than 0.7 mW was consumed with a supply voltage of 1.8 V. The coupling of BPSK modulator along with passive modulator allowed full duplex data transmission between the controller and the medical IMDs. This ensures and increases the durability of the entire system. In [37], wireless implantable medical devices required a continuous source of power supply which poses a threat to the health related issues. Here, the system was setup consisting of antenna and capable of operating at MHz frequency range. So that the large number of losses from the tissues could be compensated. This work presented an antenna having a size much smaller; and provided with the same power efficiency in the specified range of frequency of operation (Fig. 12).

2   Conclusion

Biomedical applications in the field of biomedical engineering include the development and designing of biomedical instruments capable of solving medical and health related problems. These instruments range from cosmic imaging systems like standard X ray, magnetic resonance imaging, etc. to the evolution of numerous therapeutic medical devices like EEG, MRI, etc. used for diagnosis. Here, in this paper, we have reviewed the designing and implementation of LNA using various topologies in biomedical applications like neural applications, biosensors, ECG, EEG, hearing aids, medical implantable devices, etc (Fig. 13).

References

1. Hameed, A., Oudah, A.: Improved design of low noise amplifier. Int. J. Multimedia. Ubiquitous Eng. **10**(1), 255–264 (2015)
2. Ahmed, S., Manoli, Y.: Design of low power direct conversion RF-front end with double balanced current-driven subharmonic mixer in 013 μm CMOS. IEEE Trans. Circuits Syst. **60**(5), 1322–1330 (2013)

3. Reddy, K.V., Sravani, K., Prashantha kumar, H.: A 280 μW sub-threshold Balun for LNA for medical radio using current reuse-technique. In: Proceedings of IEEE Conference on Microelectronics and Electronics (PRIME-LA), pp. 1–4 (2017)
4. Bharucha, E., Sepehrian, H., Gosselin, B.: A survey of neural front end amplifiers and their requirements toward practical neural interfaces. J. Low Power Electron. Appl., **4**, 268–291, ISSN 2079-9268
5. Chae, M., Kim, J., Liu, W.: Fully-differential self-biased bio-potential amplifier. Electron. Lett. **44**(24), 1390–1391 (2008)
6. Majidzadeh, V., et al.: A micropower neural recording amplifier with improved noise efficiency factor. In: Proceedings of the European Conference on Circuit Theory and Design, 23–27 August 2009, Antalya, Turkey, pp. 319–322 (2009)
7. Mirbozorgi, S.A., Bahrami, H., Rusch, L.A., Gosslin, B.: A low-power 2.4 GHz receiver for wireless implantable neural simulators. In: IEEE International Symposium on Circuit and Systems (ISCAS), pp. 1082–1085 (2014)
8. Holleman, J., Otis, B.: A sub microwatt low-noise amplifier for neural recording. In: Proceedings of 29th Annual Internationally Conference of IEEE EMBS, Lyon, France, 23–26 August 2007
9. Gao, H., et al.: A 96-channel full data rate direct neural interface in 0.13 μm CMOS. IEEE J. Solid Stare Circuits **47**, 1043–1055 (2012)
10. Perlin, G.E., Wise, K.D.: An ultra compact integrated front end for wireless neural recording microsystems. J. Microelectromechanical Syst. **19**, 1409–1421 (2010)
11. Jo-Yu, W., Kea-Tiong, T.: A band-tunable, multichannel amplifier for neural recording with AP/LFP separation and dual-threshold adaptive AP detector. In: Proceedings of the Annual International Conference of the IEEE Engineering in Medicine and Biology Society (EMBC 2011), Boston, MA, USA, 30 August–3 September 2011, pp. 1847–1850 (2011)
12. Dubey, D., Gupta, A.: A low power low noise amplifier for biomedical applications. In: IEEE International Conference on Electrical, Computer and Communications Technologies (ICECCT), pp. 1–6 (2015)
13. Gangopadhyay, D., Dixon, A.M.R., Nataranjan, K., Gupta, S.: Compressed sensing analog front end for bio-sensor applications. IEEE J. Solid State Circuits **49**(2), 426–439 (2014)
14. Helmy, A.A., Jeon, H.-J., Lo, Y.-C.: A self-sustained CMOS microwave chemical sensor using a frequency synthesizer. IEEE J. Solid-State Circuits **47**(10), 2467–2483 (2012)
15. Bao, S.-D., Zhang, Y.-T., Sheh, L.-F.: Physiological signal based entity authentication for body area sensor networks and mobile health care systems. In: Proceedings of 2005 IEEE Engineering in Medicines and Biology, September 2005, pp. 1–4 (2005)
16. Lee, T.-S., Hong, J.H., Cho, M.C.: Biomedical digital assistant for ubiquitous healthcare. In: Proceedings of 29 International Conference of IEEE EMBS, August 2007, pp. 1789–1793 (2007)
17. Hong, L., et al.: Fully integrated fluorescence biosensors on chip employing multi-functional nanoplasmonics optical structure in CMOS. IEEE J. Solid State Circuits **52**(9) (2017)
18. Tivatansakul, S., Ohkura, M.: Healthcare system focusing on emotional aspects using augmented reality. In: IEEE International conference on Biometrics and Kansei engineering, pp 218–222 (2013)
19. Lee, K.-H., et al.: CMOS capacitive biosensor with enhanced sensitivity for label-free DNA detection. ISSCC Dig. Tech. Papers, pp. 120–122 (2012)
20. Bailey, J.J., Berson, A.S., Garson, A., Horan, L.G., Macfarlanc, P.W., Mortara, D.W., Zywictz, C.: Recommendations for the standardization and specifications in automated electrocardiography bandwidth and signal processing. Circulation **81**, 730–739 (1990)
21. Macfarlane, P.W., Lawrie, T.D.W. (eds.): Comprehensive Electrocardiology. Theory and Practice in Health and disease, vol. 1–3. Pergamon Press, New York (1989)

22. Nemati, E., Deen, M.J., Mondal, T.: A wireless wearable ECG sensor for long term applications. In: Proceedings of IEEE Communication Magazine, January 2012, pp. 36–44 (2012)
23. Park, C., Chou, P.H., Hibbs, A.: An ultra wearable wireless low power ECG monitoring system. In: IEEE Conference of Biomedical Circuits and System, pp. 241–244 (2006)
24. Yazicioglu, R.F., Merken, P., Puers, B., Van Hoof, C.: A 200 µW EightChannel EEG acquisition ASIC for ambulatory EEG systems. In: 2010 IEEE JSSC, 8 February 2010, vol. 43, no. 12, pp. 124–126 (2010)
25. Wang, L.-H., et al.: Implementation of wireless ECG acquisition SoC for IEEE 802154 (ZigBee) application. IEEE J. Biomed. Health Inf. 19(1), 247–255 (2015)
26. Nanda, C., et.al: 1 V CMOS instrumental amplifier with high electrode offset cancellation for ECG acquisition system. In: Proceedings 2010 IEEE Students Technology Symposium, 3–4 April 2010
27. Lakshmi, M.R., et al.: Survey on EEG signal processing methods. IJARCSSE 4(1) (2014)
28. Sriaam, N., et al.: An adaptive error modelling scheme of the lossless compression of ECG signals. IEEE Trans. Inf. Biomed. 12(5) (2008)
29. Agarwal, R., Gotman, J.: Computer assisted sleep staging. IEEE Trans. Biomed. Eng. 48(12), 1412–1422 (2001)
30. Sun, Y., et al., The research of EEG analysis method based on sound of different frequency. In: 2007 IEEE Conference on Complex Medical Engineering (2007)
31. Subbulakshmi, N., Manimegalai, R.: A survey of filter bank algorithms for biomedical application (ICCCI-2014), Coimbatore, 3–5 January 2014 (2014)
32. Kim, D.-W., et al.: Implementation and verification of a platform for bluetooth linked hearing aids system with smart phone and multimedia devices. In: Conference on ICCE (2013)
33. Kim, S., et al.: An energy efficient analog front-end circuit for a sub 1–V digital hearing aid chip. IEEE J. Solid State Circuits 41(4), 876–882 (2006)
34. Halperin, D., et al.: Security and privacy for implantable medical devices. IEEE Comput. 2C(7) (2008)
35. Yamu, H.: A fully integrated low power BPSK demodulator for implantable medical devices. IEEE Trans. Circuit Syst. 52(12), 2552–2562 (2005)
36. Galbraith, D., Soma, M., White, R.L.: A wide-band efficient inductive transdermal power and data link with coupling insensitive gain. IEEE Trans. Biomed. Eng. 34(4), 265–275 (1987)
37. Poon, A.S.Y., et al.: A mm sized implantable power receiver with adaptive link compensation. In: IEEE Conference on Solid-State Circuit, pp. 294–296 (2009)
38. Cha, H.-K., et al.: A CMOS MedRadio receiver RF front-end with a complementary current-reuse LNA. IEEE Trans. Microw. Theory Tech. 59(7), 1846–1854 (2011)

Control Methods for Minimizing Cross Regulation in Single Inductor Multiple Output Converters – A Review

Md. Samiullah[1]([✉]), Atif Iqbal[2], Imtiaz Ashraf[1], Arshad Mohammad[1], and Reyaz Hussan[1]

[1] Department of Electrical Engineering,
Aligarh Muslim University, Aligarh, India
md_samiullah@zhcet.ac.in
[2] Qatar University, Doha, Qatar

Abstract. Single Inductor Multiple Output (SIMO) converter provides a cost-effective alternative to conventional power converters, which have more than one output. Its potential applications include portable devices, where multiple sources are required. SIMO on the other hand suffers with a major drawback of cross regulation among the outputs i.e. change in load at one submodule affects another. The problem of cross regulation plays a major role in stability and performance analysis of converters. In this paper the control methods ensure a good cross regulation and later the performances are summarized and collected.

1 Single Inductor Multiple Output Converter

Multiple outputs from non-isolated type of converter were normally obtained by using a linear regulator at the output of the first converter or by connecting another converter in series or parallel of the base converter [1, 2]. From isolated DC-DC converters more than one outputs were taken by using transformers with multiple secondary windings [3, 4]. These all were poorly efficient, bulky and costlier. A single Inductor multiple output (SIMO) converter was then proposed which provides multiple DC outputs with each output being regulated independently [5]. It uses a simple circuit (Fig. 1) and yields high efficiency in low complexity and cost. SIMO thus provides a cost-effective alternative to multiple switching power converters used to get different outputs for applications where many voltage levels are required simultaneously.

1.1 Modes of Operation

- Discontinuous conduction mode (DCM)
- Pseudo continuous conduction mode (PCCM)
- Continuous conduction mode (CCM)

Discontinuous conduction mode (DCM): In discontinuous-conduction-mode (DCM) of any converter, the inductor current falls to zero before the switching cycle ends and again starts from the zero value in the next cycle [6]. A typical waveform is shown in Fig. 2a.

© Springer Nature Switzerland AG 2020
A. P. Pandian et al. (Eds.): ICICCS 2019, AISC 1039, pp. 168–175, 2020.
https://doi.org/10.1007/978-3-030-30465-2_19

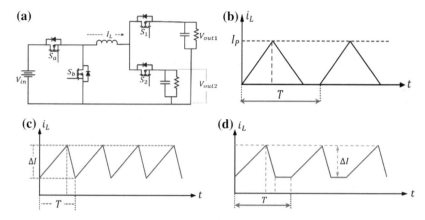

Fig. 1. (a) Single inductor multiple output converter [5]. (b) Typical inductor current in of DC-DC converter in DCM. (c) Inductor current in of DC-DC converter in CCM. (d) Inductor current in of converter during PCCM

Continuous conduction mode (CCM): In continuous conduction mode of converter (CCM), the current in the energy transferring inductor never goes down to zero during the entire switching cycle as Fig. 2b depicts. Normally a single output converter works in DCM with light loads to get high efficiency but the same converter works in CCM for heavy loads to deliver more power. However, SIMO converters have single inductor being shared among different loads. So, it may cause a severe cross regulation when the converter goes in CCM during heavy loads [6].

Pseudo continuous conduction mode (PCCM) [7–11]: There is one more mode of operation known as Pseudo continuous conduction mode (PCCM) in which the inductor current has certain minimum DC value I_{DC} as shown in Fig. 2c [11]. This base value I_{DC} can be adjusted as required by the load. During heavy loads its value is raised to allow more power to be delivered.

2 Cross Regulation

In numerous applications we need multiple power sources which is easily achievable using multioutput (MO) converters [12–15]. But the issue of cross regulation is seen in these converters. When the current or voltage of one output terminal is affected by the load variation on another, we say the system is cross regulated. This is a serious issue and quite common with MO DC-DC converters. The following example illustrates the cross regulation in SIMO converters.

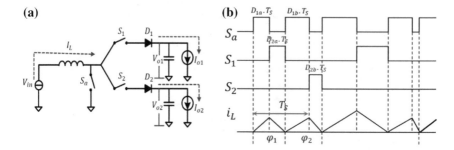

Fig. 2. (a) A Typical SIMO as proposed in [2]. (b) Inductor current and switching sequence of SIMO

The two outputs V_{o1} and V_{o2} in Fig. 2a are being regulated. Figure 2b shows that when the inductor current of one converter falls to zero, another converter activates. They are operating in an alternate fashion. The operating phases of the two converters are φ_1 and φ_2 and the switching time is T_S. These all can be modified according to the load currents I_{o1} and I_{o2}. Moreover, we have the following relations

$$I_{O1} = \frac{V_i}{2L}\left(\frac{M_1 - 1}{M_1^2}\right)\frac{\varphi_1^2}{\varphi_1 + \varphi_2} \tag{1}$$

$$I_{O2} = \frac{V_i}{2L}\left(\frac{M_2 - 1}{M_2^2}\right)\frac{\varphi_2^2}{\varphi_1 + \varphi_2} \tag{2}$$

$$\frac{Io1}{Io2} = \left(\frac{M_2}{M_1}\right)^2 \frac{M_1 - 1}{M_2 - 2}\left(\frac{\varphi_1}{\varphi_2}\right)^2 \tag{3}$$

Where, V_i is the input voltage, L is the value of the inductor, $M_1 = \frac{V_{o1}}{V_i}$ and $M_2 = \frac{V_{o2}}{V_i}$

Because of load variation let suppose there is sudden increase in the current $Io1$ at the port 1 of output, then V_{o1} will experience a voltage drop, while the voltage at the other port will remain unchanged. To cope up the voltage variation at V_{o1}, the controller will issue a larger φ_1 to deliver more current to this end (3). However, this increased value of φ_1 will decrease the value of I_{O2} (2) which ultimately affect the voltage V_{o2}. The controller finally finds some values of φ_1 and φ_2 to attain a new steady state. This is how cross regulation occurs between the outputs of multioutput converters. Moreover, if there is large variation of loads, the converter might go in unstable mode of operation.

3 Control Methods for Minimization of Cross Regulation

SIMO DC-DC converter uses only one Inductor to generate multiple voltage/current levels for different ports/sub-modules at the output. Apart from the reduced footprints area, it also gives higher power conversion efficiency [16]. However, since a single inductor accumulates the current for all different terminals, a serious design and control challenges arises to address issues like cross regulation and system stability. This section will cover up the control methods researchers have gone through so far.

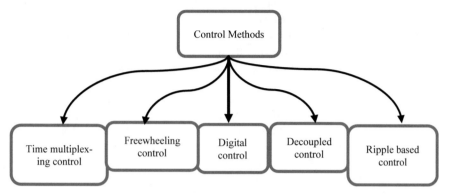

3.1 Time Multiplexing Control

This method of control is only applicable when the SIMO is operating under DCM [12, 17]. This method is an extension to the work proposed by [5], where the SIMO is operating at the boundary of DCM and CCM and does not have a fixed switching frequency. So, there is an interdependence between the timings of the two sub-converters which exhibits a severe cross regulation. However, in time multiplexing control of SIMO converter, each switching cycle is divided among the multiple outputs. For N outputs, each sub-converter is given a time multiplexed interval of 'switching cycle/N'.

A fixed frequency cycle is used to operate the converters and current in the inductor falls to zero after discharging in the outputs (Fig. 3b). The energy transfer of converts is unaffected by one another and no cross regulation is observed between the outputs.

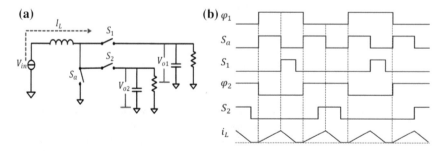

Fig. 3. (a) SIMO as proposed in [17] (b) Switching of SIMO in time multiplexing

3.2 Freewheeling Control

Converters working in DCM is not efficient enough to supply heavy loads and it is also having large current ripples along with a high value of peak inductor current [15, 18]. This ultimately affects the performance and dynamic behavior of the converter. The same converter when moves to CCM from DCM, it again suffers a significant cross regulation [19]. Moreover, when a boost converter is operated in continuous conduction mode, there exists a low frequency zero in the right half plane of the loop gain transfer function. Dominant pole compensation then affects the transient response and makes the converter with very limited bandwidth [20].

To address the issues mentioned above freewheeling control is proposed which is primarily based on new mode of conduction known as pseudo continuous conduction mode (PCCM) [11]. In this method of control, a freewheeling interval is added by using an additional switch connected across the inductor to handle heavy loads generating large currents and makes the outputs decoupled.

Figure 4 shows the working mode of converter in PCCM. The dc level I_{DC} can be easily adjusted according to the load and ripples requirement. Thus, a high value of inductor current can be maintained with very small current and voltage ripples. The inductor current instead of falling down to zero, stays at some DC value during freewheeling period and maintains the continuity as well.

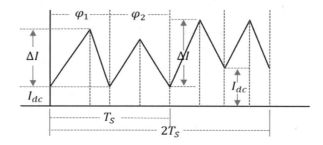

Fig. 4. SIMO operating in PCCM with freewheeling

$$\frac{di_L}{dt} = \frac{V_L}{L} \tag{4}$$

To achieve the freewheeling period and to maintain the inductor current constant at some value, V_L is made zero (4) by sorting the switch across the inductor. Ideal freewheeling is lossless with a fixed DC value. But the practical case is deviating the ideal and this becomes the main shortcoming which invites some another fellow to carry the research further.

3.3   Digital Control

Digital controllers are another mean to control the converter operation [21]. In particular, these controllers provide an effective way to control SIMO in CCM along with suppressing cross regulation [22, 23]. A voltage control based separate regulation for differential output and common mode voltage is adopted in [22] which is advantageous in minimization of cross regulation and extension of number of output ports. However, an adaptive gain compensation is required to eliminate differential mode loop gain dependent on inductor current. Following the voltage-based control, [23] also proposed a digital controller based on Phase Sequence Interchange (PSI) concept. According to the load conditions, phase sequence is arranged to minimize cross regulation.

A current control based digital controller is proposed in [21] which estimates the inductor current at the beginning of the next switching cycle. Based on this predicted value and the reference current, the controller determines the duty cycles of switches for the next switching period. Since the estimation is based on the reference load current, all current control loops are independent to each other and cross regulation is greatly minimized. However, this method of control is only effective when the inductor current is above certain threshold level.

3.4   Decoupled Control

Decoupling cross regulation transfer function is another approach which is adopted to suppress cross regulation of SIMO working in CCM [24]. In order to minimize the gain of cross regulation transfer function a cross derivative state feedback system is proposed in [25]. The converter is controlled efficiently by well managing the state feedbacks from the self and cross-output. The cross-loop compensator which is actually a lead compensator acts as a derivative feedback.

In [26] a multivariable based controller is proposed which is based on dynamic performance and decoupling requirements. Here an open loop transfer function matrix is formed and its diagonal and off diagonal elements are determined. Then by comparing this matrix with the desired one, coefficients of the controller are found out. Diagonal elements are adjusted according to the dynamic performance requirement and off diagonal elements are tuned to decouple the outputs.

3.5   Ripple Based Control

Improving the transient response by using ripple-based control is also proven good in reducing the cross regulation. Different methods are proposed to enhance the transient response of the converter. In [27], common mode peak current and differential mode ripples are adopted for enhancing transient response which in turn decrease the cross regulation. Peak current mode (PCM) control is used by [28] for a faster transient response. [29] suggested capacitor current as the control variable instead of using peak current. Capacitor current is more promising in representing the load variation and directly reflects any change in load.

4 Conclusion and Future Scope

This paper provides an overview on the control schemes adopted to minimize the occurance of cross regulation in SIMO converter. This converter with its multiple advantages and applications suffers from poor performance and instability in system dynamics due to presence of coupled outputs. Designing a controller for fast and robust system is still challenging. More accurate control variables to consider load parameter variations and change in input voltage are needed to go through which can improve converter performance and stability. Apart from SIMO, many new topologies with dual and hybrid outputs have been proposed recently which are not exempted from the issue of cross regulation and needs to explore.

References

1. Chang, J.M., Pedram, M.: Energy minimization using multiple supply voltages, IEEE Trans. Very Large Scale Integr. Syst. **5**(4), 436–443 (1997)
2. Single inductor multiple output boost regulator. April 1997
3. Zare, A.N.F., Blaabjerg, A.G.F.: Multi-output DC – DC converters based on diode-clamped converters configuration : topology and control strategy. December 2008 (2010)
4. Xing, Y., Wu, H., Zhang, J.: A family of multi-port buck-boost converters based on DC-link-inductors (DLIs), vol. 8993, no. c (2014)
5. Examiner, P., Deneke, A.: United States Patent, vol. 1, no. 12 (2001)
6. Huang, W., Qahouq, J.A.A., Dang, Z.: CCM – DCM power-multiplexed control scheme for converter with no cross regulation. IEEE Trans. Ind. Appl. **53**(2), 1219–1231 (2017)
7. Dasika, J.D., Bahrani, B., Sacedifard, M., Karimi, A., Rufer, A.: Multivariable control of single-inductor dual-output buck converters. IEEE Trans. Power Electron. **29**(4), 2061–2070 (2014)
8. Chen, B.: Digitally controlled low cross-regulation single- inductor dual-output (SIDO) buck converter, pp. 2497–2500 (2015)
9. Simo, A.S.: Control scheme for reduced cross-regulation in single-inductor multiple-output DC–DC converters. IEEE Trans. Industr. Electron. **60**(11), 5095–5104 (2013)
10. Shen, Z., Chang, X., Wang, W., Tan, X., Yan, N., Min, H.: Predictive digital current control of single-inductor multiple-output converters in CCM with low cross regulation, vol. 27, no. 4, pp. 1917–1925 (2012)
11. Ma, D., Ki, W., Tsui, C.: A pseudo-CCM/DCM SIMO switching converter. IEEE J. Solid State Circuits **38**(6), 1007–1014 (2003)
12. Modepalli, K., Parsa, L.: A scalable N-Color LED driver using single inductor multiple current output topology. IEEE Trans. Power Electron. **31**(5), 3773–3783 (2016)
13. Chen, H., Zhang, Y., Ma, D.: A SIMO parallel-string driver IC for dimmable LED backlighting with local bus voltage optimization and single time-shared regulation loop. IEEE Trans. Power Electron. **27**(1), 452–462 (2012)
14. Kim, H.-C., Yoon, C.S., Jeong, D.-K., Kim, J.: A single-inductor, multiple-channel current-balancing LED driver for display backlight applications. IEEE Trans. Ind. Appl. **50**(6), 4077–4081 (2014)
15. Chen, Y.-T., Shih, F.-Y.: New multi-output switching converters with MOSFET-rectifier post regulators. IEEE Trans. Ind. Electron. **45**(4), 609–616 (1998)

16. Zheng, Y., Ho, M., Guo, J., Leung, K.N.: A Single-inductor multiple-output auto-buck – boost DC – DC converter with tail-current control. IEEE Trans. Power Electron. **31**(11), 7857–7875 (2016)
17. Ma, D., Ki, W.H., Tsui, C.Y., Mok, P.K.: Single-inductor multiple-output switching converters with time-multiplexing control in discontinuous conduction mode. J. Solid State Circuit **38**(1), 89–100 (2003)
18. Ma, D., Ki, W.-H.: Fast-transient PCCM switching converter with freewheel switching control. IEEE Trans. Circuits Syst. II Express Briefs **54**(9), 825–829 (2007)
19. Patra, P., Patra, A., Misra, N.: A single-inductor multiple-output switcher with simultaneous buck, boost, and inverted outputs. IEEE Trans. Power Electron. **27**(4), 1936–1951 (2012)
20. Viswanathan, K., Oruganti, R., Srinivasan, D.: A novel tri-state boost converter with fast dynamics. IEEE Trans. Power Electron. **17**(5), 677–683 (2002)
21. Shen, Z., Chang, X., Wang, W., Tan, X., Yan, N., Min, H.: Predictive digital current control of single-inductor multiple-output converters in CCM with low cross regulation. IEEE Trans. Power Electron. **27**(4), 1917–1925 (2012)
22. Trevisan, D., Mattavelli, P., Tenti, P.: digital control of single-inductor multiple-output step-down DC–DC converters in CCM. IEEE Trans. Ind. Electron. **55**(9), 3476–3483 (2008)
23. Chen, B.W., Chang-Chien, L.R.: Digitally controlled low cross-regulation single-inductor dual-output (SIDO) buck converter. In: 2015 IEEE International Symposium on Circuits and Systems (ISCAS), pp. 2497–2500 (2015)
24. Wang, Y., Xu, J.: A cross regulation analysis for single-inductor dual-output CCM buck converters, J. Power Electron. kpubs.org (2016)
25. Patra, P., Ghosh, J., Patra, A.: Control scheme for reduced cross-regulation in single-inductor multiple-output DC–DC converters. IEEE Trans. Ind. Electron. **60**(11), 5095–5104 (2013)
26. Dasika, J.D., Bahrani, B., Saeedifard, M., Karimi, A., Rufer, A.: Multivariable control of single-inductor dual-output buck converters. IEEE Trans. Power Electron. **29**(4), 2061–2070 (2014)
27. Sun, W., Han, C., Yang, M., Xu, S., Lu, S.: A ripple control dual-mode single-inductor dual-output buck converter with fast transient response. ieeexplore.ieee.org (2015)
28. Wang, W., Xiaojuan Wei, J., Zhou, G.: Peak inductor current and differential-mode voltage control of single-inductor dual-output buck converters in continuous conduction mode. In: 2009 IEEE 6th International Power Electronics and Motion Control Conference, pp. 495–499 (2009)
29. Wang, Y., Xu, J., Yin, G.: Cross-regulation suppression and stability analysis of capacitor current ripple controlled SIDO CCM buck converter. IEEE Trans. Ind. Electron. **66**(3), 1770–1780 (2019)

A Web Application to Optimization of Transport in Military Operations

Hardy Aguilar[1], Patricio González[1], Matías Molina[1], Daniela Quila[1],
Gustavo Gatica[1], Jairo R. Coronado-Hernandez[2],
Javier Baron-Villamizar[3], and Jesús Silva[4(✉)]

[1] Universidad Andrés Bello, Santiago, Chile
h.aguilarvalenzuela@gmail.com,
p.gonzalezmarchant@outlook.com,
m.molinafaundes@uandresbello.edu, dp.quila@gmail.com,
ggatica@unab.cl
[2] Universidad de La Costa, Barranquilla, Colombia
jcoronad18@cuc.edu.co
[3] Escuela Naval de Cadetes "Almirante Padilla", Cartagena, Colombia
dfam@enap.edu.co
[4] Universidad Peruana de Ciencias Aplicadas, Lima, Peru
jesussilvaUPC@gmail.com

Abstract. Transport is an operation necessary to carry out any logistical mission, especially in times of war, peace or natural disasters. The distribution of the necessary demanded resources is done from a military unit, to the different locations or military bases. However, operational efficiency depends on the planners. In more than 60% of trips, shipping and return isn't efficient, even between the same units. The cause is the non-consolidation of trips and the lack of return load, coming from perimeter units. Planning is done without consolidating trips and in many cases on demand. It's presented a web application, a parametric framework to any geographical area, given the integration with applications such as Google Maps®. Computational times are reasonable, given a to hardiness to the problem. The software architecture is scalable and extensible, complying with software quality practices present in ISO 25000.

Keywords: Web application · Military logistics · Vehicle routing

1 Introduction

Logistics are vital for the survival of private companies. However, as a discipline of application and study, it has a military origin, being an indispensable operation to comply each mission [1, 2]. Its goal is to facilitate processes for timely supply necessary to support the community and military units [3]. There is a transversal problem in the armed forces and the private sector. The problem are the trips of vehicles to maximum capacity of load of going and return, avoiding trips without load [4]. The literature proposes to consider the capacity of the vehicle in return trips, being used for the collection of products [5–8]. In this scenario, two sets customer are considered, the

© Springer Nature Switzerland AG 2020
A. P. Pandian et al. (Eds.): ICICCS 2019, AISC 1039, pp. 176–185, 2020.
https://doi.org/10.1007/978-3-030-30465-2_20

first being the receiver and the second the senders. In this paper, a web application is presented for the assignment and routing of a fleet of military vehicles to distribute and collect resources. The problem in literature is known as vehicle routing problem with Backhaul and is confronted with various techniques. However, it doesn't always consider real logistical operational problems [9].

The article, in its second section presents the background to understand the problem of optimization. The third chapter presents the Solution technique and architecture of web application. The results are presented in the fourth chapter and are concluded with respect to this project under development.

2   Problem Statement

The base scenario for the development of the web application establishes a set of military units distributed in a region or country. For Example, in a Country with a large number of military units established, including brigades, regiments and training schools. Some assumptions are made to address the problem, among which are k military units to supply. Military units have unlimited capacity. A vehicle must travel through L dispatch customers and B collection customers without exceeding its Q capacity. For the cost of the route only the Euclidean distance is considered. The problem in literature is known as The Vehicle Routing Problem With Backhauls (VRPB), and is approached with exact, heuristic and metaheuristic solution techniques [10].

VRPB is NP-Hard [10]. The objective is to determine the routes that cover the demand of all customers for dispatch and collection. The routes generated have a minimum travel cost. It's represented as a non-directed complete graph $G = (V, A)$, V and A are sets of vertices and arcs. Set V is partitioned into two subsets, $L = \{1, 2, 3, ...i\}$ and $B = \{1, 2, 3, ...j\}$ of dispatch and collection clients respectively. There is a set of homogeneous k vehicles, each with a Q capacity, available in each depot or units in this case. Each arc (i,j) has a non-negative cost associated with it. In addition, 5 additional constraint must be met. First, each route must start and end in the same depot. Second, each customer must be visited only once, regardless of the route that is assigned. Third, dispatch customers must be served before pickup customers. Fourth, the total demand associated with linehaul-backhaul customers for each route must not separately exceed the capacity of the vehicle. Fifth, the number of routes associated with each depot must not exceed the k vehicles available.

2.1   Mathematical Formulation

An MILP discrete time formulation has been developed for the VRPB on transport military operation, based on [11]. The sets, all parameters and the decision variables include are listed below. The complete formulation is as follows:

Set
L Customers with product deposit requirements.
B Customers with product collection requirements.
G_L Feasible linehauling routes.

G_B Feasible backhauling routes.

Parameters

i Linehaul customers.

j Backhaul customers.

q Capacity used of the product.

c Cities covered

Q Vehicle capacity.

R Routes.

$Cost\_L_i$ Total cost of the Linehaul route.

$Cost\_B_j$ Total cost of the Backhaul route.

$Cost\_Z_{ij}$ Cost of transition from Linehaul to Backhaul route.

Variables

$$X_i = \begin{cases} 1, \textit{if the route } i \in G_L \textit{ is part of the solution.} \\ 0 \textit{ other case.} \end{cases}$$

$$Y_i = \begin{cases} 1, \textit{if the route } j \in Gj \textit{ is part of the solution.} \\ 0, \textit{other case.} \end{cases}$$

$$Z_{ij} = \begin{cases} 1, \textit{if the route } i \in GL \textit{ is connected to a route belonging to } j \in Gj. \\ 0, \textit{other case.} \end{cases}$$

Mathematical Model

$$min\, z = \sum_{i \in G_L} cost\_L_i * X_i + \sum_{j \in G_B} cost\_B_j * Y_j + \sum_{i \in L} \sum_{j \in B} cost\_Z_{ij} * Z_{ij} \tag{1}$$

s.t.

$$\sum_{l \in L_c} X_l = 1, \forall c \in L \tag{2}$$

$$\sum_{j \in B_c} Y_j = 1, \forall c \in B \tag{3}$$

$$\sum_{l \in L_c^T} X_l - \sum_{l \in L_c} z_{cj} = 0, \forall c \in L \tag{4}$$

$$\sum_{l \in B_c^l} Y_l - \sum_{l \in L_c} Z_{ic} = 0, \forall c \in B \tag{5}$$

$$\sum_{i \in L} \sum_{j \in B} Z_{ij} = R \tag{6}$$

$$Q_{min}^L \leq \sum_{i\in P} q_i \leq Q \left(resp.\ Q_{min}^B \leq \sum_{i\in P} q_i \leq Q \right) \tag{7}$$

$$Q_{min}^L = Max\left[0, \left(\sum_{i\in L} q_i\right) - (R-1)*Q\right] \tag{8}$$

$$Q_{min}^B = Max\left[0, \left(\sum_{i\in B} q_i\right) - (R-1)*Q\right] \tag{9}$$

$$X_l \in \{0,1\}, \quad Y_l \in \{0,1\}, \quad Z_{ij} \in \{0,1\} \tag{10}$$

The objective function (1) minimizes the costs of the route to be used. In Eqs. (2) and (3) it requires that each client in both Linehaul and Backhaul be visited by a single route. Equation (4) forces the solution to contain an arc starting with a client L (Linehaul). This must be fulfilled as long as the arc is part of the possible routes belonging to Linehaul customers (G_L). Equation (5) requires the solution to have an arc (i,j) with $i \in L$ and $j \in B$. The above is for the solution to be feasible for the Backhaul customers route (G_B). Equation (6) forces the solution to contain R routes with arcs (i, j),$i \in L$ and $j \in B$. Equation (7) allows to determine that the demanded capacity of the vehicle for both Linehaul and Backhaul doesn't exceed the maximum. Also so that the demanded capacity of the vehicle isn't less than the minimum permitted. To comply with the above, Eqs. (8) and (9) are derived [12].

3 Solution Technique and Architecture

The web application has three layers. The first layer is where the data entry, through the layer of Pipe and Filters. This layer is integrating the system with the logic layer of the application where VRPB is implemented. Then, the routes assigned for the dispatch of resources and for the collection of resources are presented. Finally, the generated routes are displayed on a map in the user interface layer. The algorithm used is based on ant colony optimization [13–16], working with a number of ants, number of generations and pheromones. It also uses the strategy to update the pheromones (ant-density, ant-quality or ant-cycle). This is defined internally in the algorithm. To obtain the routes each ant generates its own route, selecting each deposit according to a probabilistic rule. The probabilistic rule varies according to the selected strategy (density, quality or cycle). Then, the global update of the pheromone is carried out. This process is done by the evaporation of pheromone remaining in the roads that the ant didn't choose. Afterwards, the evaporated pheromone is deposited on the chosen route [15]. Finally, this process is iterated until the number of generations is reached, as can be seen in Algorithm 1.

Input requirement: Information given by the user.
Output: Route of vehicle
1   Initialize;
2   Func solve (graph, current_demand)
3   best_cost = float
4   best_solution = []
5   for gen in range(generations):
6   ants= [Ant (graph,type,ini_end,current:demand, graph_range) for i in range (ant_count)]
7   for ant in ants:
8   for i in range(graph – 1):
9   ant.select_next()
10   ant.total_cost += graph.matrix[ant.tabu[-1]][ant.tabu[0]]
11   if ant.total_cost < best_cost:
12   best_cost = ant.total_cost
13   best_solution = [ ] + ant.tabu
14   ant.update_pheromone_delta()
15   update_pheromone(graph, ants)
16   **Return:** best_solution, best_cost, ant.demand
Algorithm 1: A proposed algorithm in web environment

3.1   Solution Architecture

The architecture defined for the development of the application is hybrid [17]. From this layer it connects to the component where the application logic resides. In addition, a user interface (UI) module is established, which allows data to be uploaded, and results to be displayed. The Pipe and Filters layer consists of two big components. The first one receives the data from the users through a user interface module. Then the data is processed by the algorithm developed in the logic layer. The second prepares the output data so that it can be used by the Google Maps API [17, 18]. The data layer stores and manages to respond to customer requests. The architecture at the level of functional components is detailed in Fig. 1, which responds to the proposed scenario. It should be noted that the solutions are provided over Euclidean distances.

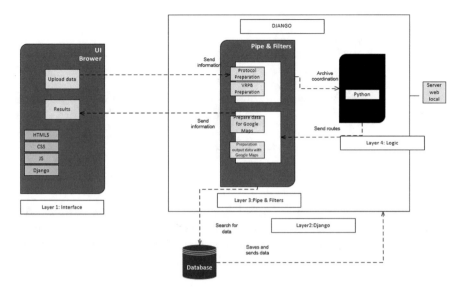

Fig. 1. Functional architecture

3.2   User Interface

The interfaces were designed to capture content and functionality, complying with the quality standard established by ISO 25000 [19, 20]. Figure 2 shows the user the options of starting a new route or submitting a request. In Fig. 3 you can see the choice to start a new route. User selects origin base, vehicle capacity, destination base and requested demand.

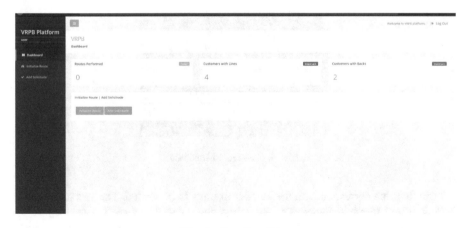

Fig. 2. Functionalities

Fig. 3. Start route

3.3 Quality Attributes

The developed application hasn't operability disadvantages, as it can be executed by the user from any web browser. It is also fault-tolerant, as it allows routing using the information available in the database. Architecture allows you to make changes to components without affecting the application logic. In this way, it can be integrated with other software that uses maps.

4 Simulation Results

The simulations carried out make it possible to model various scenarios. As an application, concurrent simulations are available, which don't affect computational performance. In addition, to minimize network usage, it is feasible to dispense with the integration with Google Maps, and work only with the routes. The summary of each simulation is made up units to be visited, origin and destination, and the total of vehicles used. Figure 4.

```
1   6
2   2
3   1
4   -32.9910894 -71.546911 0 123123
5   -33.4402301 -70.6471716 0 12
6   -33.0217758 -71.6370338 0 123 0
7   -33.561901 -70.6889039 12 0 0
8   -33.5619012 -70.6813812 1234 0 0
9   -33.5717312 -70.6896408 0 2 0
10  -33.4419038 -70.6584225 0 4 0
11  -18.4779137 -70.3113652 0 12 0
12
```

Fig. 4. Generated Instance

The first line corresponds to the number of units to be visited. The second indicates the initial and terminal depot. The third line simulates that there is one vehicle per depot previously loaded into the database. From line 4, the coordinates of the initial depot, terminal depot and the units to be visited are displayed. From line 4, the third column corresponds to the user's request of the backhaul type, collection of resources. When the value of this column is different from 0, it corresponds to the demand required to be collected. Otherwise, it means that the user's request is of the linehaul type, with the next column being the required demand. For each unit is indicated its name, a list of the units to be visited in Figs. 5 and 6.

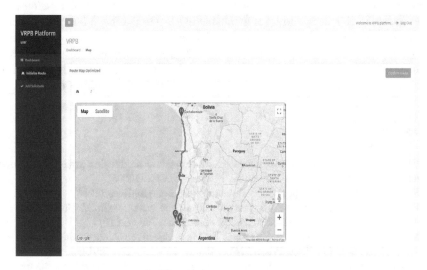

Fig. 5. Generated route

Fig. 6. Units to visit

5 Conclusions

According to the problem, it's possible to design a system that allows the consolidation of trips in the armed forces. The software architecture is portable, manageable, with reusable components, and allows for modifications and extensibility, in addition to being scalable. This is due to a set of interacting layers that have interfaces to communicate with other services. The product developed is extensible and can be implemented in similar and/or different areas. One of the aspects on which we are working, is the inclusion of time windows. A planning must exist within a time horizon corresponding to a calendar month according to daily, weekly or monthly frequency.

References

1. Fontena, H.: Logística civil y logística militar. Cuaderno de difusión académia de guerra naval no. 12. Academia de guerra naval, Valparaíso, Chile (2005)
2. Fontena-Faúndez, H.: Proposición para una definición de logística (1988)
3. Fontena, H.: Integración de los procesos logísticos para maximizar la eficacia y la disponibilidad operativa. Experiencia de la Armada de Chile (2010)
4. Molina, J., Eguia, I., Racero, J., Guerrero, F.: Multi-objective vehicle routing problem with cost and emission functions. Procedia Soc. Behav. **160**, 254–263 (2014)
5. Belloso, J., Juan, A.A., Faulin, J.: An iterative biased-randomized heuristic for the fleet size and mix vehicle-routing problem with backhauls. Int. Trans. Oper. Res. **26**(1), 289–301 (2019)
6. Lamos Díaz, H., Aguilar Imitola, K., Barreto Robles, M., Niño Niño, P., Martínez Quezada, D.: A memetic algorithm for location-routing problem with time windows for the attention of seismic disasters: a case study from Bucaramanga, Colombia. INGE CUC **14**(1), 75–86 (2018)
7. Mingozzi, A., Giorgi, S., Baldacci, R.: An exact method for the vehicle routing problem with backhauls. Transp. Sci. **33**(3), 315–329 (1999). pubsonline.informs.org
8. Ghaziri, H., Osman, I.H.: A neural network algorithm for the traveling salesman problem with backhauls. Comput. Ind. Eng. **44**(2), 267–281 (2003)
9. Vigo, D., Toth, P.: Vehicle Routing; Problems Methods, and Applications (2014)
10. Toth, P., Vigo, D.: An exact algorithm for the vehicle routing problem with backhauls. Transp. Sci. **31**(4), 372–385 (1997)
11. Baldacci, R.: Algorithms for Location and Routing Problems in Distribution Systems (1999)
12. Henao, J.P.: Efecto de la cantidad de carga en el consumo de combustible en camiones (2012)
13. Kakas, A.C., et al.: Ant colony optimization. In: Encyclopedia of Machine Learning, pp. 36–39. Springer, Boston (2011)
14. Gambardella, L., Taillard, É., Agazzi, G.: MACS-VRPTW a multiple ant colony system for vehicle routing problems with time windows. In: Corne, D., Dorigo, M., Glover, F. (eds.) New Ideas in Optimization, pp. 63–76. McGraw-Hill's, London (1999)
15. Mazzeo, S., Loiseau, I.: An ant colony algorithm for the capacitated vehicle routing. Electron. Notes Discrete Math. **18**, 181–186 (2004)
16. Romero-Conrado, A., Coronado-Hernandez, J., Rius-Sorolla, G., García-Sabater, J.: A tabu list-based algorithm for capacitated multilevel lot-sizing with alternate bills of materials and co-production environments. Appl. Sci. **9**(7), 1464 (2019)

17. Kobayashi, S., Fujioka, T., Tanaka, Y., Inoue, M., Niho, Y., Miyoshi, A.: A geographical information system using the Google Map API for guidance to referral hospitals. J. Med. Syst. **34**(6), 1157–1160 (2010)
18. Kobayashi, S., Fujioka, T., Tanaka, Y., Inoue, M., Niho, Y., Miyoshi, A.: A Geographical Information System Using the Google Map API for Guidance to Referral Hospitals. J. Med. Syst. **34**(6), 1157–1160 (2010)
19. International Organization for Standardization, Systems and software engineering – Systems and software Quality Requirements and Evaluation (SQuaRE) – System and software quality models, ISO/IEC (2011)
20. Amelec, V.: Validation of strategies to reduce exhausted shelf products in a pharmaceutical chain. Adv. Sci. Lett. **21**(5), 1403–1405 (2015)

Big Data Marketing During the Period 2012–2019: A Bibliometric Review

Amelec Viloria[1]([⊠]), Jesús Vargas[1], Ernesto García Cali[1],
David Martínez Sierra[2], Alexandra Perdomo Villalobos[3],
Osman Redondo Bilbao[3], Alberto Enrique Mercado Sarmiento[4],
and Hugo Hernández-Palma[5]

[1] Universidad de La Costa, St. 58 #66, Barranquilla, Atlántico, Colombia
{aviloria7,jvargas41,egarcia29}@cuc.edu.co
[2] Centro para la industria petroquímica Sena Regional Bolívar,
Provincia de Cartagena, Colombia
dmartinezsierra@misena.edu.co
[3] Corporación Universitaria Latinoamericana, Barranquilla, Colombia
alex-perdomo@hotmail.com, oredondo@ul.edu.co
[4] Corporación Universitaria Minuto de Dios – UNIMINUTO,
Barranquilla, Colombia
alberto.mercado@uniminuto.edu
[5] Universidad del Atlántico, Puerto Colombia, Colombia
hugohernandezp@mail.uniatlantico.edu.co

Abstract. The present study identifies the most significant trends in production of high impact scientific papers related to the Big Data Marketing variable during the period between the years 2012 and 2019 through a revision of the Scopus database, which manages to highlight the relevance of 113 indexed papers. For this purpose, the following descriptive bibliometric indicators are implemented: production volume, type of document, number of citations, and country of application. In the studied time period, the evidence suggests an annual growth in the production volume of papers related to the variable, but with a significant drop in 2017. The knowledge areas that showcases more researches about the Big Data Marketing variable are computer science, mathematics, decision-making, and engineering domain.

Keywords: Big data · Marketing · Bibliometric review · Volume of production

1 Introduction

Marketing represents a key factor in consumer behavior, so companies invest more efforts in the creation of strategies that are aimed to capture the greatest possible number of new customers and retains the current ones [1]. The design and implementation of such strategies in traditional marketing is based on the analysis of small data sets (megabytes or gigabytes, or kilobytes) with limited analytical platforms and implementation capacity [2]. However, there are tools for the management of large

© Springer Nature Switzerland AG 2020
A. P. Pandian et al. (Eds.): ICICCS 2019, AISC 1039, pp. 186–193, 2020.
https://doi.org/10.1007/978-3-030-30465-2_21

amounts of information in order to obtain enough material for the design of strategies that covers increasingly broad and complex market niches [3, 4].

Therefore, the present study aims to interpret the most significant data from a bibliometric approach on studies in marketing and big data as a starting point in the creation of strategies aimed at the achievement of new clients and loyalty of existing customers [5–7]. The papers studied have been published in the period of 2012–2019 and have been indexed in Scopus database.

The bibliometric approach allows to classify the scientific papers according to the most significant components: production volume per year, principal authors, country of origin of the publication, among others. Bibliometric indicators are statistics deducted from the different characteristics of the scientific publications, based on the important role played by them in the dissemination and transmission of knowledge generated in research [8–10].

2 Materials and Methods

The bibliometric analysis allows a quantitative approach for the scientific production related to the Big Data Marketing variable through the interpretation of charts and graphs derived from the gathered database [11–13]. The present research was developed through three phases with the purpose of answering the research questions. The following chart explains the sequence of these phases.

2.1 Database Selection

In first phase, scientific papers are selected considering if they were carried out under the Big Data Marketing variable and if they were indexed in Scopus database, centering the search in the period 2012–2019 and selecting those classified by year of publication, author, affiliation, country of origin, area of research, most relevant papers, and publication type. After applying the search patterns, a total of 114 papers were obtained to start the data classification by the categories mentioned above.

2.2 Development of the Bibliographic Analysis

The organization of the collected data starts with the processing of graphs and tables which offer a great help for the interpretation of the information obtained and their respective quantitative analysis. The categories to classify the information obtained are: Year of publication, authors and affiliation, type of investigation, area of research, country of origin of the research, most relevant articles [14, 15].

2.3 Construction and Analysis of the Final Document

After categorizing and organizing the information, the construction of the final document was performed with its subsequent analysis of tables and graphs to generate conclusions and limitations identified in the research process [16, 17].

3   Analysis and Results

113 documents were identified after applying the search parameters in the years from 2012 to 2019, demonstrating a sustained growth in the production of papers related to the variable of study, except for the year 2017 when such production was just 13 papers contrasting with 24 papers in the year 2016.

3.1   Distribution of Papers Per Year

Figure 1 shows the years with highest scientific production on the Big Data Marketing variable, being 2018 the year with greater number of papers, according to the Scopus database, with a total of 27 papers. There is a growth in productivity with the exception of the year 2017 when production reached just 13 studies. However, since it is a relatively new variable, growth will be in positive for the coming years since during the first quarter of the year 2019 this data base has registered 12 papers.

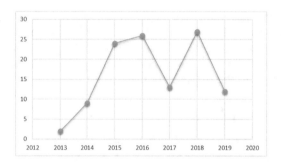

Fig. 1. Distribution of papers per year

3.2   Production by Authors and Affiliation

After reviewing the results of the gathered data, it is evident that because it is a relatively new topic in the study field and its involvement and growth is directly proportional to the new technologies, there is an evident equity in the number of publications with this topic among the authors. Figure 2 clearly shows that, in the first five places by number of publications, there are 5 authors with the same number of papers related to the Big Data Marketing variable with a total of three items during the analyzed period of time. In the following five positions there are also consistent results showing that each of the following authors registered two papers published with the variable in study during the years between 2012 and 2019.

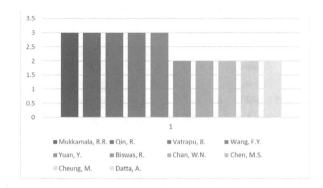

Fig. 2. Distribution of papers by author

In addition to the distribution of publications by author, the production of those papers must be analyzed by country of origin [18, 19]. Table 1 shows that most developed countries lead the production of papers related to the Big Data Marketing variable. The United States is by far the largest producer of scientific papers in this area of research, with a total of 40 papers published and indexed in Scopus database during the period 2012–2019, followed by China and India with 18 and 10 papers respectively.

Table 1. Productivity by country of origin

Country of origin	Number of papers
USA	40
China	18
India	10
Taiwan	9
Singapore	6
Australia	5
Italy	4
Japan	4
South Korea	4
Spain	4

Figure 3 shows that more institutions are producing scientific papers with the variable in study and allows to know that the vast majority of affiliations are for Asian countries. The Chinese Academy of Sciences, has a total of 6 publications. It is worth noting that, while the United States is the country with the highest production of scientific papers, once the data of the institutions with a greater number of affiliations were analyzed, it was determined that the production of that country is distributed in a large number of universities whose affiliations vary between one and two publications [20, 21].

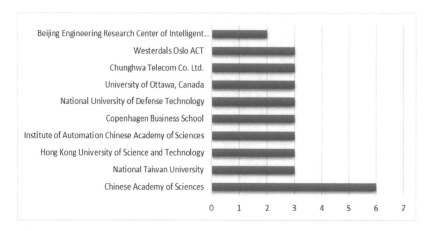

Fig. 3. Distribution of papers by affiliation.

3.3   Publications by Area of Research

Table 2 shows the researching production carried out by each knowledge area, with sciences associated with computing or systems as the ones that generate a greater volume of production with 48.2% in terms of the search criteria related to this study. It may be inferred that the studies that are conducted about Big Data are directly related to the advances in information technology. Next, the results show a percentage of 12.7% and 12.2% in mathematics and decision-making respectively, which leads to infer that the variable under study does not belong to a specific area of research. On the contrary, it can be applied to different study fields with different purposes, always starting from the efficient management of large amounts of information, and what it represents to areas such as Engineering (9.6%) and Administration (6.6%).

3.4   Most Influential Publications

Table 3a and 3b permits to see the relevance and impact of the papers studied under the variable Big Data marketing, being the article entitled *Big Data consumer analytics and the transformation of marketing* in the *Journal of Business Research* which has been cited 148 times and represents the most cited scientific paper within the database collected after the application of the search filters to develop the present research. It is worth noting that the first five papers in this ranking have been published in journals with significantly higher rating (Q1), which represents a high impact in terms of the number of citations enjoying high quality and scope in the different studies involving the Big Data marketing variable.

Table 2. Productivity per area of research.

Research area	Documents by area of research	%
Computer science	95	48.2%
Mathematics	25	12.7%
Decision-making	24	12.2%
Engineering	19	9.6%
Business administration and accounting	13	6.6%
Social sciences	8	4.1%
Economy and finance	4	2.0%
Energy	3	1.5%
Materials science	3	1.5%
Medicine	2	1.0%
Psychology	1	0.5%

Table 3a. Most influential publications

Ranking	Papers	Journal	Quartile	Citations
1	Big Data consumer analytics and the transformation of marketing	Journal of Business Research	1	148
2	The service revolution and the transformation of marketing science	Marketing Science	1	87
3	Big data reduction framework for value creation in sustainable enterprises	International Journal of Information Management	1	52
4	A study on sentiment computing and classification of sina weibo with Word2vec	Proceedings - 2014 IEEE International Congress on Big Data	1	51
5	Big Data and consumer behavior: imminent opportunities	Journal of Consumer Marketing	1	31
6	Mastering structured data on the semantic web: From HTML5 microdata to linked open data	Mastering Structured Data on the Semantic Web: From HTML5 Microdata to Linked Open Data	3	26
7	T-PICE: Twitter personality based influential communities extraction system	Proceedings - 2014 IEEE International Congress on Big Data, BigData Congress 2014	1	24

Table 3b. Most influential publications

Ranking	Papers	Journal	Quartile	Citations
8	Big Data Analytics: Security and privacy challenges	Proceedings - IEEE Symposium on Computers and Communications	1	18
9	A glimpse on big data analytics in the framework of marketing strategies	Soft Computing	4	17
10	Perspectives on big data	Journal of Marketing Analytics	2	18

4   Conclusions

After analyzing the gathered data, it can be concluded that the variable Big Data Marketing has gained interest from researchers at a global level, since the advancement of information technology allows to store and process large amounts of data that can be interpreted and used in the decision-making processes of large global companies. The present study was developed through the selection of scientific papers published and indexed in the Scopus database during the period 2012–2019 without distinction of language, country of origin, or authors. A remarkable growth is evident in the number of papers produced during these years, with the exception of the year 2017 when production significantly fell to 13 studies related to the variable, while the production during the previous year was 26, and 27 articles in the following year. Technological advances in the different platforms for the collection and analysis of information have a direct impact on current researches about issues related to Big Data applied to marketing. Therefore, the results obtained in the analysis of annual production show that the variable in study is getting stronger in research and becomes critical for the decision-making processes in big companies.

References

1. Najafabadi, M.M., Villanustre, F., Khoshgoftaar, T.M., Seliya, N., Wald, R., Muharemagic, E.: Deep learning applications and challenges in big data analytics. J. Big Data, **2**(1) (2015)
2. Minelli, M., Chambers, M., Dhiraj, A.: Big Data, Big Analytics: Emerging Business Intelligence and Analytic Trends for Today's Businesses (2013)
3. Erevelles, S., Fukawa, N., Swayne, L.: Big Data consumer analytics and the transformation of marketing. J. Bus. Res. **69**(2), 897–904 (2016)
4. Tirunillai, S., Tellis, G.J.: Mining marketing meaning from online chatter: strategic brand analysis of big data using latent dirichlet allocation. J. Mark. Res. **51**(4), 463–479 (2014)
5. Leeflang, P.S.H., Verhoef, P.C., Dahlström, P., Freundt, T.: Challenges and solutions for marketing in a digital era. Eur. Manage. J. **32**(1), 1–12 (2014)
6. Rust, R.T., Huang, M.H.: The service revolution and the transformation of marketing science. Market. Sci. **33**(2), 206–221 (2014)

7. Wedel, M., Kannan, P.K.: Marketing analytics for data-rich environments. J. Market. **80**(6), 97–121 (2016)
8. He, W., Wu, H., Yan, G., Akula, V., Shen, J.: A novel social media competitive analytics framework with sentiment benchmarks. Inf. Manage. **52**(7), 801–812 (2015)
9. Yang, Y., Pan, B., Song, H.: Predicting hotel demand using destination marketing organization's web traffic data. J. Travel Res. **53**(4), 433–447 (2014)
10. Marine-Roig, E., Anton Clavé, S.: Tourism analytics with massive user-generated content: a case study of Barcelona. J. Destination Market. Manage. **4**(3), 162–172 (2015)
11. Fan, S., Lau, R.Y.K., Zhao, J.L.: Demystifying big data analytics for business intelligence through the lens of marketing mix. Big Data Res. **2**(1), 28–32 (2015)
12. Mariani, M.M., Di Felice, M., Mura, M.: Facebook as a destination marketing tool: evidence from Italian regional destination management organizations. Tourism Manage. **54**, 321–343 (2016)
13. Power, D.J.: Using 'Big Data' for analytics and decision support. J. Decision Syst. **23**(2), 222–228 (2014)
14. Xu, Z., Frankwick, G.L., Ramirez, E.: Effects of big data analytics and traditional marketing analytics on new product success: a knowledge fusion perspective. J. Bus. Res. **69**(5), 1562–1566 (2016)
15. Christopher, M., Ryals, L.J.: The supply chain becomes the demand chain. J. Bus. Logistics **35**(1), 29–35 (2014)
16. Buhalis, D., Foerste, M.: SoCoMo marketing for travel and tourism: empowering co-creation of value. J. Destination Market. Manage. **4**(3), 151–161 (2015)
17. Rehman, M.H.U., Chang, V., Batool, A., Wah, T.Y.: Big data reduction framework for value creation in sustainable enterprises. Int. J. Inf. Manage. **36**(6), 917–928 (2016)
18. Dolnicar, S., Ring, A.: Tourism marketing research: past, present and future. Ann. Tourism Res. **47**, 31–47 (2014)
19. Gaitán-Angulo, M., Cubillos Díaz. J., Viloria. A., Lis-Gutiérrez. J.P., Rodríguez-Garnica. P. A.: Bibliometric analysis of social innovation and complexity (Databases Scopus and Dialnet 2007–2017). In: Tan, Y., Shi, Y., Tang, Q. (eds.) Data Mining and Big Data DMBD 2018. Lecture Notes in Computer Science, vol 10943. Springer, Cham (2018)
20. Lis-Gutiérrez, J.P., Henao, C., Zerda, Á., Gaitán, M., Correa, J.C., Viloria, A.: Determinants of the impact factor of publications: a panel model for journals indexed in scopus 2017. In: Tan, Y., Shi, Y., Tang, Q. (eds.) Data Mining and Big Data DMBD 2018. Lecture Notes in Computer Science, vol. 10943. Springer, Cham (2018)
21. Viloria, A., Viviana Robayo, P.: Virtual network level of application composed ip networks connected with systems - (NETS Peer-to- Peer). Indian J. Sci. Technol. **9**(46) (2016). https://doi.org/10.17485/ijst/2016/v9i46/107376

Automatic Selenium Code Generation for Testing

Nikhil Rathi, Rohith Srivathsav, Rishabh Chitlangia[✉],
and V. K. Pachghare

College of Engineering Pune, Pune, Maharashtra, India
{rathin15.it, srivathsavrs15.it, chitlangiarrl5.it,
vkp.comp}@coep.ac.in

Abstract. Developers and product teams are striving to push new features out to customers at a much faster rate and are now realizing the advantages of automation. With the advent of Agile development model, writing automate test codes is now becoming a part of the software development life cycle. Manual testing which is completely done by human hands is costly, time consuming and error prone. However, this transition from manual to automation testing is not so easy especially for the existing manual testers. It is becoming mandatory and very important for them to have programming as a skill. In this paper, we discuss about an application that automatically generates Selenium WebDriver code which can then be used for testing. We present the design, analysis, results and the benefits of using this application to automate test cases. Our application will not only remove programming as a skill for testers but also save time and money for organizations. We input the test case in the form of a simple CSV file in the application which then runs it against a rule-based template engine to produce the desired code. The results highlight the advantage of using this application over manually writing automation test codes.

Keywords: Automation · Selenium · Automation testing · Software testing · Software engineering

1 Introduction

Software testing is an essential part of software development cycle. Normally, to ensure the operation of any written piece of code, there must be a testing phase to discover whether the code reacts as intended [1]. As per IEEE Standard 610.12-1990, software testing is defined as "The process of analyzing a software item to detect the differences between existing and required conditions (that is, bugs) and to evaluate the features of the software items" [2]. Software testing provides nonpartisan and unbiased information about the quality of the software and helps the stake holders to gauge the risk of its failure. With increase in software complexity there has been an increase in the importance of efficient software testing procedures.

All authors have contributed equally.

A. P. Pandian et al. (Eds.): ICICCS 2019, AISC 1039, pp. 194–200, 2020.
https://doi.org/10.1007/978-3-030-30465-2_22

Manual Testing is a process of finding out the defects or bugs in a software program where tests are executed manually by a tester. In this method, the tester plays an important role. All the features of the application are verified and validated manually. The tester needs to manually carry out all the phases of a STLC right from test planning to test closure. Thousands of test cases may be executed before the release of a software and a lot of time would be spent in completely testing the software. It increases the risk of error and mistakes because of human dependency and has a limited scope. Testing types like performance and batch cannot be done manually.

We're working with computers after all, machines designed to perform processes repeatedly, the same way every time and so it does make sense to use them for software testing. One of the problems with automating test cases is that one needs programmers. This leads to an increase in dependence on individuals who have the required skill-set for the same.

In this paper we aim to provide a solution to the above problem. The paper provides an overview of an application that automatically generates code to automate test cases.

2 Automation Testing

Automated Software Testing involves testers to write test scripts to automate test execution. Testers are required to use appropriate automation tools to develop the test scripts. These test scripts are then executed to validate the software. This drastically reduces time spent on testing and prevents human errors that can occur while testing. The generated test report helps the tester to determine whether an application performs as expected. Automated testing is a must in case of regression, repetitive, load and dynamic testing which includes unit testing, system testing and performance testing [3]. Automation testing drastically reduces manual intervention by the tester during test execution phase.

Automating a set of manual test cases involves translating the instructions into their automatically executable counterpart (automation code). While these manual instructions are written in natural language, the automation code is typically expressed in some form of script or visual programming language. This automation code can sometimes be reused if we can keep track of similar manual test steps that have been automated before [3].

3 Drawbacks of Automation Testing

3.1 Migration from Manual to Automation Testing

Due to the transitional shift in the industry from manual to automation testing a higher importance is now given to functional automation, scripting and programming skill-set. This transition isn't as easy as flipping a switch. It comes with a learning curve and this would involve training the current testers and hiring new personnel. Both require time and money.

3.2 Scripting

Even after a successful transition from manual to automation testing methods scripting takes up a lot of time and resources. SDLC gets stuck if the testing team does not have the required test automation scripts. Since scripting is done manually it is also prone to errors and mistakes.

4 Proposed Solution

We propose a method wherein the tester will be able to generate the automation test code using the test data without having to manually write code.

The application takes in the test data as input and uses a robust Rule based Template Engine to generate code for the test case along with the report. The Template Engine has a well defined set of rules observed after analysing many test scripts.

Instance of every test case will be maintained which can be later retrieved, updated and executed. This helps the tester to create similar test cases quickly and efficiently.

This way we can generate executable automation code directly from test data thereby streamlining the testing process.

The steps to be performed by the tester will be:

1. Prepare test data
2. Upload the test data in the application
3. Get the automation code
4. Execute the code and get the test report

5 Overview of the Architecture

For any web application development, it is important to choose a correct technology stack which allows rapid prototyping, constant iteration, code reuse, maximum efficiency, and robustness. It is also important that the technology stack is easy to learn and understand by the developers working on the front-end and the back-end. Thus, we are using the two most popular JavaScript frameworks: React [4] for the frontend and Node.js [5] for the backend. For database management, we have chosen the extremely popular relational database, MySQL (Fig. 1).

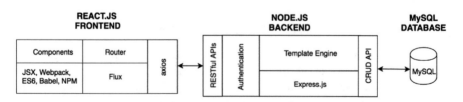

Fig. 1. Architecture of proposed solution.

The tester uploads the test data in the form of a csv file which is converted into JSON and sent to the backend Node.js server after proper authentication. Express picks up this JSON and inserts it into the MySQL database. The Template Engine takes the data from the database and asynchronously generates code for each row of the table using a rule-based template.

6 Working

6.1 Client and Server Architecture

The front end consists of a single page, interactive dashboard application created using React, which provides rich DOM manipulation and two-way data binding features. The tester is presented with a well-designed application, which follows the best responsive design principles, to carry out the code generation process. React Router is used for dynamic routing which enables routers to select paths according to real-time logical network layout changes.

We use Flux [6] with React, which is an application architecture for building user interfaces. Flux applications have three major parts: the dispatcher, the stores, and the views (React components). Flux eschews MVC in favour of a unidirectional data flow. When a user interacts with a React view, the view propagates an action through a central dispatcher, to the various stores that hold the application's data and business logic, which updates all the views that are affected. This works especially well with React's declarative programming style, which allows the store to send updates without specifying how to transition views between states.

In the server side, we use Express with Node.js. Node.js is an ideal choice for a scalable, data-driven and I/O intensive application.

All interactions between the client and the server happen with Axios [7], a promise-based HTTP client for the browser and Node.js which is similar to the Fetch API. Axios is used to make XMLHttpRequests from the browser and HTTP requests from Node.js through RESTful API calls. Authentication is carried out using the JWT auth server. To interact with the database, CRUD APIs have been implemented in Express.

6.2 Template Engine

The Template Engine is where the main business logic of the code resides. A template engine is a tool that is used to do programming tasks involving a lot of textual data. The engine takes in two inputs, the test data and a template, and generates the code. The operations in the engine occur in two phases, namely, parsing and rendering.

The bulk of the work in the engine is parsing the template and producing the necessary code that could be rendered. To help with producing the automation code, we have a Code Builder, which handles the bookkeeping as we construct the code. It adds lines of code, manages indentation, and finally gives the executable source code.

The test data which is uploaded as a csv file is persisted to the database and the engine asynchronously produces a line of code corresponding to each row of the MySQL table. Instead of using existing template engines like Jinja (Python) [8] or Apache Freemaker (Java) [9], we have created our own engine which primarily focuses on generating automation test code. A level of abstraction is created by introducing a new term "Template Engine Language" which specifies the programming language of the generated source code.

Once the test code is generated, we maintain this instance by storing the output in the database.

Here, we have given the test data for a login test scenario of a demo social media web app to generate the Java code which can be executed using Selenium WebDriver [10] (Tables 1 and 2).

Table 1. Input format for the proposed solution

Step	command	target	targetval	Val
1	open		http://localhost/twitter/	
2	type	name	email	user@test.com
3	type	id	password	password
4	click	name	login	
5	assertText	url		http://localhost/twitter/home.php

Table 2. Output code generated by proposed solution

Step	Generated code
1	`driver.get("http://localhost/twitter/");`
2	`driver.findElement(By.name("email")).sendKeys("user@test.com");`
3	`driver.findElement(By.id("password")).sendKeys("password");`
4	`driver.findElement(By.name("login")).click();`
5	`driver.getCurrentUrl().equals(http://localhost/twitter/home.php);`

7 Case Study

We created a test suite for a demo social media web application and tested it using three different approaches and got the following results (Table 3). The cost of a tester per hour is taken from a recent survey (2018) [11] and study [12].

7.1   Variables and Measurements

The independent variables recorded in this study are number of test cases (N), test cases completed per hour (t_c) and hourly rate of a tester (r). The dependent variables recorded are total number of hours required (T_n) and total cost (C).

$$T_n = N/t_c \tag{1}$$

$$C = T_n r \tag{2}$$

7.2   Assumptions and Simplifications

There was only one tester working at any given point of time. Time taken to create test reports was not taken into consideration while calculating test cases completed per hour. The total number of test cases in the test suite was 500.

Table 3. Comparison of different testing approaches on custom test suite of 500 test cases

	Manual	Automated	Proposed solution
No of test cases completed per hour	8	21	39
Total no of hours required	62.5	24	13
Hourly rate of tester (INR) [11]	250	600	250
Total cost (INR)	15625	14286	3205

7.3   Results

The results clearly show us that by using our proposed solution, maximum test cases are completed per hour at the minimum cost that can be inferred from Figs. 2 and 3. The proposed solution allows us to use manual testers to achieve better results than automation testing.

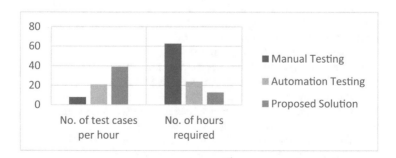

Fig. 2. Time Comparison of different testing approaches on custom test suite of 500 test cases

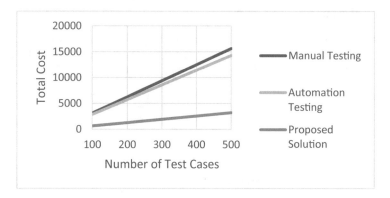

Fig. 3. Cost vs Number of Test cases for different testing approaches

8 Conclusion

In this paper, we have proposed a solution in the form of an application that will automate automation testing. This application will allow organizations to smoothly migrate from manual to automation testing. It does not require testers to have an in-depth knowledge of programming. The template engine supports multiple programming languages reducing the need for specialized personnel. The application allows the tester to store all the instances of test cases in a central place. This helps the tester to create similar test cases quickly and efficiently. In this way the proposed solution can save both time and resources spent on testing web applications.

References

1. Alaqail, H., Ahmed, S.: Overview of software testing standard ISO/IEC/IEEE 29119. IJCSNS Int. J. Comput. Sci. Netw. Secur. **18**(2), 112–116 (2018)
2. Majchrzak, T.A.: Software testing. In: Improving Software Testing. SpringerBriefs in Information Systems, pp. 11–56. Springer, Heidelberg (2012)
3. Reid, S.: The new software testing standard. In: Dale, C., Anderson, T. (eds.) Achieving Systems Safety, pp. 237–255. Springer, London (2012)
4. React. https://reactjs.org/. Accessed 20 Mar 2019
5. Node.js. https://nodejs.org/. Accessed 20 Mar 2019
6. Flux. https://facebook.github.io/flux/. Accessed 20 Mar 2019
7. Axios. https://github.com/axios/axios/. Accessed 22 Mar 2019
8. Jinja. http://jinja.pocoo.org/. Accessed 18 Mar 2019
9. Apache Freemarker. https://freemarker.apache.org/. Accessed 23 Mar 2019
10. Selenium. https://www.seleniumhq.org/. Accessed 20 Mar 2019
11. Payscale. https://www.payscale.com/research/IN/. Accessed 19 Mar 2019
12. Schroeder, P.J., Faherty, P., Korel, B.: Generating expected results for automated black-box testing. In: 17th IEEE International Conference on Automated Software Engineering, pp. 139–148 (2002)

Optimization of Parametric Model of Cumene Reactor Using Loop Shaping Methodology and Dynamic Modeling of Reactor

M. L. Vinila$^{(\boxtimes)}$, K. Aparna, and S. Sreepriya

Department of Chemical Engineering, National Institute of Technology Calicut,
Kozhikode 673601, Kerala, India
{vinila_p150083ch,
aparnak,sreepriya_p150069ch}@nitc.ac.in

Abstract. A reactor for Cumene production process has been modeled. Cumene production is a Catalytic condensation process. To yield high purity Cumene (Isopropyl benzene), it uses solid phosphoric acid (SPA) catalyst which promote the alkylation of benzene with propylene in the presence of Propane. Parametric Identification Technique has been done for the modelling of the Reactor section. Sample data is collected for a predefined time for the System Identification. Principal Component Analysis (PCA) is done to identify the Principal Components. State Space model of the Reactor is developed using System Identification Technique. Optimisation of the Parametric Model has been done by Loop shaping Methodology. Shaped Plant model is obtained using Pre and Post Compensator Design. Dynamic modelling of the Reactor has been done for the optimum values of pressure and Temperature and it is validated with Real value data.

Keywords: Cumene Reactor · Parametric model · System Identification · PCA · H∞ loop shaping · Stability robustness · Dynamic modeling · Real value data

1 Introduction

This work aims to design a Cumene Reactor that can optimize the performance of the Cumene Plant. Phenol and Acetone are the end products for most of the cumene in world wide Industries. The Catalytic Condensation process for cumene production has been commonly referred to as the poly process. The raw material for Cumene synthesis is obtained from the thermal cracking of Crude oil.

Gera et al. [1] in his study of plantwide control of Cumene process proposes best pairings with maximum throughput. Both Self-optimizing CVs and inventory control loop pairings are implemented [1]. Luyben [2] in his work developed an economically optimum design which can handle large disturbances effectively [2].

In this paper, an approach for the optimum design of Parametric model of Cumene Reactor is proposed. The approach is based on the use of Principle Component Analysis techniques [3] and the method of loop technology [12, 13]. A dynamic model

© Springer Nature Switzerland AG 2020
A. P. Pandian et al. (Eds.): ICICCS 2019, AISC 1039, pp. 201–208, 2020.
https://doi.org/10.1007/978-3-030-30465-2_23

of the Cumene Reactor section is also developed and it is validated with the real values from the plant.

This paper has the following sections. In Sect. 2, the Cumene manufacturing Process in the Cumene palnt of HOCL, Kochi has been explained. Real value data collection and modeling techniques has been illustrated in Sect. 3. An H∞ Loop shaping methodology has been used to optimize the model in Sect. 4. Section 5 narrates the design of the Cumene Reactor. Section 6 gives the dynamic simulation of the Cumene plant section with the expected Temperature and Pressure profile. Section 7 presents the Concluding remarks.

2 Process Description

In this section, the process of Cumene production has been elaborated [4, 5]. To ensure maximum yield and product quality benzene has been combined with a propane-propylene mixture in the ratio of (1:2:8 :: propylene :propane:benzene) in the overall process flow. In the presence of solid phosphoric acid catalyst alkylation takes place in the Reactor when the mixture is heated and routed into it. The reactor effluent is then allowed to pass through a series of fractionation columns. It helps to separate the desired cumene from the heavier alkylated products. It also help for the recycling of the unreacted benzene and rejection of non-reactants. Typical yields are 456.6 gms of Cumene from 303.9 gms of benzene and 172.4 gms of propylene.

The basic reaction in the Cumene Reactor is as follows:

$$C_6H_6 + C_3H_6 \rightarrow C_6H_5CH\ (CH_3)_2 + Heat \tag{1}$$

The reaction is exothermic and proceeds to essentially complete conversion of Propylene. Possible side reactions include the alkylation of cumene with propylene to form di-isopropyl benzene and the addition of propylene to yield oligomers. These reactions are suppressed by maintaining a proper ratio of benzene to propylene in the reaction zone and controlling the operating conditions [6].

Solid phosphoric acid catalyst is a reaction product of kieselguhr, a naturally occurring silicon dioxide clay, and phosphoric acid, P_2O_5. The reaction results in a material which can be described as phosphoric acid associated with a silicon phosphate base (Fig. 1).

$$2P_2O_5 + 3SiO_2 \rightarrow Si_3\ (PO_4)_4\ (Silica\ ortho\ Phosphate) \tag{2}$$

3 Real Value Data Collection and Modeling Techniques

3.1 Experimental Data

The parameters that influence the process related to Cumene production has been obtained from the Plant. The variables include Reactor inlet and outlet temperatures,

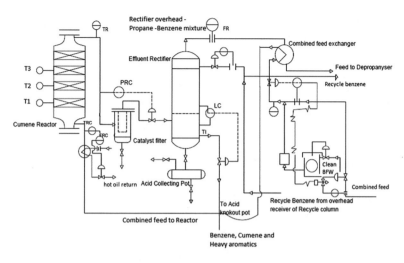

Fig. 1. Schematic diagram of cumene reactor section

Reactor effluent pressure, Liquid Hourly Space Velocity (LHSV), feed flow rate, moisture content and Reactor Bed temperatures.

The real time data of the Process has been collected which is the real reflection of the actual working of the Plant.

Table 1. Parameters identified

Parameters	Range
Temperature Reactor inlet	182 °C–192.5 °C
Temperature Reactor outlet	210 °C–216 °C
Reactor Effluent Pressure	33.99 kg/cm^2–4.01 kg/cm^2
Liquid Hourly Space Velocity (LHSV)	0.68 h^{-1}–0.72 h^{-1}
Feed flow rate	72 m^3/h–74 m^3/h
Moisture Content	149 wt.ppm–151.5 wt.ppm
Temperature Bed No: 1	210 °C–214 °C
Temperature Bed No: 2	211 °C–213 °C
Temperature Bed No: 3	190 °C–204 °C

Real data has been collected for continuous period of time with specific time duration. The Principal Components of the system can be identified from the parameters, using Supervisory Principal Component Analysis (Supervisory PCA).

3.2   Model Prediction by Principal Component Analysis

A set of predictor variables can be used for predicting the output using the method of Principal Component Analysis [3].

The transfer function of the system has been obtained from the Principal Components. The first assumption model of the system is in the form of

$$y_1 = [G_1 \quad G_2 \quad G_3 \quad G_4] \begin{bmatrix} u_1 \\ u_2 \\ u_3 \\ u_4 \end{bmatrix} \tag{3}$$

4   Loop Shaping Methodology

To deal with the uncertainties in the parametric model, loop shaping technology has been introduced. The desired open-loop shape has been obtained for the compensated plant P(s). Singular values of the Reactor transfer function has been shaped, using Pre- and post compensators W1 and W2 [8–10]

W1 = λ(1 + T1s)/(1 + T2s), where λ is a design parameter.

5   Design Example of the Cumene Reactor

Cumene Reactor has the following first assumption model from Eq. (3) considering the parameters listed in Table 1.

$$y1 = \begin{bmatrix} \dfrac{0.3053}{s+0.503} & \dfrac{-0.2737}{s+0.3651} & \dfrac{0.1256}{s+0.2801} & \dfrac{-0.04649}{s+0.3047} \end{bmatrix} \begin{bmatrix} u1 \\ u2 \\ u3 \\ u4 \end{bmatrix} \tag{4}$$

To obtain linearised and averaged model, the following post and pre compensators are used [13].

$$\text{Post–compensator}: \quad W_2 = 1 \tag{5}$$

$$\text{Pre–compensator}: \quad W_1 = \lambda \frac{(s+422)}{(s+1.10^{-4})}; \tag{6}$$

λ - design parameter.

The shaped plant should not have any unstable hidden modes. This is ensured by the proper selection of weights [11, 12, 14]. Figure 2 shows that model of the plant is shaped well.

System identification has been done considering 4 parameters identified using PCA. The time response comparison of system transfer function with the data is given below. The time response shows that the system settles in the range of 34 kg/cm^2 of pressure, which is the value of pressure to be maintained in the reactor to obtain the throughput (Fig. 3).

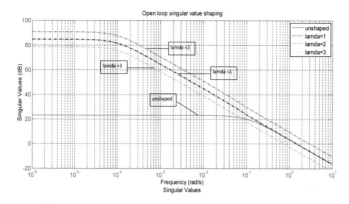

Fig. 2. Nominal plant and the shaped plant singular values

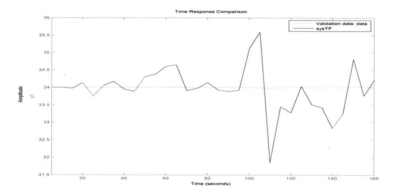

Fig. 3. The time response comparison of system transfer function with the real time data

6 Dynamic Simulation

The reliability and features of the proposed system has been examined by dynamic simulation. The performance of the Cumene reactor with the basic controllers has been examined by Aspen Hysis 8.4 (Fig. 4).

6.1 Temperature Profile in the Reactor

Combined feed to Cumene Reactor contains a mixture of Propylene, Propane and Benzene in the mole ratio of 1:2:8 and its water content is 150 ppm. The following graph shows that the Cumene Reactor maintain a temperature of 212.7 °C for a range of 1400 samples which is in line with the data from the plant. Sample data has been collected at every 4 h gap (Fig. 5).

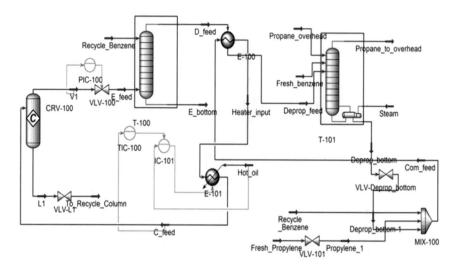

Fig. 4. Dymanic simulation of Cumene Reactor with basic controllers using Aspen Hysis 8.4

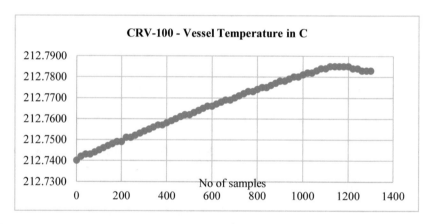

Fig. 5. Temperature profile in the Cumene Reactor for 1400 samples

6.2 Effluent Pressure Profile in Reactor

Combined feed enters the bottom of the Cumene Reactor at the pressure 34 kg/cm²g. From the Simulation the reactor effluent pressure has got a value of 3426.83 kPa which is equal to 34.943 kg/cm²g. This shows that the pressure profile is also in line with the plant data (Fig. 6).

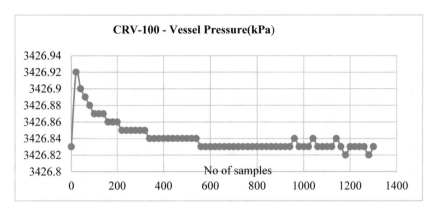

Fig. 6. The Effluent pressure profile in the Cumene Reactor for 1400 samples

6.3 The Effluent Feed Pressure at PI Controller at the Output of the Reactor

The effluent feed pressure in the PI controller, PIC-100 before is pumped to various fractionating columns is given below. At 50% output the controller maintains the required pressure (Table 2).

Table 2. Effluent feed pressure through PI controller, PIC 100

Time	Set point value	PIC-100	Output	Minimum PV value	Maximum PV value	Ramp target set point value	Ramp duration time
Seconds	kPa	kPa	%	kPa	kPa	kPa	Seconds
340	3426.87	3426.87	50	3236.2	3432.33	3426.87	300
700	3426.86	3426.86	50	3236.2	3432.33	3426.86	300
1060	3426.86	3426.86	50	3236.2	3432.33	3426.86	300

6.4 Comparison: Simulation Result with Plant Data

The Simulation result obtained has been compared with experimental data. The Reactor effluent pressure and Reactor bed Temperature have been compared (Table 3).

Table 3. Comparison of real time data and simulation result

Parameters	Real time data	Simulation result
Reactor effluent	34.00 kg/cm^2	34.943 kg/cm^2
Cumene reactor bed no. 1	214.75 °C	212.76 °C

7 Conclusion

A parametric model of the Cumene reactor plant has been developed from the real value data. Principal Component Analysis has been done for model prediction. Pre and Post compensators are designed to shape the nominal plant. Shaped Plant model is obtained using Pre and Post Compensator Design.

A dynamic simulation of the plant has been done in Aspen Hysys 8.4. The requirement in the Cumene reactor is to maintain the pressure at 34 kg/cm^2. From the simulation it is seen that it maintain a pressure in the same range and a temperature of 212.76 °C. The temperature is reduced from the actual value which will improve the performance of the system. This unique model of the plant can be further utilised for the design of controllers for the plant.

Acknowledgement. The authors are thankful to HOCL, Kochi, Kerala for their timely support for the successful completion of this study.

References

1. Gera, V., Panahi, M., Skogestad, S., Nitin, K.: Economic plant wide control of the cumene process. Ind. Eng. Chem. Res. **52**, 830–846 (2013)
2. Luyben, W.L.: Design and control of the cumene process. Ind. Eng. Chem. Res. **49**(2), 719–734 (2010)
3. Bair, E., Hastie, T., Paul, D., Tibshirani, R.: Prediction by supervised principal components. J. Am. Stat. Assoc. **101**, 119–137 (2006)
4. Chudinova, A., Salischeva, A., Ivashkina, E., Moizes, O., Gavrikov, A.: Application of cumene technology mathematical model. Procedia Chem. **15**, 326–334 (2015)
5. Maity, D., Jagtap, R., Kaistha, N.: Systematic top-down economic plantwide control of the cumene process. J. Process Control **23**, 1426–1440 (2013)
6. Coulson, J.M.: Chemical reactors. Nature **212**(5059), 236–237 (1966)
7. Walas, S.M., Ph, D.: Reaction kinetics. Chem. Eng. **72**(1), 53–61 (1999)
8. Idiarte, E.V., et al.: Analysis and design of H$_\infty$ control of non minimum phase-switching converters. IEEE Trans. Circuits Syst.-I: Fundam. Theory Appl. **50**(10), 1316–1323 (2003)
9. Doyle, J.C., Glover, K., et al.: State-space solutions to standard H2 and H$_\infty$ control problems. IEEE Trans. Automat. Contr. **34**(8), 831–846 (1989)
10. McFarlane, D., Glover, K.: A loop shaping design procedure using H$_\infty$ synthesis. IEEE Trans. Automat. Contr. **37**, 759–769 (1992)
11. Ioanniddis, G.C., Manias, S.N.: H$_\infty$ loop shaping control schemes for the buck converter and their evaluation using μ-analysis. Proc. Inst. Elect. Eng. **146**, 237–246 (1999)
12. Gahinet, P., Apkarian, P.: Linear matrix inequality approach to H$_\infty$ control. Int. J. Robust Nonlinear contr. **4**, 421–448 (1994)
13. Burl, J.B.: Linear Optimal Control. Addison-Wesley Inc, Menlo Park (1999)
14. Dileep, K., Krishnan, S., Jose, A.: Vehicular adaptive cruise control using laguerre functions model predictive control. Int. J. Eng. Technol. (IJET) **10**(6), 1719–1730 (2018). (2019)

Fractional Controller Optimization for Liquid Level System Using MATLAB

Rathod Divyani[⊠] and Deshpande Amruta

Instrumentation and Control Department, College of Engineering, Pune, India
{rathoddll7.instru, asd.instru}@coep.ac.in

Abstract. Conventional PID controller is oldest and one of the most widely used controllers in industry. In industrial development, the systems became more complex which leads to the requirement for more advanced controller. In this paper, the optimization technique for designing Fractional Order Fuzzy PI (FOFPI) controller is proposed. Controller designing is based on required values of peak overshoot and rise time. Optimal FOFPI gives better performance than PID, fuzzy logic PI and fractional order PI controller. FOFPI makes the system more robust.

Keywords: Fractional Order Fuzzy PI (FOFPI) controller · Fuzzy Logic PI (FPI) controller · Fractional Order PI (FOPI) controller · Optimization · Peak overshoot · Rise time

1 Introduction

Proportional Integral Derivative (PID) controller focuses on past, present and future error. PID controller used in various automatic process control applications in the industry to regulate the system like flow, temperature, pressure, level and many other industrial processes. In industry 95% of PI/PD/PID type control loops are used in which only 20% of control loops work well and reason behind this is bad tunning, sensor and actuator problems [1]. PI controller does not show the better response and achievement for nonlinear system, time delayed system and fractional system. The motivation behind Fractional Order Fuzzy PI (FOFPI) is to overcome the drawback of typical PI controller using optimization method.

The FOPI controller has one more parameters λ which helps to find out optimal solutions for controller. Classical optimization is not used in process because of the roughness of the objective function surface. So, for level control plant use derivative free optimization, guided by the collective behavior of social swarm [1].

Importance of FOPI and modeling method explained in [2]. The frequency domain approach which is based on expected gain margin and phase margin is mentioned in [1, 3, 4] it helps to find out the fractional parameter and sensitivity of the system. The tunning method of FOPI controller for classical feedback control system is proposed in [4, 5]. State space modeling for FOPI is viewed in [6]. The FOPI for quadruple tank and couple tank is mentioned in [5], which gives the information about cascade system

© Springer Nature Switzerland AG 2020
A. P. Pandian et al. (Eds.): ICICCS 2019, AISC 1039, pp. 209–219, 2020.
https://doi.org/10.1007/978-3-030-30465-2_24

and required tunning method. The evolution of PID and FOPI is mentioned in [7], it highlights the tunning methods, modeling, stability analysis and aspects of some simplest nonlinear system reviews.

Modeling stability analysis and aspect of some simplest nonlinear fuzzy is mentioned in [8]. Literature [9] is mentioned the difference between FPI and PID. Design, performance evolution and stability analysis of FPI controller proposed in [10]. Comparative analysis between conventional PI and fuzzy logic PI and fuzzy logic PD controller given in [11].

This paper presents a FOFPI using optimization method applied to the coupled tank system. Water level controlling of two tanks is challenging when the valve is not present between two tanks and system is of first order. FOFPI helps the system to settle down fast and to reduce overshoot.

This paper organized as follows, Sect. 1 gives the introduction about FOPI. Section 2 gives the system description in detail. Section 3 gives the mathematical modeling. Section 4 presents the tunning method of PI, FOPI, and FPI. Section 5 present the simulation and results which gives information about why optimal FOFPI is better than PI, FPI, and FOPI. Section 6 presents the conclusion of the system.

2   System Description

The coupled tank plant [12] consisting of two tanks and a pump with a water basin. Tanks are mounted above the water basin on the front plate. Water flows from first tank to the second tank through an outlet orifice located at the bottom of the tank and from second tank flows into the water basin (reservoir). Pressure-sensitive sensor used for level measurement which is available for feedback.

A single system can be configured using three types

1. Configuration 1: Single input single output system is used on one tank only. The pump feeds water into the tank.
2. Configuration 2: State coupled single input single output system consists of both tanks, tank 1 and tank 2. The pump feeds water into tank 1 and it feeds water into tank 2.
3. Configuration 3: State coupled and input coupled single input single output system consists of both tanks. The pump feeds water to tank 1 and it feeds water to tank 2 same as configuration 2 but the difference is that water to tank 2 can be feed from the pump also. The water flowing through the pump is split into two tanks (Fig. 1).

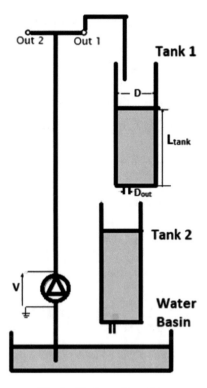

Fig. 1. Tank configuration

3 Coupled Tank Modeling

As discussed in configuration 2, tank 2 depends on tank 1 with input as voltage. Thus, the level in both the tanks is to be maintained (Table 1).

$$L_{tank2} = L_2 + L_{12} \text{ and } L_{tank1} = L_0 + L_1 \tag{1}$$

where L_0 and L_2 are equilibrium points and L_1 and L_{12} are small changes in operating points of tank 1 and tank 2 respectively.

Table 1. Common parameters

Parameter	Value	Unit
Gravitational constant (g)	981	cm/s^2
Flow constant (K)	3.3	cm^{3}v/s
Cross-section area of tank outlet hole (A_{out})	0.1781	cm^2
Area of tank (A_{tank1})	15.5179	cm^2
Peak voltage (V_p)	22	V
Orifice diameter (D_{out})	0.63	Cm
Tank maximum height (h)	30	Cm

Equation of motion for the linearized equation for tank 1 and 2 given below

$$\frac{\partial L_{12}}{\partial t} = f(L_1, L_{12}) \tag{2}$$

Applying Tylor series.

$$f(L_{tank2}, L_{tank1}) = f(L_2, L_0) + \left(\frac{\partial}{\partial L_{tank2}} f(L_2, L_0) \right) (L_{tank2} - L_2)$$
$$+ \left(\frac{\partial}{\partial L_{tank1}} f(L_2, L_0) \right) (L_{tank1} - L_0) \tag{3}$$

Transfer function is obtained from above equations

$$G_{tank1}(s) = \frac{K_{dc_{tank2}}}{\tau_{tank2}s + 1} \tag{4}$$

where

$$K_{dc_{tank2}} = \frac{A_{out1} * \sqrt{L_2}}{A_{out2} * \sqrt{L_0}} \quad \text{and} \quad \tau_{tank2} = \frac{A_{tank2} * \sqrt{2gL_2}}{A_{out2} * g} \tag{5}$$

where A_{out2} is cross-section area of tank 2 outlet hole and A_{tank2} is area of tank 2.

4   Tuning

Controller tuning is an important part which gives the specific procedure to tune. There are different methods available in literature used to tune the controller. Tuning is used to find out suitable gain values Kp, Ki and K_d. PI mainly used to eliminate the drawback of the steady state error which comes from proportional controller. PI is not used for future error it used for past and present error. So PI controller used mostly where the speed of the system is not an issue. It can be used when fast response of the system is not necessary and only one energy storage, capacitive or inductive is in process. It can be used when large transport delays are in the system and more noise and disturbances are present during operation of the process like flow control.

4.1   Proportional Integral Controller

In PID, the proportional term focuses on present error, the integral term focuses on the change in error i.e., past values and the derivative term focuses on the future error. PI mainly used to eliminate the drawback of the steady state error which comes from proportional controller. PI is not used for future error it used for past and present error. So PI controller used mostly where the speed of the system is not an issue. It can be used when fast response of the system is not necessary and only one energy storage, capacitive or inductive is in process. It can be used when large transport delays are in the system and more noise and disturbances are present during operation of the process like flow control.

Generalize equation for PID controller is

$$u(t) = K_p e(t) + K_i \int e(t)dt + K_d \frac{d}{dt} e(t) \tag{6}$$

Let us consider first order system for coupled tank, so the derivative term is zero. So the need of the tuning is to find out controller gain K_p and K_i for PI controller

$$C(s) = K_p + \frac{K_i}{s} \text{ and } P(s) = \frac{K}{\tau s + 1} \tag{7}$$

where the C(s) is the PI controller and P(s) is the first order transfer function of plant. The transfer function of the closed-loop system from reference to output is

$$\frac{Y(s)}{R(s)} = \frac{C(s)P(s)}{1 + C(s)P(s)} \tag{8}$$

substitute Eq. (7) in Eq. (8)

$$\frac{Y(s)}{R(s)} = \frac{\left(K_p + \frac{K_i}{s}\right)\left(\frac{K}{\tau s + 1}\right)}{1 + \left(K_p + \frac{K_i}{s}\right)\left(\frac{K}{\tau s + 1}\right)} \tag{9}$$

State Space Model is given by

$$X = \left[L_{21} \int L_{21} dt\right] X + [1]L_{11} \tag{10}$$

$$y(t) = [L_{21}] + D[0] \tag{11}$$

Solving Eq. (9) and comparing with characteristic equation,

$$\frac{K_p(s).K + s}{\tau} - 2\zeta\omega_n \text{ and } \frac{K_i}{\tau} - \omega_n^2 \tag{12}$$

4.2 Fractional Order Proportional Integral Controller

PI is an integer order controller whereas PI^λ is a fractional order controller. The fractional order controller provides total three adjustment parameters to tune the controller so as to get proper control action. The three parameters are K_p and K_i which are gains for proportional and integral terms whereas λ is integral order [14]. The optimization techniques on three dimensional space where, if the sytem is first order it is three dimensional space. Generalized FOPI equation is

$$C(s) = K_p + K_i s^{-\lambda} \tag{13}$$

From above equation, there are three unknown parameters K_p, K_i, and λ. For finding the range of suitable values of λ sensitivity of the system is used [4]. Sensitivity is given by

$$S(s) + T(s) = 1 \tag{14}$$

where

$$S(s) = \frac{1}{1 + C(s)P(s)} \ and \ T(s) = \frac{P(s)C(s)}{1 + C(s)P(s)} \tag{15}$$

For obtaining λ value in FOPI [3]

$$\left(\omega_{gain}^{\lambda+\mu} - \omega_{phase}^{\lambda+\mu}\right)\{a_1\left[\omega_{gain}^{\alpha}\cos\left(\frac{\pi\alpha}{2}+\phi\right) - \omega_{phase}^{\alpha}\cos\frac{\pi\alpha}{2}\right]$$
$$+ a_2\left[\omega_{gain}^{\beta}\cos\left(\frac{\pi\beta}{2}+\phi\right) - \omega_{phase}^{\beta}\cos\frac{\pi\beta}{2}\right] + a_3\left(\cos\phi - \frac{1}{A_m}\right)\}$$
$$+ \left(\cot\frac{\pi\lambda}{2} + \cot\frac{\pi\mu}{2}\right)\left(\frac{\omega_{phase}^{\lambda}\omega_{gain}^{\mu}I_p}{A_m} + \omega_{gain}^{\lambda}\omega_{phase}^{\mu}I_g\right) \tag{16}$$
$$- \left(\frac{\omega_{phase}^{\lambda+\mu}I_p}{A_m} + \omega_{gain}^{\lambda+\mu}I_g\right)\cot\frac{\pi\mu}{2} - \left(\frac{\omega_{gain}^{\lambda+\mu}I_p}{A_m} + \omega_{phase}^{\lambda+\mu}I_g\right)\cot\frac{\pi\lambda}{2} = 0$$

where ω_{gain} and ω_{phase} are gain frequency and phase frequency respectively. A_m is gain margin and ϕ is phase margin (Table 2).

$$I_p = a_1\omega_{phase}^{\alpha}\sin\left(\frac{\pi\alpha}{2}\right) + a_2\omega_{phase}^{\beta}\sin\left(\frac{\pi\beta}{2}\right) \tag{17}$$

$$I_g = a_1\omega_{gain}^{\alpha}\sin\left(\frac{\pi\alpha}{2}+\phi\right) + a_2\omega_{gain}^{\beta}\sin\left(\frac{\pi\beta}{2}+\phi\right) + a_3\sin\phi \tag{18}$$

Table 2. Tank parameters

Parameter	Tank 1 value	Tank 2 value	Unit
K_p	7.14	5.09	V/cm
K_i	9.10	1.74	V/(cm-s)
Kgf	2.38	1	V/√cm

4.3   Fractional Fuzzy Proportional Integral Controller

In the system, two input sets, one for the error given in Fig. 2(a) and another for the change in error given in Fig. 2(b). The error set was defined in two membership functions: Negative Error (NE) and Positive Error (PE) and the change in error set was defined in two membership functions: Negative Change in Error (NCE) and Positive Change in Error (PCE). The output set shown in Fig. 3 was defined in three membership functions: Negative (N), Zero (Z) and Positive (P) [8, 10]. Mathematical representation of FPI is given below

$$u(t) = K_p \frac{d}{dt} e(t) + K_i \frac{d}{dt} \int_0^t e(t) \tag{19}$$

For taking the control action the output of the fuzzy logic is converted into a relative crisp value understandable to the controller. This is done using defuzzification. There are many different methods of defuzzification out of which the fuzzy controller is designed using centroid method for defuzzification, which is also called as center of area or center of gravity.

$$u(t) = \frac{\sum x_i \mu_i}{\sum \mu_i} \tag{20}$$

where μ_i is the membership function of the fuzzy set and x_i is the value of the fuzzy set.
   Rules for FOFPI are:

1. If the error is negative (NE) and change in error is negative (NCE) then output is negative (N)
2. If the error is negative (NE) and change in error is positive (PCE) or error is positive (PE) and change in error is negative (NCE) then output is zero (Z)
3. If the error is positive (PE) and change in error is positive (PCE) then output is positive (P)

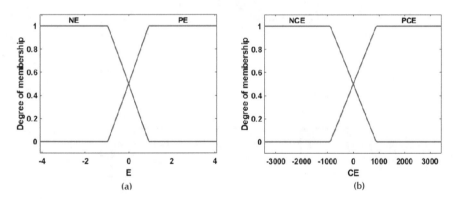

Fig. 2. Membership function for input sets

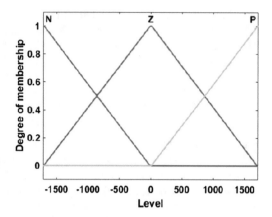

Fig. 3. Membership function for output set

5 Simulation and Results

Simulation is done using MATLAB simulink toolbox in which mathematical modeling is represented by the calculations and results shown in the graph. The Fig. 4(a) shows the block diagram for the PI controller and Fig. 4(b) shows the block diagram for the cascade system. The block diagram of fuzzy fractional order PI controller for system is shown in Fig. 5 (Table 3).

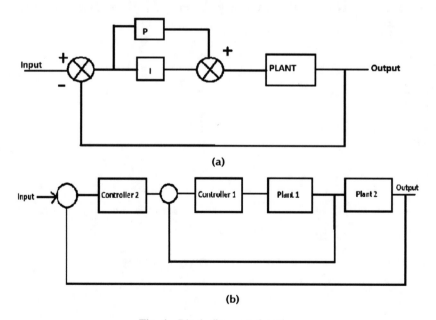

Fig. 4. Block diagram of system

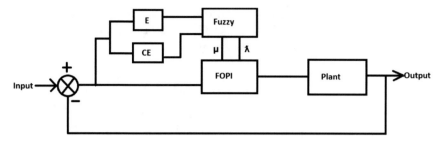

Fig. 5. Fuzzy fractional controller block diagram

Table 3. Simulation results

Controller	Percent overshoot	Settling time (sec)
PI	11	20
FOPI	8	10
Fuzzy PI	2	5
FOFPI	2	0.2

When PI controller was applied to the coupled tank the overshoot of 11% and the settling time of 20 s (5%) was observed Fig. 6(a). After applying fractional order PI controller Fig. 6(b) it was found that the overshoot was reduced to 8% whereas the settling time was reduced to 10 s (5%).

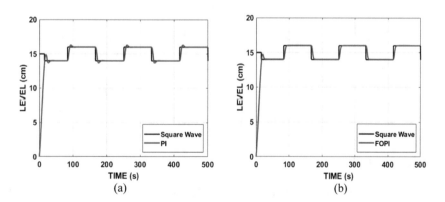

Fig. 6. Response of system

When the fuzzy controller was applied to the coupled tank with gains, the system was settled down to the setpoint Fig. 7(a) but with more settling time than fractional order PI controller. The fuzzy controller can be implemented with PI and PD controllers individually. When the fuzzy controller was used with fractional order PD controller Fig. 8(a), the settling time found was more than when the fuzzy controller was used with fractional order PI controller Fig. 7(b). Pump peak voltage and maximum continous voltage are shown in Fig. 8(b).

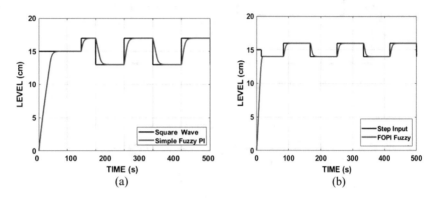

Fig. 7. Response of system for fuzzy

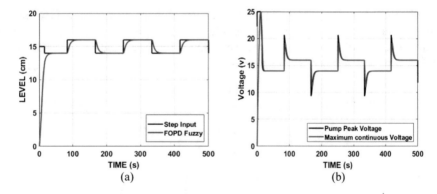

Fig. 8. Response of couple tank

6   Conclusion

In this paper, fractional order PID controller is designed using the optimization method for the coupled tank system. The system found to be robust for different inputs. The efficiency of fuzzy PI is observed greater than fuzzy PD as the settling time of the system is optimally reduced. FOFPID controller reduces the drawback of the control scheme in terms of settling time and peak overshoot. This controller is also effective for time-delayed systems.

References

1. Petras, I.: Tuning and implementation methods for fractional-order controllers. Fract. Calc. Appl. Anal. **15**, 282–303 (2012)
2. Podlubny, I.: Fractional-order systems and fractional-order controller. Inst. Exp. Phys. Slovak Acad. Sci. **12**(3), 1–18 (1994)
3. Zhao, C., Xue, D., Chen, Y.: A fractional order PID tuning algorithm for a class of fractional order plants. In: IEEE International Conference Mechatronics and Automation (2005). https://doi.org/10.1109/icma.2005.1626550
4. Zhong, J., Li, L.: Tuning fractional-order $PI^{\lambda}D^{\mu}$ controllers for a solid-core magnetic bearing system. IEEE Trans. Control Syst. Technol. **23**, 1648–1656 (2015)
5. Ionescu, C.M., Maxim, A., Copot, C., De Keyser, R.: Robust PID auto-tuning for the quadruple tank system. IFAC-PapersOnLine **49**, 919–924 (2016)
6. Shah, P., Agashe, S., Singh, A.P.: Fractional order modelling using state space theory. Int. J. Eng. Technol. **5**, 2891–2894 (2013)
7. Shah, P., Agashe, S.: Review of fractional PID controller. Mechatronics **38**, 29–41 (2016)
8. Arun, N.K., Mohan, B.M.: Modeling, stability analysis, and computational aspects of some simplest nonlinear fuzzy two-term controllers derived via center of area/gravity defuzzification. ISA Trans. **70**, 16–29 (2017)
9. Khan, A.A., Rapal, N.: Fuzzy PID controller: design, tuning and comparison with conventional PID controller. In: IEEE International Conference on Engineering of Intelligent Systems (2006). https://doi.org/10.1109/iceis.2006.1703213
10. Carvajal, J., Chen, G., Ogmen, H.: Fuzzy PID controller: design, performance evaluation, and stability analysis. Inf. Sci. **123**, 249–270 (2000)
11. Pitalúa-Díaz, N., Herrera-López, E., Valencia-Palomo, G., González-Angeles, A., Rodríguez-Carvajal, R., Cazarez-Castro, N.: Comparative analysis between conventional PI and fuzzy logic PI controllers for indoor benzene concentrations. Sustainability **7**, 5398–5412 (2015)
12. Quanser - Coupled Tanks Manual Document no. 557, rev. 04
13. Bequette, B.W.: Process Control: Modeling, Design, and Simulation. Prentice Hall, Upper Saddle River (2003)
14. Fegani, N., Charef, A.: Process step response based fractional PID controller parameters tuning for desired closed loop response. Int. J. Syst. Sci. **47**, 1–12 (2014)

Conglomerates of Bus Rapid Transit in Latin American Countries

Carmen Vásquez[1], Rodrigo Ramírez-Pisco[2], Amelec Viloria[3(✉)],
David Martínez Sierra[4], Erika Ruiz-Barrios[3], Hugo Hernández-P[5],
Jairo Martínez Ventura[5], and Juan De la Hoz Hernández[5]

[1] Departamento de Ingeniería Eléctrica,
Universidad Nacional Experimental Politécnica Antonio José de Sucre,
Vicerrectorado Barquisimeto, Barquisimeto, Venezuela
cvasquez@unexpo.edu.ve
[2] Universidad de Barcelona and Universidad Politécnica de Cataluña,
Barcelona, Spain
rramirez@unibarcelona.com
[3] Universidad de la Costa, Barranquilla, Colombia
{aviloria7,eruiz7}@cuc.edu.co
[4] Universidad Simón Bolívar, Barranquilla, Colombia
dmartinez@unisimonbolivar.edu.co
[5] Corporación Universitaria Latinoamericana (CUL), Barranquilla, Colombia
{hhernandez,academico,jdelahoz}@ul.edu.co

Abstract. Public transportation plays an indispensable role in our daily lives, an it also remains as a higher priority for local governments. Within transportation systems, the BRT transport model uses buses on segregated roads, which has been used in different cities across the country. In order to establish structures and associations to suggest statistical models, perform diagnosis and for other purposes of BRT in Latin America, the present research work has been carried out. Once the clusters are formed, an analysis is made about the fuels used in 18 systems in Latin America. Two (2) clusters were formed: Group A with the countries of Argentina, Chile, Ecuador, El Salvador, Guatemala, Peru, Uruguay, and Venezuela which serves 20% of the daily users in Latin America. In addition, cluster B, with 80% of the rest of the users, with three (3) countries: Brazil, Colombia and Mexico. It shows the advantages of the use of natural gas fuel (NGF) with respect to the use of diesel in the BRT units. The use of hybrid systems that integrate electricity and the combination of gasoline with ethanol are not considered in this work due to the variability implied by the energy matrix where BRTs are located and the type and proportion of ethanol used, respectively.

Keywords: Cluster of Latin American countries · BRT fuels · BRT · Bus rapid transit · Urban transportation

1 Introduction

Distances that separate social and economic activities from technological and organizational ones have considerably increased in the last decade. This means that a large part of time is gained due to the reduction of working hours and devoted to travel,

© Springer Nature Switzerland AG 2020
A. P. Pandian et al. (Eds.): ICICCS 2019, AISC 1039, pp. 220–228, 2020.
https://doi.org/10.1007/978-3-030-30465-2_25

bringing the consequence of an increase in the use of private vehicles [1]. However, concerns- about slowing global greenhouse gas (GHG) emissions, reducing environmental impacts, increasing the efficiency of fossil fuel use [2, 3] and guarantee quality of life, among other factors, have driven changes in some regions and countries. In this sense, urban transportation has gone from the use of private vehicles and unregulated private transport systems to other transportation means where governments have sought to make the necessary regulation and investment to ensure sustainable urban mobility.

Among the systems implemented to ensure sustainable urban mobility are the classical subway systems and, recently, the bus rapid transit (BRT). Currently, this service is used by 33,381,258 passengers per day and is available in 170 cities around the world, with an approximate length of 5,055 km, approximately 40% of the earth's diameter [4]. Of these passengers, 61.48% correspond to Latin America, followed by Asia with 28.17%.

According to [5], urban mobility is studied for several reasons: the new relationships that users have in the spaces of the transport infrastructure, the urban economy and the articulating phenomenon of criteria of order and development of the territory, the impact of externalities generated in structuring elements such as housing, land use, roads, pollution, among others. In addition to these causes, it is important to evaluate their effects, that in regions like Latin America, have caused a structural readjustment in their economy and the liberation and regulation of this sector [6].

According to [7], during the 1990s, Latin America undertakes significant projects for the construction of urban highways, which instead of reducing traffic jams produced the appearance of new and greater motorized traffic, an increase in consumption of fuels and polluting emissions, saturation of urban accesses during peak hours, traffic jams, delays and growing dissatisfaction. This new situation led to rethinking the priorities and a new appreciation of the impact and relevance of public transportation networks.

The present research is carried out to establish structures and associations to suggest statistical models, perform diagnosis, and for other purposes related to the BRT used in Latin America [8]. For obtaining the data, the BRTData base is used. The information used for the study is collected and published on the BRTdata platform of BRT + CoE and the Institute for Transportation and Development Policies (ITDP - Instituto de Políticas para el Transporte y el Desarrollo), managed and updated by WRI Brasil Ross Centro [4]. This platform has published the data from 2012 to date. For the study, those data published in October 2018 were used.

Lower costs, improvements in customer service and the provision of quality service, shorter travel times, decongestion and use of non-polluting fuels with lower environmental impact, among others, are the policies that must be established with respect to the use of urban transportation systems. Once the conglomerates are obtained, the present research analyzes the subject of fuels that emit GHG and contribute to the climate changes. In this sense, this paper is structured in an introductory section on the BRT as a public service, the description of the method, and the results of the clusters. Finally, the analysis of results discusses the issue of fuels and systems reported [4].

2 Public Transportation System and the BRT in Latin America

The concept of public service has had different definitions over the years. Initially, it was related to the service provided by the State to satisfy public needs. However, at present it still remains as responsibility of the State, but it can also be provided by private entities through granted concessions. According to [9], public passenger transportation, as a service designed to meet the needs of urban mobility of the population, is a strategic and fundamental sector for the social and economic development of the country. An efficient passenger transportation system contributes to the development of local economies, generates a significant number of direct and indirect formal jobs, allows the mobilization of the workforce from cities and regions to production centers, and fulfills an important function of inclusion and social cohesion. In this sense, it plays a fundamental and strategic role for life in society and in modernity.

The BRT began to be used in Latin America in 1972, with the pioneering experience of Curitiba (Brazil) [10], followed by the construction of the first branch in the city of Quito, later in 1995 [11]. Since 2000, the installation and operation of the BRT has registered an important growth that includes the Transmilenio of Bogotá [11], the "Interligado" system of São Paulo (Brazil, 2003) [12], the one in Mexico City (2005) designed to complement the metro, the branch of the city of Pereira in Colombia, the cases of Guayaquil (2006) and Guatemala (2007) [13], among others. Currently, there are 55 cities in 13 countries that count on this service with a number of passengers served by 20,506,977 days [4]. This number may be greater, since these statistics do not include the cases of Bolivia, Cuba, and Puerto Rico.

Table 1. Length, passengers served per day and the relationship between them of the BRT in Latin America in 2018 [4]

Country	Length (km)	Passengers per day	Passengers per day/Length (km)
ARG	76	1,717,000	22,592.11
BRA	765	10,681,654	13,962.95
CHI	105	476,800	4,540.95
COL	225	3,071,541	13,651.29
ECU	117	1,055,000	9,017.09
SAL	6	27,000	4,500.00
GUA	24	210,000	8,750.00
MEX	394	2,652,204	6,731.48
PER	25	350,000	14,000.00
URU	6	25,000	4,166.67
VEN	42	240,778	5,732.81
Max	765	10,681,654	13,962.95
Prom	162	1,864,271	11,488.50
Min	6	25,000	4,166.67

3 Results

Currently in Latin America, there are 55 cities where the BRTs are located, specifically in the countries: Argentina (ARG), Brazil (BRA), Chile (CHI), Colombia (COL), Ecuador (ECU), El Salvador (SAL), Guatemala (GUA), Mexico (MEX), Peru (PER), Uruguay (URU), and Venezuela (VEN) (Global BRT Data, 2002). As can be seen in Fig. 1, the country that carries the highest number of passengers per day is Brazil, followed by Colombia and Mexico. Additionally, Fig. 2 shows that there is a high

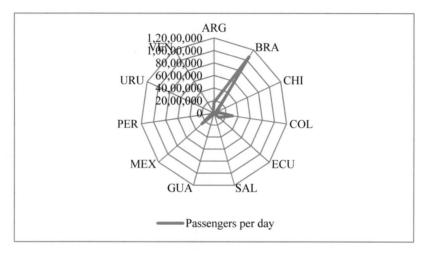

Fig. 1. Number of passengers transported per day by BRT in the countries of Latin America. Source: prepared with data from [4]

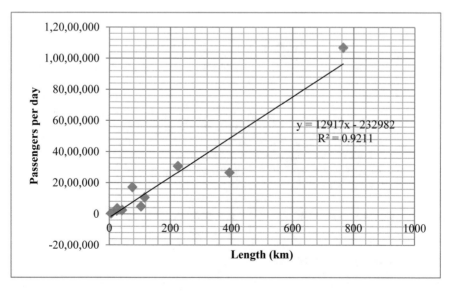

Fig. 2. Correlation of daily passengers served and length of road of the BRT in Latin America

correlation between the number of passengers served and the length (km) of 0.9211. There is no need to include page numbers or running heads; this will be done at our end. If your paper title is too long to serve as a running head, it will be shortened. Your suggestion as to how to shorten it would be most welcome. Figure 3 shows the formation of two (2) clusters, whose characteristics are defined in Table 1.

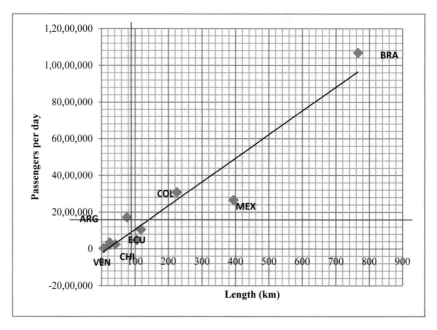

Fig. 3. Clusters in Latin America

Table 2. Definition of the cluster

Cluster	Definition	Countries	Number of countries	Number of passengers served on a daily basis
A	Countries with a BRT system with less than 172 km in length and have a daily passenger number greater than 1,864,000	Argentina, Chile, Ecuador, El Salvador, Guatemala, Perú, Uruguay y Venezuela	8 (72,7%)	4,101,578 (20%)
B	Countries with a BRT system with more than 172 km in length and have a daily passenger number greater than 1,864,000	Brazil, Colombia and Mexico	3 (27,3%)	16,405,399 (80%)

4   Analysis Results

The 61.48% of the passengers that use the BRT per day in the world correspond to users in Latin America and 80% of them live in the cities of Brazil, Colombia, and Mexico. According to Table 2, two (2) clusters of countries were formed, which are:

Cluster A, where eight (8) countries are grouped and a total of 41 corridors are managed, having 24.4 and 39% between Argentina and Chile respectively. Additionally, Cluster B formed by three (3) countries where 128 corridors are managed, 64 from Brazil, 20.3% from Colombia, and 15.6% from Mexico.

In addition to the number of passengers served, there is an important issue to consider related to the use of fuels consumed by the buses producing a significant impact on the climatic changes we are facing today. In [4], only 18 of the systems report the type of fuel they use. Among these, Tables 3 and 4 show them per year, for clusters A and B.

Table 3. Types of fuels used for Cluster A

CON	City	2009	2010	2011	2012	2013	2014	2015	2016	2017	2018
ECU	Quito				Diesel						
PAN	Panamá City							E&G			
PER	Lima				CNG						

Table 4. Types of fuels used for Cluster B

CON	City	2009	2010	2011	2012	2013	2014	2015	2016	2017	2018
BRA	Belo Horizonte					Diesel					
BRA	Fortaleza										Diesel
BRA	Guarulhos	Diesel									
BRA	Porto Alegre					Diesel					
BRA	Uberaba						Diesel				
BRA	Uberlandia						D&E				
COL	Barranquilla				Diesel						
COL	Bucaramanga				Diesel						
COL	Cali				Diesel						
COL	Cartagena								CNG		
COL	Medellín				CNG						
MEX	Juárez								CNG		
MEX	León de Los Alameda				Diesel						
MEX	México City						D&H				
MEX	Pachuca						Diesel				

Where:

D&E: Diesel and electricity;

E&G: Ethanol and gasoline;

D&H: Diesel & Hybrid (diesel-electricity);

CNG: Compressed Natural Gas.

As can be seen, since 2009, the city of Guarulhos (Brazil), clusters B, reports the use of diesel as fuel for its BRT units. Additionally, 11 of the 18 systems in total use only diesel or in combination with hybrid systems that include the use of electricity as in the case of Uberlandia (BRA), Mexico City (MEX) and Panama City (PAN). Among the policies established to reduce GHG emissions from this type of fuel increases in its performance are included [14].

The CNG reported its use since 2012, by cities such as Lima (PER) and Medellin (COL), from clusters A and B, respectively. According to the CO2, CO, NOX and PM emission factors, they tend to significantly decrease with the use of this fuel, compared to Diesel, as shown in Table 5 [15]. Although the occupancy levels of both vehicles are different, it can be observed that NOx and particulate matter (PM) emissions are approximately negligible with the use of CNG as fuel.

Table 5. Types of fuels used for Cluster B

UND	COMB	CO_2	CO	NOx	PM	UND	PASS
BRT B5	Diesel	3428,9	17,07	2,86	0,7	km/l	160
NG120	CNG	1889,6	1,79	4.6e–5	3,28E–08	km/m^3	120

The buses with hybrid technology using electricity direct their emissions to the source of electricity generation, as in the case of the BRT of Mexico which energy matrix is based on fossil fuels [16, 17]. Finally, the combination of ethanol with gasoline is one of the policies that are currently being used in a significant way to reduce emissions, not only from the BRT, but also from other types of vehicles.

5   Conclusions

Public transportation is a necessity for the population, as well as a priority for local governments, among which the BRT are used in different cities around the world to meet the needs of urban mobility. Two (2) clusters were formed, according to the number of daily passengers served in the BRT: Group A with the countries of Argentina, Chile, Ecuador, El Salvador, Guatemala, Peru, Uruguay, and Venezuela that serve 20% of the daily users in Latin America. In addition, cluster B with 80% of the rest of the users, with three (3) countries: Brazil, Colombia and Mexico. Finally, it shows the advantages of using compressed natural gas (CNG) with respect to the use of diesel in the BRT units.

Acknowledgment. The authors express their sincere thanks to the BRTdata platform and the Ibero-American Transport and Sustainable Urban Mobility Network (RITMUS - Red Iberoamericana de Transporte y Movilidad Urbana Sostenible) for their valuable contribution to the development of this work.

References

1. lLizárraga, C.: Movilidad urbana sostenible: un reto para las ciudades del siglo XXI, Economía, sociedad y territorio, pp. 283–321 (2006)
2. Sánchez, L., Vásquez, C., Viloria, A.: The data envelopment analysis to determine efficiency in Latina American countries for greenhouse gases control in electrical power generation. Int. J. Energy Econ. Policy **8**(3), 197–208 (2018)
3. Sánchez, L., Vásquez, C., Viloria, A.: Conglomerates of Latina American countries and public policies for the sustainable devoloment of electric power generation sector. In: International Conference on Data Mining and Big data, pp. 759–766 (2018)
4. ITDP: Global BRT Data [En línea] (2002). https://brtdata.org/. Último Acceso 05 Jan 2019. I. d. P. p. e. T. y. e. D
5. Grande, C.: Movilidad urbana sostenible y sistema integrado de trasnporte para el área metropolitana de San Salvador. REDIP **8**(1), 1441–1463 (2018)
6. Figueroa, O.: Transporte urbano y globalización. Políticas y efectos en Latinoamérica. Revista EURE **31**(94), 41–53 (2005)
7. Lupano, J., Sánchez, R.: Políticas de movilidad urbana e infraestructura urbana de transporte, CEPAL, Santiago de Chile (2009)
8. Rodríguez, P., Palomino, N., Mondaca, J.: El uso de datos masivos y sus técnicas analíticas para el diseño e implementación de polítias públicas en Latinoamérica y el Caribe, BID. Banco Interamericano del Desarrollo, Santiago de Chile (2017)
9. Álvarez, L., Dueñas, F.: Responsabilidad del estado por acción frente al derecho fundamental a la calidad del servicio de transporte público terrestre. Análisis de la sentencia c-066 de 1999, Universidad Santo Tómas, Bogotá (2015)
10. Duarte, F., Rojas, F.: Intermodal connectivity to BRT. a comparative analysis of Bogotá and Curitiva. J. Public Transp. **15**(2), 1–18 (2012)
11. Rodriguez, D., Vergel, E., Camargo, W.: Desarrollo Urbano orientado a los sistemas de trasnporte público masiivo tipo BRT (Bus Rapid Transit) en Quito y Bogotá, Lincoln Institute of Land Policy, Cambridge (2013)
12. Rolnik, R., Klintowitz, D.: Mobilidade na ciudade de Sao Paulo, Estudos avanzados, pp. 89–108 (2011)
13. Poole-Fuller, E.: ¿Hacia una movilidad sustentable? Desafíos de las políticas de reordenamiento del transporte público en Latinoamérica. El caso de Lima. Letras Verdes **21**, 4–31 (2017). Revista Latinoamericana de Estudios Socioambientales
14. Solíz, J., Sheinbaum, C.: Consumo de energía y emisiones de CO_2 del transporte en México y escenarios de mitigación. Rev. Int. Contam. Ambiente **32**(1), 7–23 (2016)
15. Coéllar, Y., Buitriago-Tello, L., Belalcazar-Ceron, C.: Life cycle emissions from a bus rapid transit systems and comparison with other modes of passenger transportation. CTF Cienc. Tecnol. Futuro **6**(3), 123–134 (2016)
16. Sánchez, L., Pérez, R., Vásquez, C.: Eficiencia de los países desarrollados en el control del usos de los combustibles fósiles para generar energía. Rev. Cient. Ecociencia **4**(2), 58–71 (2017)
17. Sánchez, L., Vásquez, L.M.C.: Emisiones de mercurio por lámparas fluorescentes compactas y por generación de energía eléctrica a base de combustibles fósiles. Rev. Cient. Ecociencia **5**(5), 1–18 (2017)

18. Vásquez, C., Torres-Samuel, M., Viloria, A.: Public policies in science and technology in latin American countries with universities in the top100 of web ranking. J. Eng. Appl. Sci. (2017)
19. Pérez-Esparrells, C., Cómez-Sancho, J.: Los ranking internacionales de las instituciones de educación superior y las calsificaciones universitarias en España; visión panorámica y prospectiva de futuro. Fundación de las cajas de ahorros. Documento de trabajo No. 559/2010 (2011)

Regenerative Braking Control of Induction Motor in Electric Vehicles for Optimal Energy Recovery

Konka Gurunath[1]([✉]) and A. Vijayakumari[2]([✉])

[1] Department of Electronics and Communication Engineering, Amrita School of Engineering, Coimbatore Amrita Vishwa Vidyapeetham, Coimbatore, India
kggurunath@gmail.com
[2] Department of Electrical and Electronics Engineering, Amrita School of Engineering, Coimbatore Amrita Vishwa Vidyapeetham, Coimbatore, India
a_vijayakumari@cb.amrita.edu

Abstract. This paper discusses the regenerative braking control as applied to Electric Vehicle with two control techniques along with their performance comparison. Regeneration is realized through V/F, a scalar control method and Field Oriented Control (FOC), and a vector control method for an inverter fed Induction motor. A simple Look up Table approach is proposed for a real-time implementation of V/F control, while in FOC, a methodology is proposed to shift the operating point from braking to generation region for optimal energy recovery. A model representing the commercial electric vehicle GM EV1 is developed in the present study that emulates a typical Vehicle environment. The test conditions are mainly chosen to include all possible drive ranges like acceleration, cruising and deceleration to demonstrate start-up, motoring and regeneration control through both scalar and vector control methods. The theoretical relationship between the synchronous and the rotor speeds are presented to bring out the mode change from motoring to regeneration. Furthermore, the regeneration is ascertained with the current and power flow results.

Keywords: Regenerative braking · Electric vehicle · Field Oriented Control · Induction machine

1 Introduction

Across the world, several action plans are in place to replace the IC engine vehicles with zero emission vehicles. One of the potential candidates for emission free roads is Electric Vehicles (EV), which is envisaged to decarbonize road transportation in the near future [1, 2]. EVs manifest a wide scope for retrieving power from the electric motors during decelerating periods like stops, traffic signals, downhill etc. But, the electric drive should be equipped with the control capability to recuperate power back to the source during braking periods. Thus, one among the major requirements for more efficient power-train of an EV is regenerative braking, which can improve driving range which inturn will improve battery capacity. Vehicles developed with regenerative braking (for example, Honda Insight) have proven to be the most fuel efficient [3, 4].

© Springer Nature Switzerland AG 2020
A. P. Pandian et al. (Eds.): ICICCS 2019, AISC 1039, pp. 229–238, 2020.
https://doi.org/10.1007/978-3-030-30465-2_26

However, regeneration can be executed in EVs just through software modifications in the controller without the need of any additional power electronic components.

Induction Machines (IMs) are identified as the most preferred ones due to their low cost, ruggedness, low maintenance, and good efficiency over a wide region of operation. IM exhibits higher efficiency and reliability in higher capacity machines, which is a much sought specification for traction applications and in E-Mobility [5, 6]. IM can be controlled by scalar and/or vector control techniques in both motoring and in regeneration modes. Vector control exhibits battery dynamic performance but with an overhead of laborious computations. Also their dynamic response is highly influenced by machine parameter uncertainties. Thus, robustness has to be brought into the control against parameter variations if it is to be extended for EV drive applications [7].

In [8], regenerative braking boundaries are explained through torque–speed relationships, wherein, it is inferred that by maintaining the operation in the generative region rather than in braking region, maximum energy recovery is possible. Concurrently, for modeling vehicular dynamics, system level simulation tools like ADVISOR and QSS Toolbox are to be utilized to emulate the real time vehicular environment. These tools are required to understand various power flow architectures and their dynamics in EVs and HEV's [9].

This research work proposes a methodology for maximizing the available regeneration power by shifting the operating point of IM to an optimal location whenever the brake pedal activation is received. It also involves identification of an exact torque value to recuperate the regeneration power back to the battery for instantaneous braking. A typical Europe city drive cycle is utilized for testing the proposed speed control with regeneration feature during both Motoring and braking events. The entire system is implemented in the MATLAB/Simulink platform with both scalar and vector control methods and their results are presented and analyzed.

2   Regeneration in Induction Motor

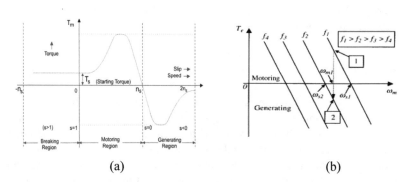

(a)   (b)

Fig. 1. (a) Typical Torque- Slip Characteristics of IM (b) Torque-Slip Characteristic of IM with variable stator frequency

IM has a torque-speed characteristic that utmost meet the expectations of traction motors, which are subjected to variable speed and variable torque condition during the operation. IMs can be operated under constant torque until base speed and above base speed maintains constant power. If IM is operated in EV applications, then frequent start-stops will be encountered and a dynamic shift in the operating point from motoring to braking and vice-versa has to be carried out by control. From the typical torque-Speed characteristics depicted in Fig. 1(a), it is observed that higher torque is possible at generation region compared to braking region. So, if the operating point of IM is shifted from braking to generation region at any given rotor speed, then maximum power can be recuperated from the machine. But shifting the operating point needs the frequency of rotating magnetic field to be brought below the rotor speed to accomplish a negative slip. Negative slip is not a normal operating scenario in motoring, so a stator frequency control is opted for accomplishing this requirement. Figure 1(b) depicts a linear slip region for various stator operating frequencies in a typical IM.

Consider a vehicle which is moving at a constant speed with its stator supplied with a frequency f_1 which corresponds to a synchronous speed of ω_{s1} as in Fig. 1(b). The rotor speed at this condition be ω_{m1} for some finite positive slip, suppose if the driver applies brake, and initiates a regeneration operation, then the proposed control algorithm is intended to generate a stator frequency f_2 which corresponds to a synchronous speed of ω_{s2}. This revised synchronous speed results a negative slip for the present rotor speed, thus making the machine to operate in generating region, because $\omega_{m1} > \omega_{s2}$. In order to operate the machine with a maximum possible negative torque an optimal f_2 has to be identified and obtained on the stator.

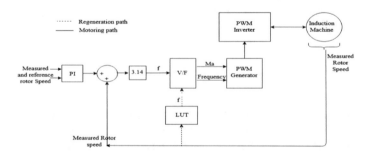

Fig. 2. V/F implementation for regeneration of IM

3   Regeneration with V/F Control Algorithm

This speed control method allows the stator voltage and frequencies to be varied independently, at the same time maintains a constant V/F ratio in order to conserve the rated flux in the machine to deliver rated torque. To accomplish regeneration with V/F control, an optimal frequency for any rotor speed has to be estimated as shown in Fig. 2. The effort in the present work is to establish a loop up table (LUT) containing the relationship between the rotor speed and the stator frequency corresponding to maximum negative torque. This LUT is formulated by experiments on an inverter fed IM in MATLAB/Simulink. The specifications of the IM used in this experiment is presented in Table 1.

Table 1. IM specifications

S no	Parameters	Used value
1	Voltage (V)	460 V
2	Power (P)	100 hp
3	Frequency (f)	50 Hz
4	Pole pairs (p)	2
5	Stator resistance (R_s)	0.4232 Ω
6	Rotor resistance (R_r)	0.4232 Ω
7	Stator reactance (X_s)	0.19044 Ω
8	Rotor reactance (X_r)	0.19044 Ω
9	Magnetizing reactance (X_m)	7.406 Ω

3.1   Optimal Frequency Identification

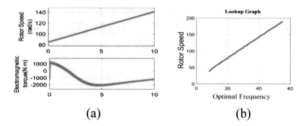

(a)   (b)

Fig. 3. (a) Identifying Optimal Frequency (b) Locus of Maximum Negative Torque

From Fig. 3(a), it is observed that the maximum negative torque occurs at 110 rad/s. It ascertains that, if this IM is used in the proposed EV, and during braking, if the rotor speed is 110 rad/s, then the induction machine should be supplied with a stator frequency of 30 Hz to recuperate optimal power from the machine. Similarly, the experiment is repeated for various frequencies and the relationship between the frequency and the speed corresponding to maximum negative torque is deduced and the same is presented in Fig. 3(b). This locus of maximum negative torque serves as the LUT for the proposed braking control.

4 Regeneration with FOC

V/F control, does not consider the machine dynamics for control loop tuning, thus leading to lack of precision in accomplishing the speed targets. But, FOC considers the dynamics of an induction machine which improves the stability during load changes, thus resulting an accurate speed tracking. Through FOC, DC machine like performance can be accomplished in IM, by decoupling the torque and speed controls. Such a decoupled control is achieved by segregating the stator currents into torque and flux producing components through coordinate transformations then by controlling them independently.

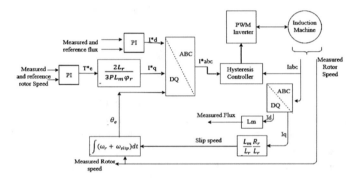

Fig. 4. IFOC implementation scheme

Coordinate transformations are utilized for the segregation of the two current components from the stator current. However, to achieve orthogonality between the two currents, the rotor flux position and magnitude is required. Rotor flux position can be either sensed or estimated and based on which FOC can be classified as Direct FOC, where sensors are used and Indirect FOC (IFOC) where rotor flux is estimated [10, 11].

Figure 4 presents the block diagram of IFOC as applied to IM, where, a hysteresis current controller is used to make the stator current to follow a reference current, while the reference current is obtained by incorporating the current magnitudes with the rotor angular position. The angular position of rotor flux is determined as the integral sum of the required rotor angle and the angle introduced due to the slip frequency i.e. slip angle. The angular rotor position is estimated as,

$$\theta_e = \int \omega_e dt = \int (\omega_r + \omega_{slip}) dt = \theta_r + \theta_{slip} \tag{1}$$

Where θ_e angle due to the synchronous speed ω_e, θ_r is angle due to the rotor speed ω_r and θ_{slip} is angle due to slip speed ω_{slip}. The rotor circuit equations of IM in synchronous frame of reference are expressed as,

$$\frac{d\varphi_{dqr}}{dt} + R_r i_{dqr} - (\omega_e - \omega_r)\varphi_{qdr} = 0 \tag{2}$$

Where φ_{dqr} rotor flux along d and q axes, i_{dqr} rotor current along d and q axes and R_r is rotor resistance/phase of IM. Now the flux linkage of rotor can be expressed as,

$$\varphi_{dqr} = L_r i_{dr} + L_m i_{dqs} \tag{3}$$

Where L_r is rotor inductance, L_m is mutual inductance of IM and $\omega_{slip} = \omega_e - \omega_r$.

For decoupled control of torque and flux, the direct axis is made to align with the rotor flux, which makes quadrature component to become zero, resulting an equation ascertaining the fact that the q-axis current solely determine the slip speed, and is expressed as,

$$\omega_{slip} = \frac{L_m}{L_r}\frac{R_r}{L_r} i_{qs} \tag{4}$$

During motoring operation, if the torque requirement of the machine is positive then the quadrature current is positive which will result the synchronous speed greater than present rotor speed as per Eq. (1). Conversely, during the regeneration if the torque requirement is negative then the quadrature current component is negative which will result the synchronous speed less than the present rotor speed as per Eq. (1). Thus, with FOC, the regeneration control is inherent upon the actuation of the brake pedal, and does not require any external control logic unlike V/F. This project aims at developing and the regeneration control with FOC as applied to IM.

5   EV System Modeling

Table 2. Vehicle parameters [12]

Parameter	Value (SI units)
Total mass of vehicle (kg)	1400 + 140
Vehicle cross section A_f (m^2)	1.8
Rolling resistance coefficient μ_u	0.0048
Drag coefficient C_w	0.19
Gradient angle θ	0
Mass increase due to inertia* (%)	5
Wheel diameter d_{wheel} (m)	0.6
Gear ratio (G)	2
Air density (Kg/m^3)	1.202

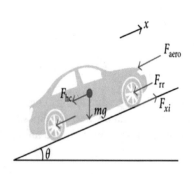

Fig. 5. Longitudinal dynamics of the vehicle

To implement the proposed regeneration control for EV applications, a European city drive cycle is considered along with the vehicle specifications. By imposing the drive cycle on the vehicle specifications, the reference speed and torque which will be exerted on the electrical motor of EV can be calculated. The power and torque imposed on the Induction motor are determined based on longitudinal forces acting on the vehicle. Where, μ_u is the rolling resistance constant, A_f is the vehicle cross section, C_w is drag coefficient, V is velocity of the vehicle, d_{wheel} is wheel diameter and G is gear ratio.

GM EV1 is considered as a sample case and Table 2 lists its parameters. Table 3 is deduced with the parameters of Table 2 so as to obtain the reference torque and speed. The IM simulation system receives these reference quantities as controller inputs and ensures the speed reference tracking as demanded by the drive cycle considered.

Table 3. Sample load calculations

Parameters	Equations	Acceleration (2 m/s2) and Velocity = 15 m/s	Deceleration (2 m/s2) and Velocity = 15 m/s
Rolling resistance (N)	$F_{rr} = mg\mu_u \cos(\theta)$	74	74
Aerodynamic resistance (N)	$F_{aero} = 0.5\rho C_w A_f V^2$	46	46
Gradient resistance (N)	$F_{hc} = mg\sin(\theta)$	0	0
Propelling force (N)	$F_{xi} = 1.05ma$	3203	−3203
Tractive force (N)	$F_{tractive} = F_{xi} + F_{hc} + F_{aero} + F_{rr}$	3323	−3083
Motor torque (N m)	$T_{motor} = 0.5d_{wheel}F_{tractive}/G$	499	−462
Motor speed (rad/s)	$\omega_{motor} = V/(0.5d_{wheel})G$	100	100
Motor power (W)	$P_{motor} = \omega_{wheel}T_{motor}$	49851	−46245

6 Validation of Regeneration Control in IM

Regeneration operation of IM is validated with (i) V/F control and (ii) IFOC. The system represented in Fig. 2 with the specifications of Tables 1 and 2 are used for V/F control while the system of Fig. 5 with the same specifications is used for IFOC. The typical Europe city drive cycle consisting of (i) acceleration during 5 s to 16 s, (ii) cruising during 16 s to 21 s and (iii) deceleration during 21 s to 31 s are utilized as input for the simulation studies. Due to the simulation time constraints a portion of the drive cycle with a duration of 35 s is only considered for the present validation.

6.1   Discussion on V/F Control Results

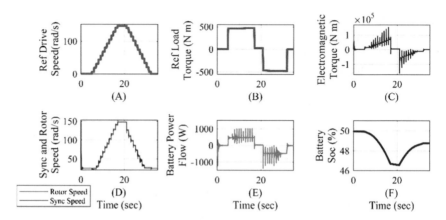

Fig. 6. (a) Reference Drive speed, (b) Reference Load Torque, (c) IM Electromagnetic Torque, (d) IM Synchronous and Rotor Speeds, (e) Battery Power flow, (f) Battery SOC waveforms with implemented V/F Control

The results of regeneration operation with V/F are presented Fig. 6(a) to (f). From Fig. 6, few observations are made (i) there is an increase in stator frequency to meet the increasing speed demand of drive cycle and vice versa, (ii) reference following capability, and (iii) reduction in power requirement during coasting. During the braking interval of 21 s to 31 s, the algorithm identifies the negative torque from the drive cycle model, rotor speed is read and accordingly optimal frequency is detected from the LUT. From Fig. 6(d), it can be comprehended that the optimal stator frequency would have been applied at every instant during braking, as the rotor speed is always found higher than synchronous speed. This ensured the negative slip region of operation of IM which in turn increases the SOC of the battery due to the power flow into it. In the entire speed range V/f ratio is found to remain constant. Also the machine is observed to follow speed and torque references as dictated by the drive cycle throughout, but with sluggishness in the torque reference tracking as seen in Fig. 6(c).

6.2   Discussion on FOC Results

The results of regeneration operation with FOC are presented Fig. 7(a) to (f) for the same input conditions. A critical observation from FOC when compared to V/F control is the substantial reduction in the torque ripple under all operating ranges, and the swiftness in the electromagnetic torque developed in Fig. 7(d).

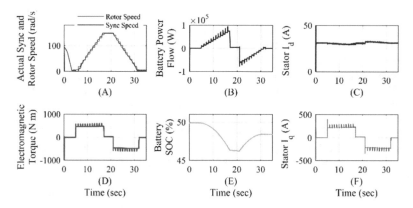

Fig. 7. (a) IM Synchronous and Rotor Speeds, (b) Battery Power flow, (c) IM Stator d-axis Current (d) IM Electromagnetic Torque, (e) Battery SOC, (f) IM Stator q-axis Current waveforms with implemented FOC Control

During the deceleration interval, when braking is initiated the torque is observed to become negative from Fig. 7(d). Further, the q component of stator current depicted in Fig. 7(f) is becoming negative during the interval between 21 s to 31 s, ascertaining a lower value of synchronous speed than the rotor speed and pushing the operation of IM to generation region. From SOC plot, it can be concluded that during acceleration the power is being fed from a battery and during braking, it is reversed, indicating the regeneration of energy.

7 Conclusion

Regenerative braking for Induction motor based electric vehicles with V/F and indirect FOC is implemented in MATLAB Simulink. Regeneration in V/F is achieved through a LUT deduced from experiments. With LUT an appropriate inverter frequency is identified for optimal energy recovery from IM during braking. On the other hand regeneration in IFOC has been achieved through its inherent logic for the shift of operating point to generation region. FOC exhibited good dynamic response with reduced torque ripple, and recommended for IM traction drives. In contrary, V/F control contends well in terms of simple implementation, acceptable response but slightly higher torque ripples and poor speed tracking at low speeds. Though FOC gives better performance for IM drives, the machine parameter uncertainties are the major barriers. But with a robust adaptive system in place, these uncertainties can be well addressed in the modern IFOC drives.

References

1. Tran, M., Banister, D., Bishop, J.D.K., McCulloch, M.D.: Realizing the electric-vehicle revolution. Nat. Clim. Chang. **2**, 328–333 (2012)

2. Vijay, B., Shanmughasundaram, R.: Control strategy for parallel hybrid electric vehicles. Procedia Comput. Sci. **143**, 678–685 (2018)
3. Chan, C.C., Wong, Y.S.: The state of the art of electric vehicles technology. In: 4th International Power Electronics Motion Control Conference, pp. 46–57 (2004)
4. Yoong, M.K., Gan, Y.H., Gan, G.D., Leong, C.K., Phuan, Z.Y., Cheah, B.K., Chew, K.W.: Studies of regenerative braking in electric vehicle. In: IEEE Conference on Sustainable Utilization and Development Engineering Technology (2010)
5. Zechmair, D., Steidl, K.: Why the induction motor could be the better choice for your electric vehicle program. World Electr. Veh. J. **5**, 546–549 (2012)
6. Zeraouila, M., Benbouzid, M.E.H., Diallo, D.: Electric motor drive selection issues for HEV propulsion systems: a comparative study. IEEE Veh. Power Propuls. Conf. **55**, 1756–1764 (2006)
7. Shah, V., Vijayakumari, A.: Field oriented control of surface mount permanent magnet synchronous machine with non linear observer for continuous rotor position estimation. In: 3rd International Conference Convergence Technology (2018)
8. Murthy, A.S., Magee, D.P., Taylor, D.G.: Vehicle braking strategies based on regenerative braking boundaries of electric machines. In: 2015 IEEE Transportation Electrification Conference and Expo (2015)
9. Sinchuk, O., Kozakevich, I., Kalmus, D., Siyanko, R.: Examining energy-efficient recuperative braking modes of traction asynchronous frequency-controlled electric drives. East.-Eur. J. Enterp. Technol. **1**, 50–56 (2017)
10. Bose, B: Modern Power Electronics and AC Drives. Prentice-Hall, Upper Saddle River (2002)
11. Krause, P.C., Wasynczuk, O., Sudhoff, S.: Analysis of Electric Machinery and Drive Systems. IEEE Press, New York (2002)
12. James, L.: Electric Vehicle Technology Explained. Wiley, Hoboken

Smart Surveillance System for Detection of Suspicious Behaviour Using Machine Learning

Aastha Joshi[(⊠)], Ninad Jagdale[(⊠)], Rajvi Gandhi[(⊠)],
and Sheetal Chaudhari[(⊠)]

Department of Information Technology, Sardar Patel Institute of Technology,
Mumbai 400058, India
aasthakj@gmail.com, ndjagdale@gmail.com,
rajvigandhi27@gmail.com, sheetalchaudhari@spit.ac.in

Abstract. A method for detecting suspicious behaviour and activities in live surveillance is presented in the following paper. Recent works in the field of convolutional neural networks have shown considerable progress in the areas of object detection and recognition, especially in images. Labels are required as learning signals in the convolutional neural networks. The proposed method finds difference between two consecutive video frames in order to find the motion history image by eliminating the use of labels. We detect suspicious behaviour in videos and live surveillance using a spatiotemporal architecture. The architecture for prediction of video frames is based on convolutional neural networks. Classification of normal and abnormal behaviour is attained for most of the scenarios. An anomalous batch of frames within the videos is identified and an alarm is generated to alert people about the malicious activities. In this way, they can be prevented on the spot instead of taking action later.

Keywords: Machine learning · Classification · Convolutional neural networks · Training · Unsupervised learning · Suspicious behaviour · Anomaly detection

1 Introduction

In a world of increasing crime, passive or dead surveillance is not enough. Active surveillance where cameras themselves can detect suspicious behaviour can narrow down the search and decrease human effort considerably. With increase in the data recorded by the videos, there is a need to identify not only objects and humans but also analyse the entire scenario to classify it and ensure that nothing suspicious is taking place. Machine learning and computer vision have performed the tasks of extracting human actions from videos [1]. Results have been obtained for activities such as pointing, walking, jogging and hand waving [2]. Important events in video surveillance, often occur with very low probability [3]. Manual detection of suspicious behaviour or activities in videos requires continuous monitoring by personnel which is a tedious and cumbersome task and does not assure accuracy.

Smart surveillance which can automatically detect mishaps are of great importance since they can reduce staffing needs to a great extent thus leading to monetary benefits to

© Springer Nature Switzerland AG 2020
A. P. Pandian et al. (Eds.): ICICCS 2019, AISC 1039, pp. 239–248, 2020.
https://doi.org/10.1007/978-3-030-30465-2_27

the organization. It also leads to the formation of a much more secure environment by detection of suspicious activities while they are taking place since measures can be taken to curb these activities. Such a system may be extremely useful in scenarios where video surveillance plays a vital role in environments like schools and colleges. However, this system requires a massive amount of configuration on every video surveillance before the analysis stage. The data in videos are unstructured in format consisting of noise, different events, high dimensionality. This makes it difficult to perform processing operations on the data and to model the data [3]. The definition of a suspicious activity also differs from scenario to scenario. For example, running in a park or a playground is definitely a normal and non suspicious activity but if it takes place on a train station or a mall it may be an indication of theft. Such differences in scenarios make it difficult for machine learning to classify a particular activity as suspicious. Some commonly considered anomalies (usually based on tracking features) are fall including motionlessness, unusual speed], wrong direction, etc. [9]. Action recognition has proved to be successful in many cases [6]. However most action recognition methods need the parts of the video that needs to be scrutinized to be labelled explicitly. This paper presents a method for extracting features from a video with the help of a spatio-temporal encoder. Statistical properties of the preprocessed data will aid in the classification of videos as normal and abnormal. Our method is tested using various videos obtained from datasets as well as using live feed. In order to obtain greater processing power, we use a Graphics Processing Unit (GPU). This can be used in two ways. One method can be to filter the abnormal frames amongst all the frames as per the trained model. The other way can be by parallel processing, the frames can be stored in the CPU and then sent for comparison to the GPU. Simultaneous comparisons can help determine abnormal frames.

2 Related Work

Extensive work is carried in the field of video processing. As we discussed in the introduction of this paper that there is an appalling need for any operations or security systems to integrate and adapt to more secure, robust and quick ways to deal with problems in the surrounding. But the real concern lies in the analysis and learning, which is not simple and whose accuracy can never match up to human analysis. Anomaly in the surrounding can be anything like, dropping a bag, kidnapping and various other cases, but some non-anomalies can also be confused by the machine as anomalies, and extent of that is usually high in such projects. In still to be developing world, we have to take care of the accuracy and efficiency of the system. Thus, in most of the projects which are based on video processing, it is rather important that the scope, surrounding and anomalies are limited. Determination of unusual activities before they are happening is very difficult. There have been a lot of practices and related work in this field. Some of the most common practice which is implemented in these type of projects is training the machine with the normal activities and patterns from the training videos [12, 13]. We later test such videos and look out for deviation in patterns. If deviation exists then it is considered as anomalous. In some procedures the method followed is: feature extraction, classification and obtaining decision. Houghs forest model is utilized for extracting features from representative pictures [1].

Motion and sequence from videos are encoded in the image. This spatial temporal voting model is then combined with the input. A 2D convolutional neural network helps here to process the image representation. Anomalous behavior based on the deviation from the trained data set is identified. In [14], We first calculate the acceleration and secondly based on this acceleration we classify crowd behaviour based on the acceleration. So, we can deduce that for an obvious anomaly to occur, violence or any kind of threatening objects like knife, gun or any abandoned objects should be involved, which is easier to detect than unknown anomaly. Some techniques involve extensive research in improving the efficiency of same. Like in [17], violence is detected in video surveillance using irregular motion information. Similarly in [18], abandoned Objects are skillfully detected using background subtraction along with human blobs.

3 Methodology

The method used is an unsupervised method which is based on extracting features from a video in order to classify certain behaviours as abnormal. The logic is that the occurrence of an abnormal event will be indicated by the video frames being significantly different from the other video frames. Our model is trained using a feature extractor for spatial features and a time based encoder-decoder which learns patterns from the input frames. The method consists of the following steps:

3.1 Architecture

See (Fig. 1).

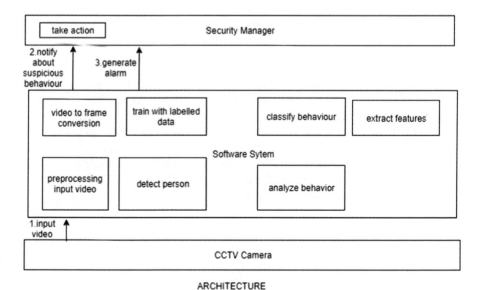

ARCHITECTURE
DIAGRAM

Fig. 1. Architecture diagram of proposed system

3.2   Preprocessing

This is the stage where the unstructured video data is converted to a uniform format to make the purpose of feature extraction simple. It should be converted and aligned in order to make it appropriate for input. Each video frame is resized so that all of them have the same size. The pixel values are further scaled and given a positive value less than 1 which is then subtracted from the mean for normalization. In order to reduce the problem of dimensionality, the image is converted to grayscale. The image that has been processed now has a mean 0 and variance 1. The input model is comprised of video frames with sets of numbers which have different gaps in them. Data is divided into smaller and smaller parts for input in the temporal encoder-decoder. Large amount of data is needed as an input to this model.

3.3   Feature Learning

We use the properties of space and time in order to learn regular patterns from the input videos supplied for training. The architecture consists of 2 parts: one for extracting features of space from each frame and an encoder-decoder to study previously extracted patterns. To regularize the extracted frames and reduce dimensionality, encoding decoding is performed on these frames. The units in the output should be lesser than input units. An error is calculated and by analysing this value, the abnormality can be determined. For the video to be normal, the error value should be as small as possible. Convolutional neural network enable us to extract features from the frames and which frames should be focused on. Convolution refers to computing the dot product between the vectors in the input. During the training process, a convolutional neural network automatically learns the features from the frames by specifying certain parameters. We use convolutional long short term memory which was invented by Shi et al. in [10]. Here, the matrix operations for computation have been replaced by convolutions. The number of weights used in Convolutional LSTM have been reduced.

3.4   Regularity Score

After the training stage has been finished, the next stage is testing. In this stage we test videos based on the model generated by training to check for anomalies. We used the same formula for regularity score which was used in [5]. An error is calculated by determining the distance between the input and reconstructed frame $e(t) = -x(t)fW(x(t))-2$ [18] where fW is the learned weights by the spatiotemporal model. A score base on abnormal frames is then calculated and scaled between 0 and 1. Subsequently, sr(t) which is the regularity can be simply derived by subtracting the abnormal score from 1: $sa(t) = e(t)e(t)min\ e(t)max$ [8] $sr(t) = 1sa(t)$ [8].

3.5   Abnormality Detection

3.5.1   Determining an Upper Limit

This is the most important event in anomaly detection. It is very easy to test if the frame we extracted from a video is abnormal or not. The simplest one is thresholding. Here, we calculate the reconstruction error like [8] i.e. our e(t). Thus our project depends upon this error. Threshold value should be carefully selected. If we set our threshold too low, then it will lead to more alarms being triggered and give rise to false positives, whereas if we set the threshold too high, there can be a chance that we dont detect some of the anomalies.

3.5.2   Number of Events

After we have followed the practice in [10], to reduce the unmeaningful as well as noisy minima which is present in the regularity score, we use persistence 1D [11] which aids in grouping local minima within a temporal window. It has been assumed that local minima within 50 frames lie along with the same abnormal event. This is a considerable length of the temporal window, an event that is abnormal should have a minimum of length 2–3 s.

4   Implementation

4.1   Dataset

We have trained our system with videos from the CAVIAR dataset [19], avenue dataset [20], subway entry and exit dataset and mall dataset [3]. Each dataset consists of certain normal as well as abnormal events. The abnormal events in the CAVIAR dataset include leaving a bag behind, leaving box behind, chasing and fighting, fighting between men and running away after fighting. Avenue dataset consists of events like running in the wrong direction loitering. Mall and subway datasets consists of anomalous events like sudden getting out of the train, running without payment for ticket and sudden stops.

4.2   Model Parameters

The model has been trained so as to minimize the reformation error of the input. An optimizer has been used for setting the rate of learning in an automatic manner based on the weight update history of the models. We make use of mini-batches that are 64 in size and train every volume for epochs or till the loss of reformation of validation data stop to decrease. We choose Hyperbolic tangent as the function of activation of spatial decoder and encoder.

5   Results

5.1   Preprocessing

In the preprocessing stage, the videos are normalized and converted to a standard format so that is acceptable for input. Frames in the form of strides are generated. The first stride consists of frames 1 to 10. Second one consists of frames of first 10 odd natural numbers. Third stride consists of frames starting from one and leaving 2 after each frame and so on (Fig. 2).

Fig. 2. Frames getting generated

5.2   Training

We train the system using our dataset to generate a model in order to learn regular patterns from the data. This model is used for future classifications (Fig. 3).

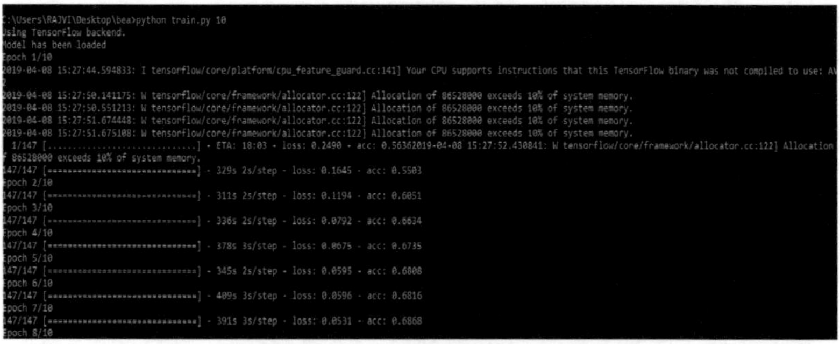

Fig. 3. Videos in dataset getting trained

5.3   Testing

The testing phase helps to classify the extracted bunch of frames as anomalous or not. This is done on the basis of the model generated by the training phase. The output of this phase determines if the frames are anomalous or not (Figs. 4 and 5).

Fig. 4. Testing results on the input videos

Fig. 5. Testing results on the input videos

5.4   Live Feed

The last phase opens the web camera of the particular device and then tests the input in a live scenario. We feed the trained model to the live feed for comparison. The output of this stage is classifying the live feed as normal or anomalous and further generating an alarm on detection of anomalous behaviour (Figs. 6, 7 and 8).

Fig. 6. Output detected anomalous by strangling a person

Fig. 7. Terminal detecting anomaly in above scenario

Fig. 8. Output detected anomalous by hitting with objects

6 Future Scope

This system can be extended to be implemented in other critical places like malls, streets, restaurants, airports, public places etc. This would require the training to be done for a wider range of activities. In each scenario, abnormal and normal are identified as they are specific to the environment. This will reduce staffing needs to a great extent and hence reducing the budget allocated to it. Also, this can be further extended to identify the criminal itself by help of face detection and comparing it with criminal records from the government.

7 Conclusion

We have implemented a smart surveillance system that detects suspicious and anomalous behaviour from video sequences. We have proposed for this system to be implemented in schools and colleges. The system helps in preventing anomalous activities as soon as they are taking place. We used Machine Learning to train the dataset and then test it to obtain results. We also integrated this approach in the live feed where real time output is generated. An alarm is generated so that the concerned authorities are alerted.

References

1. Serrano, I., Deniz, O., Espinosa-Aranda, J.L., Bueno, G.: Fight recognition in video using hough forests and 2D convolutional neural network. IEEE Trans. Image Process. **27**(6), 4787–4797 (2018)

2. Laptev, I., Lindeberg, T.: Space-time interest points. In: Proceedings of International Conference on Computer Vision, pp. 432–439 (2003)
3. Adam, A., Rivlin, E., Shimshoni, I., Reinitz, D.: Robust real-time un-usual event detection using multiple fixed-location monitors. IEEE Trans. Pattern Anal. Mach. Intell. **30**(3), 555–560 (2008). http://doi.ieeecomputersociety.org/10.1109/TPAMI.2007.70825. bibsource DBLP http://dblp.uni-trier.de
4. Dutta, J., Banerjee, B.: Online detection of abnormal events using incremental coding length (2015). http://www.aaai.org/ocs/index.php/AAAI/AAAI15/paper/view/9923
5. Hasan, M., Choi, J., Neumann, J., Roy-Chowdhury, A.K., Davis, L.S.: Learning temporal regularity in video sequences. In: 2016 IEEE Conference on Computer Vision and Pattern Recognition (CVPR), pp. 733–742, June 2016
6. Karpathy, A., Toderici, G., Shetty, S., Leung, T., Sukthankar, R., Fei-Fei, L.: Large-scale video classification with convolutional neural networks. In: 2014 IEEE Conference on Computer Vision and Pattern Recognition, pp. 1725–1732, June 2014
7. Ji, S., Yang, M., Yu, K.: 3D convolutional neural networks for human action recognition. Pami **35**(1), 22131 (2013). http://www.ncbi.nlm.nih.gov/pubmed/22392705
8. Chong, Y.S., Tay, Y.H.: Abnormal event detection in videos using spatiotemporal autoencoder. In: Computer Vision and Pattern Recognition (cs.CV) (2017)
9. Popoola, O.P., Wang, K.: Video-based abnormal human behavior recognition a review. IEEE Trans. Syst. Man Cybern. Part C: Appl. Rev. **42**(6), 865–878 (2012)
10. Dollar, P., Rabaud, V., Cottrell, G., Belongie, S.: Behavior recognition via sparse spatio-temporal features. In: Proceedings - 2nd Joint IEEE International Workshop on Visual Surveillance and Performance Evaluation of Tracking and Surveillance, VS-PETS, vol. 2005, p. 65–72 (2005)
11. Kozlov, Y., Weinkauf, T.: Persistence 1D: Extracting and filtering minima and maxima of 1D functions. http://people.mpi-inf.mpg.de/weinkauf/notes/persistence1d.html. Accessed 05 Jan 2017
12. Lu, C., Shi, J., Jia, J.: Abnormal event detection at 150 fps in matlab. In: 2013 IEEE International Conference on Computer Vision, pp. 2720–2727, December 2013
13. Cong, Y., Yuan, J., Liu, J.: Sparse reconstruction cost for abnormal event detection. In: Proceedings of the IEEE Computer Society Conference on Computer Vision and Pattern Recognition, pp. 3449–3456 (2011)
14. Chen, C., Shao, Y., Bi, X.: Detection of anomalous crowd behavior based on the acceleration feature. IEEE Sens. J
15. Mahadevan, V., Li, W., Bhalodia, V., Vasconcelos, N.: Anomaly detection in crowded scenes. In: Proceedings of the IEEE Conference on Computer Vision and Pattern Recognition (CVPR), pp. 1975–1981 (2010)
16. Medel, J.R.: Anomaly Detection Using Predictive Convolutional Long Short-Term Memory Units. Masters thesis, Rochester Institute of Technology (2016). http://scholarworks.rit.edu/theses/9319
17. Ha, J., Park, J., Kim, H., Park, H., Paik, J.: Violence detection for video surveillance system using irregular motion information. In: 2018 International Conference on Electronics, Information, and Communication (ICEIC) (2018)
18. Hargude, S., Idate, S.R.: I-surveillance: intelligent surveillance system using back-ground subtraction technique. In: International Conference on Computing Communication Control and Automation (ICCUBEA)
19. http://homepages.inf.ed.ac.uk/rbf/CAVIARDATA1/
20. http://www.cse.cuhk.edu.hk/leojia/projects/detectabnormal/dataset.html

Fog Image Classification and Visibility Detection Using CNN

Shruti Chincholkar and Manoov Rajapandy[(✉)]

Vellore Institute of Technology, Vellore, India
sschincholkar@gmail.com, manoov.r@vit.ac.in

Abstract. The paper aims to find an approach for predicting the visibility percentage of the foggy images based on the factors like image brightness, luminance, intensity, and variance. The scope of using an image for predicting weather and providing information about the other details i.e. predicting if weather is foggy has increased. The idea is to utilize the data images to firstly classify them as being foggy or not. Secondly to detect the image visibility based on the image of a particular location. Thus the paper presents the implementation of convolutional neural network for the task of detecting and classifying the images into fog and non-fog. After getting the classification output, the brightness, luminance, variance and intensity of the images is calculated to find the visibility percent of the fog images.

Keywords: Convolutional neural network · Fog classification · Visibility prediction · Multiple linear regression

1 Introduction

Weather recognition [1] using machine learning has started to gain importance as weather forecasting is emerging as one of the most important areas of research. The analysis of fog weather images aim to provide the predictions about the visibility that can be used for many purposes in the near future. By this the prediction of bad weather and changes in the atmosphere can be done. In this paper, we present the fog weather recognition and detection of visibility based on the images which will perform following operations on the fog and non-fog dataset that will be provided to the algorithm.

The algorithm will take different weather images of 2 types of weather dataset-foggy and non-foggy images. The training set and the test set created by using multiple images so as to be sure about the proper classification given by the algorithm. Then the prediction can also be done using input by camera click or from the gallery of the mobile phone. The algorithm will predict information about each image by recognizing the category of fog and also detects the visibility.

The overall system consists of the use of convolutional neural network [2] algorithm in machine learning for the classification and visibility detection of the fog weather. The use of CNN is done as it provides better accuracy in terms of an image dataset and performs multiple iterations of layers on the images to classify correctly. After getting weather the data is fog or non-fog from CNN output, the brightness,

© Springer Nature Switzerland AG 2020
A. P. Pandian et al. (Eds.): ICICCS 2019, AISC 1039, pp. 249–257, 2020.
https://doi.org/10.1007/978-3-030-30465-2_28

luminance, variance and intensity of the images is calculated to find the visibility percent of the fog images.

2 Literature Survey

This section provides an overview about the referred papers for detecting visibility from fog image and understanding the CNN algorithm working with images.

A paper published in 2017 named '[3] A Neural network approach to visibility range estimation under foggy weather conditions'. This paper uses neural network for visibility distance estimation. This paper defines different approaches which can be applied for the visibility prediction. The consideration of visibility is done into 6 classes of distance range ranging from 60 m to 400 m. The achieved accuracy rate is 90.2% for classification.

From, [4] paper published as "The Application of Deep Learning in Airport Visibility Forecast" in 2017 for predicting the airport fog visibility using DNN in which the use of temperature, humidity and dew point as the factors to predict the visibility. Here the dataset used is in a tabular form and contains all the above factors as independent variables for predicting the visibility.

[1] Mark Holmstrom, Dylan and Christopher Vo published a paper in 2016 which applies machine learning to perform weather forecasting. Here the weather forecasting is done using machine learning models i.e. the linear regression model and a variation on a functional regression model. The outcomes of this experiment were that the linear regression model outperforms the functional regression model when the data was taken for 2 days but on weather data for 4–5 days allowed the functional model to outperform the linear regression model.

A paper [5] Weather classification with deep convolutional neural networks published in the conference of 2015 which uses CNN for providing the categorization of weather into two classes i.e. the sunny and cloudy. This paper achieved 82% normalized classification. It performs classification of the weather images into 2 categories.

The above used algorithms provide great results and have been implemented on image as well as tabular dataset. But are mostly working on tabular data or if image data is used it has very limited image dataset to provide good results. In this paper we enhance the use of images for fog detection and visibility prediction [6] based on an image in a better manner by using various factors of an image to predict the visibility percentage. The factors considered are calculated for multiple images so that we can get better accuracy in terms of visibility percent.

3 Algorithm Used

The convolutional neural network [7] is an approach to classify and detect the type of image as fog or non-fog. And the multiple linear regression is an approach used to use the factors calculated from the image like brightness, intensity, luminance and variance to predict the visibility of the fog in the image as there are multiple independent variables and one single dependent variable so multiple linear regression is a considerable approach.

3.1 Convolutional Neural Network

CNN is a class of deep neural network algorithms. There are four main layers that are organized in sequence to constitute the CNN, which are Convolutional, the Pooling (max pooling in this case), Flattening and the Fully Connected layer. CNN is defined as a sequence of layers so we create an object of sequence class and then add layers of CNN.

Convolution – Convolution step takes the input image in the form of pixels and applies multiple feature detectors of different pixel values and provides a feature map as an output we get a feature detector which is extraction of features i.e. the factors which are required of an image.

Max Pooling – The input to this step is the feature detector. This step involves reducing the size of the feature map. Here we take the time take max of 2*2 blocks. It also involves reducing the complexity of the model without decreasing its performance.

Flattening – To flatten featured map into column so that we can input this into an ANN for further processing. In this step, the conversion of the output of the max pooling layer to get one huge vector of input is done. The flattening is creation of one dimensional vector. We reduce the size by convolution and pooling step so that we can keep the spatial structure of the input image. Then in this step we flatten the image so that it can be the input to the neural networks.

Full Connection – The 1D vector of the flattening layer is the input to this step.

The process of creating a convolutional neural network begins to take a more complex turn in this layer. This step involves three layers: Input layer, Fully-connected layer and Output layer.

This step involves adding ANN to the CNN. The fully connected layers are used here which provides more accurate result. Main purpose of ANN is to combine features into more attributes that predict the classes even better. Here we have 2 classes – fog and non-fog for classification based on the features.

3.2 Multiple Linear Regression

It is the relation in straight-line between two or more variables. It has multiple independent variables that will predict single dependent variable instead of a single independent variable. Multiple regression is used for estimation of the dependent variable (y) in equation.

$$y = b_0 + b_1{}^*x_1 + b_2{}^*x_2 + \dots + b_n{}^*x_n$$

Dependent variable (DV)   Independent variables (IVs)

Multiple Linear Regression

Constant   Coefficients

The x's - independent variables and y - dependent variable. Multiple regression provides the relationship amongst one dependent variable and multiple independent variables. The intercept, b0 - point is the regression plane intersects Y axis. The bi - slopes of regression plane towards direction of xi. Here the independent variables are brightness, luminance, variance, fog/non-fog, intensity and the dependent variable is visibility percentage (Fig. 1).

4   Architecture Diagram

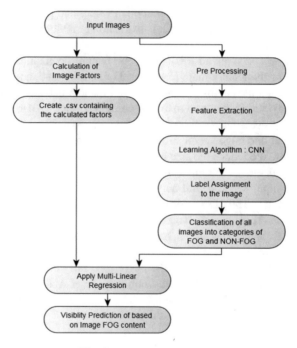

Fig. 1. Architecture diagram

Input images – The images taken as input are of two kinds – the fog data and the non-fog dataset [8]. There are two categories in which the data is divided so as to apply the CNN and the multiple linear regression. The training set and the test set. The dataset contains all images in format jpg and have numbering such as fog1.jpg. The fog and non-fog images Figs. 2 and 3.

Fig. 2. Foggy image   **Fig. 3.** Non-fog image

Convolutional neural network – The convolutional neural network classifies the data into two categories [9] – fog and non-fog. The CNN algorithm is very widely used for classification of the images into categories. Here we use CNN as we have an image dataset and it provides a very good accuracy rate in terms of classifying images. The data is taken in the form of images. The 4 layers of CNN are then applied to the images to classify the images into 2 categories.

Image factors computing – The factors computed to predict the visibility are values of brightness, luminance, variance and intensity of the images.

Brightness – The brightness [10] is the insight caused by luminance of graphic target. It is not proportional to luminance. It depends on the R, G and B values of colour appearance constraints of colour appearance models. Brightness differs to the term Lightness. Brightness in this case is a measure of red, green and blue values of an image based on which we calculate brightness of that particular image.

$$Brightness = 0.2126^*R + 0.7152^*G + 0.0722^*B$$

Luminance – It is a measure of the amount of light that falls on a surface in the image. It describes the amount of light that permits through or that is reflected from a specific area in an image, and falls within a given angle. In calculative terms it is the average value of the highest and the least intensity of an image from an RGB image [11].

$$Luminance = ((1/2)^* (\max(r, g, b) + \min(r, g, b)))$$

Intensity – Intensity refers to amount of light or numerical value of a pixel. For example, in colored images, it's depicted by the red, blue and green color level value at each pixel and then finding the mean of all the pixels to get resultant value of intensity.

Variance – Foggy/misty images have very low horizontal variance than the other non-foggy day images. Variance ($\sigma 2$) is a measure of how distant each number in set is from mean. It is measured by taking each pixel and checking how much it varies from the neighboring pixel. In case of fog, the atmosphere color is similar in most of the regions in image i.e. white pixels is more so variance value is less [12].

The variance (σ2) is a measure of how distant each value in the data set is from the mean.

After getting the values of all these factors we compute the visibility using the multiple linear regression. The values of the computed factors of each image is first stored in a csv so that we can take all the values corresponding to every image one by one to predict the visibility of the image in case of the foggy images. The CNN makes it easier by classifying the images into fog and non-fog so that we can compute the visibility values of only the foggy images which are necessary for the project.

5 Proposed Method

ALGORITHM

Step 1: Collect a dataset containing images which are downloaded from the internet. There are two different classes of dataset- fog and non-fog. The data is divided into training set and test set.

Step 2: Propose optimization of Convolutional neural network architecture for solving the task of classifying the data into two categories fog and non-fog.

Step 3: Then secondly take the same images and compute the values of brightness, luminance, variance and intensity of the images.

Step 4: Take the above computed values and insert it into a csv file.

Step 5: Use the computed values and the output of the CNN i.e. fog or non-fog to calculate the dependent variable visibility percent using the Multiple Linear Regression.

Step 6: The final output is the calculated the visibility of the image based on the above factors.

The CNN algorithm is used to classify the data and multiple linear regression is used in predicting the visibility.

The criteria followed to predict the visibility is based on –

1. Brightness – the value of brightness increases on increase in fog content in the image.
2. Luminance – the value of luminance increases on increase in fog content in the image.
3. Variance – the value of variance decreases on increase in fog content in the image.
4. Intensity – the value of intensity increases on increase in fog content in the image.

6 Result

As a preliminary experiment has been performed on two kinds of weather for checking the classification of weather into two categories non-foggy and foggy applying convolutional neural network by having 3 convolution steps and epoch value of 25. In the preliminary study, the dataset of training and test set is divided into two categories - foggy and non-foggy. The experiment was performed with 200 images in training set 80 in test set and also 160:80 which is 50% of both training and test set which provided an accuracy of 76% with loss of 0.9 which is high. The training dataset has 160 images- 80 of each types and test set contains 50 images – 25 of each type. But the image ratio of 16:5 provided an accuracy of 82% over the test set. This is the result of the CNN which classifies into fog and non-fog. The output of this data is used to predict the visibility of images using multiple linear regression.

Fog visibility percentage table in Fig. 4 -

Brightness	Luminance	Intensity	Visibility percentage
bright>=150	luminance>=60	intensity>=150	20%
150>bright>=100	60>luminance>=50	150>intensity>=100	40%
100>bright>=80	50>luminance>=40	100>intensity>=85	60%
80>bright>=60	40>luminance>=20	85>intensity>=60	100%

Fig. 4. Fog visibility

The output of the multiple linear regression is visibility of the fog images [13].

The multiple linear regression has 4 independent variables which are brightness, luminance, intensity and variance which are used to predict the dependent variable visibility percentage in values from 20% to 100% based on the values of the independent variables.

The first variable is brightness comparison to predict visibility percent (Fig. 5) –

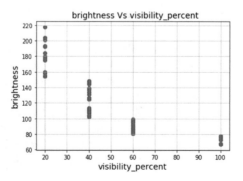

Fig. 5. Brightness vs visibility_percent

The second variable is luminance comparison to predict visibility percent (Fig. 6) –

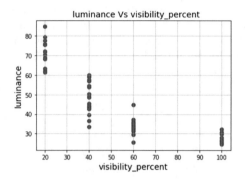

Fig. 6. Luminance vs visibility_percent

The third variable is intensity comparison to predict visibility percent (Fig. 7) –

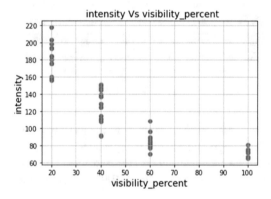

Fig. 7. Intensity vs visibility_percent

7 Conclusion and Future Work

The use of CNN to predict the class of the images leads to better classification. Then multiple linear regressions are used to calculate the predicted values of visibility. The predicted values of visibility are in range from 20% to 100% based on the values of brightness, luminance and intensity. The paper helps in visibility prediction from an image by considering various factors of the image. The further work can include the use of various other image processing techniques to perform deeper analysis of the image for finding the exact visibility distance. The use of larger dataset can be done to achieve accurate results, which is also an approach for future enhancement to this paper.

References

1. Holmstrom, M., Liu, D., Vo, C.: Machine learning applied to weather forecasting. Meteorol. Appl. (2016)
2. Convolutional neural network based feature extraction for IRIS recognition **10**(2), 65–78 (2018)
3. Chaabani, H., Kamoun, F., Bargaoui, H., Outay, F., Yasar, A.U.H.: A neural network approach to visibility range estimation under foggy weather conditions. Procedia Comput. Sci. **113**, 466–471 (2017)
4. Zhu, L., Zhu, G., Han, L., Wang, N.: The application of deep learning in airport visibility forecast. Atmos. Clim. Sci. **07**, 314 (2017)
5. Elhoseiny, M., Huang, S., Elgammal, A.: Weather classification with deep convolutional neural networks, July 2015
6. Jeevan, S., Usha, L.: Estimation of visibility distance in images under foggy weather condition. Int. J. Adv. Comput. Electron. Technol. **3**, 11–16 (2016)
7. Dey, A.: Machine learning algorithms: a review. Int. J. Comput. Sci. Inf. Technol. **7**(3), 1174–1179 (2016)
8. Carlos, J., Guerra, V., Khanam, Z., Ehsan, S., Stolkin, R., Mcdonald-Maier, K.: Weather classification : a new multi-class dataset, data augmentation approach and comprehensive evaluations of convolutional neural networks (2018)
9. Lu, C., Lin, D., Jia, J.: Two-class weather classification. IEEE Trans. Pattern Anal. Mach. Intell. **39**(12), 2510–2524 (2017)
10. Bezryadin, S., Bourov, P., Ilinih, D., Francisco, S.: Brightness Calculation in Digital Image Processing
11. Jyothi, G.: Luminance based conversion of gray scale image to RGB image. Int. J. Comput. Sci. Inf. Technol. Res. **3**(3), 279–283 (2015)
12. Kumar, V., Gupta, P.: Importance of statistical measures in digital image processing. Int. J. Emerg. Technol. Adv. Eng. **2**(8), 56–62 (2012)
13. Chaabani, H., Werghi, N., Kamoun, F., Taha, B., Outay, F., Yasar, A.U.H.: Estimating meteorological visibility range under foggy weather conditions: a deep learning approach. Procedia Comput. Sci. **141**, 478–483 (2018)

A Fuzzy-PSO Based Control of DSTATCOM for Voltage Regulation in Distribution Grids

B. Durga Hari Kiran[(✉)], Y. Hareesh Kumar, and T. Navya

Department of Electrical and Electronics Engineering,
Vaagdevi College of Engineering, Bollikunta(V), Warangal 506005, India
harikiran_bd@vaagdevi.edu.in,
bdurgaharikiran@yahoo.co.uk

Abstract. Voltage regulation is one of the major power quality issues in Low voltage distribution grids. With strict voltage regulation grid code in place, the usage of voltage-controlled DSTATCOM has increased. This paper deals with modelling and simulation of Fuzzy-Particle Swarm Optimization (Fuzzy-PSO) based control of DSTATCOM to mitigate power quality issues in Low voltage distribution grids. The proposed control is a temporary solution for voltage regulation in order to meet grid code. The model consists of two stages, power stage and control stage. The concept Fuzzy-PSO is used to operate the DSATCOM in order to minimize distribution of unnecessary reactive power compensation. Simulation results establish the voltage regulation features of Fuzzy-PSO algorithm for DSTATCOM for both linear and nonlinear loads. The results are obtained from the MATLAB/SIMULINK environment.

Keywords: DSATACOM · Distribution grids · Fuzzy · PSO · Voltage regulation

1 Introduction

Consumers wired to distribution grid end are affected with low voltages and may encounter inadequate voltage regulation. Energy companies confined to due dates (A fortnight to Three months) need to replenish the voltage levels at PCC (If the voltages are outside the admissible levels). This duration is essential for enduring remedies like grid restructuring, providing reactive power support etc. If the provider failed to confine voltage regulation to certain level, then they need to refurbish amount to the consumer during poor voltages levels. Voltage regulator devices can be utilized in order to avoid such refunds. However, the faster control of such devices has become an issue in present energy markets.

The proposed method aid in improving the grid power quality to meet the grid code, along with improving the voltage at Point of Common Coupling (PCC). The major cause of poor voltage regulation in low voltage distributions grids is, farther distance of PCC from grid transformer and absence of voltage information of the grid. In [1–13] voltage regulation is been achieved by using a voltage controlled DSTAT-COM. Configuration proposed in literature is given in Fig. 1. One advantage of shunt connection is that, it minimizes the interruptions due to plugging in or out of voltage regulator. Further, grid or load unbalances can be met at PCC with voltage regulation

© Springer Nature Switzerland AG 2020
A. P. Pandian et al. (Eds.): ICICCS 2019, AISC 1039, pp. 258–265, 2020.
https://doi.org/10.1007/978-3-030-30465-2_29

by using DSTATCOM. The swift load changes in grid can be nullified at PCC by controlling voltage waveform with DSTATCOM. Further, voltage-controlled DSTATCOM provides low impedance path which minimizes harmonic content by decoupling the grid and the loads.

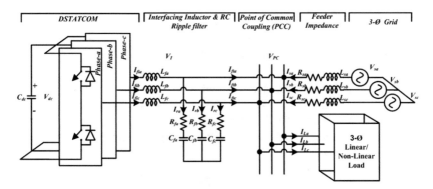

Fig. 1. Distribution system along with DSATCOM connected at PCC.

In [1] a new control algorithm for DSTATCOM was proposed for distribution system. A generalized Neural Network is used to give switching gate pulses to DSATCOM. Performance of the proposed control techniques are tested on a 3 Phase distribution system. The proposed method was more flexible to load dynamics. In [2] authors proposed Adaptive Volterra second-order filter to control DSTATCOM in distribution network. The proposed control algorithm successfully could able to reduce harmonics and improve power factor for both linear and non-linear loads. In [3] authors proposed a learning based anti Hebbian control algorithm to control DSTATCOM in a 3 Phase distribution network. The control algorithm was tested against reactive power compensation and harmonics reduction. In this work, with suitable load perturbations DC bus voltage was regulated. In [4] variable forgetting factor recursive least square based control algorithm is proposed. In this work weighted values of active and reactive power components are estimated for effective operation of DSATCOM. For Voltage control DSATCOM angular position reference is required according to [5]. Real time implementation of Adaptive Neuro Fuzzy Inference System Least Mean Square Based Control algorithm for DSTATCOM was proposed in [6]. A new online control strategy for DSTATCOM based on reinforced learning algorithm was proposed in [7]. The new controller reduces harmonics and provides reactive power compensation in Microgrids. In [8] a SPV-DSTATCOM was controlled by a hybrid control technique involving variable step size–least mean square–least mean fourth algorithm. In this work power quality improvement was considered as an objective. In [9] abnormal grid conditions were addressed with improved proportionate normalized least mean square algorithm to control DSTATCOM. The cases of steady state and load unbalancing were studied on a test system. In [10] a neural network based control algorithm for DSTATCOM under non-ideal source voltage and varying load conditions is proposed.

The proposed methods in the literature, few are mathematical and few are heuristic methods. The basic drawback of these methods is that methods like neural network needs huge data set, which needs good training. Methods like PSO and Genetic Algorithm (GA) often fall in local minimum. Hence, there is need of intelligent control which combines the optimum solution search method of PSO and Fuzzy classifier to control DSTATCOM.

2   Voltage Regulation with DSTATCOM

DSTATCOM is a device connected in shunt at PCC for the improvement of power quality, the basic structure of DSTATCOM is given in Fig. 2. With the help of DSTATCOM connected to PCC in distribution system real and reactive power can be accustomed by changing the inverter voltage magnitude as well as difference of angle between bus and inverter output.

$$P = \frac{V_{PC} V_I Sin\ \theta}{X} \tag{1}$$

$$Q = \frac{V_{PC}(V_{PC} - V_I Cos\ \theta)}{X} \tag{2}$$

Angle θ is zero for steady state operation. Let V_{PC} is less than inverter voltage V_I then reactive power is injected in to the bus from DSTATCOM. By adjusting V_I reactive power flow can be regulated from DSTATCOM.

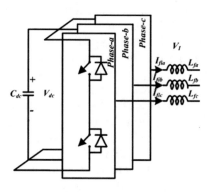

Fig. 2. DSATCOM

In this work, alternating voltage is generated using voltage source converter. A constant phase angle is maintained to generate AC voltage in phase with bus voltage.

With system voltage as reference, Phase Locked Loop (PLL) synchronizes the pluses to DSTATCOM.

3 Fuzzy-PSO

Particle Swarm Optimization (PSO) refers to the activities of bird flocking [14]. Optimization problem consists of individuals which contribute latent solution. The classical PSO (explained in [15]) fails due to its sensitivity to the control parameters which includes inertia weight and learning factors. PSO employs a searching process, which makes it difficult to take the control parameters dynamically. So, these control parameters are often taken as constants. In Fuzzy-PSO, the control variables get adjusted adaptively. In FPSO, control variables are updated by Fuzzy controller after each iteration. The fitness change between the next particle to the local best particle (X_i) as well the difference between current particle to the global best particle (Y_i) serves as an input to the fuzzy controller. local best particle and global best particle can be normalized using Eqs. 3 and 4.

$$X_i^{best} = \frac{f(x_i) - f(X_i)}{f(Y^w) - f(X_i)} \tag{3}$$

$$Y_i^{best} = \frac{f(x_i) - f(Y)}{f(Y^w) - f(Y)} \tag{4}$$

Where Y^w is the global worst particle. A Gaussian membership function [16] is used in this work.

4 Simulation Results

An improved control algorithm is evaluated and tested using MATLAB/SIMULINK on a three phase DSTATCOM to regulate voltage at PCC. The control logic is tested for both Linear and Non-Linear loads. The proposed work consists of a Fuzzy-PSO based control strategy for DSTATCOM. Distorted load currents are considered.

Below cases are studied to establish the importance of the proposed method.

1. Without DSTATCOM with balanced linear load with ideal grid condition.
2. With DSTATCOM with balanced Non-Linear Load with ideal grid condition.
3. With DSTATCOM with balanced Non-Linear Load with ideal grid condition and PI control of DSTATCOM.
4. With DSTATCOM with balanced Non-Linear Load with ideal grid condition and Fuzzy-PSO control of DSTATCOM.

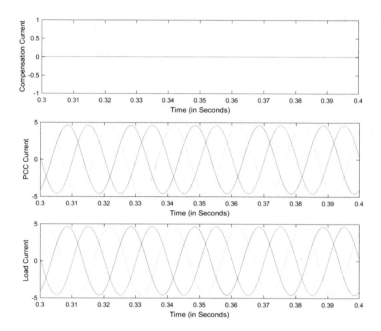

Fig. 3. Case 1: Various currents without DSATCOM with linear load

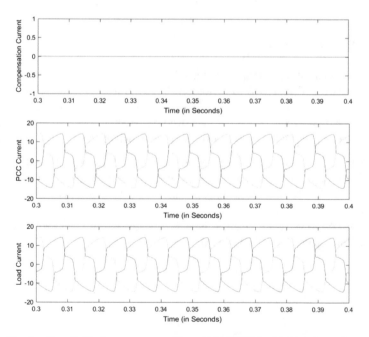

Fig. 4. Case 2: Various currents without DSATCOM with Non-linear load

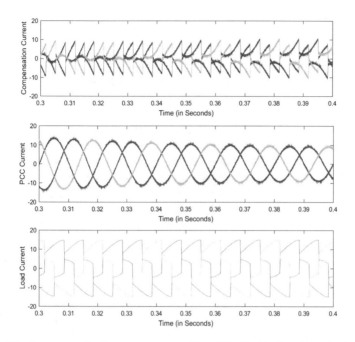

Fig. 5. Case 3: Various currents with DSATCOM with Non-linear load

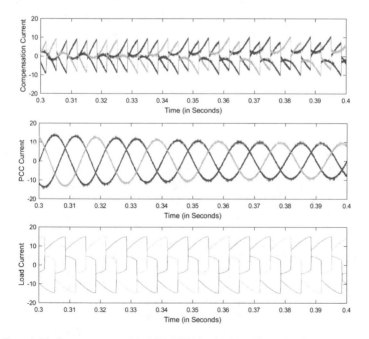

Fig. 6. Case 4: Various currents with DSATCOM with Non-linear load and with Fuzzy-PSO

Table 1. THD comparison.

Case study	THD
With PI control for DSTATCOM	23%
With proposed method Fuzzy-PSO	10%

The above table shows that the switching with Fuzzy-PSO is smoother when compared with PI controller as it is evident from the values of THD. The THD has been reduced to 10% in Fuzzy-PSO when compared with PI controller which is at 23%. By reducing THD presence of harmonics can be reduced which improves voltage regulation by eliminating the effect of non-linear loads on the system. The proposed method outperforms the classical method. As a future scope this proposed controller can be further explored to mitigate other power quality issues like voltage sag, swell etc.

5 Conclusion

In this paper, a new control strategy to control a three phase DSTATCOM to regulate voltage at PCC has been proposed. The proposed work comprises of control strategy using Fuzzy-PSO, traditional loop and dc bus regulation loop. Simulation results extend the voltage regulation capability, balanced voltages at PCC due to DSTAT-COM. The proposed Fuzzy-PSO control algorithm could able to control amplitude loop to meet grid code. The proposed solution for voltage regulation is a shunt connected one which is connected to low voltage distribution grids to fore see all smaller interruptions. The no information of grid voltage and impedance information is simulated. Also, the proposed solution provides balance and low THD voltages to the customers.

References

1. Ahmad, M.T., Kumar, N., Singh, B.: Generalised neural network-based control algorithm for DSTATCOM in distribution systems. IET Power Electron. **10**(12), 1529–1538 (2017)
2. Ahmad, M.T., Kumar, N., Singh, B.: AVSF-based control algorithm of DSTATCOM for distribution system. IET Gener. Transm. Distrib. **11**(13), 3389–3396 (2017)
3. Arya, S.R., Singh, B., Chandra, A., Al-Haddad, K.: Learning-based anti-hebbian algorithm for control of distribution static compensator. IEEE Trans. Ind. Electron. **61**(11), 6004–6012 (2014)
4. Badoni, M., Singh, A., Singh, B.: Variable forgetting factor recursive least square control algorithm for DSTATCOM. IEEE Trans. Power Deliv. **30**(5), 2353–2361 (2015)
5. Ledwich, G., Ghosh, A.: A flexible DSTATCOM operating in voltage or current control mode. IEE Proc. – Gener. Transm. Distrib. **149**(2), 215–224 (2002)
6. Badoni, M., Singh, A., Singh, B.: Adaptive neurofuzzy inference system least-mean-square-based control algorithm for DSTATCOM. IEEE Trans. Ind. Inform. **12**(2), 483–492 (2016)
7. Bagheri, M., Nurmanova, V., Abedinia, O., Salay Naderi, M.: Enhancing power quality in microgrids with a new online control strategy for DSTATCOM using reinforcement learning algorithm. IEEE Access **6**, 38986–38996 (2018)

8. Beniwal, N., Hussain, I., Singh, B.: Hybrid VSS–LMS–LMF based adaptive control of SPV-DSTATCOM system under distorted grid conditions. IET Renew. Power Gener. **12**(3), 311–322 (2018)
9. Beniwal, N., Hussain, I., Singh, B.: Implementation of the DSTATCOM with an i-PNLMS-based control algorithm under abnormal grid conditions. IEEE Trans. Ind. Appl. **54**(6), 5640–5648 (2018)
10. Jayachandran, J., Murali Sachithanandam, R.: Neural network-based control algorithm for DSTATCOM under nonideal source voltage and varying load conditions. Can. J. Electr. Comput. Eng. **38**(4), 307–317 (2015)
11. Yada, H.K., Murthy, M.S.R.: An improved control algorithm for DSTATCOM based on single-phase SOGI-PLL under varying load conditions and adverse grid conditions. In: 2016 IEEE International Conference on Power Electronics, Drives and Energy Systems (PEDES), Trivandrum, pp. 1–6 (2016)
12. Yada, H.K., Murthy, M.S.R.: Reference current extraction for three-phase four-wire h-bridge DSTATCOM using SOGI-PLL algorithm. In: 2014 Annual IEEE India Conference (INDICON), Pune, pp. 1–6 (2014)
13. Yada, H.K., Murthy, M.S.R.: A new topology and control strategy for extraction of reference current using single phase SOGI-PLL for three-phase four-wire shunt active power filter. In: 2014 IEEE International Conference on Power Electronics, Drives and Energy Systems (PEDES), Mumbai, pp. 1–6 (2014)
14. Kennedy, J., Eberhart, R.: Particle swarm optimization. In: Proceedings of IEEE International Conference on Neural Networks, vol. 4, pp. 1942–1948, November 1995
15. Clerc, M., Kennedy, J.: The particle swarm - explosion, stability, and convergence in a multidimensional complex space. IEEE Trans. Evol. Comput. **6**(1), 58–73 (2002)
16. Zheng, Y., Dong, Z.Y., Luo, F.J., Meng, K., Qiu, J., Wong, K.P.: Optimal allocation of energy storage system for risk mitigation of discos with high renewable penetrations. IEEE Trans. Power Syst. **29**(1), 212–220 (2014)

Efficient Techniques of Transformation from Biogas to Bio-methane for IoT Based Photovoltaic Biogas Hybrid System

Amandeep Kaur Bhangal and Jyotsna Sengupta[✉]

Department of Computer Science, Punjabi University,
Patiala 147002, Punjab, India
amandeepbhangal@ymail.com, jyotsna.sengupta@gmail.com

Abstract. Biogas is formed when organic biodegradable substances are degraded through bacteria under some anaerobic conditions. Methane is considered as the major substance of biogas but it also includes various number of impurities for instance carbon dioxide (CO_2), carbon monoxide (CO), oxygen (O_2), hydrogen sulfide (H_2S), water (H_2O) and dust particles. Transformation is a proper way to remove these contaminates. This conversion offers advantages to improve the quality for a number of anaerobic digestion plants by increasing its life validity. Two technologies are mainly used to get pure Bio-Methane. Cleaning is a primary technique, which has removal of impurities and upgrading process regulates the calorific value of gas. Techniques for raw biogas cleaning are dissimilar in functioning, quality conditions and the efficiency of incoming gas. Upgrading technology is usually performed in the scientific way to achieve the standard value to utilize as fuel for electrification and vehicles. This paper presents a systematical review of major analysis evolution on technologies for conversion of raw biogas to Bio-Methane.

Keywords: Water scrubbing · Raw biogas · Carbon dioxide · Hydrogen sulfide · Pressure swing adsorption · Biogas upgrading · Gas liquid adsorption

1 Introduction

Photovoltaic Biogas Hybrid power generation system for rural areas is purely based on renewable resources, which are solar energy and biogas energy. However, according to the current state of affairs, world's energy requirements heavily depend on precious and finite resources such as coal, natural gas, and petroleum crude oil. These fossil fuels take immeasurable years within the earth. Nowadays, some interesting topics such as reduction in fossil fuels, energy security, and emission of greenhouse gases have encouraged renewable energy studies. A number of renewable energy resources which are existing in the form of solar energy, hydro power, biogas power and wind energy. Until now, biogas has not become a central energy source, but its production is a valuable for waste management treatment that is increasingly developing into an environmentally way and cost effective sustainable process especially in rural areas. The production of biogas is done by the anaerobic digesters using biological degradation of organic material. Organic compounds, for instance animal manure, food wastes and agriculture waste are generally used

© Springer Nature Switzerland AG 2020
A. P. Pandian et al. (Eds.): ICICCS 2019, AISC 1039, pp. 266–273, 2020.
https://doi.org/10.1007/978-3-030-30465-2_30

in the biogas production process. Biogas is a mixture of methane (CH_4, 50% to 70%) [1], carbon dioxide and other impurities for instance water hydrogen, carbon monoxide, ammonia, oxygen, nitrogen, siloxanes and dust particles as given in Table 1 [10]. However, direct use of raw biogas can produce a number of hardware problems such as corrosion and low calorific value, which can reduce the efficiency of hybrid system. Transformation is a conversion of biogas to Bio-Methane to remove the impurities and improve the quality of biogas. The Bio-Methane production process consists of two steps which are listed under biogas cleaning and biogas upgrading as shown in Fig. 1. The treatment of untreated biogas is usually done under cleaning technology, which is used for elimination of water and dangerous gases to keep away from corrosion in combustion engine and pipes. In upgrading process, the raw biogas is split into two gas streams, which is compounded of methane (95% to 97%) and carbon dioxide (1% to 3%) [1]. Moreover, carbon dioxide removal is essential to regulate the calorific value of biogas, but only for vehicle. Bio-Methane has need of more than 97% purity for vehicle use [14]. However, when biogas is used for heating and electricity generation, only water and H_2S removal is required. Each and every technology has its own specific advantages and drawbacks. This paper reviews of some transformation techniques and it shows that no technology is the optimal solution of biogas transformation. The selection of the best suitable technique depends on the quality of the raw biogas.

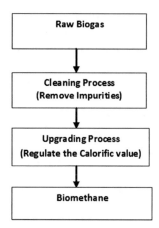

Fig. 1. Flow chart of transformation of biogas to Bio-Methane

2   Water Removal Techniques from Biogas

Biogas contains approximately 10% water and removal of water is the simplest task in contrast of CO_2 and H_2S. Generally, water quantity in raw biogas usually depends on the temperature which can be altered by a few parameters. For example water content is 6% at 30 °C but decrement in temperature can reduce the water quantity in biogas [9]. Three technologies are now in trends for removal of water; cooling and compression, absorption and adsorption.

2.1 Cooling and Compression

The simplest way of removing water from biogas is cooling of biogas. This method has limit of dew points to 0.5 °C and to get lower dew points the raw biogas to be compressed before freezing. The lower dew point and higher pressure are required in this method. The condensed water are entrapped and removed. To achieve compression with cooling of the biogas, demisters and cyclones to the biogas pipe system is applied [10].

2.2 Absorption

Absorption of water from biogas is done in two ways; absorption through glycol solution and absorption with hydroscopic salt. Diethylene glycol, triethelene glycol and ethylene glycol are commonly used as a glycol solution. The glycol solution requires regeneration unit, where a temperature of 200 °C is used [7]. In case of hydroscopic salt, salt is dissolved as it absorbs water from the raw biogas. However, salt is not regenerated and new salt is to be used for further process.

2.3 Adsorption

Generally, water and carbon dioxide (CO_2) react with each other and produce different type of acids, which can become harmful for equipments. Under this method, molecules of liquid are gathered on the top surface of element, which is utilized as adsorbent. This process is slightly dissimilar from absorption of water, where fluid is dissolved. Moreover, silica, aluminum oxide, magnesium oxide, zeolite or activated carbon elements are used for adsorption [9]. These elements are regenerated with the help of changes in different parameters such as pressure and temperature.

Table 1. Biogas impurities and their range, average and impact

Impurities of biogas	Range	Average	Impact
Carbon dioxide (CO_2)	25–55%	35%	High volume of CO_2 reduces to calorific value
Water (H_2O)	0–10%	3.1%	Water and CO_2 reaction to produce acids. Due to lower energy and high pressure, calls to freezing and condensation problems
Nitrogen (N_2)	0.01–5%	1%	It reduces to flame propagation speed
Hydrogen (H_2)	0–1%	<1%	Combustion leads to corrosion in engines
Oxygen (O_2)	0.01–2%	0.3%	Explosive mixture due to high volume of O_2 in biogas
Hydrogen sulphide (H_2S)	0–30 000 mg/m^3	500 mg/m^3	It is caused to occurring corrosion in pipe. If H_2S available high amount on produced biogas, it may produce to sulphuric acid from utilization equipment during their combustion. This gas is heavily toxic and causing serious health risks
Ammonia (NH_3)	0.01–2.5 mg/m^3	0.7 mg/m^3	The high volume of ammonia can cause to inhibition of methane formation at the digester
Siloxanes (R_2SiO)	0–0.02%	0.02%	Combustion in process causes to microcrystalline quartz and SiO_2. It also leads to deposition at spark plugs

3 Removal of H_2S, CO_2 and Other Impurities

CO_2 can reduce the heat of biogas and H_2S produce harmful toxic which arise corrosion in the system equipment. Biogas upgrading process regulates to volumetric energy of gas through a removal of carbon dioxide from biogas to produce methane. Some desulphurization processes and upgrading processes remove oxygen and other impurity components as given impurities in Table 1, up to some extent. Hence, the selection of upgrading process for biogas should be done with further analysis. As hydrogen sulfide is removed during the cleaning process, but some techniques are used for removal of carbon dioxide as well as hydrogen sulfide and other pollutants [9].

3.1 Water Scrubbing

Water scrubbing is constructive and most popular method for removal of pollutant such as carbon dioxide (CO_2) and hydrogen sulfide (H_2S). H_2S is naturally present in natural gas, crude petroleum and degradation process of organic material. CO_2 and H_2S have no color and efficiently soluble in water rather than methane [9]. Water absorption or scrubbing technology is a totally physical process and during this process, raw biogas is cleaned up with water. This process is pressurized to raw biogas and fed in scrubber column, which is also filled with water. CO_2 and H_2S are dissolved with water and processed water can be recycled to recover more amount of methane. Regeneration or recycling is achievable with the help of de-pressuring or stripping with air. If the volume of H_2S is very high, then stripping with air is not recommended because it can cause of any operational problem [1]. However, it has a requirement of a large quantity of water which is a major drawback of water scrubbing. So, waste water that used in treatment plants can be utilized to make it cost efficient technology.

3.2 Polyethylene Glycol Absorption

Polyethylene Glycol Scrubbing is largely similar to water scrubbing, but the dissimilarity is that impurities are absorbed through organic solvent. Selexol and many more acids are used in the form of solvent to remove impure gases such as H_2S and CO_2. H_2S and CO_2 are more easily soluble in selexol where outcomes in a less demand of solvent and decreased pumping. Moreover, adsorption with selexol is also removed to halogenated hydrocarbons, which are always intended with recirculation [7]. Raw biogas should be compressed up to 6–7 bars and maintained temperature within 10–20 °C. A separate dehydration step requires for producing biogas mixture because of hygroscopic characteristics and desulphurization during this method. Typically the production of gas from this technology has 93–98% methane [10]. The major advantage of this absorption is that it has no need of any special chemical for removal of impurities. However, large quantities of water, regeneration problem, limited removal of H_2S, are under the main drawbacks. Moreover, CO_2 decrease pH values of the solvent and corrosion to the equipment caused by H_2S.

Table 2. Advantages and drawbacks of different transformation techniques

Method	Advantages	Disadvantages
Water scrubbing	Removal of H_2S and CO_2 It is cheap when easier availability of water It has low losses of CH_4	It required high volume of water
Polyethylene glycol	It removes to CO_2 and S elements such as H_2O, H_2S, HCN and NH_3 Purity of methane is up to 95–98% It is more energetic and favorable than water It is also regenerative	Expensive investment Heater is necessary for regeneration Reduce the operation when glycol dilution with water
Pressure swing adsorption	Removal of H_2S and CO_2 It uses low energy and regenerative technique It is useful for small capacities	Expensive technique To save CH_4 losses, extensive process control needed
Membrane separation	Removal of H_2S and H_2O It is easy to use Without any increase in cost, it treats little gas flows	Membrane selectivity is low It compromises with purity of CH_4 and quantity of improved biogas It requires a number of processes to gain high purity More losses of CH_4
Cryogenic separation	Remove to H_2S, N_2, O_2 Purity of methane is up to 95–98% CO_2 and CH_4 in great purity	It is too expensive and complicated process and CO_2 can remain in the CH_4.

3.3   Pressure Swing Adsorption

Pressure swing adsorption technique is one of the most practical technologies to produce Bio-Methane and impure gases are separated by using adsorption. Under this process, raw biogas penetrates through a top surface of solid or adsorbent and unwanted pollutants are trapped with the help of volume of molecular sieve [2] and [12]. Usually, a number of activated carbons are used for this adsorption such as zeolites and so on. Normally, different types of properties are changed to improve adsorption for example high pressure and low temperature. This material selectively absorbs CO_2, H_2S, NH_3 and H_2O, and small part of N_2 and O_2 are also adsorbed. However, the most of the applications, H_2S and H_2O are separated before the gas mixture entering into the adsorption columns [13]. Only a little volume of methane is absorbed and the greater part of methane goes through adsorbents without adsorption. Generally, the Bio-Methane gas is released from the adsorption tower with an uninterrupted rate. Adsorbent material is regenerated by stepwise decreased pressures and flushes out with raw biogas or produced Bio-Methane. During this process, a gas which is emitted as the byproduct of a chemical process is leaving the absorber. Afterwards, the pressure is increased again with raw biogas or Bio-Methane. The Bio-Methane gas has methane with purity over 96% as shown in Table 2 [4].

3.4   Membrane Separation

Membrane separation technique is used to separate pollutants such as carbon dioxide (CO_2) and hydrogen sulfide (H_2S). It is a new technique which is rapidly growing in the production of Bio-Methane. The continuing advancement in the research field of membrane separation technique is justified to make it better technique in the cleaning of biogas. Basically, two methods are used for the membrane separation technique, which are high pressure gas separation and gas liquid adsorption membrane [10].

3.4.1   High Pressure Gas Separation

This technique is used to remove hydrogen, hydrogen sulfide and moisture of raw biogas through membrane separation technique. Here, pressurized gas is cleaned through carbon filtering that applies a bed of activated carbons for impurities. The bed of carbon is pursued by a filter and heater. Untreated biogas is passed out from three states for transformation of biogas to achieve 96% production of methane [12]. Moreover, the impure gas, which gets from two states, is recycled to recover methane gas. The waste gas of third state is used in steamer to get more methane, which is approximately 10–20%.

3.4.2   Gas Liquid Absorption Membranes

Gas liquid absorption through membrane separation is a new technique for cleaning of raw biogas. Micro-porous Hydrophobic Membranes (MHMs) is the main element of this physical process which is a flat sheet made of polymeric material. It separates the gasses from a liquid. Gas molecules flow only in a single direction, which can be easily diffused with the help of the membranes. These molecules can be absorbed with the help of liquid on the opposite side which is in counter current [7]. This technique reduces construction cost because it works at atmospheric pressure. Membrane separation technique is the best method for treatment of landfill gasses.

3.5   Cryogenic Separation

CH_4 and CO_2 both are dissimilar due to condensing temperatures and these can be separated from the mixture by their condensation and distillation. The cryogenic separation uses different boiling points and sublimation points of gasses from the point of separation. Before, applying cryogenic technique, water and H_2S should be separated to conquer numerous problems for example freezing and so on. During this process when CH_4 is condensed, N_2 and O_2 can also separate [12]. The cryogenic technique requires compressing untreated raw biogas to a high pressure, just liking as 200 bars. A great quantity of energy, approximately from 5 to 10% of the Bio-Methane should be generated during this process [4]. However, the cryogenic separation technique leads to enormous advantages to produce liquid and great purity of Bio-Methane. The loss of CH_4 is small in volume, generally less than 1% and purity is approximately 99% [13]. Nowadays, the technique of cryogenic separation is going on progress, but few profitable plants are previously in operation. Cryogenic technique takes little volume of energy cost to achieve liquid Bio-Methane.

4 Conclusion

Biogas is utilized as energy fuel for combustion engine for power generation. However, direct supply of raw biogas may cause much higher losses, so clean biogas in the form of Bio-Methane should be used as a fuel for engine. This paper is a scientifically analysis of cleaning and upgrading techniques, requirement of gas quality for a variety of utilization purposes. To perform these two jobs, a number of different techniques have been discussed in this paper. The better technique can be selected according to the maximum achievement of methane volume, investment cost, operation cost and maintenance cost. Generally, condensation method is used in cleaning process to remove water to avoid corrosion. Water absorption is a cheap method when water is available to remove H_2S and CO_2. The pressure swing adsorption technique uses less energy and produces a high efficient density of methane, which is approximately 95-98%. It is an efficient method for removal of CO_2 and useful technique for small capacity impurities such as H_2S, NH_3 and H_2O in raw biogas. The Polyethylene glycol process removes H_2S and CO_2 as well as S components such as H_2S, NH_3, HCN and H_2S. It is more energetic and favorable than water adsorption, and lose of CH_4 is less than other technologies. The construction of membrane separation is trouble free and operation is also uncomplicated and constructive to remove H_2S and H_2O. Cryogenic Separation process treats little gas flows, without any extra addition in construction cost. It is capable of removing H_2S, N_2, O_2, and efficiency level of methane is approximately 99.5% which is the highest quality. Thus an appropriate technology for the conversion of biogas to Bio-Methane can be selected according to need. All techniques have their limitations and advantages, but stability, reliability, and efficiency for removal of impurities are main challenges in this field.

References

1. Beil, M., Beyrich, W.: Biogas upgrading to biomethane. In: Wellinger, A., Murphy, J., Baxter, D. (eds.) The Biogas Handbook, pp. 342–377. Woodhead Publishing Limited, Germany (2013)
2. Lien, C.C., Lin, J.L., Ting, C.-H.: Water scrubbing for removal of hydrogen sulfide (H_2S) in biogas from hog farms. J. Agric. Chem. Environ. **3**, 1–6 (2014)
3. Al Mamun, M.R., Torii, S.: Removal of hydrogen sulfide (H_2S) from biogas using zero-valent iron. J. Clean Energy Technol. **3**(6), 428–432 (2015)
4. Mondal, M.K., Balsora, H.K., Varshney, P.: Progress and trends in CO_2 capture/separation technologies: a review. Energy **46**(1), 431–441 (2012)
5. Nallamothu, R.B., Teferra, A., Rao, B.V.: Biogas purification, compression and bottling. Glob. J. Eng. Des. Technol. (GJEDT) **2**(6), 34–38 (2013)
6. Nguyen, D.M.K., Imai, T., Dang, T.-L.T., Kanno, A., Higuchi, T., Yamamoto, K., Sekine, M., Grad, M.: Response surface method for modeling the removal of carbon dioxide from a simulated gas using water absorption enhanced with a liquid-film-forming device. J. Environ. Sci. (JES) **65**, 116–126 (2017). JES-01008

7. Peiris, A.P.T.S.: Feasibility study of production of bio methane from bio wastes in Sri Lanka and develop cost model for the production process. KTH School of Industrial Engineering and Management Energy Technology EGI-2016: OUSL Division of Heat & Power SE-100 44 STOCKHOLM (2016)
8. Pellegrinia, L.A., Guid, G.D., Consonnib, S., Bortoluzzib, G., Gattib, M.: From biogas to biomethane: how the biogas source influences the purification costs. Chem. Eng. Trans. (CET) **43**, 409–414 (2015). Italian Association of Chemical Engineering
9. Pirola, C., Galli, F., Bianch, C., Manenti, F.: Biogas to biomethane upgrading by water absorption column at low pressure and temperature. Technology **3**(2 & 3), 99–103 (2015)
10. Ryckebosch, E., Drouillon, M., Vervaeren, H.: Techniques for transformation of biogas to biomethane. Biomass Biogas **35**(5), 1633–1645 (2011)
11. Sha, D.R., Nagarsheth, H.R.: Biogas up gradation using water scrubbing for its use in vehicular applications. Int. Adv. Res. J. Sci. Eng. Technol. **2**(6), 46–48 (2015)
12. Sun, Q., Li, H., Yan, J., Liu, L., Yu, Z., Yu, X.: Selection of appropriate biogas upgrading technology-a review of biogas cleaning, upgrading and utilization. Renew. Sustain. Energy Rev. **51**, 521–532 (2015)
13. Truong, L.V.A., Abatzoglou, N.: A H_2S reactive adsorption process for the purification of biogas prior to its use as a bioenergy vector. Biomass Bioenergy **29**(2), 142–151 (2005)
14. Warren, K.: A techno-economic comparison of biogas upgrading technologies in Europe. M. Sc. Thesis, University of Jyvaskyla, Finland (2012). Accessed 17 Nov 2016
15. Xiao, Y., Yuan, H., Pang, Y., Chen, S., Zhu, B., Zou, D., Ma, J., Yu, L., Li, X.: CO_2 removal from biogas by water washing system. Chin. J. Chem. Eng. **22**(8), 950–953 (2014)
16. Zulkeflia, N.N., Masdara, M.S., Jahima, J., Majlanb, E.H.: Overview of H_2S removal technologies from biogas production. Int. J. Appl. Eng. Res. **11**(20), 10060–10066 (2016)

An Algorithm Implementation for Harmonics Reduction in Power Electronic Transformer Based Electric Locomotives

Sachin Gee Paul and C. S. Ravichandran$^{(\boxtimes)}$

Sri Ramakrishna Engineering College, Coimbatore, India
{sachingeepaul, eniyanravi}@gmail.com

Abstract. Power Electronic Transformer (PET) technology is one of the future technologies for high power conversion drives. Over the past years, various research and studies are conducted to explore the technical challenges for high-speed rails. The traditional locomotive transformer experience from a massive weight with poor efficiency. Thus power electronic transformers (PETs) is one of the promising solutions in the field of rolling stock with greater advantage of the reduction in size and weight of the locomotive transformer. This paper aims to describe high-frequency transformer merged DC to DC power converter feeding to a traction converter coupled traction motor is utilized in electric locomotives. This paper also suggests a dual phase algorithm method for the reduction of higher order harmonics during power conversion stages. The IGBT based dc to dc converter with traction converter for power electronic transformer has been verified by the simulation results with MATLAB/SIMULINK.

Keywords: Power electronic transformer (PET) · High frequency transformer (HFT) · Line frequency transformer (LFT) · Dual phase analysis (DPA)

1 Introduction

Railway transportation is one of the busiest and fast growing network in the world. In literatures, many methods have been made to replace line frequency transformer by high-frequency transformer units integrated with various power converters, thus Power Electronic Transformer (PETs) plays a key role. A Power electronic transformers offer significant advantages in reduced size and weight over conventional line frequency transformer [1]. One of the major problem facing on conventional electric locomotives is due to heavy weight its difficult to propels the axle load of the locomotive. This problem can be overcome by using PETs topology with proper converter configuration is addressed. The scheme of converters in various series-parallel combinations with medium frequency transformer is analyzed in previous literature. The line frequency transformer (LFT) used conventional electric locomotives is the major hurdle for achieving high-speed technology on locomotives. As per design criteria of transformer frequency is inversely proportional to cross section area, due to low operating frequency of LFT, overall weight tends to increase. In addition to this nearly one third weight of entire locomotive is only because of bulky LFT. The high growth of railway

© Springer Nature Switzerland AG 2020
A. P. Pandian et al. (Eds.): ICICCS 2019, AISC 1039, pp. 274–281, 2020.
https://doi.org/10.1007/978-3-030-30465-2_31

system ensures for railway vehicles executes the dimension constraints with efficiency of the traction equipment in precised manner [2]. Earlier literature discussed about transformer less [3] operation in electric locomotive which eliminate transformer and intermediate conversion stages and hence offers an even higher potential in efficiency improvement as well as a reduction in weight and cost [4].

In this paper, harmonics reduction is obtained by calculating the switching angle of the inverters thereby total harmonics can be reduced. This paper propose an algorithm based technique by which the harmonics with higher order generated in the traction converters as well as on the power electronic transformer are reduced. More over instead of transformer less, a transformer with high frequency is used in the intermediate stage of the dc to dc converter in an electric locomotive. Power electronic transformer (PETs) is the combination of various power converters with intermediate transformer having high frequency. A high voltage 25 kV, 50 Hz single phase ac catenary voltage is directly converted to dc using cascade H bridge converter. After converting, the high voltage dc is further reduced to lower voltage by using dc to dc converter of PETs, which then fed to the traction converter connected to the traction load. Thus the weight of the transformer is considerably reduced by increase in frequency of transformer used in converter units, thus overall weight of the locomotive is reduced.

2 Architecture for PET in Rolling Stock

2.1 Conventional Electric Locomotives

According to conventional system, an overhead line called as high voltage catenary 25 kV, 50 Hz single phase that run over the top of the main train tracks. The Line Frequency Transformer [LFT] or main line transformer mounted inside the locomotive is step down to suitable voltage for converters. This line frequency transformer inside the locomotive will be the root cause of the weight. Because as the line frequency 50 Hz is inversely proportional to the size of the transformer results increases burden of transformer. The reduced voltage from the transformer is fed to AC to DC converter, after that traction inverters which converters DC to AC. Finally this ac is then applied to traction motors, which is coupled in the axle of locomotives.

2.2 Power Electronic Transformer Based Traction Drive

Power electronic transformer is the combination of various sets of power converter with medium or high frequency transformer. The weight of the LFT itself carries around 1/3 of the total weight of locomotive for 25 kV, 50 Hz systems. Figure 1 shows the block diagram of Power Electronic transformer (PET) with traction converter feeding traction motors.

The merging of various power electronic converters with high-frequency transformer thereby it can also achieve the isolation from high voltage side to load end. Initially overhead line catenary voltage is collected from pantograph is fed to a cascade H bridge front end converter. In this front end converter, the 25 kV high voltage is

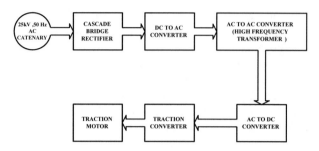

Fig. 1. Power electronic transformer fed traction drive

directly rectified. Each H- bridge splits the high voltage into various cell units, thereby voltage stress on each H bridge cell is reduced finally converted dc voltage is fed to DC to DC converter through High-Frequency Transformer (HFT). By using high-frequency concept the weight volume ratio of the transformer is considerably reduced. The essential features of PET-based traction drive system as follows [5]:

1. The transformer is operating at high frequency (100 Hz to 1 kHz).
2. The terminal voltages of the HFT generated are pulsed signals not sinusoidal, because of the heavy switching frequency of high power converter units.
3. The converter at incoming side is directly connected to the catenary of 25 kV, this converters should be capable of withstanding the highest voltage.

The connecting power electronic converters to grid is the most challenging task in power electronic transformer based electric locomotives [6]. The major difference when compared to conventional locomotives is the elimination of line frequency transformer at the front end. The coupling procedure is done by connecting all the multilevel H bridge converter-elements in cascade manner where each bridge termed as cell [7]. In this way, the grid voltage can divide into different cells thereby reduction of switching stress on each device is achieved. By cascade connection scheme of different converter-cells, maximum overall voltage is generated [8]. The major challenge in front-end converter is the capability of withstanding direct high voltage from the catenary since there is no transformer connected at the front end of the locomotive [9]. The recent development of power semiconductor devices like the introduction of 6500 V IGBTs for currents of 200 A, 400 A, and 600 A will meet this challenge [8].

3   Harmonic Reduction in Locomotive

The generation of harmonic in the output voltage of PWM or multilevel inverter in locomotive is greater because of several switching sections and due to various power converters. For several reasons such that nonlinearities in filters, modulation algorithm, dead times, voltage drops across power switches and dc bus modulation, the harmonic generation is more predominant Especially in the field of electric locomotives, a large number of power switches and converters are operated at higher switching frequency. So generation of harmonics is greater than predetermined values. Normally there are several mitigation techniques adopted in power converters. In most of the cases, lower order harmonics can be easily filtered out but higher order harmonics are still there.

3.1 Conventional Methods for Reduction of Harmonics

Typical by FFT analysis scheme, harmonic spectrum are obtained by using the following equation,

$$Hi = \sum_{p=0}^{n-1} R^{pi} s_p \tag{1}$$

Where, H_i = ith harmonic (i = 0....n−1), R = Transform parameter and s_p = Input sample.

In FFT transform the value tends to became zero as the input increases, so it called as a decaying function. The continuously decreasing outputs as the integers of FFT are going up. As the harmonic value is increases then the transformation value is negligible. Thus for higher order harmonics monotonous spectrum with small magnitude is formed. Because of this reason equalization of harmonics may not be correctly. So this paper focusing to find out higher order harmonics with aid of dual phase analysis.

3.2 Dual Phase Analysis

Dual phase analysis mainly carried out by micro phase analysis and macro phase analysis [10]. In micro phase is the analysis mainly focusing on harmonic spectrum with small scale magnitude. It also defined as the analysis for which measurement of lower order harmonics are carried out with respect to base frequency. In this method, the harmonics up to a pre-set value are carried out with respect to reference of base. The total harmonics is not consider into account but the level of harmonics is chosen with reference to the total number of stages considered and the accuracy. In the case 11th order harmonics measurement analysis considered as in the micro stage. The analysis of frequencies up to 550 Hz with reference to initial base frequency 50 Hz termed as micro phase stage but this stage is not tolerable for the complete process. The increase in harmonics order will results in accuracy to be decreased for micro phase analysis.

In the case of macro phase, analysis for harmonic spectrum of greater magnitude considering frequencies higher than micro phase analysis is measured. For this method measurement of higher order harmonics are considered. In this method harmonics of higher order is obtained by a interim shift of base frequency. The frequency base is shifted to harmonic that are nearly resembles to frequency fundamental. In these ways analysis of next stage of execution is carried out. At this instant the micro phase process is executed with the base frequency of new.

3.3 Normalization Procedure on Harmonic Spectrum

In these way analysis of higher order harmonics are measured and scaled with reference to lower order harmonics [10–12]. To do a normalization, compare harmonics of different magnitude of higher order with at a different scale of magnitude with original. The main advantage for normalization is to remove irregularities that arise when change in magnitude scales.

Execute the next stage of micro phase analysis carried out before selecting nearest matching base frequecy. This execution process is repeated until maximum frequency. The macro phase analysis is also followed by the micro phase by shifting the base. Now, the execution is turn back to the initial level and the next nearest frequency is selected. Repeated all the steps until expected frequencies reached. By taking all the normalized maximum values from all the available sources, repetition frequency can be avoided and turn back to original. During repeated analysis, more number of values obtained on that case, the value close to the harmonic base is consider as a final value. In these way total harmonics generation can be reduced in power converters used on PET based power converters for electric locomotives. The flow chart representation of dual phase analysis as shown in Fig. 2.

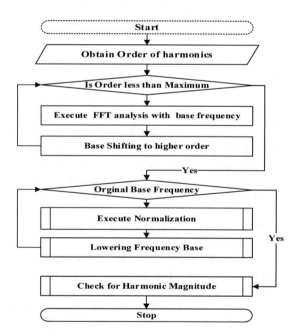

Fig. 2. Flowchart of dual phase analysis

3.4   Challenges for Traction Inverters

Balancing of voltage from traction inverter is the crucial task. Since there are two sets of traction motors are coupled in the axle of the front end as well as at the back end of the locomotive. When both the traction motors rotate synchronously, then only loco-motive drives the axle load. For this, in order to feed traction motor, it should ensure that both the traction inverter have same output voltage. In order to achieve, certain control technique is adopted for traction inverter with some controllers.

4 Result and Discussion

Simulation of dc to dc converter with traction converter feeding traction motor for 25 kV input parameter is carried out using MATLAB. For the simulation, high-frequency transformer is used with traction converter is also simulated by providing with suitable PWM. Since 25 kV is feeding from catenary is directly converted to dc by using cascade H bridge converter.

Before applying to high-frequency transformer its frequency is altered because of high-frequency switching occurs on the power converter. Since the dc to dc converter output is same as that of input of traction inverter. The traction converter output as load voltage of 2100 V is obtained from the simulation output.

The traction terminal voltage of the motor is vary based on the type of locomotives. Generally terminal load voltage of traction motor lie in between 850 V to 2500 V. The output load voltage of traction converter as shown in Fig. 3. Since because of several switching, unwanted noise pulse will distort the waveform. To avoid noises, a first order filter is coupled between traction inverter and load. Since motor is acting as load which is connected to the traction inverter. The simulation waveform of the speed of traction motor as shown in Fig. 4. The FFT analysis for converters was also carried out in this paper when compare to converters bridges without filter configuration the THD level can drastically reduced and reach around 4.26% as shown in Fig. 5.

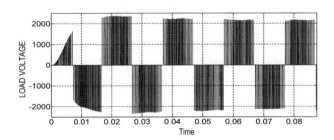

Fig. 3. Load voltage of traction converter

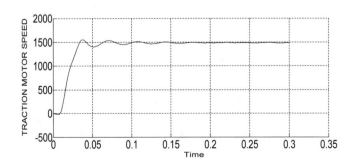

Fig. 4. Speed of traction motor

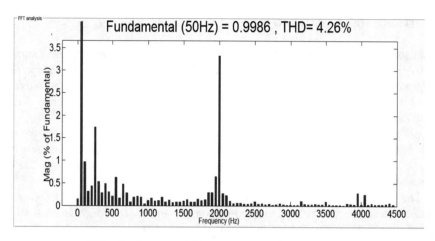

Fig. 5. Converter bridge harmonics level with filter

5 Conclusion

In this paper, Traction converter coupled with dc to dc converter in power electronic transformer for electric locomotives is proposed. More over harmonic reduction by using dual phase analysis is also presented. By replacing bulky line frequency transformer with high-frequency transformer, inside the converter unit of power electronic transformer has been simulated using MATLAB/SIMULINK. The load end also coupled with induction motor feed from traction inverter has discussed. The proposed scheme results in excellent method for weight reduction and also improves the power transfer capability of converters in electric locomotives for modern drives.

References

1. Gu, C., Zedong, Z., Li, Y.: A Power electronic transformer with multiport bidirectional resonant DC–DC converters for electric traction applications (2015)
2. Ronanki, D., Williamson, S.S.: Evolution of power converter topologies and technical considerations of power electronic transformer-based rolling stock architectures. IEEE Trans. Transp. Electr. **4**(1), 211–219 (2018)
3. Dieckerhoff, S., Bernet, S.: Power loss-oriented evaluation of high voltage IGBTs and multilevel converters in transformer less traction applications. IEEE Trans. Power Electron. **20**(6), 1328–1336 (2005)
4. Zhao, C., Dujic, D.: Power electronic traction transformer-medium voltage prototype. IEEE Trans. Ind. Electr. **61**(7), 3257–3268 (2014)
5. Feng, J., Chu, W.Q., Zhang, Z., Zhu, Z.Q.: Power electronic transformer-based railway traction systems: challenges and opportunities. IEEE J. Emerg. Sel. Top. Power Electron. **5**(3), 1237–1253 (2017)
6. Kouro, S., Malinowski, M., Gopakumar, K., Pou, J., Franquelo, L.G., Wu, B., Rodriguez, J., Perez, M.A., Leon, J.I.: Recent advances and industrial applications of multilevel converters. IEEE Trans. Ind. Electron. **57**(8), 2553–2580 (2010)

7. Cecati, C., Dell'Aquila, A., Liserre, M., Monopoli, V.G.: Design of H-bridge multilevel active rectifier for traction systems. IEEE Trans. Ind. Appl. **39**(5), 1541–1550 (2003)
8. Bernet, S.: Recent developments of high power converters for industry and traction applications. IEEE Trans. Power Electron. **15**(6), 1102–1117 (2000)
9. Rufer, A., Schibli, N., Chabert, C., Zimmermann, C.: Configurable front-end converters for multicurrent locomotives operated on 16 2 = 3 Hz AC and 3 kV DC systems. IEEE Trans. Power Electron. **18**(5), 1186–1193 (2003)
10. Vijayalakshmi, V.J., Ravichandran, C.S., Amudha, A.: Predetermination of higher order harmonics by dual phase analysis. Appl. Mech. Mater. **573**, 13–18 (2014)
11. Chiasson, J.N., Tolbert, L.M., Mckenzie, K.J., Du, Z.: A complete solution to the harmonic elimination problem. IEEE Trans. Power Electron. **19**(2), 491–499 (2004)
12. Enjeti, P.N., Ziogas, P.D., Lindsay, J.F.: Programmed PWM techniques to eliminate harmonics: a critical evaluation. IEEE Trans. Ind. Appl. **26**, 302–316 (1990)

Evolutionary Algorithm Based Robust Fixed Structure Controller for pH in Sodium Chlorate Process

S. Sreepriya[✉], K. Aparna, and M. L. Vinila

Department of Chemical Engineering,
National Institute of Technology, Calicut, India
{sreepriya_p150069ch, aparnak,
vinila_p150083ch}@nitc.ac.in

Abstract. PID Controllers are used in almost 95% of control loops in the most process industry. Since the design of a controller for an industrial system is complex due to the presence of model uncertainties, nonlinearities, and disturbances, these controllers are usually not tuned optimally and performance is compromised. Robust fixed structure control is a remedy to the problem, and such a controller design is an optimization problem, characterized by nonlinear multimodal large pursuit space, tight constraints, and expensive objective function. The classic optimization method is not appropriate here as a solution set is to be computed. Evolutionary algorithm, a robust search, and optimization method can handle this problem. Covariance Matrix Adapted Evolutionary strategy is successfully utilised to design robust PID controller for Sodium chlorate process. The performance in terms Integral Square Error (ISE) for set point tracking, robust stability and disturbance attenuation was found satisfactory.

Keywords: Evolutionary algorithm · Sodium chlorate process · Fixed structure controller · Robust control · pH control

1 Introduction

Robustness of design is a key aspect of a controller as most practical systems/processes are subjected to disturbances, measurement noises, un-modeled plant dynamics, and plant parameter variations. A robust control strategy tends to be able to cope with the differences between the true system and modelled system. Traditional robust controller designs lead to a controller with higher order dynamics, whose practical implementation is difficult. The structure specified controller design is a solution to this issue. But the drawback of structure specified controller design is characterized by nonlinear multimodal large pursuit space, tight constraints, and expensive objective function. Evolutionary algorithm is a widely used optimization technique, which can handle ill-behaved problems with features like multimodality, discontinuity, randomness, etc. This paper focuses on EA based controller design for Sodium chlorate process [1].

Sodium chlorate manufacturing process is one of the biggest energy intensive electrochemical process, where the pH of the electrolysis cell has a major influence on

© Springer Nature Switzerland AG 2020
A. P. Pandian et al. (Eds.): ICICCS 2019, AISC 1039, pp. 282–291, 2020.
https://doi.org/10.1007/978-3-030-30465-2_32

the current efficiency and hence the power consumption [2]. Sodium chlorate is manufactured by industrial electrolysis of hot acidulated brine in undivided cells [3]. The reactions taking place in the chlorate cell depends on pH and active chlorine concentration. Several methods are proposed in the literature to improve current efficiency [4–6]. However pH control to improve efficiency, is not proposed so far. pH is one of the most commonly used process variable and it is the most difficult process variable to control [7, 8] as it is nonlinear, highly sensitive, and exhibit high variation in its behavior as a function of varying concentrations of the reactants. [8] Different pH control strategies for pH neutralization process is available in the literature [8–15]. From recent kinds of literature, it is found that Artificial intelligence techniques are found to be superior to traditional methods in control of such non-linear parameters. But industries prefer PID controllers, as they are simple, fewer variables to tune and of low order. However, in complex industrial systems, PID controller design is complicated due to the presence of various uncertainties such as modeling errors, disturbances, plant parameter variations, etc. In practice, it is very important to design low order-fixed structure controllers since it requires on-site controller tuning. Due to this reason, fixed order/fixed structure controller design approaches have received considerable attention. Various design methods satisfying robust stability and H_∞ specifications have been proposed [16]. However, these methods depend on the specific structure and so they cannot be applied to a broader class of processes. Most approaches in conventional H_∞ controllers utilize LMI (Linear Matrix Inequality) method [17–19]. But this method demands deep knowledge of semidefinite programming theory. Contrary to conventional methods, probabilistic methods based on randomized algorithms and mixed methods were later proposed in the literature [20, 21]. Many pieces of research have employed a PID controller structure for mixed H2/H_∞ optimal design. Chen et al. [22] suggested a mixed H_2/H_∞ PID controller design using simple GA for a SISO system. The performance criteria were minimizing ISE in face of constraints, disturbance attenuation, and robust stability constraints [22]. Chen also proposed H_∞ PID controller using GA for multiple input multiple output system for optimizing a combination of disturbance rejection and robust stability [23]. A. Krohling et al. proposed GA based PID controller with disturbance rejection for a system with servomotor. Kitsios et al. illustrated a PID controller design by considering disturbance attenuation and robustness as objectives [24]. Tan et al. presented the procedure for optimal PID tuning in a multivariable process using Ricatti and H-infinity loop-shaping [25]. Ho et al. presented a controller for a MIMO system using Integral Square Error and performance criteria combining robust stability and disturbance rejection as performance criteria [1]. They also proposed a mixed H_2/H infinity PID controller design based on IGA for a MIMO system [26]. Estimation of distribution algorithms (EDA) is used to alleviate the issues in GA while applied for non-separable function. As compared to GA, instead of crossover and mutation, EDAs utilize probability distribution obtained from objective function to create search points. Rest of the algorithms are the same. Here, robust optimal PID controller design scheme for sodium chlorate process using CMAES is presented. CMAES is a robust method, which is capable of managing problems featuring multimodality, discontinuity, time-variance, randomness, and disturbances [27]. The learning part of CMAES algorithm performs the search in such a way that it improves convergence rate [28]. Literature shows the utilization of CMAES algorithm in myriads of optimization problems [29, 30].

2   Robust Fixed Structure Controller

The schematic of a control system with perturbation is given in Fig. 1 .The system has p_i inputs and p_0 outputs.

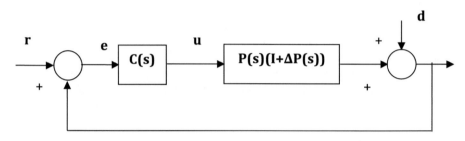

Fig. 1. Control system with plant perturbation

The plant perturbation $\Delta P(s)$ of plant $P(s)$ is bound by a weighing function $W_1(s)$ given below [1]

$$\bar{\sigma}(\Delta P(j\omega)) \leq \bar{\sigma}(W_1(j\omega)), \forall_\omega \in [0, \infty) \tag{1}$$

This work suggests a robust fixed structure PID controller design for pH in sodium chlorate process. CMAES is used for minimizing the performance criteria, the sum of Integral Square Error, robust stability, and disturbance attenuation.

2.1   Structure of a Robust PID Controller

Let the specification of a fixed structure controller be of the form [1]

$$C(s) = \frac{N_c(s)}{D_c(s)} = \frac{B_m s^m + B_{m-1} s^{m-1} + \cdots + B_0}{s^n + a_{n-1} s^{n-1} + \cdots + a_0} \tag{2}$$

where m and n are the orders. This controller is used to minimize the performance index, where [1]

$$B_k = \begin{bmatrix} b_{k11} & \cdots & b_{k1n_i} \\ \vdots & \ddots & \vdots \\ b_{kn_01} & \cdots & b_{kn_0n_i} \end{bmatrix} \tag{3}$$

for k = 0,1,2....m. The PID and lead lag combinations that are the most common industrial controllers are the special cases of this controller.

3 Evolution Strategy: Covariance Matrix Adapted Evolution Strategy (CMAES)

In CMAES, population candidates are generated from samples of a probability distribution which is generated while optimization is being done. This method uses the knowledge of correlations among parameters to accelerate the convergence. Recombination in CMAES as compared to GA is finding a new mean value for distribution and mutation is adding a random vector. In CMAES, two time evolution paths of distribution mean are utilised, one path for adaptation of covariance matrix C and another one global step size control [30, 31].

3.1 CMAES Algorithm

In CMAES new population is obtained by sampling of a multivariate nominal probability distribution. CMAES uses the learning of correlation among variables to accelerate the convergence. The detailed steps to perform algorithm is given below [28].

1: Create an initial solution, generation g, which is random [32].
2: The offspring/candidate for next generation($g + 1$ generation), x_k^{g+1} is obtained by sampling a Gaussian distribution. This is done using C and the global step size at g [32].

$$x_k^{(g+1)} = z_k, \quad z_k = N\left(\langle x \rangle_\mu^{(g)}, \sigma^{(g)^2} \mathbf{C}^{(g)}\right) \quad k = 1, \ldots, \lambda \tag{5}$$

where $\langle x \rangle_\mu^{(g)} = \sum_{i=1}^{\mu} x_i^{(g)}$ with m, as the best individuals obtained in population.

The number of decision variables helps in finding the values of c_c, cov, c_s, and d for the next calculations [32] [38]:

$$c_c = \frac{4}{n+4}, \quad c_\sigma = \frac{10}{n+20}, \quad d = \max\left(1, \frac{3\mu}{n+10}\right) + c_\sigma,$$

$$c_{cov} = \frac{1}{\mu} \frac{2}{(n+\sqrt{2})^2} + \left(1 - \frac{1}{\mu}\right) \min\left(1, \frac{2\mu - 1}{(n+2)^2 + \mu}\right) \tag{6}$$

The c and ccov helps the adaptation of step size and C [33].

$$\mathbf{P}_\sigma^{(0)} = \mathbf{P}_c^{(0)} = \mathbf{0} \text{ and } \mathbf{C}^{(0)} = \mathbf{I}$$

3: $\mathbf{P}_c^{(g+1)}$ evolution path is calculated using [32]

$$\mathbf{P}_c^{(g+1)} = (1 - c_c) \cdot \mathbf{P}_c^{(g)} + \sqrt{c_c(2 - c_c)} \cdot \frac{\sqrt{\mu}}{\sigma^{(g)}} \left(\langle x \rangle_\mu^{(g+1)} - \langle x \rangle_\mu^{g}\right) \tag{7}$$

$$\mathbf{C}^{(g+1)} = (1 - c_{cov}) \cdot \mathbf{C}^{(g)} + c_{cov}$$
$$\cdot \begin{pmatrix} \frac{1}{\mu} \mathbf{P}_c^{(g+1)} (\mathbf{P}_c^{(g+1)})^T + \\ (1 - \frac{1}{\mu}) \frac{1}{\mu} \sum_{i=1}^{\mu} \frac{1}{\sigma^{(g)^2}} (x_i^{(g+1)} - \langle x \rangle_\mu^{(g)})(x_i^{(g+1)} - \langle x \rangle_\mu^{(g)})^T \end{pmatrix} \quad (8)$$

The parameter $c_{cov} \in [0, 1]$ decides the rate of variation of C [33]. The covariance matrix C of the distribution is updated so that the likelihood of previous successful search steps is increased. Both updates are natural gradient descent. Then CMA performs an iterated principal components analysis (PCA) of successful search steps while retaining all principal axes.

4: The second time evolution path, conjugate evolution path $\mathbf{P}_\sigma^{(g+1)}$ is used for controlling global step size $\sigma^{(g+1)}$ as per the equation given below [33]

$$\mathbf{P}_\sigma^{(g+1)} = (1 - c_\sigma) \cdot \mathbf{P}_\sigma^{(g)} + \sqrt{c_\sigma(2 - c_\sigma)}$$
$$\cdot \mathbf{B}^{(g)} (\mathbf{D}^{(g)})^{-1} (\mathbf{B}^{(g)})^{-1} \frac{\sqrt{\mu}}{\sigma^{(g)}} \left(\langle x \rangle_\mu^{(g+2)} - \langle x \rangle_\mu^{g} \right) \quad (9)$$

The B(g) and D(g) are obtained by PCA [34]:

$$\mathbf{C}^{(g)} = \mathbf{B}^{(g)} (\mathbf{D}^{(g)})^2 (\mathbf{B}^{(g)})^T \quad (10)$$

B(g) have its columns as the normalized eigen vectors of C(g) and D(g) [33] $\sigma^{(g+1)}$ is given by

$$\sigma^{(g+1)} = \sigma^{(g)} \exp\left(\frac{c_\sigma}{d} \left(\frac{\|\mathbf{P}_\sigma^{(g+1)}\|}{E(\|N(0, \mathbf{I})\|)} \right) - 1 \right) \quad (11)$$

5: 2–4 are repeated until the stopping criteria is reached.

Constraint handling is done using Penalty parameter-less constraint-handling Fitness function is obtained using the equation [35]

$$F(\bar{x}) = \begin{cases} f(\bar{x}) & \text{if } g_j(\bar{x}) \geq 0 \quad \forall j = 1, 2 \cdots, m \\ f_{max} + \sum_{j=1}^{m} (g_j(\bar{x})) & \text{otherwise} \end{cases} \quad (12)$$

$f(\bar{x})$ is the objective function
$g_j(\bar{x})$ is the jth normalized absolute constraint violations value
f_{max} is the value of the objective function for the worst solution in the population [35].

Solution is selected based on fitness function and constraint violation. if two feasible solutions are there, the solution with better f value is opted. If one feasible and one infeasible solution is there, then feasible solution is selected. If there is two infeasible solutions, the solution with lesser violation of constraint is selected [34].

4 Sodium Chlorate Process

Sodium chlorate is manufactured by the electrolysis of hot acidulated brine (NaCl) and can be described by the reaction

$$NaCl + 3H_2O \rightarrow NaClO_3 + 3H_2 \tag{13}$$

The quantity and quality of the $NaClO_3$ crystal depend on a lot of factors like pH of bulk, temperature, concentration, the current density of electrodes, types of electrodes, cell voltage, etc. [2, 36]. A major issue confronting the industry is power consumption. Hence cell efficiency is a major factor to be considered. When power is applied, the primary reaction occurring in the chlorate cell is the formation of chloride ions and H^+ ions to form Cl_2 and H_2 respectively.

Chlorine molecules are hydrolyzed as per the reaction is given by Eq. (14)

$$Cl_2 + H_2O \leftrightarrow Cl^- + H^+ + HClO \tag{14}$$

Hypochlorous acid is a weak acid and is partially dissociated into hypochlorite ions (H^+ and ClO^-) called Active Chlorine.

$$HClO \leftrightarrow H^+ + ClO^- \tag{15}$$

If the pH of the bulk is favorable, the hypochlorous ion produced reacts with more hypochlorous acid to form chlorate ion as per the following reaction

$$2HClO + ClO^- \rightarrow ClO_3^- + 2H^+ + 2Cl^- \tag{16}$$

and thus efficiency is improved. If pH is not favorable, a parasitic reaction given by Eq. (17) takes place and efficiency decreases to 66% [2]. This necessitates the control of pH in the cell.

$$ClO^- + 3H_2O \rightarrow 2ClO_3^- + 3O_2 + 6H^+ + 4Cl^- + 6e^- \tag{17}$$

The sodium chlorate process is identified from a database collected during industrial production of sodium chlorate in the plant of Travancore Cochin Chemicals, India. This plant has a designed production capacity of 5000 kg of dry sodium chlorate crystals per day. Measured the input and output parameters on a daily basis during stable industrial production and collected the database of 9000 samples.

5   Simulation Results

The model and controller of the process is simulated using MATLAB 2015b software on a 64 bit PC with 4 GB RAM. The stste space plant model for pH from HCl Flow rate, NaOH Flow rate, Electrolyte temperature, DC Load current, and pH of feed input is identified from data collected from the plant,

$$\dot{x} = Ax + Bu$$
$$y = Cx \tag{17}$$

Using Automatic Weight selection algorithm [37] weights for the design is selected. The weighing function to represent plant uncertainties $W_1(s)$ is given by

$$W_1(s) = \frac{100s + 1}{s + 100} I_{2\times2} \tag{18}$$

For disturbance rejection, a weighting function is considered which consist of an LPF and is represented as $W_2(s)$,

$$W_2(s) = \frac{s + 1000}{1000s + 1} I_{2\times2} \tag{19}$$

The controller structure used is

$$K(x) = 10^{x1}(1 + \frac{1}{10^{x2}s} + \frac{10^{x3}}{1 + 10^{(x3-x4)}s}) \tag{20}$$

where $x = (x_1, x_2, x_3, x_4)$ are the parameters. The initial pursuit space is set as

$$-2 \leq x_i \leq +200 \, i = 1, 2, 3, 4$$

The population is fixed at 50, and ten independent runs conducted. The sampling time is 0.01 s. The optimal controller obtained is given below

$$K(s) = 1848.84(1 + \frac{1}{0.1709s} + \frac{0.1802s}{1 + (8.7679 \times 10^{-4})s}$$

The performance indices like best value, mean value and standard deviation in Integral Square Error along with the performance criteria of disturbance attenuation and robust stability for 10 trials are reported in the Table 1.

Robust stability critiriea is to satisfy the inequality

Table 1.   Norms of closed loop plant

J_2	J_∞	$J = J_2 + J_\infty$
0.0104	0.9808	0.9912

$$F_1 = \|W_1(s)T(s)\|_\infty \prec 1 \tag{21}$$

Disturbance attenuation obeys

$$F_2 = \|W_2(s)S(s)\|_\infty \prec 1 \tag{22}$$

where $S(s)$ is sensitivity function and $T(s)$ is complementary sensitivity function. So to minimize both F_1 and F_2 value a combined value J_∞ is used here. $J_\infty = F_1 + F_2$. From the results obtained the performance function J_2 and J_∞ values are less than one, the performance is good. It indicates good disturbance rejection and robust stability. ISE value (J_2) is small indicating good set point tracking. Figure 2 shows the system response with the controller. Statistical performance of the algorithm is given in Table 2.

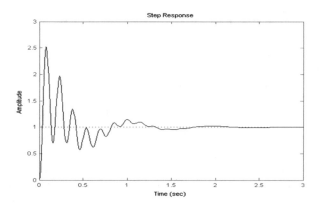

Fig. 2. System step response with the controller.

Table 2. Statistical Performance

Best value	Mean value	Standard deviation	Average Functional Evaluation (AFE)
0.9912	0.9931	0.0058	3017

6 Conclusion

The application of CMAES for optimal robust PID controller design for Sodium chlorate cell pH has been demonstrated. The objective of design is to minimize ISE and attain robust stability and attenuation of disturbances. For performance analysis, ten runs with independent population initialization is done. The performance function J_∞ value is less than one, which indicates good disturbance rejection and robust stability. ISE value (J2) is small indicating good set point tracking. Hence the design of pH controller using CMAES is satisfactory based on set point tracking, robust stability, and disturbance attenuation. The values best, mean and standard devistion of objective got in 10 trials indicates that CMAES method is able to give good performance consistently.

References

1. Ho, S.-J., Ho, S.-Y., Shu, L.-S.: IEEE Trans. Syst. Man Cybern. Part A Syst. Hum. **34**(5), 588–600 (2004). https://doi.org/10.1109/tsmca.2004.832834
2. Viswanathan, K.: J. Electrochem. Soc. **131**(7), 1551 (1984). https://doi.org/10.1149/1.2115908
3. Vogt, H.: J. Electrochem. Soc. **128**(2), 29C–32C (1981). https://doi.org/10.1149/1.2127407
4. Endrődi, B., Sandin, S., Smulders, V., Simic, N., Wildlock, M., Mul, G., Mei, B.T., Cornell, A.: J. Clean. Prod. **182**, 529–537 (2018). https://doi.org/10.1016/j.jclepro.2018.02.071
5. Gordon, G., Tachlyashlki, S.: Environ. Sci. Technol. **25**(3), 468–474 (1991). https://doi.org/10.1021/es00015a014
6. Kriem, L.S.: Thesis (2017)
7. Tadeo, F., López, O.P., Alvarez, T.: IEEE Trans. Control Syst. Technol. **8**(2), 236–246 (2000). https://doi.org/10.1109/87.826795
8. Hermansson, A.W., Syafiie, S.: Control Eng. Pract. **45**, 98–109 (2015). https://doi.org/10.1016/j.conengprac.2015.09.005
9. Lakshmi Narayanan, N.R., Krishnaswamy, P.R., Rangaiah, G.P.: Chem. Eng. Sci. **52**(18), 3067–3074 (1997). https://doi.org/10.1016/s0009-2509(97)00130-9
10. Li, M., Wang, F., Gao, F.: Ind. Eng. Chem. Res. **40**(12), 2660–2667 (2001). https://doi.org/10.1021/ie990715e
11. Shabani, R., Sedigh, A.K., Salahshoor, K.: 497–500 (2010). https://doi.org/10.1109/iccas.2010.5669969
12. Mwembeshi, M.M., Kent, C.A., Salhi, S.: Comput. Chem. Eng. **28**(9), 1743–1757 (2004). https://doi.org/10.1016/j.compchemeng.2004.03.002
13. Kumbasar, T., Eksin, I., Guzelkaya, M., Yesil, E.: ISA Trans. **51**(2), 277–287 (2012). https://doi.org/10.1016/j.isatra.2011.10.007
14. Mota, A.S., Menezes, M.R., Schmitz, J.E., Da Costa, T.V., Da Silva, F.V., Franco, I.C.: Chem. Eng. Commun. **203**(4), 516–526 (2016). https://doi.org/10.1080/00986445.2015.1048799
15. Kambale, S.D., George, S., Zope, R.G.: Int. Res. J. Eng. Technol. (1), 2395–2456 (2015)
16. Ho, M.T., Lin, C.Y.: IEEE Trans Autom. Control **48**(8), 1404–1409 (2003). https://doi.org/10.1109/TAC.2003.815028
17. Apkarian, P., Noll, D., Tuan, H.D.: Int. J. Robust Nonlinear Control. **13**(12), 1137–1148 (2003). https://doi.org/10.1002/rnc.807
18. Saeki, M.: IFAC Proc. **16**, 415–420 (2005). https://doi.org/10.1016/j.automatica.2005.07.006
19. Saeki, M.: Automatica **42**(1), 93–100 (2006). https://doi.org/10.1016/j.automatica.2005.07.006
20. Tempo, R., Calafiore, G., Dabbene, F.: Algorithms for Analysis and Control of Uncertain Systems. Springer (2004)
21. Fujisaki, Y., Oishi, Y., Tempo, R.: Mixed Deterministic/Randomized Methods for Fixed Order Controller Design. IEEE Trans. Autom. Control **53**(9) (2008)
22. Chen, B., Cheng, Y., Lee, C.: Genetic approach to mixed H2/H, optimal PID control. IEEE Control Syst., 0272–1708 (1995)
23. Sen Chen, B., Cheng, Y.M.: IEEE Trans. Control Syst. Technol. **6**(6), 707–718 (1998). https://doi.org/10.1109/87.726532
24. Kitsios, I., Pimenides, T., Groumpos, P.: (3), 1196–1201 (2002). https://doi.org/10.1109/cca.2001.974035

25. Tan, W., Chen, T., Marquez, H.J.: Asian J. Control. **4**(4), 439–451 (2010). https://doi.org/10.1111/j.1934-6093.2002.tb00085.x
26. Shu, L.-S., Ho, S.-Y., Ho, S.-J., Huang, H.-L., Hung, M.-H.: IEEE Trans. Control Syst. Technol. **13**(6), 1119–1124 (2005). https://doi.org/10.1109/tcst.2005.857403
27. Iruthayarajan, M.W., Baskar, S.: Expert Syst. Appl. **36**(5), 9159–9167 (2009). https://doi.org/10.1016/j.eswa.2008.12.033
28. Mohaideen Abdul Kadhar, K., Baskar, S.: Appl. Soft Comput. J. **34**, 337–348 (2015). https://doi.org/10.1016/j.asoc.2015.05.022
29. Bhaskar, V., Gupta, S.K., Ray, A.K.: Rev. Chem. Eng. **16**, 1–54 (2000). https://doi.org/10.1515/REVCE.2000.16.1.1
30. Büche, D., Hansen, N., Ocenasek, J., Koumoutsakos, P., Kern, S., Müller, S.D.: Nat. Comput. **3**(3), 355–356 (2004). https://doi.org/10.1023/b:naco.0000036904.41423.1c
31. Sadeghi, M., Kalantar, M.: J. Renew. Sustain. Energy. **6**(6), 1–24 (2014). https://doi.org/10.1063/1.4901087
32. Manoharan, P.S., Kannan, P.S., Baskar, S., Willjuice Iruthayarajan, M., Dhananjeyan, V.: Eng. Optim. **41**(7), 635–657 (2009). https://doi.org/10.1080/03052150902738768
33. Sreepriya, S., Baskar, S., Willjuice Iruthayarajan, M.: In: 2010 2nd International Conference on Computing, Communication and Networking Technologies, ICCCNT 2010, no. (LMI), pp. 1–4 (2010). https://doi.org/10.1109/icccnt.2010.5591817
34. Jeyadevi, M.W.I.S., Bhaskar, S.: Eur. Trans. Electr. Power **21**(E1), 1343–1360 (2011). https://doi.org/10.1002/etep
35. Manoharan, P.S., Kannan, P.S., Bhaskar, S., Iruthayarajan, M.W.: Gener. Transm. Distrib. IET **1**(2), 324 (2007). https://doi.org/10.1049/iet-gtd
36. Adam, L.C., Gordon, G.: Inorg. Chem. **38**(6), 1299–1304 (1999). https://doi.org/10.1021/ic980020q
37. Nair, S.S.: Int. J. Eng. Sci. Technol. **3**(1), 122–138 (2011)

Solution of Fuzzy Differential Equation of Order 2 by Intuitionistic Fuzzy Numbers (IFS)

A. Rajkumar[1] and C. Jesuraj[2,3(✉)]

[1] Department of Mathematics, Hindustan Institute of Technology and Science, Chennai, India
arajkumar@hindustanuniv.ac.in
[2] Hindustan Institute of Technology and Science, Chennai, India
jesu2112@gmail.com
[3] Department of Mathematics, IFET College of Engineering, Villupuram, India

Abstract. Any real life problems can be modelled by differential equation. In this paper second order fuzzy differential equation is converted to fuzzy differential equations and solved by fuzzy sumudu transforms method with initial values as intuitionistic triangular fuzzy numbers. To understand the proposed method we solved a numerical problem with graphical representations.

Keywords: Differential equation · Fuzzy differential equation · Fuzzy sumudu transforms and IFS

1 Introduction

Zadeh, introduces fuzzy logic and a generalized theory of uncertainty [3] Dubois and Prade worked on fuzzy differential calculus in 1982 [2]. Later 1987 Kaleva also worked on Fuzzy differential equations by initial boundary conditions [6]. Analyzed a system of linear fuzzy differential equations, and Dubey applied triangular intuitionistic fuzzy numbers to solve the linear programming model and to solve system of fuzzy differential equation using triangular fuzzy numbers by a, Sadeghi. Kumar apply new way to solve FDE using fuzzy numbers [7]. Finding Solution for linear fuzzy ordinary differential equation by fuzzy boundary values by Gasilov, Amrahov and Fatullayev [8]. Fuzzy Intuitionistic set was introduced by Atanassov, and he worked on both membership and non-membership [10, 11]. Intuitionistic fuzzy theory was used for medical diagnoses by De, Biswas, and Roy [12, 13]. Mondal, used intuitionistic fuzzy number to solve linear differential equations by GH methods [14, 15, 19]. In 2015 Nirmala and Pandian worked on Solving Intuitionistic Fuzzy Differential Equation under Generalised Differentiability Concept [16], solving fuzzy differential equation by successive approximations by Ettoussi, Melliani, Elomari and Chadli [17, 18]. In this paper we used intuitionistic TFN as initial states to solve second order DE by sumudu transform method.

© Springer Nature Switzerland AG 2020
A. P. Pandian et al. (Eds.): ICICCS 2019, AISC 1039, pp. 292–298, 2020.
https://doi.org/10.1007/978-3-030-30465-2_33

2 Premilinearies

Definition 2.1: Intuitionistic fuzzy number:

An IFS $\tilde{A}^i = \{r, \mu_{\tilde{A}^i}(r), \vartheta_{\tilde{A}^i}(r)/r \in R\}$ is called an IFN if (a) $\tilde{A}^i$ is normal and convex, i.e., its convex (b) $\mu_{\tilde{A}^i}, \vartheta_{\tilde{A}^i}$ are upper semi continuous and lower also. (c) Supp $A = \{r \in X/\vartheta_{\tilde{A}^i}(r) < 1\}$ is bounded.

Definition 2.2: TIFN: Graphical representation and membership and non-membership functions are follows (Fig. 1),

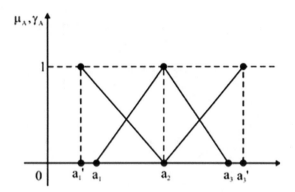

Fig. 1. ITFN

$$\mu_{\overline{A}}(x) = \begin{cases} \dfrac{x - r_1}{r_2 - r_1}, r_1 \leq x \leq r_2 \\ \dfrac{r_3 - x}{r_3 - r_2}, r_2 \leq x \leq r_3 \\ 0, \qquad otherwise \end{cases}, \mu_{\overline{B}}(x) = \begin{cases} \dfrac{r_2 - x}{r_2 - q_1}, q_1 \leq x \leq r_2 \\ \dfrac{x - r_2}{q_2 - r_2}, r_2 \leq x \leq q_2 \\ 0, \qquad otherwise \end{cases}$$

2.1 Theorem: Let f: R → F(R) be a real valued function. If $g(bt)\Theta e^{-t}$ be improper integral on [a, ∞), then FST is $\int\limits_{0}^{\infty} g(bt)\Theta e^{-t}dt$

$$G(u) = S[g(t)] = \int_{0}^{\infty} g(ut)\Theta e^{-\tau}dt, u \in [\tau_1, \tau_2]$$

$$= \left[\int_{0}^{\infty} \underline{g}_k(ut).e^{-\tau}dt, \int_{0}^{\infty} \overline{g}_k(ut).e^{-\tau}dt\right]$$

Thus $S[g(t)] = \left[s\left[\underline{g}_k(t)\right], s[\overline{g}_k(t)]\right]$

3   Main Results

Now consider the second order fuzzy differential equation with initial value problems using TFN, $j''(t) = j(c)$, $j(0) == (\alpha - 1, 1 - \alpha; 1 - \beta, \beta - 1), j'(0) = (\alpha - 1, 1 - \alpha; 1 - \beta, \beta - 1)$

By fuzzy sumudu transforms, we have

$$j_A(t) = \left(\underline{j}(t, \alpha), \bar{j}(t, \alpha) \right), j'_A(t) = \left(\underline{j}(t, \beta), \bar{j}(t, \beta) \right)$$
$$s[j_A(t)] = \left(s\left[\underline{j}(t, \alpha) \right], s\left[\bar{j}(t, \alpha) \right] \right), s[j'_A(t)] = \left(s\left[\underline{j}(t, \beta) \right], s[\bar{j}(t, \beta)] \right)$$

The following four cases are explains the solution procedure.

Case: 1 Let j(t) and $j'(t)$ be(i)-differentiable. Then

$$\left[\frac{S[j(t)]}{u_1^2} \ominus \left(\frac{j(0)}{u_1^2} \right) \ominus \left(\frac{j'(0)}{u_1} \right) \right] = S[j(t)]$$

Thus:

$$S\left[\underline{j}(t, \omega) \right] = \frac{1 - \omega}{1 + u_1}, S[\bar{j}(t, \omega)] = \frac{\omega - 1}{1 + u_1}$$
$$S\left[\underline{j'}(t, \varepsilon) \right] = \frac{\varepsilon - 1}{1 + u_1}, S[\bar{j'}(t, \varepsilon)] = \frac{1 - \varepsilon}{1 + u_1}$$

Then we get the alpha and beta cut representation of solution is

$$\underline{j}(t, \omega) = (1 - \omega)e^{-t}, \bar{j}(t, \omega) = (\omega - 1)e^{-t}$$
$$\underline{j'}(t, \varepsilon) = (\varepsilon - 1)e^{-t}, \bar{j'}(t, \varepsilon) = (1 - \varepsilon)e^{-t}.$$

Case: 2 let j(t) is (i) and $j'(t)$ is (ii)-differentiable, then

$$\frac{-S[j(t)]}{u_1^2} \ominus \frac{-j(0)}{u_1^2} \ominus \frac{-j'(0)}{u_1} = S[j(t)]$$

Thus:

$$S\left[\left(\underline{j}(t, \omega) \right) \right] = (1 - \omega) \left[\left(\frac{1}{u_1^2 + 1} \right) + \frac{u_1}{u_1^2 + 1} \right], S[\bar{j}(t, \omega)] = (\omega - 1) \left[\left(\frac{1}{u_1^2 + 1} \right) + \frac{u_1}{u_1^2 + 1} \right]$$
$$S\left[\left(\underline{j'}(t, \varepsilon) \right) \right] = (\varepsilon - 1) \left[\left(\frac{1}{u_1^2 + 1} \right) + \frac{u_1}{u_1^2 + 1} \right], S[\bar{j'}(t, \varepsilon)] = (1 - \varepsilon) \left[\left(\frac{1}{u_1^2 + 1} \right) + \frac{u_1}{u_1^2 + 1} \right]$$

Then we get the alpha and beta cut representation of solution is

$$\underline{j}(t,\omega) = (1-\omega)(\sinh t + \sin t); \quad \bar{j}(t,\omega) = (\omega-1)(\sinh t + \sin t).$$
$$\underline{j'}(t,\varepsilon) = (\varepsilon-1)(\sinh t + \sin t); \quad \bar{j'}(t,\varepsilon) = (1-\varepsilon)(\sinh t + \sin t).$$

Case: 3 let $j'(t)$ is (i) and $j(t)$ is (ii)-differentiable, then

$$\left[\frac{-S[j(t)]}{u_1^2} \ominus \left(\frac{-j(0)}{u_1^2} \right) \ominus \left(\frac{j'(0)}{u_1} \right) \right] = S[j(t)]$$

Thus:

$$S\left[\underline{j}(t,\omega)\right] = (\omega-1)\left[\left(\frac{1}{u_1^2+1} \right) - \left(\frac{u_1}{u_1^2+1} \right) \right]$$

$$S\left[\bar{j}(t,\omega)\right] = (1-\omega)\left[\left(\frac{1}{u_1^2+1} \right) - \left(\frac{u_1}{u_1^2+1} \right) \right]$$

$$S\left[\underline{j'}(t,\varepsilon)\right] = (1-\varepsilon)\left[\left(\frac{1}{u_1^2+1} \right) - \left(\frac{u_1}{u_1^2+1} \right) \right]$$

$$S[\bar{j'}(t,\varepsilon)] = (\varepsilon-1)\left[\left(\frac{1}{u_1^2+1} \right) - \left(\frac{u_1}{u_1^2+1} \right) \right]$$

Then we get the alpha and beta cut representation of solution is

$$\underline{j}(t,\omega) = (\omega-1)(\sinh t - \sin t); \quad \bar{j}(t,\omega) = (1-\omega)(\sinh t - \sin t).$$
$$\underline{j'}(t,\varepsilon) = (1-\varepsilon)(\sinh t - \sin t); \quad \bar{j'}(t,\varepsilon) = (\varepsilon-1)(\sinh t - \sin t).$$

Case: 4 let $j'(t)$ is (i) and $j(t)$ is (ii)-differentiable, then

$$\frac{S[j(t)]}{u_1^2} \ominus \frac{j(0)}{u_1^2} \ominus \frac{-j'(0)}{u_1} = S[j(t)]$$

Thus:

$$S\left[\underline{j}(t,\omega)\right] = (\omega-1)\left[\frac{1}{u_1+1} \right], S\left[\bar{j}(t,\omega)\right] = (1-\omega)\left[\frac{1}{u_1+1} \right]$$

$$S\left[\underline{j'}(t,\varepsilon)\right] = (1-\varepsilon)\left[\frac{1}{u_1+1} \right], S[\bar{j'}(t,\varepsilon)] = (\varepsilon-1)\left[\frac{1}{u_1+1} \right]$$

Then we get the alpha and beta cut representation of solution is

$$\underline{j}(t, \omega) = (\omega - 1)e^{-t}, \bar{j}(t, \omega) = (1 - \omega)e^{-t}.$$
$$\underline{j}'(t, \varepsilon) = (1 - \varepsilon)e^{-t}, \bar{j}(t, \varepsilon) = (\varepsilon - 1)e^{-t}.$$

4 Numerical Examples

Solve the fuzzy differential equation of order 2 by fuzzy sumudu transform. Take initial values as intuitionistic triangular fuzzy numbers $j''(t) = g(t, j(t), j'(t))$ where $j(0) = (7, 8, 9; 9, 8, 6)$ & $j'(0) = (8, 9, 10; 9, 10, 7)$

Solution

We know that $\frac{S[j(t)]}{u_1^2} \ominus \frac{j(0)}{u_1^2} \ominus \frac{j'(0)}{u_1} = S[j(t)]$

Taking fuzzy sumudu transform and inverse fuzzy sumudu transform we get the following results (Table 1 and Fig. 2).

$$S(\underline{\varphi}(t, \omega)) = \frac{t^2}{2!}(k) + (7 + \alpha) + t(8 + \alpha)$$

$$S(\overline{\varphi}(t, \omega)) = \frac{t^2}{2!}(k) + (10 - 2\alpha) + t(12 - 3\alpha)$$

$$S(\underline{\varphi}'(t, \beta)) = \frac{t^2}{2!}(k) + (8 + \beta) + t(10 + \beta)$$

$$S(\overline{\varphi}'(t, \beta)) = \frac{t^2}{2!}(k) + (8 + 2\alpha) + t(10 - 3\beta)$$

Table 1. Values of $S(\underline{\varphi}(t, \omega)); S(\overline{\varphi}(t, \omega)); S(\underline{\varphi}'(t, \beta)); S(\overline{\varphi}'(t, \beta)$

(α, β)	$S(\underline{\varphi}(t, \omega))$	$S(\overline{\varphi}(t, \omega))$	$S[\underline{\varphi}'(t, \beta)]$	$S[\overline{\varphi}'(t, \beta)]$
0	47.0000	66.0000	56.0000	56.0000
0.1	47.5000	64.6000	56.5000	55.0000
0.2	48.0000	63.2000	57.0000	54.0000
0.3	48.5000	61.8000	57.5000	53.0000
0.4	49.0000	60.4000	58.0000	52.0000
0.5	49.5000	59.0000	58.5000	51.0000
0.6	50.0000	57.6000	59.0000	50.0000
0.7	50.5000	56.2000	59.5000	49.0000
0.8	51.0000	54.8000	60.0000	48.0000
0.9	51.5000	53.4000	60.5000	47.0000
1	52.0000	52.0000	61.0000	46.0000

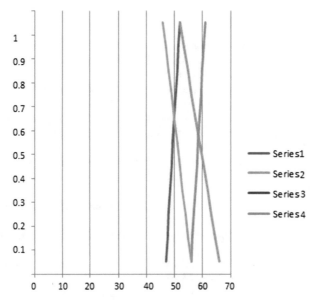

Fig. 2. Graphical representations

5 Application

1. In Electrical Engineering, movement of electricity in a circuit can be described by fuzzy differential equation by fuzzy numbers.
2. In Mechanical Engineering, we can find the flow of uniform axial load in any road can be identifying by fuzzy intuitionistic numbers, etc.

6 Conclusion

We solved fuzzy differential equation of order 2 by fuzzy sumudu transform method; here we used two point initial conditions as intuitionistic fuzzy numbers, we obtained best solution. This will helpful for those who are using intuitionistic fuzzy environment to any real life problems.

References

1. Atanassov, K.T.: Intuitionistic fuzzy sets. VII ITKR's session, Sofia (deposited in Central Science and Technical Library of the Bulgarian Academy of Sciences 1697/84) (1983)
2. Dubois, D., Parade, H.: Operation on fuzzy number. Int. J. Fuzzy Syst. **9**, 613–626 (1978)
3. Atanassov, K.T.: Intuitionistic fuzzy sets. Fuzzy Sets Syst. **20**, 87–96 (1986)
4. Zadeh, L.A.: Toward a generalized theory of uncertainty (GTU) an outline. Inf. Sci. **172**, 140 (2005)

5. Dubois, D., Prade, H.: Towards fuzzy differential calculus: part 3. Differ. Fuzzy Sets Syst. **8**, 225–233 (1982)
6. Kaleva, O.: Fuzzy differential equations. Fuzzy Sets Syst. **24**, 301–317 (1987)
7. Chalco-Cano, Y., Romn-Flores, H.: On the new solution of fuzzy differential equations. Chaos, Solitons Fractals **38**, 112–119 (2008)
8. Gasilov, N., Amrahov, S.E., Fatullayev, A.G.: Solution of linear differential equations with fuzzy boundary values. Fuzzy Sets Syst. **257**, 169–183 (2014)
9. Lata, S., Kumar, A.: A new method to solve time-dependent intuitionistic fuzzy differential equation and its application to analyze the intutionistic fuzzy reliability of industrial system. Concurrent Eng.: Res. Appl. **20**, 1–8 (2012)
10. Atanassov, K.T.: Intuitionistic fuzzy sets. Fuzzy Sets Syst. **20**(1), 87–96 (1986)
11. Atanassov, K.T.: Operators over interval valued intuitionistic fuzzy sets. Fuzzy Sets Syst. **64** (2), 159–174 (1994)
12. De, S.K., Biswas, R., Roy, A.R.: An application of intuitionistic fuzzy sets in medical diagnosis. Fuzzy Sets Syst. **117**(2), 209–213 (2001)
13. Kharal, A.: Homeopathic drug selection using intuitionistic fuzzy sets. Homeopathy **98**(1), 35–39 (2009)
14. Mondal, S.P., Roy, T.K.: First order linear non homogeneous ordinary differential equation in fuzzy environment. Math. Theory Model **3**(1), 85–95 (2013)
15. Mondal, S.P., Banerjee, S., Roy, T.K.: First order linear homogeneous ordinary differential equation in fuzzy environment. Int. J. Pure Appl. Sci. Technol. **14**(1), 16–26 (2013)
16. Nirmala, V., Pandian, S.C.: Numerical approach for solving intuitionistic fuzzy differential equation under generalised differentiability concept. Appl. Math. Sci. **9**(67), 3337–3346 (2015)
17. Ettoussi, R., Melliani, S., Elomari, M., Chadli, L.S.: Solution of intuitionistic fuzzy differential equations by successive approximations method. Notes on Intuitionistic Fuzzy Sets **21**(2), 51–62 (2015)
18. Melliani, S., Elomari, M., Atraoui, M., Chadli, L.S.: Intuitionistic fuzzy differential equation with nonlocal condition. Notes on Intuitionistic Fuzzy Sets **21**(4), 58–68 (2015)
19. Mondal, S.P., Roy, T.K.: System of differential equation with initial value as triangular intuitionistic fuzzy number and its application. Int. J. Appl. Comput. Math. **1**, 449–474 (2015)
20. Stefanini, L., Bede, B.: Generalized Hukuhara differentiability of interval-valued functions and interval differential equations. Nonlinear Anal. **71**, 1311–1328 (2009)
21. Rajkumar, A., Jesuraj, C.: Mathematical model for dengue virus infected populations with fuzzy differential equations. In: ICAICR 2018, CCIS 955, pp. 206–217. Springer, Singapore (2019)

DeepTrackNet: Camera Based End to End Deep Learning Framework for Real Time Detection, Localization and Tracking for Autonomous Vehicles

Dinesh Kumar Amara[1]([✉]), R. Karthika[1], and K. P. Soman[2]

[1] Department of Electronics and Communication Engineering, Amrita School of Engineering, Coimbatore, Amrita Vishwa Vidyapeetham, Coimbatore, India
`dineshkumar.amara@gmail.com`
[2] Centre for Computational Engineering and Networking (CEN), Amrita School of Engineering,Coimbatore, Amrita Vishwa Vidyapeetham, Coimbatore, India

Abstract. Vehicle detection and tracking in real time remains as the most challenging tasks for the autonomous vehicles and mobile robots. The designed algorithms should have less latency and with high accuracy and performance. This paper proposed a unified end to end deep learning framework for real-time detection, localization and tracking for autonomous vehicles called Deep-TrackNet. In the proposed DeepTrackNet architecture Mobilenet SSD is used for vehicle detection, localization and deep regression network is used for vehicle tracking. Based on the experimental analysis performed we infer that the proposed DeepTrackNet architecture has produced satisfactory results for real-time vehicle detection and tracking on an Nvidia Jetson embedded computing platform using monocular camera with a reasonable latency. Mobilenet SSD took 84 (ms) median detection time and deep regression tracker has the least overall average failure frames at 4.388 on the VTB TB-100 dataset (http://cvlab. hanyang.ac.kr/tracker_benchmark/datasets.html). The code along with detailed results are available in the GitHub repository (https://github.com/dineshresearch/ DeepTrackNet).

Keywords: Vehicle detection · Localization · Single Shot Multi Box Detection · Offline tracking · DeepTrackNet · Deep regression networks

1 Introduction

Real-time vehicle detection and tracking are the challenging tasks applied in many applications of computer vision domains like Autonomous vehicles, Advanced Driver Assistance systems, Robot navigation and Surveillance [1, 22, 23]. Vehicle detection and tracking are generally tackled by using RADAR sensors and LIDAR sensors but because of the cost and the poor ability in estimating the size and the shape of vehicles, camera related solutions have recently gained attention in autonomous vehicles research community. Availability of large datasets [2–4], embedded processing power [5], led to an

© Springer Nature Switzerland AG 2020
A. P. Pandian et al. (Eds.): ICICCS 2019, AISC 1039, pp. 299–307, 2020.
https://doi.org/10.1007/978-3-030-30465-2_34

interest in applying deep learning architectures for real-time detection, localization and tracking of the vehicles are performed superior to the prior algorithms [6, 7].

Our main contributions from this research are

1. Developed an end to end framework using Single Shot Multi Box Detector (SSD) architecture for the detection, localization and deep regression network for tracking vehicles and implemented it on Nvidia Jetson tx1 embedded board in real-time.
2. SSD Mobilenet with Faster-RCNN, R-FCN Convolutional Neural Network (CNN) architectures are compared for vehicle detection and localization.
3. Compared and evaluated the performance of deep learning based offline trained CNN architecture called deep regression network and feature based online trackers on the Visual tracker benchmark (VTB) [10].

The paper is organized as follows: Sect. 2 includes the related work. Section 3 discuss different vehicle detection and localization algorithms, Sect. 4 describes about the various online tracking algorithms and Sect. 5 includes the offline deep regression network tracking algorithm. Section 6 contains the experimental analysis and the results and Sect. 7 is conclusion.

2   Related Work

Extensive research was done using conventional image processing techniques and feature extraction methods for vehicle detection and localization. HOG and SIFT are generally used for feature extraction from the target image along with Support Vector Machine (SVM is commonly used [8]). Different convolutional neural networks architectures were proposed for real-time object detection, localization and widely used ones are You look only once (YOLO), Tiny YOLO, Fast Region convolutional network (RCNN), Faster-RCNN, SSD [9].

For vehicle tracking popular track by detection algorithms uses discriminative appearance based model which consists of a classifier that classifies object presence in an image patch [11]. Supervised learning algorithms in machine learning like SVM, Random forest [12], Naive Bayes classifier and other boosting algorithms are widely used in predicting the target in the image patch. But the machine learning algorithms need offline prior training with a training dataset but the biggest disadvantage is they perform well only for the testing data that is similar to training data.

3   Detection and Localization Algorithms

3.1   Mobile Net

Mobile Nets are network architectures constructed for the purpose of running efficiently on mobile embedded based vision application. Mobile Nets achieve this with the following techniques:

1. Rather than a standard convolution perform depth wise convolution followed by a 1×1 convolution. If 1×1 convolution is following a depth wise convolution then it is called a point wise convolution. A separable depth wise convolution is the combination of depth wise convolution followed by a point wise convolution.
2. Reducing the size of input/output channels by using a width multiplier, set to a value between 0 and 1.
3. Reducing the size of original input using a width multiplier, set to a value between 0 and 1.

The size of cumulative parameters and the computation required can be reduced using the above techniques. Usually models produce higher accuracy and were optimized for mobile devices.

3.1.1   Single Shot Multi Box Detection (SSD)

Single Shot Detection (SSD) is a popular algorithm in object detection which is faster than RCNN (Region convolution neural network) and Region Proposal Network (RPN). The detailed architecture and layers are showed in the Fig. 1.

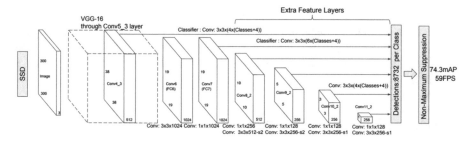

Fig. 1. Single shot multi box detector (SSD) architecture

3.2   Faster Region Convolutional Neural Network (Faster RCNN)

Faster R-CNN consists of two networks: network using these proposals to detect objects and region proposal network (RPN) for producing region proposals. When it shares the most computation with the object detection network, the time cost of generating region proposals are smaller in RPN than the selective search, RPN ranks region boxes (anchors) and proposed the ones most likely containing objects. An anchor is basically a box and it play crucial role in Faster R-CNN (Fig. 2).

RPN determine which regions of image to be selected. A pre-trained model is then combined with RPN. The SSD is a single convolutional network learn to predict bounding box locations and arrange the locations in one pass. Faster RCNN cannot be trained end to end but SSD can. SSD architecture includes various convolution layers and a base network.

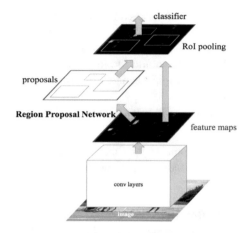

Fig. 2. Faster R-CNN architecture

3.3 Region Based Fully Convolutional Networks (R-FCN)

R-FCN computes the feature maps from the whole image. It derive the region proposals (ROIs) from the feature maps directly. For every ROI, no more feature extraction is needed. By reducing the work needed for finding each ROI R-FCN increases the speed. The region-based feature maps are independent of ROIs and can be computed outside each ROI.

4 Online Tracking Algorithms

Online trackers are the feature based trackers which tracks the target by extracting the features from the previous frame [18–20] and comparing them with the current frame without any prior offline training [13–17].

4.1 Boosting Tracker

Tracking is classified as a binary classification problem distinguishing the object from the background. It is a online Ada boost based tracking algorithm where a combination of weak classifiers is ensembled into the strong classifier to achieve the robustness against variations. A voting weight is assigned to each of the weak classifiers based on the prior estimated error of the classifier [12].

Online boosting involved the random initialization of the selectors which have feature pool of different weak classifiers. It consists of Haar-like features, orientation of histograms and local binary patterns. A strong classifier is obtained by a linear combination of several weak classifiers.

$$h^{strong}(x) = sign\left(\sum_{n=1}^{N} \alpha_n . h_n^{sel}(x)\right) \tag{6}$$

4.2   Multiple Instance Learning (MIL)

It is an online discriminative classifier for separating the object from the background. Compared to conventional boosting algorithms an adaptive appearance model that adjusts itself as the object moves and changes its shape can perform accurate tracking. MIL is analogous to the voila Jones method which is used for detection. It consists of three parts: image representation, appearance and motion models. It is based on MIL boosting algorithm and online ada boost algorithm.

Following a greedy strategy for updating the tracker location as

$$l_t^* = l(\operatorname{argmax} p(y|x)) \tag{7}$$

$$x \varepsilon X^s$$

4.3   Kernalised Correlation Filter *(KCF)*

Discriminative learning methods are used by considering the object as positive and the background environment as negative in the sequence of images thus converting tracking it into a classification problem. Because of the computational constraints, the number of negative samples is limited but KCF uses the Kernel Ridge Regression which is similar to linear correlation filter with the kernel [13]. Kernel trick transforms from the linear space to the non-linear space using the kernel function [13].

$$w = \sum_i \alpha_i \varphi(x_i) \tag{8}$$

4.4   Median Flow

Tracking is done both forward and backward and the differences between the trajectories is measured. It is similar to normalized cross-correlation (NCC), minimizing this forward and backward error leads to reliable tracking. It involves reverse sequence applying to the same tracking algorithm.

4.5   Discriminative Correlation Filter (Dlib)

Discriminative Correlation Filter uses the correlation filters to estimate the translation and the scale correlating the previous frame and the present frame by minimizing the cost function of the optimal correlation filter (h). H^l is one filter per feature dimension and desired correlation output is g.

5   Offline Deep Regression Network Tracking Algorithm

The proposed deep regression network algorithm first crop the current input frame and get the search region and pass it to a stack of convolutional layers. Crop the previous frame based on region of interest and pass it to a stack of convolutional layers. Finally, both the convolutional layer outputs are passed through three fully connected layers that output the predicted location of the target in the search region of the frame. It is highly inspired from Goturn [21] tracker (Fig. 3).

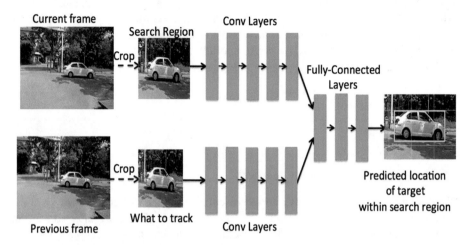

Fig. 3. Deep regression network architecture for vehicle tracking

6   Experimental Analysis and Results

The block diagram shown in Fig. 4 explain the proposed DeepTrackNet architecture. The Deep regression network is an offline tracker and track vehicles at 100 Frames per Second (FPS) during test time. The network is pre trained with the 314 video sequences from ALOV300++ dataset and 13,082 still images of 251 objects and some additional images from Imagenet database. The offline training and one pass regression features makes the tracker very fast to use it in real time reliably. The Video stream is converted to individual frames and they are passed further for detection and tracking. Nvidia Jetson TX1 embedded controller board with specifications as central processing unit (CPU) Quad ARM® A57/2 MB L2, Graphical Processing Unit (GPU) NVIDIA Maxwell™, 256 Compute Unified Device Architecture (CUDA) cores and 4 GB 64 bit LPDDR4 memory, ZED camera with 720 p resolution operated at 30 FPS. OpenCV, Dlib, Tensorflow, Keras libraries with CUDA acceleration software libraries were used for algorithm development and implementation.

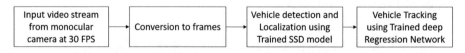

Fig. 4. Block diagram of the proposed DeepTrackNet architecture

By performing the detection for 10 times and by calculating the detection time's box plot in Fig. 5 is plotted. The boxplots illustrates that out of the RFCN, Faster RCNN, SSD algorithms SSD performs faster detection with least latency of 84 ms.

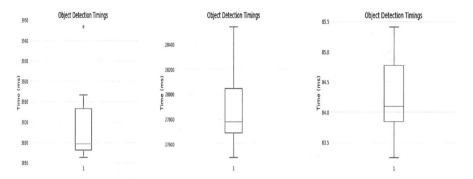

Fig. 5. Boxplot illustrating detection timing for RFCN (Left), faster RCNN (Middle), SSD (Right) algorithms in milliseconds

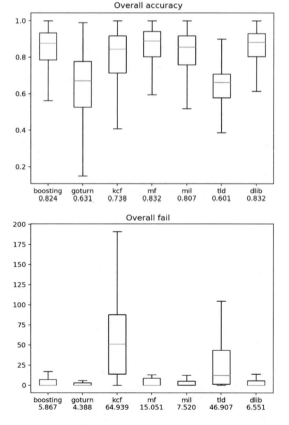

Fig. 6. Boxplot illustrating overall average accuracy (in 1000 frames) and overall average failure frames at the overlap threshold 0.5 (in 1000 frames)

Figure 6 shows the overall accuracy and the failure rate for the trackers with box plot. The detailed results are shown in the Table 1 on VTB dataset (TB-100 Sequences). The proposed deep regression tracker has the lowest Overall Average Failure frames at 4.388. The tracker achieved almost real time tracking with low latency and computational power because of the offline training process.

Table 1. Results on VTB dataset (TB-100 Sequences) for the trackers

	Boosting	Proposed tracker	KCF	MF	MIL	TLD	Correlation tracker
Overall average overlap (in 1000 frames)	82.4	63.1	73.8	83.2	80.7	60.1	83.2
Overall Average Failure frames at the overlap threshold 0.5 (in 1000 frames)	5.867	4.388	64.939	15.051	7.520	46.907	6.551

7 Conclusion

Various deep learning based algorithms were compared and evaluated for vehicle detection and localization on embedded platform. Calculated the performance of the online feature trackers and deep regression network on the Visual tracker benchmark (VTB) TB-100 video sequences. Evaluated and compared the vehicle detection timing for SSD Mobile net with Faster Region convolutional network (Faster-RCNN), Region-based Fully Convolutional Networks (R-FCN). Mobile net SSD outperformed others with least median latency of 84 ms. The reliability and robustness of the framework can be improved from the perturbation and adversarial attacks. Future work can be extended by reducing the model size by network optimization, pruning and quantization. The code is available open for further research.

References

1. Geronimo, D., et al.: Survey of pedestrian detection for advanced driver assistance systems. IEEE Trans. Pattern Anal. Mach. Intell. **32**(7), 1239–1258 (2010)
2. Geiger, A., et al.: Vision meets robotics: the KITTI dataset. Int. J. Robot. Res. **32**(11), 1231–1237 (2013)
3. Geiger, A., Lenz, P., Urtasun, R.: Are we ready for autonomous driving? the KITTI vision benchmark suite. In: 2012 IEEE Conference on Computer Vision and Pattern Recognition (CVPR). IEEE (2012)
4. Cordts, M., et al.: The cityscapes dataset for semantic urban scene understanding. In: Proceedings of the IEEE Conference on Computer Vision and Pattern Recognition (2016)
5. Meyer, G.G., Främling, K., Holmström, J.: Intelligent products: a survey. Comput. Ind. **60**(3), 137–148 (2009)

6. Udalski, A.: The optical gravitational lensing experiment. Real time data analysis systems in the OGLE-III survey. arXiv preprint arXiv:astro-ph/0401123 (2004)
7. Leonard, J.J., Durrant-Whyte, H.F.: Mobile robot localization by tracking geometric beacons. IEEE Trans. Robot. Autom. **7**(3), 376–382 (1991)
8. Sapankevych, N.I., Sankar, R.: Time series prediction using support vector machines: a survey. IEEE Comput. Intell. Mag. **4**(2), 24–38 (2009)
9. Lienhart, R.: Reliable transition detection in videos: a survey and practitioner's guide. Int. J. Image Graph. **1**(03), 469–486 (2001)
10. Ma, C., et al.: Hierarchical convolutional features for visual tracking. In: Proceedings of the IEEE International Conference on Computer Vision (2015)
11. Li, X.: A survey of appearance models in visual object tracking. ACM Trans. Intell. Syst. Technol. (TIST) **4**(4), 58 (2013)
12. Kulkarni, V.Y., Sinha, P.K.: Random forest classifiers: a survey and future research directions. Int. J. Adv. Comput. **36**(1), 1144–1153 (2013)
13. Li, Y., Zhu, J.: A scale adaptive kernel correlation filter tracker with feature integration. In: European Conference on Computer Vision. Springer, Cham (2014)
14. Smeulders, A.W.M., Chu, D.M., Cucchiara, R., Calderara, S., Dehghan, A., Shah, M.: Visual tracking: an experimental survey. IEEE Trans. Pattern Anal. Mach. Intell. (1), 1 (2013)
15. Yang, H., Shao, L., Zheng, F., Wang, L., Song, Z.: Recent advances and trends in visual tracking: a review. Neurocomputing **74**(18), 3823–3831 (2011)
16. Zhang, K., Zhang, L., Yang, M.-H.: Real-time compressive tracking. In: European Conference on Computer Vision, pp. 864–877. Springer, Heidelberg (2012)
17. Kalal, Z., Mikolajczyk, K., Matas, J., et al.: Tracking-learning-detection. IEEE Trans. Pattern Anal. Mach. Intell. **34**(7), 1409 (2012)
18. Babenko, B., Yang, M.-H., Belongie, S.: Robust object tracking with online multiple instance learning. IEEE Trans. Pattern Anal. Mach. Intell. **33**(8), 1619–1632 (2011)
19. Saffari, A., Leistner, C., Santner, J., Godec, M., Bischof, H.: On-line random forests. In: 2009 IEEE 12th International Conference on Computer Vision Workshops (ICCV Workshops), pp. 1393–1400. IEEE (2009)
20. Grabner, H., Grabner, M., Bischof, H.: Real-time tracking via on-line boosting. In: BMVC, vol. 1, p. 6 (2006)
21. Held, D., Thrun, S., Savarese, S.: Learning to track at 100 fps with deep regression networks. In: European Conference on Computer Vision. Springer, Cham (2016)
22. Deepika, N., Variyar, V.V.S.: Obstacle classification and detection for vision based navigation for autonomous driving. In: 2017 International Conference on Advances in Computing, Communications and Informatics (ICACCI). IEEE (2017)
23. Jose, A., Thodupunoori, H., Nair, B.B.: A novel traffic sign recognition system combining Viola–Jones framework and deep learning. In: Soft Computing and Signal Processing, pp. 507–517. Springer, Singapore (2019)

Real-Time Speed Bump Detection Using Image Segmentation for Autonomous Vehicles

J. Arunpriyan[1]([⊠]), V. V. Sajith Variyar[2], K. P. Soman[2], and S. Adarsh[1]

[1] Department of Electronics and Communication Engineering,
Amrita School of Engineering, Amrita Vishwa Vidyapeetham, Coimbatore, India
`arunpriyan.j@gmail.com, s_adarsh@amrita.edu`
[2] Center for Computational Engineering and Networking (CEN),
Amrita School of Engineering, Amrita Vishwa Vidyapeetham, Coimbatore, India
`{vv_sajithvariyar, kp_soman}@amrita.edu`

Abstract. Autonomous vehicle technology, which is evolving at a faster pace than predicted is promising to deliver higher safety benefits. Detecting the obstacles accurately and reliably is important for safer navigation. Speed bumps are the obstacles installed on the roads in order to force the vehicle driver to reduce the speed of the vehicle in the critical road areas, such as hospitals and schools. Autonomous vehicles have to detect and slower the speed appropriately to drive safely over the speed bump. In this paper, we propose a novel method to detect the upcoming speed bump by using a deep learning algorithm called SegNet, which is a deep convolutional neural network architecture for semantic pixel-wise segmentation. The trained model will give segmented output from the monocular camera feed placed in front of the vehicle.

Keywords: Autonomous vehicle technology · Obstacle avoidance · Speed bump · Deep learning · Semantic segmentation · SegNet · Monocular camera

1 Introduction

According to the definition of Society of Automotive Engineers (SAE) J3106, Autonomous vehicles are the vehicles that are equipped with technology for reducing vehicle crashes, congestion energy consumption and pollution meanwhile improving transport accessibility [1]. Development in this field has progressed long way after the vision guided robotic Van is developed by Mercedes-Benz in the year of 1980 [2]. Currently, various universities and institutions all over the world are working on research projects based on Autonomous vehicles [2]. The modern research and development are based on implementing High Automation (Level 4) and even Full Automation (Level 5) in autonomous vehicles [3]. The modern autonomous driving techniques are based on the highly expensive components, such as Global Positioning (GPS) for localization, Light Detection and Ranging (LIDAR) for navigation, and Laser Range Finder (LRF) for obstacle detection [4].

The acceptance and implementation of any new technology mainly depend on cost and technology constraints. With the advancement in deep learning based image

A. P. Pandian et al. (Eds.): ICICCS 2019, AISC 1039, pp. 308–315, 2020.
https://doi.org/10.1007/978-3-030-30465-2_35

processing techniques, normal monocular vision cameras, which are cheaper than the expensive sensors can be used to understand the environment and avoid obstacles in the navigation path [5]. In the current scenario of unstructured road infrastructure, obstacle detection is very important for autonomous vehicles to safely navigate the vehicle. The cost, slow scanning speed and low spatial resolution of active sensors like Lidar and radar hindered the development of the obstacle detection system [5]. Monocular cameras proven to be more efficient in obstacle detection and also can be used for obstacle identification [6].

Speed bumps are obstacles installed on the roads to constrain the vehicles to slower speeds near the key areas where a higher number of pedestrians could potentially cross the roads, such as Hospitals and Schools. IRC099 of Indian Road Congress (IRC) suggests guidelines for constructing speed bumps also makes it mandatory to install the Speed bump warning signboards [7]. The Indian road scenarios do not follow the guidelines seriously and the speed bumps are not with a uniform structure. We have also observed that many speed bumps are maintained poorly and warning signboards are missing.

2   Literature Review

Earlier research works carried out described in these articles [8–12] uses the accelerometer and GPS for detecting speed bump. These techniques are not real time and detect speed bump by processing the accelerometer data after the vehicle passes the speed bump. Techniques described in [13,14] uses cloud storage to store information about speed bumps. Vehicle driver will be alerted only about known speed bumps about which the information is already available in the cloud storage. The systems cannot alert the driver about unknown speed bumps and network connectivity issues could alert the driver at the wrong location.

Few research works were carried out using vision based techniques. Devapriya et al. [15, 16] proposed a real-time speed bump detection by analyzing the images from the road and applying computer vision enhancement based on a Gaussian filter and a connected component approach. Srimongkon et al. at [17] show bump detection using a Gaussian mixture model with segmentation, dilation, and erosion operations. The 3D vision-based approach at [18] have used single road image and it suits for images with a single vanishing point only. Deep learning based techniques have proven to be efficient than conventional image processing techniques by a large margin.

In the work done at [19] uses deep learning based image processing technique to detect the speed bump in real-time. The system uses the stereovision camera to estimate the speed bump's distance and alert is if the speed bump is within the specified distance limit. This approach uses deep learning based object detection technique, which predicts only the approximate region of the speed bump. For safer navigation of autonomous vehicles, obstacle detection needs to be accurate and reliable.

Our contribution in this research work is to propose a novel deep learning based image pixel segmentation technique to detect the speed bump in real time using a monocular camera. We are proposing to use the semantic segmentation algorithm called SegNet [20], which is a deep encoder-decoder architecture for multi pixel-wise segmentation. SegNet has been proved to be an efficient architecture for real-time semantic segmentation over the other algorithms benchmark by CamVid dataset [20]. SegNet was implemented as multiclass pixel-wise segmentation for segmenting 11 classes such as road, building, cars, poles etc. and speed bump is not one of the predefined class. In this work, the SegNet architecture is exclusively trained to detect speedbump and the remaining pixels in the image are grouped to *non-speedbump class*.

This paper is organized as follows. The following Sect. 2 gives the details of the data and methods used for pixel-wise segmentation. Results of our experiments are detailed in Sect. 3. Section 4 concludes the paper and explains about the scope for the future work.

3   Data and Methods

3.1   Dataset Description

Speed bump dataset is created by capturing the images of various sized marked speed bumps available in and around Coimbatore city. In order to have the image dataset with various lighting conditions, they are ought to be captured at different times of the day. Figure 1 shows sample images present in our dataset. We have captured it to include various sized speed bumps found around the city.

(a)   (b)

Fig. 1. (a) Big speed bump with wide stripes (b) Small speed bump with small stripes

3.2   Preprocessing of Data

The collected data set for speed bump segmentation includes other background information such as trees, pedestrians etc. To remove the objects that are not in interest, we need to crop the images to a uniform size of 360 by 480. The pre-processed images will be annotated to highlight the pixels that belong to *speedbump class* and all remaining pixels that belong to the *non-speedbump class*.

To increase the training samples and include different varieties of training data, image augmentation is performed on the data set. Data augmentation techniques comprised rotation, horizontal flip, Zooming, brightness, contrast change and HSV colour space conversion. Table 1 provides the statistics of the speedbump image dataset.

Table 1. Speed bump image dataset statistics

S. No.	No. of captured images	Images obtained from five types of augmentation	Total No of images
1	217	1085	1302

3.3 Proposed Method

Semantic Segmentation. Semantic segmentation is the process of associating each pixels of an image with the available classes. Semantic segmentation is extensively used for object localization, tracking, detection, classification, and other scene understanding tasks [21]. For autonomous vehicles, it is vital to understand the driving scenery to make a proper decision at the right time and navigate the car safely [21].

SegNet architecture, which is a multiclass pixel-wise segmentation algorithm, consists of a series of encoders and decoder set, which followed by a final pixel-wise classifier follows. A uniqueness of SegNet is that it uses max-pooling indices in the decoders for performing upsampling of low-resolution feature maps. This technique has the advantage of keeping high frequency details in the segmented images and reduces the number of trainable parameters in the decoders. Figure 2 illustrates the layers of SegNet architecture.

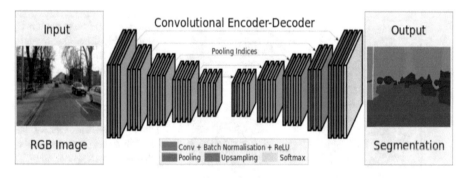

Fig. 2. Architecture of SegNet - Layers of SegNet illustrated [20]

4   Experimental Results and Discussion

SegNet is implemented for road scene understanding by the authors [20] for segmenting 11 classes. Since segmenting speed bump is a binary classification problem we have modified the SoftMax layer output parameter as '2'. The SegNet is trained for 38,000 iterations using GPU Tesla K40C. Out of total speedbump image dataset, 1000 images are used for training and 300 are used for testing. We have used the optimum hyperparameters such as fixed learning rate of 0.01 and momentum of 0.9 to avoid overfitting. Figure 3 explains the block diagram of speed bump segmentation system.

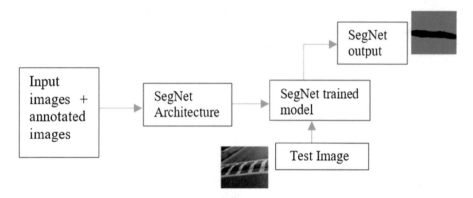

Fig. 3. Block diagram explaining speed bump segmentation system

To test the performance of the system the trained model is tested with different scenarios. Table 2 discusses in detail about the performance of the system in various scenarios.

Table 2. Performance of the trained SegNet model for test speed bump image dataset

S. No.	Input image	Output image	Output description
1.			Speed bump image captured during day time. Our Model is able to detect the speed bump with high accuracy
2.			Speed bump image captured during night time. Our Model is able to segment the pixels that belong to speedbump class
3.			Speed bump image captured and augmented by modifying brightness and contrast is predicted precisely
4.			Captured speed bump image is augmented to change the orientation. Our model is able to segment around 70% of the pixels correctly to speedbump class

(continued)

Table 2. (*continued*)

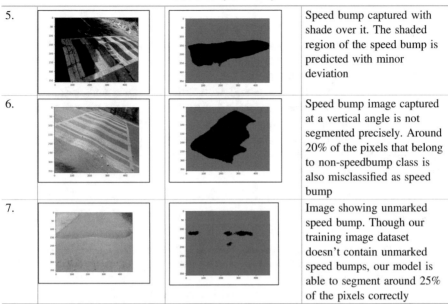

5.			Speed bump captured with shade over it. The shaded region of the speed bump is predicted with minor deviation
6.			Speed bump image captured at a vertical angle is not segmented precisely. Around 20% of the pixels that belong to non-speedbump class is also misclassified as speed bump
7.			Image showing unmarked speed bump. Though our training image dataset doesn't contain unmarked speed bumps, our model is able to segment around 25% of the pixels correctly

To evaluate the performance of the trained model we have used commonly used performance measures like global accuracy, class average accuracy, and mIoU, which is also known as Jacard Index. Global accuracy specifies the percentage value of the pixels that are correctly classified in the dataset. Class average accuracy is the average of the predictive accuracy of all classes. Mean intersection over union (mIoU) is the average of IoU calculated for each class. IoU is defined as follows:

$$IoU = Area\,of\,overlap/Area\,of\,Union. \tag{1}$$

Table 3 shows the performance measures obtained by testing the model against our test image dataset. In a captured image, pixels of speed bump occupies a relatively low area of the total image when compared to the pixels that belong to non-speedbump class that creates a variation in class average accuracy.

Table 3. Test performance measures of trained SegNet Model

S. No.	Global accuracy	Class average accuracy	mIOU
1.	91.781%	68.003%	48.872

5   Conclusion and Future Work

This paper presents a method to detect the speed bump with the help of SegNet, using a monocular camera. The above results have proved that the proposed method helps to detect the speed bump in autonomous vehicles with minimum resources and low implementation cost. The model accuracy is improved by training the model with more speed bump images, which are captured with different angles and various lighting condition. The overall accuracy can be improved by using stereo vision camera with depth information instead of a monocular camera.

This work can be extended to detect the unmarked speed bump, rubber rumble strips. Furthermore, this work can be implemented in embedded hardware in order to test their real time performance.

References

1. Emani, S., Soman, K.P., Sajith Variyar, V.V., Adarsh, S.: Obstacle detection and distance estimation for autonomous electric vehicle using stereo vision and DNN. Adv. Intell. Syst. Comput. **898**, 639–648 (2019)
2. Bimbraw, K.: Autonomous cars: past, present and future. In: 2015 12th International Conference on Informatics in Control, Automation and Robotics. vol. 01, pp. 191–198 (2015)
3. Zhou, C., Li, F., Cao, W.: Architecture design and implementation of image based autonomous car: THUNDER-1. Multimed. Tools Appl., 1–17 (2018)
4. Okuyama, T., Gonsalves, T., Upadhay, J.: Autonomous driving system based on deep Q learnig. In: 2018 International Conference on Intelligent Autonomous Systems, ICoIAS 2018, pp. 201–205 (2018)
5. Deepika, N., Sajith Variyar, V.V.: Obstacle classification and detection for vision based navigation for autonomous driving. In: 2017 International Conference on Advances in Computing, Communications and Informatics, ICACCI 2017, pp. 2092–2097 (2017)
6. Prabhakar, G., Kailath, B., Natarajan, S., Kumar, R.: Obstacle detection and classification using deep learning for tracking in high-speed autonomous driving. In: TENSYMP 2017 - IEEE International Symposium Technology Smart Cities, pp. 3–8 (2017)
7. IRC099: Tentative Guidelines on the Provision of Speed Breakers for Control of Vehicular Speeds on Minor Roads. Indian Roads Congress January 1988. https://archive.org/details/govlawircy1988sp99_0. Accessed 15 Mar 2019)
8. Mednis, A., Strazdins, G., Zviedris, R., Kanonirs, G., Selavo, L.: Real time pothole detection using android smartphones with accelerometers. In: 2011 International Conference on Distributed Computing in Sensor Systems and Workshops (DCOSS), pp. 1–6 (2011)
9. Rishiwal, V., Khan, H.: Automatic pothole and speed breaker detection using android system. In: 2016 39th International Convention on Information and Communication Technology, Electronics and Microelectronics (MIPRO), pp. 1270–1273 (2016)
10. Chen, Q., Ding, D., Wang, X., Liu, A.X., Munir, A.: A speed hump sensing approach to global positioning in urban cities without GPS signals. In: 2017 IEEE International Conference on Smart Computing (SMARTCOMP), pp. 1–8 (2017)
11. Mohan, P., Padmanabhan, V., Ramjee, R.: Nericell: rich monitoring of road and traffic conditions using mobile smartphones. In: Proceedings of the 6th ACM Conference on Embedded Network Sensor Systems, pp. 323–336 (2008)

12. Gunawan, F.E., Yanfi, Soewito, B.: A vibratory-based method for road damage classification. In: 2015 International Seminar on Intelligent Technology and Its Applications (ISITIA), pp. 1–4 (2015)
13. Kiran, M.V.K., Vimalkumar, K., Vinodhini, R.E., Archanaa, R.: An early detection-warning system to identify speed breakers and bumpy roads using sensors in smartphones. Int. J. Electr. Comput. Eng. **7**, 1377–1384 (2017)
14. Pooja, P.R., Hariharan, B.: An early warning system for traffic and road safety hazards using collaborative crowd sourcing. In: 2017 International Conference on Communication and Signal Processing (ICCSP), pp. 1203–1206 (2017)
15. Devapriya, W., Babu, C.N.K., Srihari, T.: Real time speed bump detection using Gaussian filtering and connected component approach. In: 2016 World Conference on Futuristic Trends in Research and Innovation for Social Welfare (Startup Conclave), pp. 1–5 (2016)
16. Devapriya, W., Babu, C.N.K., Srihari, T.: Advance driver assistance system (ADAS) - speed bump detection. In: 2015 IEEE International Conference on Computational Intelligence and Computing Research (ICCIC), pp. 1–6 (2015)
17. Srimongkon, S., Chiracharit, W.: Detection of speed bumps using Gaussian mixture model. In: 2017 14th International Conference on Electrical Engineering/Electronics, Computer, Telecommunications and Information Technology (ECTI-CON), pp. 628–631 (2017)
18. Geetha Kiran, A., Murali, S.: Automatic bump detection and 3D view generation from a single road image. In: 2014 International Conference on Advances in Computing, Communications and Informatics (ICACCI), pp. 2232–2238 (2014)
19. Varma, V.S.K.P., Adarsh, S., Ramachandran, K.I., Nair, B.B.: ScienceDirect real time detection of speed speed bump/Bump and distance distance estimation estimation with deep learning using GPU stereo camera with deep learning using GPU and ZED stereo camera. Procedia Comput. Sci. **143**, 988–997 (2018)
20. Badrinarayanan, V., Kendall, A., Cipolla, R.: SegNet: a deep convolutional encoder-decoder architecture for image segmentation. IEEE Trans. Pattern Anal. Mach. Intell. **39**, 2481–2495 (2017)
21. Naresh, Y.G., Little, S., O'Connor, N.E.: A residual encoder-decoder network for semantic segmentation in autonomous driving scenarios. In: 2018 European Signal Processing Conference, pp. 1052–1056, September 2018

Demand in the Electricity Market: Analysis Using Big Data

Amelec Viloria[1(✉)], David Martínez Sierra[2],
James Frasser Camargo[1], Karina Batistas Zea[3],
Jorge Pacheco Fuentes[4], Hugo Hernández-Palma[4],
and Sadhana J. Kamatkar[5]

[1] Universidad de la Costa, St. 58 #66, Barranquilla, Atlántico, Colombia
{aviloria7, jfrasser}@cuc.edu.co
[2] Universidad Simón Bolívar, Barranquilla, Colombia
dmartinez@unisimonbolivar.edu.co
[3] Corporación Universitaria Latinoamericana, Barranquilla, Colombia
kbatistazea@hotmail.com
[4] Universidad del Atlántico, Puerto Colombia, Colombia
jorgepacheco30@hotmail.com,
hugohernandezp@mail.uniatlantico.edu.co
[5] University of Mumbai, Mumbai, India
sjkamatkar@mu.ac.in

Abstract. The traditional business model of energy companies is changing in recent years. The introduction of smart meters has led to an exponential increase in the volume of data available, and their analysis can help find consumption patterns among electric customers to reduce costs and protect the environment. Power plants generate electricity to cover peak consumption at specific times. A set of techniques called "demand response" tries to solve this problem using artificial intelligence proposals. This document proposes a method for processing large volumes of data such as those generated by smart meters. Both for the preprocessing and for the optimization and realization of this analysis big data techniques are used. Specifically, a distributed version of the k-means algorithm and several indices of internal validation of clustering for big data in Spark. The source data correspond to the consumption of electric customers in Bogota, Colombia during the year 2018. The analysis carried out in this study about consumers helps their characterization. This greater knowledge about consumer habits and types of customers can enhance the work of utilities.

Keywords: Big data · Response to demand · Clustering · Smart meters · Electricity consumption

1 Introduction

During most of the 20th century, the relationship between electricity users and distribution companies remained unchanged. No suppliers were chosen and, therefore, there was no need to treat consumers as customers. However, deregulation, the green agenda and the continuous technological leaps have changed this relationship. New

© Springer Nature Switzerland AG 2020
A. P. Pandian et al. (Eds.): ICICCS 2019, AISC 1039, pp. 316–325, 2020.
https://doi.org/10.1007/978-3-030-30465-2_36

constraints such as security of supply, competitiveness, and sustainability are the three priority axes to change the current energy model, which can be achieved through objectives such as reducing emissions and improving renewable energy generation and energy efficiency. An essential tool in this new model are the so-called "smart meters" which should not be understood only as devices that measure consumption but as true sensors for an electrical network. These sensors facilitate a highly flexible and adaptable network that intelligently integrates the actions of users who connect to it to achieve an efficient, safe, and sustainable supply [1, 2].

One of the main problems of the electricity sector is the need to have generation capacity and an oversized network to cover the peaks of high consumption of customers at certain times. However, there are solutions based on adapting demand to available energy instead of increasing supply to meet demand. This is called "Demand Response" and its objective is to change customers' electricity consumption habits in response to changes in supply prices. The main drawback is the large volume of information available in these networks, since it can only be handled with big data techniques [3, 4].

The proposal is based on the processing of these large datasets in a distributed and parallel way. In particular, the application of data mining techniques to better understand the consumption patterns of customers. On the one hand, HDFS [1] is used for distributed data storage, while the processing will be done with Spark [2], a distributed and parallel computing platform. In particular, the k-means algorithm of the MLlib library [3] of Spark will be implemented, as well as four indices of clustering validation for big data [4]. This study could help to plan the connections of renewable energy sources to the network, with a double objective: reducing prices and generating environmental sustainability.

2   Theoretical Review

The irregularity of electricity demand is one of the main problems of the sector. This is because power companies must have both an oversized generation capacity and network redundancy to cope with large amounts of demand that only take a few hours a year. Normally, a threshold of 20% is established for the generation of latent electricity, which should cover approximately 5% of the network's service time (peak demand) [5]. Some of the resources to solve this problem need the involvement of users. These solutions are studied under the name of 'demand response DR' [6]. In contrast to the conventional ideas of increasing supply to meet demand, solutions aim to satisfy it with the available energy.

The objective is to change the patterns of energy consumption of customers in response to changes in the prices offered. This will allow utilities to better manage demand with a better fit of predictions and a reduction in the cost of energy for customers. There are multiple initiatives of possible pricing schemes, which in some cases, even maintain the benefits for the suppliers [7]. One of the main advantages of the response to demand is to offer a sustainable option with more volatile energy generation, especially in Spain, where there is a high presence of renewable generation sources. To implement demand response mechanisms, electricity networks must evolve to an infrastructure that allows the flow of information between the different participants of the electrical system. In this field, big data becomes an essential technology to analyze this flow of information and turn it into useful knowledge.

These customer consumption data obtained through smart meters are nothing but multiple time series. The analysis of time series can be understood as a sequence of values observed over time and ordered chronologically [8]. Since time is a continuous variable, samples are recorded at successive equally spaced points. Therefore, time series are a sequence of discrete time data.

In the context of time series data mining, the main challenge is the way to represent the data. The most common approach is to transform the time series into another area for reducing the dimensionality and developing an indexing mechanism. The measurement of the similarity between the sub-sequences of time series and segmentation are the two main tasks in the mining of time series that correspond to the classical tasks of data mining. The increasing use of time series data has led to a large number of attempts at research and development in the data mining field [9].

In this paper, the focus will be on clustering, a data mining method for grouping unlabeled instances of datasets. The idea is that the instances collected in the same group will have a similar behavior [10]. The clustering of time series emerges as a useful approach to undermine common patterns based on time-dependent data [11] that are characterized by having a high dimensionality and a large size.

Focusing on clustering from energy consumption data, there are many proposals framed in this field. In [12], the effect of the measures of similarity in the application of the grouping to discover the energetic patterns of the buildings is presented. To obtain typical customer load profiles [13], a stability index is proposed to choose the grouping algorithm that best suits this pattern recognition issue. In addition, another priority index (based on the stability index) is proposed to determine the priority range of the groupings. In [14], the authors develop a partitioning clustering technique to extract useful information from electricity prices. While in [15], clustering techniques are used with the objective of grouping and labeling the samples of a data set to predict the behavior of time series based on the similarity of the pattern sequences.

In relation to the intelligent management of electricity demand, in [16], the authors propose a Virtual Power Player as a manager to meet the demand and reserve of electrical energy required. In [17], an analysis of customer smart meter data is presented to better understand the maximum demand and the main sources of variability in their behavior.

In addition to classic data management methods, the big data approach has recently emerged due to the availability of a large amount of data, distributed file systems, and powerful distributed processing engines. This has led many of the data mining algorithms to adapt to the big data environment, such as the clustering algorithms. In terms of the field of energy consumption, several large data solutions have emerged today, such as optimizations of smart networks in [18], and energy consumption patterns in [19].

3   Data and Methods

The original data is stored in 35 tables, divided into hundreds of CSV files. These tables contain extensive information on consumption, rates, meters, geographic or personal data of Endesa's electric consumers in Bogota between 2012 and 2018 for a total size of 1.8 TB.

The clients considered in this study are those that have rates 2.0A and 20DHA, both from the free market of Endesa for contracted powers under 10 kW where immense majority of homes and small stores are located. The 2.0A rate maintains a fixed price throughout the year, while the 20DHA has a two-period time discrimination. In the latter, the peak and valley periods mark two different prices according to the hour in which the energy is consumed: tip from 12 h to 22 h in winter and from 13 h to 23 h in summer; valley from 10 pm to 12 am in winter, and from 11 pm to 1 pm in summer.

Table 1 shows the distribution of customers according to their rate. Due to the great variety of powers contracted, Table 1b shows a distribution of clients for each power range. These ranges are based on the standard power values that can currently be contracted by having smart meters.

Table 1.   Distribution of clients

Rate	Clients
2.0A	124.254
2DHA	6,754

a. By contracted rate

Power	Clients
[0.1-2.3)	8,914
[2.3-3.45)	21,942
[3.45-4.60)	38,547
[4.60-5.75)	22.785
[5.75-6.90)	8,457
[6.90-8.05)	3,248
[8.05-9.20)	4,954
[9.20-10.0)	84

b. By contracted power

4   Results and Discussions

The following environments have been used to carry out the experiments:

- Own cluster: 72 processors Intel Xeon E7-4820, 128 GB RAM and 8 TB of storage.
- EMR Cluster (Elastic Map Reduce) from AWS: five m3.2xlarge instances with 16 Intel Xeon E5-2670 v2 (Ivy Bridge) processors, 30 GB RAM and 2 80 GB SSDs each.

4.1 Data Processing

The first objective is to get a set of data that can be mined. First, the large amount of data obtained will be stored in a distributed file system (HDFS) configured in our cluster. To reduce the data size, the CSV files are transferred to Parquet, a column-oriented data format that compresses and encodes them. Once stored and formatted, all the processing of this data is carried out in a distributed way and in parallel with Spark. Because the data will also be processed with Amazon Web Service online tools, they are also stored in the S3, the Amazon's online storage system. Subsequently, the necessary attributes are studied and selected to build a first dataset. The processed tables are shown below (Table 2).

Table 2. Processed table

Table	Elements (millions)	Size (MB)
Clients	21.3	716
Contracts	41.2	560
Master contracts	31.5	666
Load curves	2,157.75	312.885

Once the first dataset is built, users are selected with: a contracted power equal to or less than 10 kW, a 2.0A or 20DHA rate, and having all consumption readings for the year 2018. Finally, instances with null values are discarded. This dataset is made up of 46,752,124 instances corresponding to the 365 load curves of the year 2018 from 144,019 customers. For each consumer, there are 24 hourly consumption readings for each of its 365 instances.

To build the 2018 time series, it is necessary to transform this dataset of daily instances into one of annual instances. In this way, the new dataset would consist of 131,039 instances, one per client, with: 8760 (365 $\times$ 24) hourly readings of 2018, the Universal Supply Point Code (CUPS), the rate and power contracted.

Next, a new attribute is generated to categorize the consumers according to the contracted power. Finally, an alternative dataset with normalized consumption differences is built to find the difference between each pair of consecutive consumption values and divide it by the average consumption of that day. In this way, a dataset with the hourly consumption of the year 2018 is obtained and another one with the normalized consumption differences in the same period. These datasets will be used both jointly and individually in the following sections in order to find patterns in customers' electricity consumption habits.

4.2 Determination of the Optimal Number of Clusters

Before applying clustering algorithms to our datasets, it is necessary to determine the optimal number of clusters (k) to obtain. To do this, four clustering validation indices for big data are applied to each of the datasets (BD-CVIs) [4]: BD-Silhouette [4], BD-Dunn [4], Davies-Bouldin [20], and Within Set Sum of Square Errors (WSSSE) [21].

Figure 1a shows the graphical representation of BD-Silhouette index results. For this index, the optimal values of k are their maximums, 6 and 9. These values coincide with the maxima of the graph corresponding to the BD-Dunn index (Fig. 1b). In the case of the Davies-Bouldin index, the optimal values are at the minimum, which coincide again in 6 and 9, as shown in Fig. 1c. Finally, the results of the WSSSE index represented in Fig. 1d do not yield a clear value. In this index, a stabilization of values is wanted and, as can be seen, there is no concrete value in it. After analyzing these results, the values 6 and 9 were obtained as optimal for making the clustering.

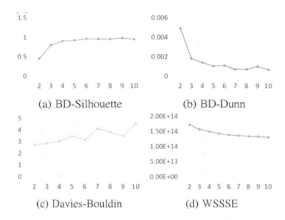

Fig. 1. Big data indices for clustering validation

As for the commodity dataset, these indices were re-applied to the standardized difference dataset discussed at the end of section IV-A. In this case, the results for the optimal value of k were 5 and 7.

4.3   Clustering

Once the optimal number of clusters was calculated, the version implemented in Spark of the k-means algorithm is applied to each dataset. This implementation was developed to be able to extract patterns in parallel and distributed systems. At the time of executing the algorithm, the object RDD (Resilient Distributed Dataset) and the k obtained previously will be given as an input. As a result, a series of clusters with elements from each of the datasets was obtained.

For the consumption dataset, two of the clusters obtained with k = 9 had less than 5 elements, so it was decided to work with the other optimal value obtained from k = 6. In the case of the standardized difference dataset, two of the clusters for k = 5 contained a single element. For this reason, the value k = 7 was chosen. The distribution of elements in the 6 clusters of the consumption dataset is shown in Table 3a. In the same way, Table 3b shows how the elements corresponding to the 7 clusters of the normalized difference dataset are distributed. Figure 2 presents the hourly consumption curves formed by the centroids of each of the clusters during a week of January 2018. It highlights that most consumers are grouped.

Table 3. Clusters of datasets for optimal k

Cluster	Elements
0	12.021
1	50.587
2	1012
3	138
4	1117
5	43.789

Cluster	Elements
0	42.147
1	8147
2	2578
3	18.945
4	28.462
5	7191
6	1029

a.Dataset of consumption for k=6   b. Standardized difference dataset for k = 7

In clusters 1 and 5, they present consumptions less than 1 kWh. It can also be noted that a small group of customers, which make up cluster 3, has a very high consumption at night (from 19 to 8 h) and practically zero during the day.

Figure 3 shows the curves of the same centroids during a week in July. Note that the consumers of clusters 0, 1 and 5 maintain a consumption practically equal to that of January, although the maximum peaks are closer to midnight. However, those of the cluster descend radically to match those of 0. The consumption curves of cluster 3 also stand out, where the hours of high consumption are reduced to 6 (00 to 6 h). On the other hand, Fig. 4 shows the curves of the normalized time differences in the same period of time as Fig. 2. It can be observed that the largest peaks of consumption differences between hours belong to the elements of clusters 1, 2, 5 and 6. These clusters turn out to be the least numerous, representing 19% of the total of consumers. This indicates that the consumption of most of the users throughout the day maintains a certain uniformity.

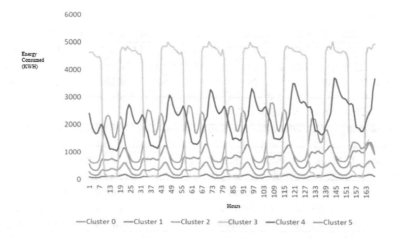

Fig. 2. Curves of hourly consumption of the centroids during a week of January.

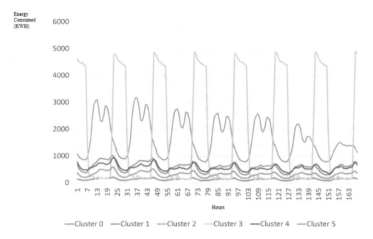

Fig. 3. Curves of hourly consumption of the centroids during a week of January

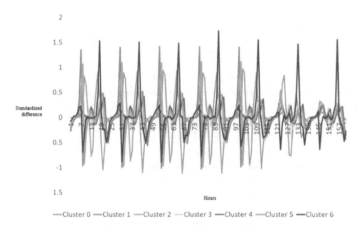

Fig. 4. Normalized hourly standard curves of centroids

5 Conclusions

This study presents the application of big data techniques for the analysis of electrical consumer data. The obtained characterization of these clients leads to the following conclusions:

- More than 86% of customers have consumption load curves where the maximum values do not exceed the kWh.
- There is a group of 126 customers with a very high night consumption. And, although the number of consumption hours during the summer is much less than in winter, they always occur in valley periods. The 95.3% of these users have contracted a rate adapted to their consumption, with time discrimination.

- In its vast majority, the clients with a low consumption contracted a low power and vice versa.
- Users with a medium-high consumption contracted a medium-high power. However, 78.42% of the clients that contracted these power levels consumed energy values that did not reach 1 kWh. So more than 3/4 of these clients have contracted powers well above what they need.
- Throughout the year, the consumption peaks reached in the mornings and midday are produced in valley hours. The same happens with the night peaks in summer, since they appear between 23 and 1. It indicates that hourly discrimination rates could be beneficial for customers with these consumption habits. However, only 6% (4,548) contracted this type of rate.
- In future studies, consumers will be characterized according to their consumption and rates. In addition, the rates and optimal powers to be hired in a personalized way for each type of client will be analyzed and recommended.

References

1. Sánchez, L., Vásquez, C., Viloria, A.: The data envelopment analysis to determine efficiency in Latina American countries for greenhouse gases control in electrical power generation. Int. J. Energy Econ. Policy 8(3), 197–208 (2018)
2. Sánchez, L., Vásquez, C., Viloria, A.: Conglomerates of Latina American countries and public policies for the sustainable development of electric power generation sector. In: International Conference on Data Mining and Big data, pp. 759–766 (2018)
3. Lom, M., Pribyl, O., Svitek, M.: Industry 4.0 as a part of smart cities. In: 2016 Smart Cities Symposium Prague (SCSP), pp. 1–6 (2016)
4. Góngora, G.P.M.: Revisión de literatura sobre ciudades inteligentes: una perspectiva centrada en las TIC. Ingeniare 19(19), 137–149 (2016)
5. Arnal, J.C.: 'Smart cities': Oportunidad económica y desafío urbano. Econ. Aragon. 48, 79–93 (2012)
6. Ontiveros, E., Vizcaíno, D., López Sabaer, V.: Las ciudades del futuro: inteligentes, digitales y sostenibles futuro: inteligentes, digitales y sostenibles (2016)
7. Andrés, C., Andrade, D., Hernández, J.C.: Smart Grid: Las TICs y la modernización de las redes de energía eléctrica – Estado del Arte. Sistemas Telemática 9, 53–81 (2011)
8. Papastamatiou, I., Marinakis, V., Doukas, H., Psarras, J.: A decision support framework for smart cities energy assessment and optimization. Energy Procedia 111, 800–809 (2017)
9. Lucena, M., Sánchez, L., Vásquez, C., Viloria, A.: Regulatory framework and environmental management of the compact fluorescent lamps. J. Eng. Appl. Sci. 12(13), 3495–3498 (2016)
10. Araujo, G.: Eficiencia técnica de los niveles de electrificación de países latinoamericanos. Revista Digital de Investigación y Postgrado (REDIP) 5(4), 977–993 (2015). Universidad Nacional Experimental Politécnica "Antonio José Sucre", Venezuela
11. Feitosa, M., Carvalho, A., Mendes, M., Marques, M.: Receita tributária e qualidade dos serviços públicos no Brasil e nos países membros da OECD (2017). https://www.occ.pt/dtrab/trabalhos/xviicica/finais_site/117.pdf
12. Vázquez-Barquero, A., Surgimiento y transformación de clusters y milieus en los procesos de desarrollo. Revista Eure, XXXII(95), 75–92 (2006). Chile

13. Mombeini, H., Yazdani-Chamzini, A.: Modelling gold price via artificial neural network. J. Econ. Bus. Manage. **3**(7), 699–703 (2015)
14. Kulkarni, S., Haidar, I.: Forecasting model for crude oil price using artificial neural networks and commodity future prices. Int. J. Comput. Sci. Inf. Secur. **2**(1), 81–89 (2009)
15. Bontempi, G., Ben Taieb, S., Borgne, Y.A.: Machine learning strategies for time series forecasting. In: M.-A., Zimányi, E. (eds.) Lecture Notes in Business Information Processing, Aufaure, vol. 138, no. 1, pp. 70–73. Springer, Heidelberg (2013)
16. Duan, L., Xu, L., Liu, Y., Lee, J.: Cluster-based outlier detection. Ann. Oper. Res. **168**(1), 151–168 (2009)
17. Abhay, K.A., Badal, N.A.: Novel approach for intelligent distribution of data warehouses. Egypt. Inform. J. **17**(1), 147–159 (2015)
18. Savasere, A., Omiecinski, E., Navathe, S.: An efficient algorithm for data mining association rules in large databases. In: Proceedings of 21st Very Large Data Base Conference, vol. 5, no. 1, 432–444 (1995)
19. Stolfo, S., Prodromidis, A. L., Tselepis, S., Lee, W., Fan, D. W.: Java agents for metalearning over distributed databases. In: Proceedings of 3rd International Conference on Knowledge Discovery and Data Mining, vol. 5, no. 2, pp. 74–81 (1997)
20. Organización Latinoamericana de Energía (OLADE), Balance Energético-Metodología OLADE (2016)
21. Varela Izquierdo, N., Cabrera H.R., Lopez Carvajal, G., Viloria, A., Gaitán Angulo, M., Henry, M.A.: Methodology for the reduction and integration of data in the performance measurement of industries cement plants. In: Tan, Y., Shi, Y., Tang, Q. (eds.) Data Mining and Big Data, DMBD 2018. Lecture Notes in Computer Science, vol. 10943. Springer, Cham (2018)

Design and Implementation of a Low Cost Slender Visual Surveillance System

Dhruv Tyagi[✉], Rishabh Shroff, Raj Khatri, Aakash Thakkar,
and Nirmal Thakur

Department of Mechatronics Engineering, MPSTME,
SVKM's NMIMS University, Mumbai, India
{dhruv.tyagi48, rishabh.shroff42, raj.khatri01,
aakash.thakkar45}@nmims.edu.in,
nirmal.thakur@nmims.edu

Abstract. This paper presents the design and implementation of affordable small size visual surveillance system. The proposed system is less than 8 mm in thickness, allowing it to go underneath common household doors. The system is controlled by handheld mobile devices, such as a smart phone or laptop, which provide good quality images/video wirelessly using Wi-Fi. The single board computer actuates the sub micro motors on the basis of commands provided by the user through the smartphone. The system can be further improved to provide better resolution images and long range operation using Radio-frequency protocols.

Keywords: Raspberry-Pi · Remote controlled robot · Slim robot · Spy surveillance system · Visual surveillance

1 Introduction

Current trends in small scale surveillance systems bear various limitations such as noisy operation, noticeable size and a limited access form factor. Additionally, majority low cost spy systems cannot enter or exit a closed room on command until a significantly wide opening is available. Also, the spy systems which do have a small form factor, either have limited mobility or a high cost.

To address these issues, the paper suggests a slim surveillance system which can comfortably enter a closed room environment through the narrow opening beneath commonly used household doors. Using its slim configuration and mobile operation, the system can enter and exit a closed room on demand. Moreover, the system avoids detection due to its small size and silent operation.

Commonly used household doors are made from wooden material and have a clearance designed to account for wood expansion due to moisture. This feature is exploited in the proposed surveillance system.

A. P. Pandian et al. (Eds.): ICICCS 2019, AISC 1039, pp. 326–335, 2020.
https://doi.org/10.1007/978-3-030-30465-2_37

2 Related Work

A thin robot which can cross from underneath a door was built as the Sprawl Tuned Autonomous Robot (STAR), developed at the Biomimetic Millisystems Lab, UC Berkeley [1]. The system works on the concept of Hexapod configuration. Although this robot has a slightly configurable height, the minimum height extends a little more than 1 cm, which is more than the clearance offered by many household doors.

Another research paper describes the use of Raspberry Pi Single Board Computer along with Virtual Network Computing (VNC) to build a bot for military applications [2]. The VNC application allows the bot to be controlled wirelessly through a mobile device. This provides easy control and debugging of the bot. Raspberry Pi has also been successfully used as a video server [3].

Similarly, one research describes Wi-Fi controlled Raspberry Pi robot system, which uses an On-Board camera to provide live video feed [4]. The Wi-Fi connection provides a stable data feed for the live video stream.

Raspberry Pi Single Board Computer has also been used to control spy robots [5], in which a satisfactory real time operation is observed.

A spy robot has been developed, capable of detecting fires and controlling a water sprinkler system to help extinguish the fire [6].

Several systems have seen the use of wireless control and monitoring for robotic spy applications [7, 8]. The reliable and extensive use of wirelessly controlled spy robots can be noted in a wide range of systems and applications.

Web-based control for surveillance and spy robots, which are running on Raspberry Pi, have been shown to exhibit convenience of control with motion detection and IoT [9].

A system was employed with a night vision camera, for efficient surveillance in dark environments as well [10]. This ensures operation during both, day and night time.

A surveillance robot was designed in such a way that it can detect leakage of gas from LPG cylinders [11]. This allows the robot to be operated not just inside a house but also some factories.

A surveillance system was designed which detects motion, captures the image at that instant and mails the image to the user [12]. This keeps the user updated of any activity, even if the user is a long distance away from the place.

3 System Configuration

3.1 Structure

The body of the surveillance system is made using a thin Medium Density Fiberboard (MDF) to help keep the frame of the bot light-weight and strong. The dimensions of the cutout used are $155 \times 155 \times 5$ mm. Figure 1 illustrates the structure, clearly highlighting the narrow features.

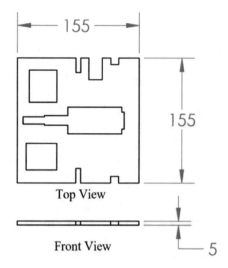

Fig. 1. MDF frame

3.2   Single Board Computer

The system makes use of Raspberry Pi Zero W Single Board Computer (SBC) module. The board was selected keeping in mind its versatility, small size and low power consumption, which helps in keeping the bot under restrictions and operate for a longer duration. Table 1 shows the SBC's specifications.

Table 1. Raspberry Pi Zero W specifications

Detail	Specification
CPU	1 GHz ARM11 Broadcom CPU, Single Core
GPU	Videocore IV
RAM	512 MB
Storage	Micro SD card slot
Interface	Camera Serial Interface
Power	Micro-USB
Connectivity	Micro-USB, mini-HDMI, 802.11n wireless LAN, Bluetooth 4.0

3.3   Camera

Raspberry Pi Spy Camera is used in the proposed surveillance system. The camera has a resolution of 5 megapixels and is capable of recording high-definition videos at 30 frames per second. The selection of this camera ensures good quality and offers flexibility.

The camera is interfaced to the SBC using Camera Serial Interface (CSI), capable of dedicated, high-rate pixel data transfer for better visualization. The specifications of the mentioned camera are provided in Table 2.

Table 2. Camera specifications

Detail	Specification
Resolution	5 megapixel
Module dimensions	8.6 mm × 8.6 mm × 5.2 mm
Lens diameter	6.9 mm
Cable length (excluding module)	52 mm
Weight	1.1 gms

3.4   Motors and Wheels

A challenging part of this system is to provide the required torque for movement. The actuating component should be small in size, cheap and should be able to efficiently maneuver the system. Most of the cheap motors available are not compatible with the size or the torque requirements of the proposed system.

Therefore, Sub-Micro DC Geared Motors, with matching specifications, are used for this bot. As per the requirement, it contains a gearbox with planetary gear assembly, with a reduction ratio of 136:1. Having a free running speed of 500 rpm, these are sufficient for the system's operational use. The dimensions of the motor, with a maximum body diameter of 6 mm, allow us to keep the bot as narrow as possible while providing sufficient torque. The system makes use of 2 such motors, fitted on either side, close to the center of the opposite edges. Table 3 provides detailed specifications of the motors used.

Table 3. Motor specifications

Detail	Specification
Gear ratio	136:1
No load speed at 6 V	500 RPM
No load current at 6 V	45 mA
Stall current at 6 V	400 mA
Stall torque at 6 V	8 oz.in
Weight	1.25 gms

The wheels are custom designed and 3D printed with a maximum outer diameter of 7.5 mm, as shown in Fig. 2. 6 such wheels are printed, out of which 2 are fitted on the central driving motors, whereas the other 4 are fitted freely on each of the 4 corners of the bot. These 4 don't provide any active movement and are merely present for stability.

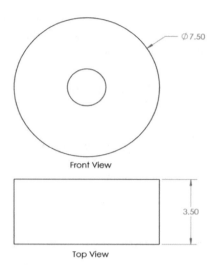

Front View

Top View

Fig. 2. Wheel diagram

3.5   Electronics

Power Demand. The SBC requires 4.75–5.25 V to operate, while a typical lithium ion battery provides, at full charge, only 4.25 V. Further, the SBC consumes a maximum 250 mAh during operation and less than 80 mAh in idle state. To overcome this problem, a 5 V 1 A boost converter of 85% efficiency was used which can work with inputs as low as 1 V.

The motor, along with gearbox, weighs only 1.25 g. Motor is powered directly through the SBC, hence it operates at 3.3 V. At this voltage level it consumes less than 30 mA for nominal torque and less than 200 mA while stalling.

Battery. The battery was selected based on 2 criteria; total current consumption and width of the battery. Current consumption of the components was calculated by adding maximum SBC current (I_{SBC}), with motor current (I_{MOT}), as follows:

$$I_{Total} = I_{SBC} + (2 * I_{MOT}) \tag{1}$$

$$I_{Total} = 250\,mA + 2 * (30\,mA) = 310\,mA \tag{2}$$

Note that this current should be provided by the boost converter and not the battery. Therefore, the battery current (I_B) must be calculated, using the Power output of Boost converter (P_{OUT}), Power input of Boost converter (P_{IN}) and Boost converter efficiency (η) as follows:

$$P_{OUT} = (P_{IN} * \eta) \tag{3}$$

$$(5\,V * 0.310\,A) = (3.7\,V * I_B * 0.85) \tag{4}$$

$$I_B = 0.49\,A \tag{5}$$

Considering a minimum operation time of 1 h, a single cell lithium ion battery was chosen along with a Battery Management System (BMS) and a boost converter to output a stable 5 V to the SBC. The battery width is 5 mm, ideal for the application. The battery specifications are mentioned in Table 4.

Table 4. Battery specifications

Detail	Specification
Nominal voltage	3.7 V
Current rating	800 mAh
Dimensions	$45 \times 35 \times 5$ mm

Battery Management System. Lithium ion batteries are delicate in nature. They may get damaged if over-discharged or over-charged. They also may damage if discharge rate exceeds twice its capacity (also known as C-rating of a battery where C refers to its capacity. For lithium ion, C-rating is typically 2C). To protect against above situations, a Battery Management System is used. We selected a BMS based on TP4056 chip which offers 3 V discharge cut-off, 4.2 V overcharge cut-off, 1 A continuous discharge (which is less than 2C for our battery) and short circuit protection. Battery charges through a micro USB type-b port installed on the BMS itself. Figure 3 shows the complete electronic circuit block diagram.

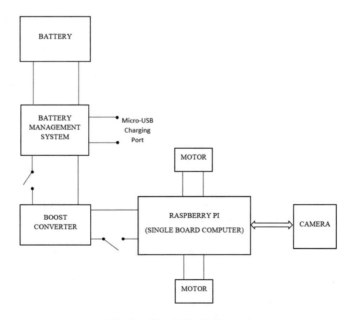

Fig. 3. Circuit block diagram

4 3D Physical Model

The physical model of the proposed surveillance bot was created first, with the help of software. Individual parts were designed or imported and then assembled together for better visualization of the system. This helped in providing guidelines for the placement and positioning of the components throughout the frame. Figure 4. shows the 3-D model of the bot.

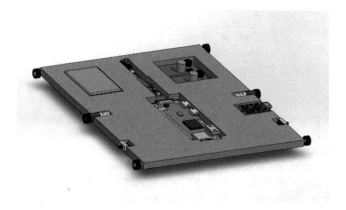

Fig. 4. 3D model of surveillance system

5 Software

The surveillance system is controlled wirelessly using a mobile device like a smart phone or a laptop. The communication is established between the SBC and the controlling device over Wi-Fi with the help of a Wi-Fi chip, present on the SBC.

The Raspberry Pi board runs on a Linux based, Raspbian operating system, which provides camera interfacing and low level hardware control. The software part of this bot consists of a Python script which interfaces with the camera, takes in real-time video/image data, performs an auto-white balance function. This function helps in securing accurate image data under changing light conditions.

Virtual Network Computing (VNC), is a system for graphical desktop sharing [2]. Using the VNC system, we can run and edit the python source code of the bot. After running the script, this screen sharing system is used to input commands for the motors, along with the display of the live video feed from the onboard camera.

The system can also be configured to store the image/video data in the micro SD card connected to the SBC, with a storage capacity of 16 gb. The SBC and the operating system are capable of Image Processing (IP) functionalities. These IP operations can be used to detect, observe and automatically take images of certain colors and shapes of interest.

For the purpose of experimentation, the video feed resolution is kept as 100×100 pixels, with a frame rate of 10 frames per second.

6   Experimental Results

Through the mobile device and the VNC system, the user provides the directional guiding values for movement of the bot. Also through the connection, the onboard SBC shares live video/image data to the controlling device's screen. The resolution and refresh rate of this data can be altered in real time with minor tweaks in code, without needing to extract the bot itself from its current position.

Figure 5(a–d) shows the movement of the surveillance system as seen from the onboard camera. These images were captured by the onboard camera during its traversal underneath the door and relayed back to the handheld mobile device.

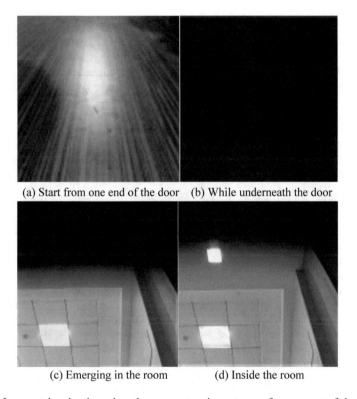

(a) Start from one end of the door   (b) While underneath the door

(c) Emerging in the room   (d) Inside the room

Fig. 5.   Images taken by the onboard camera at various stages of movement of the system

7   Conclusion

The paper presented the implementation and the details of a smart phone controlled slim visual surveillance system with a maximum thickness of 8 mm. Because of its narrow and thin features, the system has the ability to traverse from underneath a door. The smartphone control of this system is employed using an established Wi-Fi connection between the mobile device and the surveillance system. This system can be applied for the use of wireless surveillance in scenarios with locked doors in which the user needs visual surveillance of the inaccessible room. The experimental results exhibited an operation time of 1 h with a stable 10 frames per second live video feed. The system had a traversing speed of 5 cm/s with minimal noise during operation. The system can be modified for various applications in scenarios such as accessing otherwise inaccessible rooms during fire hazard scenarios. This is possible using fire resistant material or spray to ensure reliable operation in such scenarios. To improve video quality and operation range, Radio-Frequency protocols can be implemented.

References

1. Zarrouk, D., Pullin, A., Kohut, N., Fearing, R.S.: STAR, a sprawl tuned autonomous robot. In: 2013 IEEE International Conference on Robotics and Automation, Karlsruhe, pp. 20–25 (2013)
2. Kumar, D.N.S.R., Kumar, D.: VNC server based robot for military applications. In: 2017 IEEE International Conference on Power, Control, Signals and Instrumentation Engineering (ICPCSI), Chennai, pp. 1292–1295 (2017)
3. Salih, F., Mysoon Omer, S.A.: Raspberry Pi as a video server. In: 2018 International Conference on Computer, Control, Electrical, and Electronics Engineering (ICCCEEE), Khartoum, pp. 1–4 (2018)
4. Güleçi, M., Orhun, M.: Android based WI-FI controlled robot using Raspberry Pi. In: 2017 International Conference on Computer Science and Engineering (UBMK), Antalya, pp. 978–982 (2017)
5. Abdalla, G.O.E., Veeramanikandasamy, T.: Implementation of spy robot for a surveillance system using Internet protocol of Raspberry Pi. In: 2017 2nd IEEE International Conference on Recent Trends in Electronics, Information & Communication Technology (RTEICT), Bangalore, pp. 86–89 (2017)
6. Kolambe, K., Pote, R., Jadhav, A., Chennur, V.: Spy robot with fire detection and water sprinkling. In: 2018 Second International Conference on Electronics, Communication and Aerospace Technology (ICECA), Coimbatore, pp. 1844–1848 (2018)
7. Balakrishnan, M., Gowthaman, S., Kumaran, S.J., Sabhapathy, G.R.: A smart spy robot charged and controlled by wireless systems. In: 2015 International Conference on Innovations in Information, Embedded and Communication Systems (ICIIECS), Coimbatore, pp. 1–4 (2015)
8. Patoliya, J., Mehta, H., Patel, H.: Arduino controlled war field spy robot using night vision wireless camera and Android application. In: 2015 5th Nirma University International Conference on Engineering (NUiCONE), Ahmedabad, pp. 1–5 (2015)
9. Jayakumar A.J.K., Muthulakshmi, S.: Raspberry Pi-based surveillance system with IoT. In: Thalmann, D., Subhashini, N., Mohanaprasad, K., Murugan, M. (eds.) Intelligent Embedded Systems, Lecture Notes in Electrical Engineering, vol. 492. Springer, Singapore (2018)
10. Salman, H., Acheampong, S., Xu, H.: Web-based wireless controlled robot for night vision surveillance using shell script with Raspberry Pi. In: Barolli, L., Javaid, N., Ikeda, M., Takizawa, M. (eds.) Complex, Intelligent, and Software Intensive Systems, CISIS 2018, Advances in Intelligent Systems and Computing, vol. 772. Springer, Cham (2019)
11. Suryavamsi, P.S.N., Arockia Selvakumar, A.: IoT controlled mobile robot for home security and surveillance. In: Hemanth, J., Fernando, X., Lafata, P., Baig, Z. (eds.) International Conference on Intelligent Data Communication Technologies and Internet of Things (ICICI)

2018, ICICI 2018, Lecture Notes on Data Engineering and Communications Technologies, vol. 26. Springer, Cham (2019)

12. Premkumar, S., Arthi, T.S.: Distance-based IoT surveillance alert system. In: Abraham, A., Dutta, P., Mandal, J., Bhattacharya, A., Dutta, S. (eds.) Emerging Technologies in Data Mining and Information Security, Advances in Intelligent Systems and Computing, vol. 814. Springer, Singapore (2019)

NoSQL Database for Storing Historic Records in Monitoring Systems: Selection Process

Amelec Viloria[1(✉)], David Martínez Sierra[2], Laura de la Hoz[1],
Mario Orozco Bohórquez[1], Osman Redondo Bilbao[3],
Alberto Roncallo Pichón[4], Jorge Pacheco Fuentes[5],
and Hugo Hernández-Palma[5]

[1] Universidad de la Costa, St. 58 #66, Barranquilla, Atlántico, Colombia
{aviloria7,ldelahoz35,morozco5}@cuc.edu.co
[2] Universidad Simón Bolívar, Barranquilla, Colombia
dmartinez@unisimonbolivar.edu.co
[3] Corporación Politécnico de la Costa Atlántica, Barranquilla, Colombia
osmanenrique29@hotmail.com
[4] Corporación Universitaria Minuto de Dios – UNIMINUTO,
Barranquilla, Colombia
alberto.roncallo@uniminuto.edu
[5] Universidad del Atlántico, Puerto Colombia, Colombia
jorgepacheco30@hotmail.com,
hugohernandezp@mail.uniatlantico.edu.co

Abstract. The storage of historical information consists of the registration of large concentrations of data based on samples of information issued by a given system. Historical databases play a key role in industrial control systems that process and provide the information needed for the tasks of supervision, event records, management, and maintenance in the industrial sector. The purpose of the present research is to analyze the current state of monitoring and control systems, recognizing their advantages and disadvantages, and reaching the state of the art about the selection of a NoSQL database for the recording of time series. The development of the work focused on the comparison of column-oriented storage managers as *HBase*, *Cassandra*, and *Big Table*. After designing and implementing two cases of tests for comparing the behavior of both managers in the scenarios of Intensive Readings and Reading/Writing, it was concluded that both of them present excellent performance for storing historic records.

Keywords: Database · Historic record · NoSQL · Supervision and control

1 Introduction

Among the main challenges of the historical databases, the handling of large volumes of information stands out, so as to perform functions such as the recording, reading, processing, and compression of the information with appropriate levels of scalability. Scalability is the property of computer systems that study their behavior during its continued growth [1].

© Springer Nature Switzerland AG 2020
A. P. Pandian et al. (Eds.): ICICCS 2019, AISC 1039, pp. 336–344, 2020.
https://doi.org/10.1007/978-3-030-30465-2_38

For control systems, this quality is inherent to the process of maintaining and updating the platform according to the requirements of the business [2–4]. The work of integration of new control devices, as well as the infrastructure and administrative restructuring in the operational areas, create a complex scenario for the evolution of computer systems at the industrial field [5].

The historical storage has been done since the 80 s, primarily with technologies of relational databases. However, the relational model introduces limitations resulting from their own conception that contribute to an inefficient behavior when the systems grow, causing a significant reduction in performance and availability for the management of time series [6–8].

In the last decade, new trends and storage technologies that address these issues have arisen, including the well-known NoSQL Databases. These databases propose a more versatile structure of persistence, although in some cases, they lose certain functionalities as transactions and the inability to make connections of tables. Among its main attributes are the absence of scheme in the data records, easy horizontal scalability, and speed [9].

This paper conducts a research about the NoSQL technologies with the aim of selecting a database manager according to the requirements for storing historic records in industrial supervision and control systems in critical environments where soft real-time systems are needed [10].

The development includes the analysis of the various structures of NoSQL databases, the definition of attributes that will guide the selection process, the comparison, and its theoretical background.

2 Theoretical Review

2.1 HBase

The data storage is oriented to columns and comprises a set of tables containing rows and columns. Each table must have a primary key and a column represents the attribute of an object. *HBase* allows many attributes to be grouped together in what is known as family of columns, so the elements of a family column are stored together. With *HBase,* the table scheme must be previously defined and the families of columns are specified [8, 11].

HBase is like a clone of Big *Table* developed in *Java* as part of *Hadoop Apache-MapReduce framework*, providing a "high availability mode to store large amounts of scattered data" [12].

HBase depends on a Distributed File System (HDFS). In this system, there is a need for compacting, data compression, as well as filters for reducing the disk access. *HBase* can be a source or a destination for *MapReduce tasks* executed through *Hadoop*. The outstanding use of *HBase* is the system of real-time messaging on *Facebook,* built on *HBase* from 2010 [13, 14].

2.2 Big Table

It is a distributed storage system to manage structured data and can scale to a very large size as *Petabytes* of data across thousands of servers in commodities. Google uses it in its projects from 2006 [15] obtaining relevant results to date as broad applicability, scalability, high performance, and high availability. For *Google*, the implementation of *Big Table* has proven to be advantageous because the flexibility with which it has been developed is unpredictable [1].

It has other characteristics such as: distributed, high efficiency, and proprietary. It is built on GFS (Google File System), Chubby Lock Service, and some other Google services and programs [17].

It started in 2004. This manager stores information in multidimensional tables. The cells have temporary versions of their values, so that historical values can be tracked [18].

2.3 Cassandra

As a database manager, it is determined by its scalability and high availability without compromising performance. It also has great linear scalability and proven hardware fault tolerance. It also provides support for replication across multiple data centers and is the best in its class, providing lower latency and users can be sure that will survive regional cuts.

Cassandra has features such as column indexes with the performance of structured updates, firm support for materialized views, and powerful cache storage capacity [19].

It allows the efficient use of many applications beyond simple key/value. The reading and writing show great performance in linear form as new machines are added, without any application downtime or interruption. In addition, it is suitable for applications that cannot lose data, even when a data center stops working [20].

Cassandra has integration Hadoop, MapReduce with support. It is made up of column families that contain rows and columns and each row is identified by a key. Each row has multiple columns, each of them with a name, value, and time tag [21].

In order to achieve high scalability and durability of a *Cassandra cluster*, data is replicated in a number of nodes that can be defined as a replication factor by *Cassandra* example. Replication is managed by a coordinator node of the particular key that is being modified. The coordinator node for any key is the first node in the hash consistent ring that is visited by walking clockwise from the position of the key [22].

Compared to *Big Table,* and *HBase*, *Cassandra* stores data into local files instead of a distributed file system. However, the representation of the data in memory and on disk, as well as the processing of reading and writing operations is taken from *Big Table*. Writing Operations first go through a persistent confirmation record and then to an in-memory data structure.

3   Materials and Methods

The NoSQL databases implement different storage structures like the Key-value, Columns, documents, and graphs [23]. Table 1 presents a summary of its main features, the development of this research will focus on the databases aimed to columns, which are the most used elements for the registration of monitoring activities and time series.

Table 1. Categories of NoSQL databases

Category	Description
Key-value	Using a hash table where each unique key will reference a particular data, this structure makes it the simplest and easiest model to implement. It provides fast searches and its weaknesses arise when working with just a part of the value
Columns	Created to store and process large amounts of data distributed in several stations, the use of keys remains, but they reference multiple columns, which in turn are organized into families. It provides fast searches and makes a good distribution of stored information
Documents	Consists of versioned documents which are collections of key-value pairs, improves the problem of managing only a part of the value
Graphs	It uses a flexible model based in graphs allowing shortest paths in searches as well as the increase in relationships

Taking the requirements of historical databases for monitoring and control systems as a basis, the features that will be the center of the analysis were identified to make the comparative matrix that will guide the selection process of technology. Table 2a and b show the main factors and their description.

Once the attributes that will guide the comparison were defined, a study of the state of the art was carried out about the main existing managers aimed at columns [24, 25], selecting *HBase*, *Cassandra*, and *Big Table*. The description of their main characteristics to conform the comparative table of the technologies can be observed in paragraph 2.

Table 2. Attributes for comparison

Attribute	Description
Type of storage	It is the way in which the data is stored in a database. It is a collection of bits that describes both the contained data from the database and its associated metadata.
Management of time series	A time series is a set of numerical data obtained at regular periods of time. Potentially large volumes of data can be managed and accessed in a way of high performance.
Type of License	A software license is a contract between the licensor and the licensee of the software to use the software to fulfill a series of terms and conditions established within its clauses.
Implementati on Language	It is an artificial language designed to express processes that can be carried out by machines such as computers.
Data Storage	Allows to read or write data to media or storage media, and forms the secondary storage or the computer memory.
Concurrency Control	Ensures the integrity of data in spite of accesses of concurrent users.

Attribute	Description
Transactions	In a Database Management System (DBMS), it is a set of commands that are executed by forming a unit of work, that is to say, in an indivisible or atomic form.
Availability	Ability of the system to run continuously. Ensures a reduced time between failures.
Fail tolerance:	Ability of the system to react and recover from failures or exceptional situations.
Consistency	Guarantees the consistency of data and transactions.
Complexity	Analyzes the curve of familiarization with the technology for installation, configuration, and administration.
Documentati on and Community	Quality of documentation, updates, and attention of the community.

4   Analysis and Results

After a first analysis on the basis of the characteristics described in paragraph 2 (see Table 3a and b), *Big Table* is discarded due to its proprietary license. The comparison of the factors for *HBase* and *Cassandra* shows similar results without major differences to decide the selection. Therefore, a deeper analysis is carried out on the behavior of these managers in the case of certain scenarios of use.

4.1   Comparative Advantage (Benchmark)

This section presents a more detailed analysis of the *Hbase* and *Cassandra* technologies. Two of the most commonly applied cases of use for the management of historical data are Writing and Writing/Reading. Following, the applied performance tests and the results are described in detail.

Table 3. Cassandra vs Big Table vs HBase

Characteristic	Cassandra	Big Table	HBase
Type of License	Apache Open Source	Proprietary	Apache Open Source
Implementation Language	Java	C C++	Java
Examples of use	Banking, Finance	Gmail, Google Book Search	Messaging, Monitoring Systems
Best Uses	Write often, read Less	Designed to scale through hundreds or thousands of machines	Random Reading when writing in large databases

Characteristic	Cassandra	Big Table	HBase
Concurrency Control	Multi-versioning of Concurrency and control	Logged In	Logged In
Transactions	Local	Local	Local
Data Storage	Disk	GFS	Hadoop
Management of time series	Excellent complement		The consultations on this series of time could retrieve a range of versions.
Availability	High Availability	High Availability	The availability is acceptable, have given priority to raise levels of consistency
Fail tolerance:	High	High	High
Consistency	Acceptable consistency, prioritized to raise Availability	High Consistency	High Consistency
Complexity	Easy installation and Configuration. Family structure.	Complex installation, configuration and administration	Complex installation, configuration and administration.
Documentation and Community	Very good documentation and community active development	Excellent documentation	Excellent documentation and development community

4.2 Reading

This scenario aims to test the behavior of the two managers in the case of intensive reading. It implements a test case for 95% of the readings and 5% of writing during 10 min. 10 million records will be used for each node, taking the research as a reference [26]. The tests will be limited to 1 and 3 stations that are the minimum and maximum of nodes that are usually used in the sector for the supervision and control tasks. Table 4 shows the main factors and their description.

From the results, it can be concluded that Cassandra provides better performance than HBase for intensive reading. In the case of the Transfer and Reading Rate, it shows higher values around 10 times. However, the opposite occurs in writing, although HBase samples are excellent, in this sector the Cassandra are suitable for the application domain of interest which is limited to soft real-time applications [28].

Table 4. Comparative advantages for intensive reading

Nodes	A factor	HBase	Cassandra
1	Transfer rate (operations/second)	2.5K	25K
3	Transfer rate (operations/second)	4.5K	50K
1	Latency in reading (milliseconds)	55	6
3	Latency in reading (milliseconds)	95	7
1	Latency in reading (milliseconds)	0.42	6
3	Latency in reading (milliseconds)	0.25	7

4.3 Reading/Writing

This scenario aims to test the behavior of the two managers in the case of intensive reading. It implements a test case for the 50% of reading and 50% of writing. Table 5 shows the results obtained.

Table 5. Comparative advantages for Reading/Writing

Nodes	A factor	HBase	Cassandra
1	Transfer rate (Operations/second)	8	30K
3	Transfer rate (Operations/second)	10K	60K
1	Latency in reading (milliseconds)	80	8
3	Latency in reading (milliseconds)	99	10
1	Latency in reading (milliseconds)	0.07	6
3	Latency in reading (milliseconds)	0.07	7

Again, in the case of Read/Write testing, similar results are obtained to the precedent test. Cassandra far exceeds the rate of transfer and reading, it being below the latency during writing, although with very good levels according to the requirements of the application domain.

5   Conclusions

After the development of the numerals 2, 3 and 4 it could be argued that *HBase* and *Cassandra* are excellent technologies to manage large volumes of data and the development of monitoring and control applications.

Both are great features and implement key features for their activity as the high levels of consistency and availability, respectively. Similarly, their capabilities stand out for replication, fault tolerance, and their levels of horizontal scalability.

Regardless of their benefits. *Cassandra* stands out in tests of intensive reading and reading-writing. In addition, it was easy to assimilate its structure and work scheme, which led to the streamlining of the process of installation and configuration of the tests, including productive experiences to extend the platform and integrate new features. It also stressed its scalability in small clusters.

References

1. Rashid, A.M., Albert, I., Cosley, D., Lam, S.K., McNee, S.M., Konstan, J.A., Riedl, J.: Getting to know you: learning new user preferences in recommender systems. In: Proceedings of the 7th International Conference on Intelligent User Interfaces, ACM, pp. 127–134 (2002)
2. Gómez, S., Zervas, P., Sampson, D.G., Fabregat, R.: Context-aware adaptive and personalized mobile learning delivery supported by UoLmP. J. King Saud Univ. Comput. Inf. Sci. **26**(1), 47–61 (2014)
3. Lops, P., De Gemmis, M., Semeraro, G.: Content-based recommender systems: state of the art and trends. In: Recommender systems handbook. Springer, Boston (2011)
4. Malik, F., Baharudın, B.: Analysis of distance metrics in content-based image retrieval using statistical quantized histogram texture features in the DCT domain. J. King Saud Univ. Comput. Inf. Sci. **25**(2), 207–218 (2013)
5. Su, X., Khoshgoftaar, T.M.: A survey of collaborative filtering techniques. Adv. Artif. Intell. **4**(1), 1–12 (2009)
6. De Campos, L.M., Fernández-Luna, J.M., Huete, J.F., Rueda-Morales, M.A.: Combining content-based and collaborative recommendations: a hybrid approach based on Bayesian networks. Int. J. Approximate Reasoning **51**(7), 785–799 (2010)
7. Isinkaye, F.O., Folajimi, Y.O., Ojokoh, B.A.: Recommendation systems: principles, methods and evaluation. Egypt. Inform. J. **16**(3), 261–273 (2015)
8. Buder, J., Schwind, C.: Learning with personalized recommender systems: a psychological view. Comput. Hum. Behav. **28**(1), 207–216 (2012)
9. Balaguera, M.I., Vargas, M.C., Lis-Gutierrez, J.P., Viloria, A., Malagón, L.E.: Architecture of an object-oriented modeling framework for human occupation. In: Tan, Y., Shi, Y., Tang, Q. (eds.) Advances in Swarm Intelligence, ICSI 2018, Lecture Notes in Computer Science, vol. 10942. Springer, Cham (2018)

10. Taneja, A., Arora, A.: Cross domain recommendation using multidimensional tensor factorization. Expert Syst. Appl. **92**(1), 304–316 (2018)
11. Frolov, E., Oseledets, I.: Tensor methods and recommender systems. Wiley Interdisc. Rev. Data Min. Knowl. Discov. **7**(3), 1–12 (2017)
12. Kolda, T.G., Bader, B.W.: Tensor decompositions and applications. SIAM Rev. **51**(3), 455–500 (2009)
13. Panniello, U., Tuzhilin, A., Gorgoglione, M.: Comparing context-aware recommender systems in terms of accuracy and diversity. User Model. User-Adap. Interact. **24**(2), 35–65 (2014)
14. Rafailidis, D., Kefalas, P., Manolopoulos, Y.: Preference dynamics with multimodal user-item interactions in social media recommendation. Expert Syst. Appl. **74**(1), 11–18 (2017)
15. Karatzoglou, A., Amatriain, X., Baltrunas, L., Oliver, N.: Multiverse recommendation: n-dimensional tensor factorization for context-aware collaborative filtering. In: Proceedings of the fourth ACM conference on Recommender systems, ACM, pp. 79–86 (2010)
16. Zheng, C., Haihong, E., Song, M., Song, J.: CMPTF: contextual modeling probabilistic tensor factorization for recommender systems. Neurocomputing **205**(1), 141–151 (2016)
17. Hidasi, B., Tikk, D.: Fast ALS-based tensor factorization for context-aware recommendation from implicit feedback. In: Machine Learning and Knowledge Discovery in Databases, pp. 67–82 (2012)
18. Lee, J., Lee, D., Lee, Y.C., Hwang, W.S., Kim, S.W.: Improving the accuracy of top-n recommendation using a preference model. Inf. Sci. **348**(1), 290–304 (2016)
19. Balaguera, M.I., Lis-Gutierrez, J.P., Gaitán-Angulo, M., Viloria, A., Portillo-Medina, R.: An ontological framework for cooperative games. In: Tan, Y., Shi, Y., Tang, Q. (eds.) Advances in Swarm Intelligence, ICSI 2018, Lecture Notes in Computer Science, vol. 10941. Springer, Cham (2018)
20. Varela Izquierdo, N., Cabrera, H.R., Lopez Carvajal, G., Viloria, A., Gaitán Angulo, M., Henry, M.A.: Methodology for the reduction and integration of data in the performance measurement of industries cement plants. In: Tan, Y., Shi, Y., Tang, Q. (eds.) Data Mining and Big Data, DMBD 2018, Lecture Notes in Computer Science, vol. 10943. Springer, Cham (2018)
21. Hepner, G.F.: Artificial neural network classification using a minimal training set. Comparison to conventional supervised classification. Photogram. Eng. Remote Sens. **56**(4), 469–473 (1990)
22. Agarwal, B., Mittal, N.: Text classification using machine learning methods-a survey. In: Proceedings of the Second International Conference on Soft Computing for Problem Solving (SocProS 2012), 28–30 December 2012, pp. 701–709. Springer, India (2014)
23. Larrañaga, P., Inza, I., Moujahid, A.: Tema 6. Clasificadores Bayesianos. Departamento de Ciencias de la Computación e Inteligencia Artificial (En línea: http://www.sc.ehu.es/ccwbayes/docencia/mmcc/docs/t6bayesianos.pdf. Acceso: 9 de enero de 2016), Universidad del País Vasco-Euskal Herriko Unibertsitatea, España (1997)
24. Quinlan, J.R.: C4. 5: Programs for Machine Learning. Elsevier, Burlington (1993)
25. García, D.A.: Algoritmo de discretización de series de tiempo basado en entropía y su aplicación en datos colposcopicos. Tesis de Maestría en Inteligencia Artificial. Universidad Veracruzana. México. (En línea: http://cdigital.uv.mx/bitstream/123456789/32352/1/garcialopezdaniel.pdf, acceso: 10 de agosto de 2015 (2007)
26. Corso, C.L.: Alternativa de herramienta libre para la implementación de aprendizaje automático. (En línea: http://www.investigacion.frc.utn.edu.ar/labsis/Publicaciones/congresos_labsis/cynthia/Alternativa_de_herramienta_para_Mineria_Datos_CNEISI_2009.pdf, acceso: 10 de agosto de 2015), Argentina (2009)

Water Quality Monitoring for Horticulture and Aquaculture

Rishika Anand and Monika Choudhary[✉]

Department of CSE, IGDTUW, Delhi, India
rishikaanand30@gmail.com, monika.ch13@gmail.com

Abstract. Recently, water pollution is considered as the greatest threat for green globalization. In past scenario, water quality was measured by chemical tester laboratory test. This was a manual and time consuming system. Different methodologies have been evolved to detect and improve water quality. In this paper, a new model is proposed to continuously monitor the water quality for horticulture and aquaculture. Several sensors are used to detect various parameters which are significant in determining the quality of water. Sensor values are stored and analyzed in Thingspeak cloud. Users can view the data collected from sensors via an android application. This improvised system for measuring water quality in horticulture and aquaculture can be implemented in running water. It monitors more number of parameters and users get timely alerts if the value of any parameter(s) is above threshold value. Pushbullet is used to get timely notification.

Keywords: Water quality monitoring · Thingspeak · Wemos D1 mini · Temperature · Turbidity

1 Introduction

Ensuring water safety is a challenge due to excessive sources of pollutants. The rapid jump of industrialization and greater emphasis on agricultural growth combined with agricultural fertilizers and non-enforcement of laws have led to water pollution to a large extent. Other sources of water pollution include floods and droughts and lack of awareness and knowledge among people.

Poor water quality spreads disease like diarrhoea, cholera, etc. Water contaminations can pass into the food chain and accumulate until they reach toxic levels, eventually killing mammals, fish and birds. Deteriorating water quality is a serious threat for aquaculture. All the above issues make water quality monitoring highly essential.

Water quality can be detected by using different sensors like turbidity, Total Dissolved Solids (TDS), Dissolved Oxygen (DO), Temperature, pH, flow etc.

Due to heavy rain, a large amount of suspended particles may come from mountains which increases the level of TDS. Industrial waste, silt, sewage, plankton, road salts used on street during the winter, fertilizers and pesticides used on lawns and farms also increases the TDS level. TDS increases the turbidity level of water hindering the sunlight to penetrate inside water. Due to higher TDS and turbidity water plants are not

© Springer Nature Switzerland AG 2020
A. P. Pandian et al. (Eds.): ICICCS 2019, AISC 1039, pp. 345–356, 2020.
https://doi.org/10.1007/978-3-030-30465-2_39

able to release oxygen inside water which reduces dissolved oxygen level. Low dissolved oxygen level makes water harmful for aquatic animals. As TDS increases, temperature also increases due to a presence of different suspended particles (lead, cadmium, nitrate, arsenic and others) in the water. High temperature is harmful for aquatic organisms. At high pH or low pH water becomes alkaline or basic which is hazardous for aquatic life. There is a permissible limit for the various parameters measured from these sensors which is discussed in Table 1.

Section 2 comprises of literature review and limitations of previous models. Section 3 presents the proposed model and lays down the hardware and Software requirements of the proposed system. Implementation and Result is discussed in Sects. 4 and 5 contain Conclusion and Future Scope.

2   Related Work

As monitoring and regulating water quality is of significant importance, several researchers have worked in this domain. It may not be specific to water pollution monitoring but similar concepts are evolved. Various models are developed based on sampling water to help in monitoring aquatic environments. They have contributed by measuring different parameters, working on different type hardware catering to the particular needs of the system.

Elamparithi et al. in [1], monitors two water quality parameters (pH and turbidity) with the help of raspberry pi as a microcontroller. Short message service (SMS) is sent as an alert to the user. Temperature and pH parameters are monitored by Mitar et al. in [11]. Data is stored in IBM cloud and MQTT (Message Queuing Telemetry Transport) algorithm for wireless communication. Similar parameters are monitored in [16].

Saravanan et al. in [17] monitored colour with temperature and flow. Arduino was used as a microcontroller and LCD screen to show the measured values over it. Das and Jain in [3], took conductivity into account along with temperature & pH and used thingspeak cloud to store the measured data. Pande et al. in [9], used combination of previously monitored four parameters (temperature, turbidity, pH, level) to detect water quality. Arjun et al. in [5], monitored pressure as an additional parameter. Raspberry pi 3 was used. These systems were used for monitoring water level in tank of society. Kafli et al. in [15], monitored carbon monoxide as an additional parameter.

Geetha et al. in [2], monitored four parameters (pH, Conductivity, Turbidity, and Level) with the help of T1 CC3200 microcontroller. It used ubidots cloud to store data and sent SMS alerts. Kamble et al. in [6], took DO into account with pH, Conductivity, Turbidity and web GUI is made to access the data. Similar parameters were monitored with conductivity as in [18, 19] and temperature as an additional parameter in [21]. Four parameters (turbidity, pH, temperature, flow) were used to monitor water quality by Vaishnavi et al. in [8]. Prasad et al. in [12], addition parameter redox is monitored. Similar parameters were monitored with conductivity by Theofanis et al. in [20]. Light intensity and CO2 was monitored along with turbidity and temperature by Chi et al. in [13]. Zigbee was used as a communication module.

Shah et al. in [7], monitored five parameters (pH, conductivity, temperature, flow, pressure). The model was used for water distribution. Similar parameters were monitored with turbidity by Gopavanitha et al. in [4]. Srivastava et al. in [10], monitored pH, TDS, oxidation and reduction potential (ORP), salinity and conductivity. Bluetooth was used for communication, google firebase cloud was used to store data and android application was made to access data. Raju et al. in [14], monitored six parameters (Bicarbonates, Ammonia, DO, salt, CO3, pH) to detect water quality.

3   Proposed Model

In the proposed model, real time water quality monitoring for horticulture and aquaculture in IOT environment is presented. Several sensors (Temperature, Total Dissolved Solids, pH, Rain) are used to monitor parameters (Turbidity, DO, Temperature, Rain, pH, TDS) to detect water quality. This will ensure proper growth of crops and survival of aquatic organisms.

Sensors are connected to Wemos D1 mini board with the help of multiplexer. Parameters are monitored using sensors and the collected data is send to Thingspeak cloud with the help of Wi-Fi which is in-built communication module in Wemos D1 mini. Android App (ThingView) is used so that farmers can easily access the data. Pushbullet application is used to receive notification if the values of parameter are greater than the threshold value and farmers can take actions accordingly.

3.1   Architecture of Proposed System

Hardware Architecture

Sensors (Total Dissolved Solids, rain, temperature and pH) are directly interfaced to the multiplexer. Multiplexer is connected to core controller i.e. Wemos D1 mini. The sensor parameters such as Total Dissolved Solids, turbidity, rain, DO, Alkalinity, pH and temperature are measured by placing the sensor into water.

WeMos D1 mini is used as a microcontroller: ESP-8266EX. Its operating Voltage is 3.3 V & 5 V. It has 11 digital I/O pins and 1 analog pin [23]. Cost of the Wemos D1 mini is INR 237 [24]. It has an in-built communication module i.e. Wi-Fi which is used to send data from microcontroller to cloud. Multiplexer is used to connect all the analog sensors to the microcontroller. Solar Panel is used to give the power to the system so that the user need not to change its battery time to time.

pH sensor is used to measures alkalinity or acidity, or the base and caustic present in a solution. It is generally expressed with a numeric scale ranging from 0–14. High pH leads to high alkalinity and this reduces the amount of light entering into the water so turbidity increases which is harmful for aquaculture. Its cost is INR 2400 [26].

Temperature sensor is used to detect the temperature of water. More temperature of water means more turbidity is there and less dissolved oxygen which is harmful for aquaculture. Its cost is INR 115 [28].

Rain sensor is used to detect the rainfall. Due to rain, more suspended particles will flow in the water which causes more turbidity and more no. of TDS. Its cost is INR 103 [30].

Total Dissolved Solid sensor is used to detect the suspended particles present in the water. TDS (Total Dissolved Solids) indicates that how many milligrams of soluble solids dissolved in one liter of water. High TDS means high turbidity is there which decrease the penetration of light into water and thus DO decreases. Less DO causes death of aquatic organisms. Its cost is 1029 [32].

DO is monitored with the help of salinity and temperature. Formula to monitor DO is given in Eq. 1 (Figs. 1 and 2) [37].

$$\ln C_G = A_1 + A_2(100/T) + A_3\ln(T/100) + A_4(T/100)$$
$$+ S\left(B_1 + B_2(T/100) + B_3(T/100)^2\right) \tag{1}$$

where: T = Absolute temperature (°K)
      S = Salinity

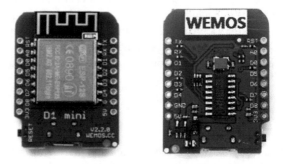

Fig. 1. Wemos D1 mini front end & back end [25]

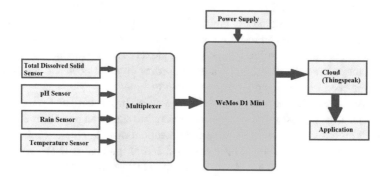

Fig. 2. Hardware architecture of system

Software Architecture
After the development of system hardware architecture we have to focus on software part that is needed to run the application.

ThingSpeak cloud is used because it has many features. It collect data in private channels, share data with public channels, MATLAB analytics and visualizations are there, give alerts to the user, event scheduling and app integration.

Arduino language is used to interface sensors with WeMos D1 mini to get the values of parameters monitored through sensors. Values are then sent to the cloud through wireless communication (Wi-Fi) using Arduino language.

User can access the data which is stored in thingspeak cloud by downloading an android application (ThingView) from google play. Figure 3 shows how to access thingspeak data on android application.

Pushbullet android application is used to get notification whenever the parameter value is greater than threshold value.

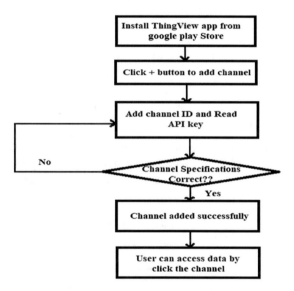

Fig. 3. Flow chart of ThingView App

3.2   System Flow

After giving power to the system, the system gets started and connection between different modules (Microcontroller, Sensors, multiplexer & Cloud) is established. Sensors are able to monitor the quality of water. Microcontroller will get the values monitored by sensors through multiplexer. Values monitored by sensors will be send to Thingspeak cloud with the help of wireless communication (Wi-Fi). Farmer can easily access the values of different parameters through android application (ThingView). If sensor value is greater than the threshold value than notification as an alert on push-bullet application is sent to the farmer and farmers can take actions accordingly (Fig. 4).

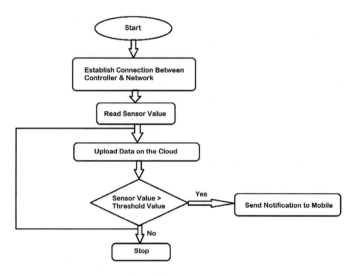

Fig. 4. Flow chart of system

3.3 Parameters Monitored

Different parameters (Turbidity, Rain, pH, Temperature, Total Dissolved Solids, Dissolved Oxygen, and Location) are monitored to know the quality of water. Threshold value of these parameters for aquaculture and horticulture are mentioned in Table 1.

Table 1. Range of parameters for horticulture and aquaculture

S. No.	Parameters of water	Range for horticulture	Range for aquaculture
1	Turbidity	1280 NTU (derived 3 ntu = ppm) [35]	167 NTU (derived 3 ntu = ppm) [35]
2	pH	5 < pH < 7 [36]	4 < pH < 11 [34]
3	Temperature	12 < temp < 30 [33]	13 < temp <30 [31]
4	Total dissolved solids	<=3840 ppm [29]	<=500 ppm [27]
5	DO	Not applicable	3–8 mg/L [22]

3.4 Proposed Deployment of System

In proposed deployment system, microcontroller unit i.e. WeMos D1 mini can be put inside the metal pipe so that we can save it from running water. All the sensors (temperature, total dissolved solids, rain, and pH) are connected to multiplexer. Multiplexer is connected to microcontroller unit. Sensors can be submerged partially in running water to get the values of different parameters (Temperature, pH, Dissolved Oxygen,

Turbidity, Total Dissolved Solids, rain). Metal pipe can be installed in the pillar of the cemented bridge so that we can get the different parameter readings of running water (Fig. 5).

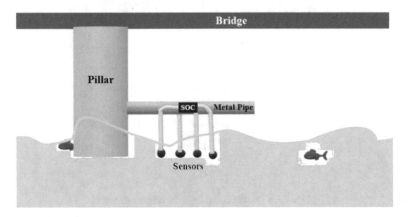

Fig. 5. Proposed deployment of system in running water

4   Implementation and Results

The proposed model is made to detect different parameters (Turbidity, Rain, pH, Temperature, Total Dissolved Solids, Dissolved Oxygen, and Location) to monitor water quality for aquaculture and horticulture.

Figure 6 shows the plot of pH with respect to time. pH is monitored continuously with the help of pH probe. If pH is more than maximum threshold value or less than minimum threshold value then notification is sent to pushbullet application.

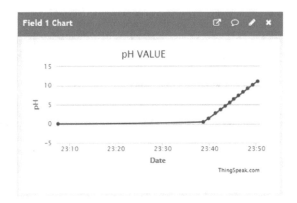

Fig. 6. pH variation with respect to time

Figure 7 shows the plot of rain with respect to time. Rain is monitored continuously with the help of rain sensor. 0 is represented as no rain and 1 is represented for rain.

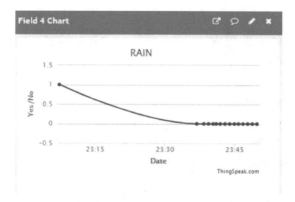

Fig. 7. Variation of rain with respect to time

Figure 8 shows the plot of turbidity with respect to time. Turbidity is monitored continuously with the help of TDS sensor.

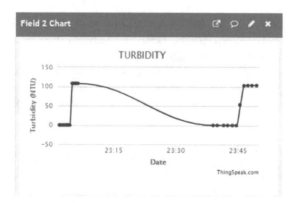

Fig. 8. Variation of turbidity with respect to time

Figure 9 shows the plot of TDS with respect to time. TDS is monitored continuously with the help of TDS sensor.

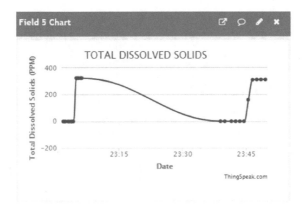

Fig. 9. Variation of TDS with respect to time

Figure 10 shows the plot of temperature with respect to time. Temperature is monitored continuously with the help of Temperature sensor.

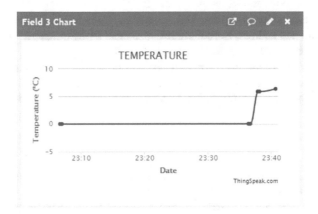

Fig. 10. Variation of temperature with respect to time

Figure 11 shows the plot of Dissolved Oxygen with respect to time. Dissolved Oxygen is monitored continuously with the help of Temperature sensor and salinity value which is measured by TDS sensor.

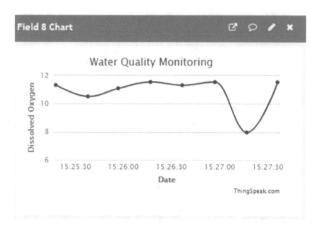

Fig. 11. Variation of DO with respect to time

5   Conclusion and Future Work

In this paper, a new model is proposed for real time water quality monitoring. Parameters (temperature, pH, rain, DO, Location, TDS, turbidity) are monitored for horticulture and aquaculture. Sensor values are stored in the thingspeak cloud and user can access those values by android app.

In the future, more number of parameters can be monitored such as oxidation and reduction potential (ORP), Nitrates, ammonia, etc. Application can be made for the use of different purposes like smart cities, water quality monitoring authorities, etc.

References

1. Elamparithi, P., Adhlin Esther, S., Jeya Gowri, A.: Water quality monitoring system using IoT. Int. J. Technol. Res. Eng. **5**(7) (2018)
2. Geetha, S., Gouthami, S.: Internet of things enabled real time water quality monitoring system. Smart Water **2**(1), 1 (2016)
3. Das, B., Jain, P.C.: Real-time water quality monitoring system using Internet of Things. In: 2017 International Conference on Computer, Communications and Electronics (Comptelix), IEEE, pp. 78–82 (2017)
4. Gopavanitha, K., Nagaraju, S.: A low cost system for real time water quality monitoring and controlling using IoT. In: 2017 International Conference on Energy, Communication, Data Analytics and Soft Computing (ICECDS), IEEE, pp. 3227–3229 (2017)
5. Arjun, K., Latha, C.A., Prithviraj: Detection of water level, quality and leakage using Raspberry Pi with internet of things. Int. Res. J. Eng. Technol. (IRJET) **04**(06) (2017)
6. Kamble, R., Kakade, S., Mahajan, A., Bhosale, A.: Automatic water quality monitoring system using Arduino. Int. J. Recent Innov. Eng. Res. (2017)
7. Shah, J.: An internet of things based model for smart water distribution with quality monitoring. Int. J. Recent Innov. Eng. Res. **6**(3) (2017)
8. Daigavane, V.V., Gaikwad, M.A.: Water quality monitoring system based on IoT. Int. J. Electron. Eng. Res. **10**(5), 1107–1116 (2017)

9. Pande, A.M., Warhade, K.K., Komati, R.D.: Water quality monitoring system for water tanks of housing society. Int. J. Electron. Eng. Res. **9**(7), 1071–1078 (2017)
10. Srivastava, S., Vaddadi, S., Sadistap, S.: Smartphone-based system for water quality analysis. Appl. Water Sci. **8**(5), 130 (2018)
11. Simić, M., Stojanović, G.M., Manjakkal, L., Zaraska, K.: Multi-sensor system for remote environmental (air and water) quality monitoring. In: 2016 24th Telecommunications Forum (TELFOR), IEEE, pp. 1–4 (2016)
12. Prasad, A.N., Mamun, K.A., Islam, F.R., Haqva, H.: Smart water quality monitoring system. In: 2015 2nd Asia-Pacific World Congress on Computer Science and Engineering (APWC on CSE), IEEE, pp. 1–6 (2015)
13. Chi, Q., Yan, H., Zhang, C., Pang, Z., Da Xu, L.: A reconfigurable smart sensor interface for industrial WSN in IoT environment. IEEE Trans. Ind. Inform. **10**(2), 1417–1425 (2014)
14. Raju, K.R.S.R., Varma, G.H.K.: Knowledge based real time monitoring system for aquaculture using IoT. In: 2017 IEEE 7th International Advance Computing Conference (IACC), IEEE, pp. 318–321 (2017)
15. Kafli, N., Isa, K.: Internet of Things (IoT) for measuring and monitoring sensors data of water surface platform. In: 2017 IEEE 7th International Conference on Underwater System Technology: Theory and Applications (USYS), IEEE, pp. 1–6 (2017)
16. Srishaila Mallikarjuna Swamy, P.M., Mahalakshmi, G.: Real time monitoring of water quality using smart sensor. J. Emerg. Technol. Innov. Res. (2017)
17. Saravanan, K., Anusuya, E., Kumar, R.: Real-time water quality monitoring using Internet of Things in SCADA. Environ. Monit. Assess. **190**(9), 556 (2018)
18. Bhatt, J., Patoliya, J.: IoT based water quality monitoring system. IRFIC, 21 February 2016
19. Faustine, A., Mvuma, A.N., Mongi, H.J., Gabriel, M.C., Tenge, A.J., Kucel, S.B.: Wireless sensor networks for water quality monitoring and control within lake victoria basin: prototype development. Wirel. Sensor Network **6**(12), 281 (2014)
20. Lambrou, T.P., Anastasiou, C.C., Panayiotou, C.G., Polycarpou, M.M.: A low-cost sensor network for real-time monitoring and contamination detection in drinking water distribution systems. IEEE Sens. J. **14**(8), 2765–2772 (2014)
21. Nikhil, R., Rajender, R., Dushyantha, G.R., Khadri, M.N.S., Kalshetty, J.N.: Smart water quality monitoring system using IoT environment. Int. J. Innov. Eng. Technol. (2018)
22. Towers, L.: How to achieve good water quality management in aquaculture (2008). https://thefishsite.com/articles/how-to-achieve-good-water-quality-management-in-aquaculture. Accessed 20 Mar 2019
23. Battista, T.: Wemos D1 Mini, ESP8266 Getting started guide with Arduino (2017). https://www.hackster.io/innovativetom/wemos-d1-mini-esp8266-getting-started-guide-with-arduino-727098. Accessed 29 Sept 2018
24. Calcutta Electronics: Wemos D1 mini (2018). https://calcuttaelectronics.com/product/wemos-d1-mini-esp826/ Accessed 29 Jan 2019
25. Wemos: D1 mini V2.2.0 (2017). https://wiki.wemos.cc/products:retired:d1_mini_v2.2.0. Accessed 29 Sept 2018
26. Amazon: pH Sensor Module (2019). https://www.amazon.in/Sensor-Module-Probe-Shield-MSP430/dp/B01LZUENX2?tag=googinhydr18418-21&tag=googinkenshoo-21&ascsubtag=_k_CjwKCAjwtYXmBRAOEiwAYsyl3CkoI5K6HhIAz3jn0RIlg5fLbY5jLR5YohOUB3_e520BrnKGRHlY0hoCHGwQAvD_BwE_k_&gclid=CjwKCAjwtYXmBRAOEiwAYsyl3CkoI5K6HhIAz3jn0RIlg5fLbY5jLR5YohOUB3_e520BrnKGRHlY0hoCHGwQAvD_BwE Accessed 25 Apr 2019
27. Water Filters Online: Total Dissolved Solids (2018). http://www.waterfiltersonline.com/tds-sources.asp. Accessed 20 Mar 2019

28. Robu.in: Temperature Sensor (2019). https://robu.in/product/ds18b20-water-proof-temperature-probe-black-1m-2/. Accessed 25 Apr 2019

29. Kamthe, B.: TDS levels in irrigation water for traditional crops - Need recommendations for TDS-resistant crops (2007). https://www.indiawaterportal.org/questions/tds-levels-irrigation-water-traditional-crops-need-recommendations-fortds-resistant-crops. Accessed 20 Mar 2019

30. Robu.in: Rain Sensor (2019). https://robu.in/product/raindrops-detection-sensor-module-rain-weather-humidity/. Accessed 25 Apr 2019

31. Neospark. http://neospark.com/images/waterqua.pdf. Accessed 20 Mar 2019

32. MG Superlabs: TDS Sensor (2019). https://www.mgsuperlabs.co.in/estore/Gravity-Analog-TDS-Sensor-Meter-for-Arduino. Accessed 25 Apr 2019

33. Climatic Requirements. https://www.kzndard.gov.za/images/Documents/Horticulture/Veg_prod/climatic_requirements.pdf. Accessed 20 Mar 2019

34. Yokogawa India Ltd. (2019). https://www.yokogawa.com/in/library/resources/application-notes/ph-in-fish-farming/. Accessed 20 Mar 2019

35. Lynge, D.: PPM into NTU (2017.) https://www.quora.com/How-do-I-convert-NTU-to-PPM-of-turbidity. Accessed 20 Mar 2019

36. Mosaic Company: Soil pH (2019). https://www.cropnutrition.com/efu-soil-ph. Accessed 20 Mar 2019

37. Chemical Oceanography. http://www.soest.hawaii.edu/oceanography/courses/OCN623/Spring2012/Non_CO2_gases.pdf. Accessed 20 Mar 2019

A Neural Network Based Overvoltage Prediction System for Long Cable Issue

Anju Joseph[1], K. V. Vineetha[1](✉), Dhanesh G. Kurup[2], and R. Mini[3]

[1] Department of Computer Science and Engineering, Amrita School of Engineering, Bengaluru, Amrita Vishwa Vidyapeetham, Bengaluru, India
anjusen.george@gmail.com, jain_vineetha@blr.amrita.edu
[2] Department of Electronics and Communication Engineering, Amrita School of Engineering, Bengaluru, Amrita Vishwa Vidyapeetham, Bengaluru, India
dg_kurup@blr.amrita.edu
[3] Department of Electrical and Electronics Engineering, Amrita School of Engineering, Bengaluru, Amrita Vishwa Vidyapeetham, Bengaluru, India
mini_sujith@blr.amrita.edu

Abstract. In most electrical applications, the motor is fed by Pulse Width Modulated(PWM) inverters and by design they are usually in separate locations, requiring long motor leads or cables. Overvoltage is a common phenomenon in AC motor drives that are fed by long cables from PWM inverters, as they use Insulated Gate Bipolar Transistors (IGBT) with small rise time and fall time. This paper proposes an Artificial Neural Network based system to predict overvoltage at the motor terminal with long cable using the parameters cable length and motor capacity.

Keywords: Overvoltage · Artificial Neural Network (ANN) · Pulse Width Modulation (PWM) · Insulated Gate Bipolar Transistors (IGBT)

1 Introduction

Developments in power electronics technology has improved the switching frequency of electronic devices. Insulated Gate Bipolar Transistor (IGBT) devices currently available have very high switching frequency and rise time in the range of nano seconds [1,2]. PWM inverters with IGBT are extensively used in drives for variable speed induction motors. The induction motor and PWM inverter are mostly located away from each other, especially in industrial applications and hence require long cables or motor leads (Fig. 1).

The output voltage pulses from PWM inverter operates like travelling waves on these cables. Full reflection of the wave will occur at motor terminal, if the propagation delay of the wave is greater than half (0.5) the rise time of the inverter output pulse resulting in overvoltage at the motor terminal. Lossless switching and high switching frequency improves the performance of PWM

© Springer Nature Switzerland AG 2020
A. P. Pandian et al. (Eds.): ICICCS 2019, AISC 1039, pp. 357–364, 2020.
https://doi.org/10.1007/978-3-030-30465-2_40

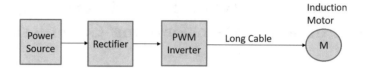

Fig. 1. Representation of induction motor driven by PWM inverter using long cable

inverters. But very high rate of rise in voltage(dv/dt), of the order of zero to six hundred volt in less than 0.1 μs, will cause damage to the insulation and bearings of the motor [2]. This paper proposes an ANN based system to predict the overvoltage at the motor terminal connected via long motor cable or leads. The proposed system provides inputs for design of application or devices to control overvoltage scenarios. The key factors that influence the occurrence of voltage reflection are rise time of inverter output voltage pulses, the length of the connecting cable between the motor terminal and inverter, the impedance of cable and motor. Simulink tool of MATLAB is used for creating simulation models. Dataset is generated from these simulation models. A Multilayer Perceptron (MLP) based neural network is used to predict the overvoltage at the motor terminal. FANN (Fast Artificial Neural Network) library is utilized for this implementation.

1.1   Theory of Voltage Reflection

The speed of an electromagnetic pulse is approximately 150–200 m/μs [1]. If the electromagnetic pulses take longer than half the rise time to travel over the cable from inverter output to motor, then a full reflection of the pulse will happen at the motor terminal. This will cause doubling of the amplitude of voltage pulse. This reflection of the pulse caused by fast switching transients can be lessened by increasing the rise time as well as the fall time of voltage output pulses of the inverter. The factors affecting voltage reflection are the rise time of the output voltage pulses of inverter (t_r), cable length used for connecting the inverter and motor which affects the propagation time of the pulse (t_p) and surge impedance of cable & motor [12]. If $t_p > t_r/2$, then a full reflection of the pulse occurs and cause doubling of the amplitude of voltage at the motor terminal [1,2,7,10]. The dv/dt gradient of the inverter output voltage can be reduced by using a passive dv/dt filter [2]. The major drawback of such filters is the large power loss. The reflected voltage amplitude depends on the co-efficient of voltage reflection(Γ_m) of motor [1] and is given by equation below:

$$\Gamma_m = \frac{Z_m - Z_c}{Z_m + Z_c} \tag{1}$$

where Z_m is characteristic impedance of the motor and Z_c is the characteristic impedance of the cable. The following equation gives the peak voltage value (V_m) at the motor terminal:

$$V_m = V_s(1 + \Gamma_m) \tag{2}$$

In Eq. (2), V_s represents source voltage. It can be concluded from the equation that voltage reflection phenomenon causes overvoltage at the motor terminal.

2   Related Work

Many researchers have worked on overvoltage in transmission lines [3–5]. Some research work focuses on the overvoltage in power lines and others on overvoltage occurrence at motor terminals connected via long leads from inverter fed systems [1]. The researchers have identified the factors causing overvoltage. ANN based solutions are proposed by few researchers to predict overvoltage on power lines [4,5]. Support Vector Machine (SVM) based classification of supply voltage to induction motor was also explored by few researchers [6] in the past. von Jouanne and et al. examines how long motor leads affect motor insulation of AC motor drives [1] that are fed by high frequency PWM inverters. Anirudh Acharya et al. designed a dv/dt output filter [2] specifically for drives of motors. They proposed a new procedure to design an LC clamp filter to reduce voltage doubling at the terminal of the motor. Pspice simulation is used to verify the effectiveness of the design. Couto and et al. proposes an alternate strategy to model a transmission line by analyzing [3] the three phase conductors separately. The results from the simulation clearly shows that there are overvoltage occurrences along the line although the values of voltage at both ends of the line are at acceptable levels. Need for new procedures for insulation co-ordination of power transmission lines is highlighted by the researchers. Thukaram et al. presents an ANN based tool to predict the peak over voltage [4] caused by switching transients during line energization. The computation of temporary overvoltage as well as switching overvoltage is done using the Electro Magnetic Transients Program (EMTP) tool. The Multi-Layer Perceptron (MLP) is trained using the Levenberg-Marquardt method. The aspects considered to influence the transient overvoltage are transmission line length, the switching angle, the strength of the source and the reactor at the receiving end. Taher et al. uses ANN to predict temporary overvoltage [5] due to transmission line energization. In Extra High Voltage lines(EHV) lines, for insulation co-ordination, the primary parameter to be considered is switching overvoltage. Simulink tool contains Power System Blockset (PSB) which is utilized for calculation of overvoltage in this approach. MLP is trained using second order of Levenberg-Marquardt method for obtaining small mean square error (MSE).

3   Methodology

Block diagram shown in Fig. 2 represents the overvoltage prediction system. The first step is to generate training data from simulation models. The simulation models are created using Simulink tool. The key components of the model are three phase power source, inverter, converter, cable, motor and a load. The next step is to train the ANN using the gathered data. This trained Neural Network is then used to predict overvoltage based on the cable and motor parameters fed by the user.

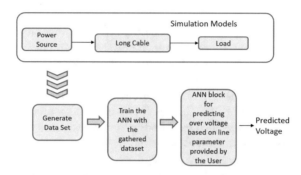

Fig. 2. Block diagram of overvoltage prediction system

4 Implementation

The proposed system uses Artificial Neural Network (ANN) to predict the overvoltage. ANN is implemented using FANN(Fast Artificial Neural Network) library, which is a C++ based library. It has functions that makes it easy to create, train and test Neural Networks. Training is carried out via back propagation in FANN.

4.1 Artificial Neural Networks

Artificial Neural Networks [8,9,11] are computing systems that draw inspiration from the biological neural networks that comprise of the brain in human beings and animals. The neural network is a framework for many different machine learning algorithms which process complex inputs of data. Such systems "learn" to perform the task of data processing by considering examples rather than being programmed with a set of task-specific rules.

The overvoltage prediction system is a feed forward ANN with 3 layers as depicted in Fig. 3. The motor capacity and cable length are the 2 inputs that are fed into the input layer. Overvoltage value is given by the output layer. This implementation uses 2 layers in the hidden layer. Sigmoid activation function is used in this model.

Simulink model for 5.4 HP motor is shown in Fig. 4. '1000 V-Multi-Conductor UL Flexible Motor Supply Cable (29501)' specification is used for modelling the cables. Dataset is generated from Simulink models with different lengths of the cable and motor capacity.

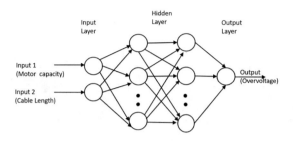

Fig. 3. Neural network diagram

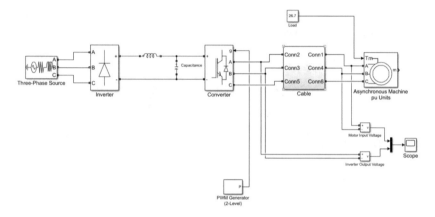

Fig. 4. Simulink model for 5.4 HP motor connected to three phase supply

5 Results

Figure 5 depicts the voltage waveform at the motor terminal connected by long cable interposed with actual output voltage of the inverter which is represented by square pulses.

The distribution of overvoltage applied at the motor terminal of 5.4 HP and 10 HP motors connected via long cable is shown in Fig. 6. The cable length (between inverter and motor) considered is between 0 and 1000 m. It is observed that there is a nonlinear relationship between the length of the cable and overvoltage. An ANN based system is suitable for making predictions in such nonlinear systems.

Neural network is trained over multiple trials to identify the structure that fit best for the prediction of overvoltage. To improve the efficiency in training, data is scrambled and fed to the neural network. Learning rate of 0.7 is used in this implementation.

Figure 7 shows the effect of using single and two layers of neurons in hidden layer. It is identified that using 2 hidden layers instead of one layer with the same number of neurons offers lesser Means Square Error.

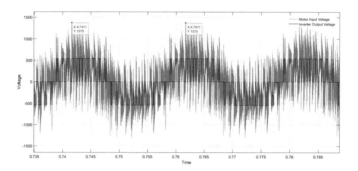

Fig. 5. Voltage signal at the motor terminal for 100 m cable

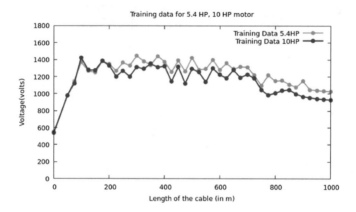

Fig. 6. Overvoltage values of 5.4 and 10 hp motor for cable length (0–1000 m)

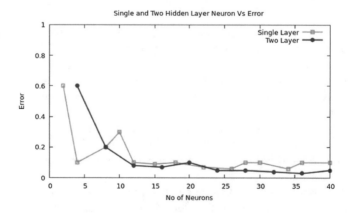

Fig. 7. Error (MSE) for single and two hidden layer neurons

Table 1. Error for various network structures.

Epoch	Neurons inlayer 1	Neurons inlayer 2	Error
1000	4	4	0.18
2000	5	5	0.09
3000	6	6	0.08
4000	8	8	0.06
6000	12	12	0.06
8000	13	13	0.06
10000	14	14	0.05
12000	15	15	0.04
15000	16	16	0.036
15000	18	18	0.031
20000	20	20	0.03
25000	20	20	0.03

Table 1 gives the error rates by changing the epochs and number of neurons. It is observed that for epoch in the range of 15000 to 20000, Mean Square Error is converging at 0.03. Hence it was concluded that epoch of 15000 with 18 neurons each in hidden layer offers the least error and is the best fit for this implementation.

6   Conclusion and Future Work

Artificial neural network is utilized in this implementation for predicting overvoltage values in a PWM inverter fed motor with long cable, for given cable parameters and motor capacity. Predicting the overvoltage would help in estimating the level of insulation required for motor and also aid in the design of filters to control the overvoltage scenarios. This implementation gives high level of accuracy for the input parameters considered. The implementation is done on a particular cable type and can be enhanced by adding more cable types on the input side. The future scope of work is the real time implementation on a hardware platform like Raspberry Pi, which can be trained for different motors and cable parameters to predict the overvoltage.

References

1. von Jouanne, A., Enjeti, P., Gray, W.: The effect of long motor leads on PWM inverter fed AC motor drive system. In: IEEE Applied Power Electronics Conference and Exposition, vol. 2, pp. 592–597 (1995)
2. Anirudh Acharya, B., John, V.: Design of output dv/dt filter for motor drives. In: IEEE International Conference on Industrial and Information System, pp. 562–567 (2010)

3. Couto, V.F., Cabral, S.H.L, Mustafa, T.I.A.H., Puchale, L.H.B.: Study of occurrence of overvoltage along power transmission lines with acceptable end voltage. In: IEEE Electrical Insulation Conference, pp. 29–32 (2015)
4. Thukaram, D., Khincha, H.P., Khandelwal, S.: Estimation of switching transient peak overvoltages during transmission line energization using artificial neural network. In: Electric Power Systems Research, vol. 76, pp. 259–269 (2006)
5. Taher, S.A., Sadeghkhani, I.: Estimation of magnitude and time duration of temporary overvoltages using ANN in transmission lines during power system restoration. Simul. Model. Pract. Theory **18**, 787–805 (2010)
6. Prez, R., Aguila, A., Vsquez, C.: Classification of the status of the voltage supply in induction motors using support vector machines. In: IEEE/PES Transmission and Distribution Conference and Exposition (T&D), pp. 1–5 (2016)
7. Mini, R., Manjiri Joshi, B., Satheesh, H., Dinesh, M.N.: Active LC clamp dv/dt filter for voltage reflection due to long cable in induction motor drives. Int. J. Electric. Comput. Eng. **6**, 1456–1469 (2016)
8. Vineetha, K.V., Kurup, D.G.: Direct demodulator for amplitude modulated signals using artificial neural network. In: International Symposium on Intelligent Systems Technologies and Applications, Advances in Intelligent Systems and Computing, vol. 683, pp. 204–211. Springer (2017)
9. Purushothaman, A., Vineetha, K.V., Kurup, D.G.: Fall detection system using artificial neural network. In: International Conference on Inventive Communication and Computational Technologies, pp. 1146–1149 (2018)
10. Mini, R., Resna, S.R., Dinesh, M.N.: LC clamp filter for voltage reflection due to long cable in induction motor drives. In: IEEE International Conference on Emerging Research Areas: Magnetics, Machines and Drives, pp. 1–6 (2014)
11. Bharadwaja, V., Ananmy, R., Nikhil, S., Vineetha, K.V., Shah, J., Kurup, D.G.: Implementation of artificial neural network on Raspberry Pi for signal processing applications. In: International Conference on Advances in Computing, Communications and Informatics, pp. 1488–1491 (2018)
12. Desmet, J., Devos, B., Stockman, K., Belmans, R.: Influencing parameters on overvoltages at the terminals of inverter supplied induction motors. In: International Conference on Electrical Machines, Proceedings, pp. 854–858 (2000)

IoT Based Real Time Vehicle Vital Parameter Monitoring and Analytics

Harish Thodupunoori[2] and Binoy B. Nair[1(✉)]

[1] SIERS Research Laboratory, Department of Electronics and Communication Engineering, Amrita School of Engineering, Coimbatore, Amrita Vishwa Vidyapeetham, Coimbatore, India
b_binoy@cb.amrita.edu
[2] Department of Electronics and Communication Engineering, Amrita School of Engineering, Coimbatore, Amrita Vishwa Vidyapeetham, Coimbatore, India
harish.thodupunoori@gmail.com

Abstract. A low cost and reliable IoT solution for real time parameter monitoring of a farming vehicle is presented. This paper provides a solution for various detection mechanisms such as vehicle speed, current active gear, coolant temperature, fuel level, battery voltage etc. and collecting multiple sensor data to cloud database and monitor through a web application and an android mobile application to observe the behavior of autonomous harvester. This system consists of a central micro controller which can connect to the local Wi-Fi network. The experimental outcomes disclose that the IoT framework is able to capture and report vital parameters of the vehicle to cloud server with accurate data and a programmed warning is sent to operator when any of the concerned parameters abnormality is identified.

Keywords: IoT · Parameter monitoring · Fuel level · Gear position · Battery voltage · Coolant temperature

1 Introduction

This project is implemented on an autonomous onion harvester [15] designed to harvest the onion field. It is an unmanned vehicle which has the embedded system that can drive the vehicle and collect onions in the farm field. It has an object detection and path planning system which provides direction to travel and tells the system about location of the onions using GPS co-ordinates and image processing technique so that it can pluck using the harvesting tray. This paper presents an IoT solution to monitor the autonomous harvester remotely.

As the automotive technology is advancing towards autonomous vehicles, remote monitoring of the vehicle is an important aspect to predict the health of the vehicle prior to damage. Many researches have been done for developing intelligent transport systems and web based vehicle condition monitoring systems like Telematic unit for advanced fuel level monitoring system [2], IoT-Based Traction Motor Drive Condition Monitoring in Electric Vehicles [4], Smart Fleet Monitoring System using Internet of Things (IoT) [5], An IoT Framework for Intelligent vehicle monitoring system [6].

© Springer Nature Switzerland AG 2020
A. P. Pandian et al. (Eds.): ICICCS 2019, AISC 1039, pp. 365–373, 2020.
https://doi.org/10.1007/978-3-030-30465-2_41

The aim of this project is to communicate the operator about the various vital parameters status of autonomous harvester [15] such as engine extreme heat conditions [4], fuel level indication [2, 5], battery voltage indication [7], current active gear indication [8], the vehicle speed [9] etc. Proper monitoring of the vehicle increases reliability, service range, maintenance effectiveness and proper control on operating the vehicle reduces maintenance cost and assists for a safe drive. Combining various techniques for parameter monitoring and IoT concept [1, 3] with a low cost and reliable system is the main objective of this paper.

This project consists of three phases where the first phase is for detecting the vital parameters of the vehicle using various sensing mechanisms discussed in the upcoming sections. Second is updating the parameter status to the cloud server using IoT framework [1, 2] and the third phase is sending the latest status data to android application [10] and an alert is sent to the operator based on the sensor data and parameter status (Table 1).

Table 1. Harvester specifications

System	Specifications
Engine	8hp, 3000 rpm diesel engine
Gear box	Forward, neutral and reverse
Fuel capacity	5.3 L

2   Methodology

Proposed system uses ESP8266 Node MCU as micro controller for interfacing with various sensors in the autonomous harvester vehicle as shown in the Fig. 1. Node MCU itself has an in-built Wi-Fi module which provides way to connect to the internet. Software segment of this framework consists of Arduino IDE which is used for programming the ESP8266, an IoT cloud API [3] that is used for communicating the operator about the vehicle status information and android application which fetches the latest status data updated in the cloud and alerts the operator if any abnormality exists. According to the hardware segment concerned, ESP8266 NodeMCU [11] is the most vital element which empowers the working of the whole circuit. Power required for the operation of ESP8266 NodeMCU is supplied by a DC power supply. The readings are continuously updated on the webpage and cloud database through the inbuilt Wi-Fi. A programmed warning is sent to the operator through an android application whenever any deviations appear in data compared to the set thresholds.

2.1   ESP8266 – Node MCU

NodeMCU is an open source embedded development board that is assembled around a very low cost SoC called the ESP8266. The ESP8266, designed and produced by Expressif Systems, contains all crucial components of modern computer: CPU, RAM, networking, and a modern operating system and SDK. The board has a built-in USB port that is already wired up with the chip, hardware reset button, Wi-Fi antenna, LED

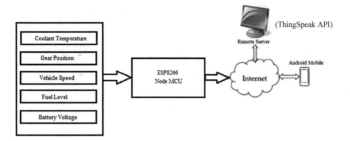

Fig. 1. Proposed system

lights and a standard-sized GPIO pins that can plug into a bread-board. NodeMCU incorporates 14 digital GPIO pins and a single analog pin (A0). For interfacing multiple analog sensors to NodeMCU, here a 16 channel analog demultiplexer/multiplexer is used.

2.2 16-Channel Analog Multiplexer/Demultiplexer

CD74HC4067 is a 16-channel digitally controlled analog switch that uses silicon-gate CMOS technology and has low ON resistance of 70 Ω (@V_{cc} = 4.5 V), fast switching and propagation speeds. It has bidirectional switches allowing any analog input to be used as an output and vice versa. The demultiplexer/multiplexer is used in this project for multiplexing the analog based sensor to single analog pin A0 on NodeMCU.

2.3 Gear Position Detection

In this proposed idea, gear position is detected using Inductive Proximity Sensor [14]. Inductive Proximity sensor output will be LOW state (i.e. 0 V) during inactive; when a metal object comes into its proximity the output will change to HIGH (i.e. 5 V). Its HIGH state changes according to the supply voltage. If the sensor is supplied with 5 V, then the output level for HIGH state is 5 V. 5 V to 36 V can be supplied to inductive proximity sensor. A setup as shown in the Fig. 2 is constructed to detect the gear position.

Fig. 2. Gear position detection configuration

A metal poll is attached to gear lever which moves based on the gear position. The metal poll comes into proximity of the inductive proximity sensor whenever the particular gear is active. The state of the inductive proximity sensor is sent to the micro controller as the active gear position. The autonomous harvester has only forward, reverse and neutral gears. The gear positions for Forward, Reverse and Neutral positions are enumerated as 16, 5 and 10 respectively for visualizing on the graph (Fig. 3).

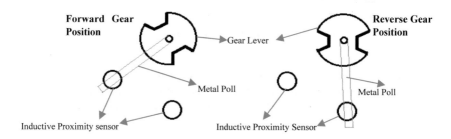

Fig. 3. Gear position representations

2.4   Vehicle Speed Detection

Vehicle Speed is calculated using the RPM of the vehicle wheel. An inductive proximity sensor is utilized for calculating the RPM of the vehicle wheel. As discussed earlier, inductive proximity sensor changes its state from LOW to HIGH whenever it detects a metal object. This mechanism is used in RPM calculation.

As shown in Fig. 4, there are 6 metals polls arranged such that they cover the circumference of the vehicle wheel. The inductive proximity sensor's face is placed such that every time the metal poll comes into sensors proximity, sensor changes its state. Then timers and interrupts in micro controller are used to calculate the time taken for complete rotation of the wheel. Here a high priority interrupt is used to detect RPM and the configuration is such that whenever sensor output goes LOW to HIGH, a function *pollcounter* will be executed. Once the time taken is known we can calculate the RPM by using the below formula.

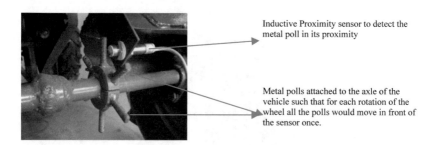

Fig. 4. Wheel RPM and vehicle speed detection configuration

$$rpm = \frac{60}{2} \times \frac{1000}{(millis(.) - passedtime(.))} \times \frac{pollcounter}{polls} \qquad (1)$$

After calculating RPM, the formula to calculate Vehicle Speed is:

$$\text{Vehicle Speed} = \frac{2 \times \pi \times r \times 60 \times rpm \times 0.0254}{1000} \qquad (2)$$

Where $\pi = 3.14$ and Wheel Radius (r) = 13.5 in.
Vehicle Speed is in KMPH.

2.5 Fuel Level Detection

Float based fuel level sensor is interfaced with NodeMCU. NodeMCU module reads the float sensor output value in the voltage form and calculates the fuel level based on the sensed voltage. Here the output voltage of the float sensor is variation in the voltage. As the float level on the fuel changes based on the level of the fuel, the float sensor resistance changes and based on that the voltage available at the NodeMCU analog input (F_MON) will be varied. NodeMCU maps the voltage available at various positions of the float and calculates the percentage of the fuel in fuel tank. The calculated fuel level in percentage is sent to the web server using the in-built Wi-Fi module (Fig. 5).

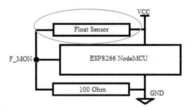

Fig. 5. Fuel level detection model circuit diagram

Here the two terminals of the float level sensor are assumed to be the two terminals of a variable resistor. The resistance between the terminals of float level sensor is minimum of 0 Ω and maximum of 100 Ω. When the float is at the top, i.e. when the fuel tank level is full the resistance across the float sensor will be approximately 0 Ω. As the float moves to the down position, the resistance reaches to max of 100 Ω which indicates the fuel tank is empty. The voltage available on the F_MON (pin A0) pin of ESP8266 NodeMCU is varied as the float moves from top to the bottom of the fuel tank. We can observe the voltage (V_{F_MON}) available at top and bottom points of the fuel tank and map accordingly to represent the fuel tank level in percentage. The following formula has been used for calculation and mapping of the fuel level with the voltage available at the F_MON pin of micro controller.

Formula for Fuel level detection:

$$V_{F\_MON} = V_c \times \frac{100}{100 + F_{SensRes}} \quad (3)$$

Where V_{cc} = 3.3 V
$F_{SensRes}$ = 0 to 100 Ω
V_{F_MON} Mapping with Fuel Level:
Fuel Tank Level @0% → V_{F_MON} = 1.65 V
Fuel Tank Level @100% → V_{F_MON} = 3.3 V

2.6 Battery Voltage Monitoring

Battery Voltage is an important factor for ignition and for all the other electronic components which make the harvester autonomous. Here the battery voltage is monitored through a resistor divider network as shown in Fig. 6. Node MCU has a single 10-bit ADC which means that it will map input voltages from 0 to 3.3 V into integer values from 0 to $(2^{10} - 1)$ i.e. equal to 1023 which gives a resolution of 3.22 mV per unit. 0 will correspond to 0 V, 1 to 4.9 mV, 2 to 9.8 mV and so on till 1023.

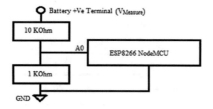

Fig. 6. Resistor divider for battery voltage measurement

$$V = V_{Measure} \times \frac{3.3}{1024} \times 10 \quad (4)$$

Where 'V' is voltage at analog pin A0 and $V_{Measure}$ is the battery voltage to be measured.

2.7 Coolant Temperature Monitoring

A temperature sensor DS18B20 probe [12] which can measure temperatures in the range of −55 °C to +125 °C is utilized for coolant temperature measurement. It has a capacitive humidity sensor and a thermistor. This probe is inserted in the coolant chamber of the vehicle and the continuous temperature readings are sent to NodeMCU. Whenever a sudden change in temperature is observed, an alert is sent to the ThingSpeak API [13] and Android app so that the operator can take the appropriate action.

Fig. 7. Autonomous harvester and sensors setup

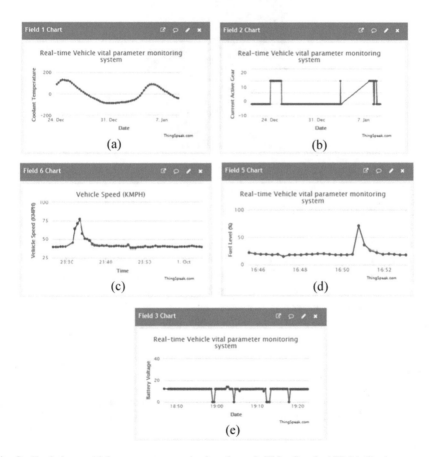

Fig. 8. Real time vehicle parameter monitoring through ThingSpeak API (a) Coolant temperature (in °C) (b) Current active gear (c) Vehicle speed (in KMPH) (d) Fuel level (in %) (e) Battery voltage (in V)

3   Experiments and Results

The discussed mechanisms are used to detect various parameters of the autonomous harvester and the sensors are interfaced with ESP8266 Wi-Fi module and the status data is sent to ThingSpeak API. Parameter information is sent every 15 s as non-commercial account of ThingSpeak is used in this project (Figs. 7, 8 and 9).

Fig. 9. Android mobile application for monitoring the autonomous harvester parameters

4   Conclusions and Future Scope

Coolant temperature, gear position, vehicle speed, fuel level and battery voltage of the autonomous harvester are captured and monitored through ThingSpeak cloud and also through an android app. Running median is taken while sending the data to the ThingSpeak to suppress noise due to disturbances and vibration.

Future scope is to analyze the collected data (ThingSpeak) using machine learning algorithms to predict reliability and maintenance time of the vehicle.

References

1. Daliya, V.K., Ramesh, T.K.: A survey on enhancing the interoperability aspect of IoT based systems. In: International Conference on Smart Technologies for Smart Nation (SmartTech-Con), Bangalore, 2017, pp. 581–586 (2017). https://doi.org/10.1109/smarttechcon.2017. 8358438

2. Dhivyasri, G., Mariappan, R., Sathya, R.: Telematic unit for advanced fuel level monitoring system. In: 2015 IEEE 9th International Conference on Intelligent Systems and Control (ISCO), pp. 1–7. IEEE (2015)
3. Nair, B.B., Keerthana, T., Barani, P.R., Kaushik, A., Sathees, A., Nair, S.A.: A GSM-based versatile unmanned ground vehicle. In: International Conference on Emerging Trends in Robotics and Communication Technologies, INTERACT-2010, Chennai, pp. 356–361 (2010)
4. Kunthong, J., Sapaklom, T., Konghirun, M., Prapanavarat, C., Ayudhya, P.N.N., Mujjal-invimut, E., Boonjeed, S.: IoT-based traction motor drive condition monitoring in electric vehicles: part 1. In: 2017 IEEE 12th International Conference on Power Electronics and Drive Systems (PEDS), pp. 1–184. IEEE (2017)
5. Penna, M., Arjun, B., Goutham, K.R., Madhaw, L.N., Sanjay, K.G.: Smart fleet monitoring system using Internet of Things (IoT). In: 2017 2nd IEEE International Conference on Recent Trends in Electronics, Information & Communication Technology (RTEICT), pp. 1232–1236. IEEE (2017)
6. Pendor, R.B., Tasgaonkar, P.P.: An IoT framework for intelligent vehicle monitoring system. In: International Conference on Communication and Signal Processing (ICCSP), Melmaruvathur, pp. 1694–1696 (2016). https://doi.org/10.1109/iccsp.2016.7754454
7. Anisi, M.H., Abdul-Salaam, G., Abdullah, A.H.: A survey of wireless sensor network approaches and their energy consumption for monitoring farm fields in precision agriculture. Precis. Agric. 16(2), 216–238 (2015)
8. Engelbrecht, J.: Survey of smartphone-based sensing in vehicles for intelligent transportation system applications. IET Intell. Transp. Syst. 9(10), 924–935 (2015)
9. Hilliard, S.R.: Vehicle speed estimation using inductive vehicle detection systems. U.S. Patent No. 6,999,886, 14 February 2006
10. Joorabchi, M.E., Mesbah, A., Kruchten, P.: Real challenges in mobile app development. In: 2013 ACM/IEEE International Symposium on Empirical Software Engineering and Measurement, IEEE (2013)
11. Singh, K.J., Kapoor, D.S.: Create your own internet of things: a survey of IoT platforms. IEEE Consum. Electron. Mag. 6(2), 57–68 (2017)
12. Gang, L.I., Zhao, Y.: Principle and Application of 1Wire bus digital thermometer DS18B20. Mod. Electron. Tech. 21, 027 (2005)
13. Maureira, M.A.G., Oldenhof, D., Teernstra, L.: ThingSpeak–an API and web service for the Internet of Things. Retrieved7/11/15World WideWeb (2011)
14. Kejík, P.: A low-cost inductive proximity sensor for industrial applications. Sens. Actuators Phys. 110(1), 93–97 (2004)
15. AUTOCAR Prof. Mag. 14(21), 37 (2018)

Dropout-Permanence Analysis of University Students Using Data Mining

Amelec Viloria[1(✉)], David Martínez Sierra[2], Martha García Samper[1],
Wilmer Orlando Cadavid Basto[3], Alberto Roncallo Pichón[4],
Hugo Hernández-Palma[5], Victoria Diago Orozco[1],
and Sadhana J. Kamatkar[6]

[1] Universidad de la Costa, St. 58 #66, Barranquilla, Atlántico, Colombia
{aviloria7,mgarcia2,vdigo2}@cuc.edu.co
[2] Universidad Simón Bolívar, Barranquilla, Colombia
dmartinez@unisimonbolivar.edu.co
[3] Corporación Universitaria Latinoamericana, Barranquilla, Colombia
wcadavid@ul.edu.co
[4] Corporación Universitaria Minuto de Dios – UNIMINUTO,
Barranquilla, Colombia
alberto.roncallo@uniminuto.edu
[5] Universidad del Atlántico, Puerto Colombia, Colombia
hugohernandezp@mail.uniatlantico.edu.co
[6] University of Mumbai, Mumbai, India
sjkamatkar@mu.ac.in

Abstract. Dropout is a rejection method present in every educational system, related to the various selection processes, academic performance, and the efficiency of the system in general, that is, the result of the combination and effect of different variables. In this sense, the dropout of university students related to their academic performance is a matter of concern since several years ago. Academic information is analyzed in order to identify factors that influence students´ dropout at the University of Mumbai, India, by using a data mining technique. The data source contains information provided to the entrance (personal and educational background) and that is generated during the study period. The data selection and cleansing are made using different criteria of representation and implementation of classification algorithms such as decision trees, Bayesian networks, and rules. the following factors are identified as influential variables in the desertion: approved courses, quantity and results of attended courses, origin and age of entry of the student. Through this process, it was possible to identify the attributes that characterize the dropout cases and their relationship with the academic performance, especially in the first year of the career.

Keywords: University dropout · Data mining · Classification algorithms · University studies

© Springer Nature Switzerland AG 2020
A. P. Pandian et al. (Eds.): ICICCS 2019, AISC 1039, pp. 374–383, 2020.
https://doi.org/10.1007/978-3-030-30465-2_42

1   Introduction

Data mining (DM) tools, based on intelligent techniques, facilitate advanced data processing and allow an automatic in-depth analysis [1]. Its strength is due to the fact that it is part of the process called Knowledge Discovery in Database (KDD), which objective is the search of data patterns that are valid, novel, potentially useful, and understandable [2]. Data mining in education (DME) is not a new concept, and its study and application became more relevant in recent years [3]. The use of DM techniques allows to deduce phenomena within the educational field. In this way, it is possible to determine the probability of dropping out or continue with the students' studies, as well as their performance during the course. The final product of the model benefits students, teachers, parents, and education managers, not just to inform about the situation of students whose performance could be associated with a particular characteristic (positive or negative), but also as advice for the decision-making processes. In other words, it is intended that these final models facilitate reflection and self-regulation during studies [4–6].

In other researches, DME techniques have been successfully used to create predictive models of student performance [7]. A case of application [8] shows the results of a study on academic performance of the Computer Science studies (technical engineering and higher education) of eight public Spanish universities. In this research, new students and the total number of students in a course were studied separately, disaggregated by sex, age, admission grade, and origin. Globally, new students in Computer Science presented differentiated profiles according to the technical or higher education career. Age became a determining factor to explain the dropout during the first year of studies, and the qualification at the entrance of the career reflects different average durations of the students (those students who begin studies with a higher grade, take less time to graduate, and the opposite). [9] analyzed the dropout factors in the first year of the career based on data gathered through a specific questionnaire considering internal and external factors of the university. They used four techniques and conclude that the level of approach in the estimation of the dropout factors depend, to a large extent, on the university, since they obtained different results among the three institutions considered in the study [10, 11].

Another study [12] states that, based on a research of retention and graduation rate in postgraduate careers, data mining algorithms working with large databases achieve better results in terms of graduation time determination than identifying student retention factors. Considering statements such as those of Planck Barahona (2014) which indicate that several studies have managed to determine that the qualifications obtained at the beginning of the university are closely related to the subsequent academic performance. The analysis of the academic performance of students in the first academic year and calendar was included as part of the process. As stated by [13], the factors or variables that affect academic performance are diverse and correspond to multiple interactions of students and their environment.

This work focuses on the analysis of variables directly related to the student's academic results and their interaction with the university, based on the data obtained from the student's career. The use of the DM classification technique is proposed to detect the characteristics and factors of greater incidence in the students at the University of Mumbai (UM) in relation to the suspension or dropout of their studies. To this purpose, the use of three classification algorithms is proposed for greater reliability in the results. The problem addressed is complex because the data can have a high dimensionality (many variables or characteristics that can influence) and are often unbalanced (many students tend to approve and just a few drop out). The objective is to detect beforehand which are the students that present characteristics related to the possibility of dropout and thus provide containment or special help and thus avoid and/or decrease the cases of student dropout. The choice of the career is due to the fact that it has the highest number of dropout cases compared to other careers (related to administration, tourism, laws, nutrition, among others) offered at the university.

2   Materials and Methods

This research analyzes the academic situation of university students based on the data of their studies at the UM. The selected sample corresponds to the students of Engineering, traditional on-ground modality, with 5 years of studies, and final thesis project. The time period selected for the study corresponds to students admitted from 2009 to 2018, for a total of 20,124 cases analyzed.

The attribute to determine the dropout cases, that is to say if the students leave their studies or not is of dichotomous type, deserts ("Des") and does not desert ("NoDes"). For this purpose, the attribute was the final condition of a nominal type, which can adopt one of the four possible values: graduated, in progress, temporary dismissal, and definitive dismissal. The definitive suspended indicates that the student has dropped out the career ("Des") and the other conditions, for processing, were grouped as non-desertion ("NoDes"). It should be clarified that the temporary dismissal refers to the temporary suspension of the student's activity. Academic performance is considered the degree of students' success related to obtaining good grades, few failed exams, few or no recourse subject, completed and approval without delay with respect to the curriculum of the career [14, 15].

The attributes selected for the study after the integration, gathering, and filtering of the data (the details are described later) are presented in Table 1, which includes the description and the standardized denomination of each attribute, as well as the type of data and possible values for the nominal ones.

As an analysis method, the KDD Process is implemented, consisting of the following phases: Data Integration and Gathering, Data Filtering, Data Mining and Evaluation and Interpretation of Results. During the development of the KDD process, as a consequence of the intermediate results, it is common to interrupt the sequence of phases of the process, to return to the previous steps, thus being an iterative and interactive process necessary to achieve a high quality of the knowledge to discover [16–18].

In data mining, there is a need to determine the level of maturity of the processes and models, and if they are adequate to solve the problem(s), so they should be reviewed, interpreted and evaluated, and finally conclude if it is possible to extract meaningful knowledge. Based on the operational data from the University database, an

Table 1. Selected and standardized attributes

Selected attributes	Standardization	Data type
Condition of Dropout	Dropout	Nominal Des, NoDes
Total of Approved Finals in 1st year	1°Apr	Numeric
Proportion of attended courses in 1st year (calendar)	Curs1	Numeric
Proportion of failed courses in 1st year (calendar)	FracC1	Numeric
Number of Approved Finals in 1st year (calendar)	Apro1	Numeric
General average in 1st year	PromA1°	Numeric
Average of Approved Courses in 1st year	PromG1°	Numeric
Age of Entry	EdadI	Numeric
Educational Institution (previous)	Est	Nominal Bachi, EscEdMed, Cen, Tec, Com, Inst, Col, EdSup, Nor, Otros.
Geographical Location (origin)	Loc	Nominal Pdas, IntProv, Otras

interpretation of the application domain related to the recorded information of academic nature (integration and data gathering phase) was carried out. Due to the large number of available attributes, more than 50 tables stored in a relational database, which have personal information of students, attendance to classes, grades, etc., a compilation of attributes was carried out to determine the most relevant ones regarding the dropout condition [19].

In the data filtering phase, an exhaustive control and debugging of the data was carried out to find a complete coherence of the tables and subsets of data to be used. Two techniques of attribute selection available in the Weka tool (described later) were applied. The first technique uses algorithms that are distinguished by their way of evaluating attributes, classified into: filters, where attributes are selected and evaluated independently of the learning algorithm, and wrappers, which use the performance of some classifier (learning algorithm) to determine the desirability of a subset. Another technique applied is the so-called 'attribute selection', used to identify, based on a particular attribute, those that most affect the object attribute (in this case, the condition of desertion). It allows to optimize subsequent tests and results to be obtained with the classification technique, mainly to avoid very complex classifications, such as large decision trees and which are difficult to interpret [20, 21].

The evaluation method applied is CfsSubsetEval and the search method is Best-First, which offers a selection of attribute subsets of higher quality according to [22]. Alternatives to the algorithms for each method have been tested, but for practical purposes, no significant variations have been found in the final results. The evaluation mode used in the algorithms of attribute selection and classification is the cross validation, which divides n times the same data set mutually exclusive and of equal size.

n − 1 sets are used to build the classifier and it is validated with the remaining set (stratified partitions), so the test partitions do not overlap. The final classifier is constructed with all subsets of data and precision is obtained from the total average. The number of subsets or cross-validation folds used in the study is 10, which makes the evaluation slow but accurate [23, 24].

As an attribute-indicator, the following parameters were considered: average of approved courses, general average, final condition of course completion (if the student regularizes the subject or not at the end of the course), grades obtained in final exams, final graduation (obtaining the title), abandonment (dropout), among others, selecting those shown in Table 1. Once the data set to be processed is defined, arranged, and adapted (minable view), the application of DM techniques and algorithms is carried out (data mining phase).

The technique used is classification, based on a model designed to predict the category of instances based on a series of input attributes, from which the classifier learns a classification scheme of the data. The first algorithm used is C4.5, which generates a decision tree from the available variables, through partitions performed recursively, according to the first in depth strategy, which implementation in WEKA is called J48 [25].

The second algorithm used is called Naïve Bayes augmented to tree (Tree Augmented Network (TAN)). As all Bayesian classifiers, they are based on the Bayes theorem, known as the cause probability formula. Naïve Bayes (NB) is a simplification that has demonstrated high accuracy and speed when applied to large volumes of data. The TAN model generally obtains better results than NB, maintaining the computational simplicity and robustness. The set of parents of the attribute to be classified C is empty, while the set of parent variables of each of the predictor attributes Xi necessarily contains the attribute to be classified, and another attribute [26, 27].

The implementation of the TAN algorithm in WEKA is called BayesNet. As the last algorithm to be evaluated, OneR was chosen, which is one of the simplest and fastest classification algorithms, since it simply identifies the attribute that best explains the output class. If there are numeric attributes, it looks for the thresholds to make rules with better success rate [28]. When using a classifier, its accuracy and reliability depends mainly on cases correctly classified from the total number of elements (evaluation phase and interpretation of results). The DM tool used for research is WEKA, which is characterized by GNU license, and was designed specifically to be used in research and educational purposes. The WEKA package contains a collection of visualization tools, algorithms for data analysis, predictive and descriptive modeling, together with a graphical user interface for easy access to its functionalities [29].

3   Analysis and Results

3.1   Algorithm Classifier C4.5 - Decision Tree

The results of the tool are presented in outline and graphical form. From the input attributes, a series of conditions represented in written form is obtained by means of a set of rules, conditions of the if-else type and graph by means of a decision tree.

Figure 1 shows the set of rules generated to classify the cases of dropout ("Des") and permanence ("NoDes"). The node or initial condition represents the number of final exams approved, corresponding to the first year of the career ("1°Apro"), where they are divided into two sub-classifications, one for approved courses less than or equal to seven and for more than seven. In both cases, the second criterion of classification is the geographical location where the students come from ("Loc"), which are discriminated against in Posadas ("Pdas") (provincial capital), inland of the province ("IntProv") or others (another province or country).

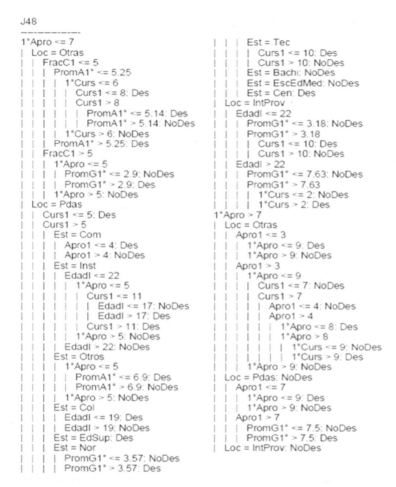

Fig. 1. Classification of Algorithm J48 (C4.5) to predict cases of dropout and permanence.

The third classification criterion varies according to the combination of previous conditions, for instances with a number of courses less than or equal to seven, for students coming from rural areas. At this level, the attributes that refer to the age of entry are presented ("AgeI"), on the other hand, for students from developed cities, the number of courses attended in the first calendar year ("Curs1") and for those from "other" geographical locations, the number of subjects that the student has not been able to regularize or promote ("FracC1"). On the other hand, for a number of subjects approved in the first year of more than seven, it was identified as relevant for students coming from "other" locations or from developed cities, to the number of courses approved in the first calendar year ("Apro1"), for "Other" locations the cut-off point is three, for "Pdas" seven, and for those from rural areas ("IntProv"), it was directly predicted that they will not drop out.

3.2   OneR Rules Classifier Algorithm

For the prediction of the desertion condition, the quantity of approved courses in the first year of the career ("1°Apro") is identified as a conditioning attribute, with a cut-off point in the value 7, indicating that below this number it is classified as cases of dropout, as happens with cases where no course was approved ("?": indicates that the field is empty) and for instances with 7 or more approved courses, as cases of permanence (this occurred in 75.8% of cases).

Table 2 shows the detail of the result evaluations obtained by the three classification algorithms (J48, TAN, and OneR). For each algorithm, the percentage of correctly classified instances (ICC) can be seen in the second column. It can be observed that the J48 algorithm obtained a higher percentage of ICC (79.8%) and, in contrast, the algorithm with the lowest percentage of ICC is for the OneR algorithm (75.8%). These values are largely due to the robustness of the algorithm [30].

Table 2. Evaluations of results of the classification algorithms

Algorithms	ICC	Dropout	VP	FP	Precision
J48 (C4.5)	79.8%	Des	77.9%	19.3%	78.6%
		NoDes	82.2%	22.6%	81.5%
BayesNet (TAN)	77.9%	Des	82.1%	23.4%	74.1%
		NoDes	74.4%	20.2%	83.5%
OneR	75.8%	Des	84.6%	31.1%	73.4%
		NoDes	71.3%	15.4%	83.4%

From the third column, the precision detailed by classes for the two possible values that the target variable "Dropout" can take ("Des", "NoDes") are evaluated. In the fourth column, for each of the values ("Des", "NoDes"), the true positives (TP) represent the proportion of instances that are classified within a class among all the elements that really belong to the class, for example, an instance is classified as a dropout case ("Des") and effectively corresponds as such. In the opposite case to VP, FP are located in the fifth column, representing the proportion of cases that were classified within a class, but belong to a different class (FP: false positives). Following the previous example, they are the cases where they are classified as non-deserters and correspond to dropout cases.

Finally, the last column indicates the precision, that is, the proportion of cases that really belong to a class among all the elements that were classified. Regarding the accuracy obtained for cases that drop out ("Des") or do not drop out ("NoDes"), it can be seen as the general way in which the 3 algorithms classify with more accuracy the cases that do not drop out, where the OneR algorithm obtained the higher percentage (83.4%) in relation to the others, and for the cases of dropout, the algorithm that obtained greater precision is J48 (78.6%).

4 Conclusions

The DM tools provide results that must be interpreted and translated into diagnoses and consequences of the real field (in this case the university). It implies that the results of the application of techniques was used to explain part of the behavior of the situation in terms of permanence and its determination based on the academic performance of the students. The possible consequences and actions tending to the taking of specific decisions is subject to considerations of other members of the academic body of the educational institution. In the KDD process, the preparation and conditioning of the data is the most extensive and, at the same time, fundamental stage because the subsequent results depend on it. In intermediate stages, it is critical to carry out analyzes and interpretations of partial results, since the process is resumed and the purification and refinement of the extracted knowledge continues.

Through the application of data mining algorithms carried out in the present study, it was possible to identify that it is during the first year of the career where the actions of containment, support, tutoring, and all those activities that improve the academic situation of the student upon admission to the university gain more importance. Some attributes were detected which, when processed and associated with specific criteria, are strongly related to dropout and permanence. The main attribute is the number of courses approved in the first year, because it marks a notable trend over the rest of the career. Other attributes that stand out are: the number of courses attended, the cases where the student does not regularize the course when attending them, the age of entry, and the origin. The combination of these criteria obtained percentages of correct answers from 76% to 80% of the correctly classified cases.

In the analysis of the results obtained from the classification algorithms C4.5 (J48), TAN (BayesNet) and OneR, it was possible to observe percentages of similar successes. However, they do not exactly identify the same attributes. Even within the same model (for example, decision tree C4.5), not all attributes have the same importance. There are some not considered significant by the method and that could be considered important (for example, the previous educational institution of the students).

References

1. Vasquez, C., Torres, M., Viloria, A.: Public policies in science and technology in Latin American countries with universities in the top 100 of web ranking. J. Eng. Appl. Sci. **12** (11), 2963–2965 (2017)

2. Aguado-López, E., Rogel-Salazar, R., Becerril-García, A., Baca-Zapata, G.: Presencia de universidades en la Red: La brecha digital entre Estados Unidos y el resto del mundo. Revista de Universidad y Sociedad del Conocimiento **6**(1), 1–17 (2009)
3. Torres-Samuel, M., Vásquez, C., Viloria, A., Lis-Gutiérrez, J.P., Borrero, T.C., Varela, N.: Web visibility profiles of Top100 Latin American Universities. In: Tan, Y., Shi, Y., Tang, Q. (eds.) Data Mining and Big Data, DMBD 2018, Lecture Notes in Computer Science, vol. 10943, pp. 1–12. Springer, Cham (2018)
4. Viloria, A., Lis-Gutiérrez, J.P., Gaitán-Angulo, M., Godoy, A.R.M., Moreno, G.C., Kamatkar, S.J.: Methodology for the design of a student pattern recognition tool to facilitate the teaching – learning process through knowledge data discovery (big data). In: Tan, Y., Shi, Y., Tang, Q. (eds.) Data Mining and Big Data, DMBD 2018, Lecture Notes in Computer Science, vol. 10943, pp. 1–12. Springer, Cham (2018)
5. Caicedo, E.J.C., Guerrero, S., López, D.: Propuesta para la construcción de un índice socioeconómico para los estudiantes que presentan las pruebas Saber Pro. Comunicaciones en Estadística **9**(1), 93–106 (2016)
6. Mazón, J.N., Trujillo, J., Serrano, M., Piattini, M.: Designing data warehouses: from business requirement analysis to multidimensional modeling. In: Proceedings of the 1st Int. Workshop on Requirements Engineering for Business Need and IT Alignment, Paris, France (2005)
7. Vásquez, C., Torres-Samuel, M., Viloria, A., Lis-Gutiérrez, J.P., Crissien Borrero, T., Varela, N., Cabrera, D.: Cluster of the Latin American Universities Top100 according to webometrics 2017. In: Tan, Y., Shi, Y., Tang, Q. (eds.) Data Mining and Big Data, DMBD 2018, Lecture Notes in Computer Science, vol. 10943, pp. 1–12. Springer, Cham (2018)
8. Haykin, S.: Neural Networks a Comprehensive Foundation, 2nd edn. Macmillan College Publishing, Inc., USA (1999). ISBN 9780023527616
9. Isasi, P., Galván, I.: Redes de Neuronas Artificiales. Un enfoque Práctico, Pearson (2004). ISBN 8420540250
10. Haykin, S.: Neural Networks and Learning Machines. Prentice Hall International, New Jersey (2009)
11. Zhang, G.P.: Time series forecasting using a hybrid ARIMA and neural network model. Neurocomputing **50**(1), 159–175 (2003)
12. Kuan, C.M.: Artificial neural networks. In: Durlauf, S.N., Blume, L.E. (eds.) The New Palgrave Dictionary of Economics. Palgrave Macmillan, UK (2008)
13. Jain, A.K., Mao, J., Mohiuddin, K.M.: Artificial neural networks: a tutorial. IEEE Comput. **29**(3), 1–32 (1996)
14. Sevim, C., Oztekin, A., Bali, O., Gumus, S., Guresen, E.: Developing an early warning system to predict currency crises. Eur. J. Oper. Res. **237**(1), 1095–1104 (2014)
15. Sekmen, F., Kurkcu, M.: An early warning system for Turkey: the forecasting of economic crisis by using the artificial neural networks. Asian Econ. Financ. Rev. **4**(1), 529–543 (2014)
16. Singhal, D., Swarup, K.S.: Electricity price forecasting using artificial neural networks. IJEPE **33**(1), 550–555 (2011)
17. Mombeini, H., Yazdani-Chamzini, A.: Modelling gold price via artificial neural network. J. Econ. Bus. Manage. **3**(7), 699–703 (2015)
18. Kulkarni, S., Haidar, I.: Forecasting model for crude oil price using artificial neural networks and commodity future prices. Int. J. Comput. Sci. Inf. Secur. **2**(1), 81–89 (2009)
19. Bontempi, G., Ben Taieb, S., Borgne, Y.A.: Machine learning strategies for time series forecasting. In: Aufaure, M.-A., Zimányi, E. (eds.) Lecture Notes in Business Information Processing, vol. 138, no. 1, pp. 70–73. Springer, Heidelberg (2013)
20. Duan, L., Xu, L., Liu, Y., Lee, J.: Cluster-based outlier detection. Ann. Oper. Res. **168**(1), 151–168 (2009)

21. Abhay, K.A., Badal, N.A.: Novel approach for intelligent distribution of data warehouses. Egypt. Inform. J. **17**(1), 147–159 (2015)
22. Savasere, A., Omiecinski, E., Navathe, S.: An efficient algorithm for data mining association rules in large databases. In: Proceedings of 21st Very Large Data Base Conference, vol. 5, no. 1, pp. 432–444 (1995)
23. Stolfo, S., Prodromidis, A. L., Tselepis, S., Lee, W., Fan, D.W.: Java agents for metalearning over distributed databases. In: Proceedings of 3rd International Conference on Knowledge Discovery and Data Mining, vol. 5, no. 2, pp. 74–81 (1997)
24. Prodromidis, A., Chan, P.K., Stolfo, S.J.: Meta learning in distributed data mining systems: issues and approaches. In: Kargupta, H., Chan, P. (eds.) Book on Advances in Distributed and Parallel Knowledge Discovery, AAAI/MIT Press (2000)
25. Parthasarathy, S., Zaki, M.J., Ogihara, M.: Parallel data mining for association rules on shared-memory systems. Knowl. Inf. Syst. Int. J. **3**(1), 1–29 (2001)
26. Grossman, R.L., Bailey, S.M., Sivakumar, H., Turinsky, A.L.: Papyrus: a system for data mining over local and wide area clusters and super-clusters. Proceedings of ACM/IEEE Conference on Supercomputing, Article **63**, 1–14 (1999)
27. Chattratichat, J., Darlington, J., Guo, Y., Hedvall, S., Kohler, M., Syed, J.: An architecture for distributed enterprise data mining. In: Proceedings of 7th International Conference on HighPerformance Computing and Networking, Netherlands, 12–14 April, pp. 573–582 (1999)
28. Wang, L., Tao, J., Ranjan, R., Marten, H., Streit, A., Chen, J., Chen, D.: G-Hadoop: MapReduce across distributed data centers for data-intensive computing. Future Gener. Comput. Syst. **29**(3), 739–750 (2013)
29. Butenhof, D.R.: Programming with POSIX Threads. Addison-Wesley Longman Publishing Company, USA (1997)
30. Bhaduri, K., Wolf, R., Giannella, C., Kargupta, H.: Distributed decision-tree induction in peer-to-peer systems. Stat. Anal. Data Min. **1**(2), 85–103 (2008)

A Review on EEG Based Epileptic Seizure Prediction Using Machine Learning Techniques

Vibha Patel[1($\boxtimes$)], Sanjay Buch[1], and Amit Ganatra[2]

[1] Department of Computer Engineering and Information Technology, CGPIT, Uka Tarsadia University, Bardoli 394350, Gujarat, India
vibha.patel@utu.ac.in
[2] Department of Computer Engineering, Chandubhai S Patel Institute of Technology, Charusat, Changa 388421, Gujarat, India

Abstract. Epilepsy is a typical neurological disorder which influences the person with epilepsy both socially and culturally, especially in India. Epileptic seizures are caused by abnormal activities in brain that can affect patient's health.

If the occurrence of seizure could be predicted well in advance, it could be prevented through medication or proper actions. Electroencephalogram (EEG) is generally used to detect epilepsy as EEG is capable of capturing the electrical activity of brain. In literature, many machine learning techniques were used to extract features from EEG recordings and predict the occurrence seizures. However, techniques with competent performance and clinical applicability are still desirable. We can use the power of modern machine learning techniques to improve the results of seizure prediction which would help to warn the patients for upcoming seizure.

Keywords: Epilepsy · Prediction · Machine learning

1 Introduction

Epilepsy is a condition wherein an individual has repetitive seizures. A seizure is a sudden surge of electrical activity in the brain. The electrical activity is caused by complex chemical changes that occur in nerve cells. A seizure usually affects how a person appears or acts for a short time. As indicated by the World Health Organization (WHO), of the 50 million individuals with epilepsy around the world, 80% live in developing nations. Epilepsy was assessed to represent 0.5% of the worldwide burden of illness, representing 7,307,975 disability adjusted life years (DALYs) in 2005. It is estimated that there are more than 10 million persons with epilepsy (PWE) in India. Its prevalence is about 1% in our population. The prevalence is higher in the rural (1.9%) compared to urban population (0.6%) [1].

It is possible to deal with epileptic seizure in two-thirds of the patients using medication, while another 8% can be cured using resected surgery. Seizures of about 25% of patients with epilepsy cannot be managed sufficiently by any available therapy. Therefore, the early anticipation of seizures could be very valuable for those patients, caregivers, or

© Springer Nature Switzerland AG 2020
A. P. Pandian et al. (Eds.): ICICCS 2019, AISC 1039, pp. 384–391, 2020.
https://doi.org/10.1007/978-3-030-30465-2_43

family members to save patients and others from possible hazards. An effective seizure prediction approach would improve the quality of patients' daily lives [2].

Electroencephalogram (EEG) is the most often used diagnostic tool for brain disorders, specifically for epilepsy. Through electrodes, it measures the voltage fluctuations which results from ionic current within the neurons of brain. There are two types of EEGs: intracranial EEG (iEEG) and scalp EEG (sEEG). Intracranial EEG is the invasive technique in which electrodes are put legitimately on the uncovered surface of the brain to record the electrical signal. Whereas scalp EEG is the noninvasive method in which electrodes are placed on the scalp area according to certain placement specifications, such as the international 10–20 system to capture the electrical signals [2].

Stage transition takes place during the cycle of seizures. The main stages of seizure are preictal, ictal, postictal and interictal. The period of time before seizure onset is known as preictal state. The interval during which seizure occurs is called ictal state. The period immediately after seizure occurrence is called postictal state and the period between two consecutive seizures is known as interictal state. Physiological evidences suggest that there is a state transition between the interictal and the ictal state. There are numbers of clinical indications that support this theory. These indications include increases in cerebral blood flow, cerebral oxygenation, cortical excitability, highly significant blood-oxygen-level-dependent signal on fMRI studies, and variations in heart rate. Based on this theory, researchers have done much work during last decades to predict epileptic seizures based on the features extracted from sEEG and iEEG data signals. As scalp EEG is noninvasive method of data collection, it is widely used and more convenient to apply clinically.

The period of time exactly before onset of seizure defines preictal state and thus the identification of preictal state would predict the occurrence of seizure. So, the aim of seizure prediction techniques is to effectively detect the preictal state. There are many stages involved in the seizure prediction process which includes EEG signal gathering, preprocessing of the EEG signal, feature extraction, feature selection, and classification of seizure states. The principle goal of seizure prediction models should be to recognize preictal state adequate time before the onset of seizure in order to give warning to the person with epilepsy such that precautions to prevent the seizure could be taken. Enough time for the predictive pre-ictal state and maximum sensitivity are important, and they remain as a performance issue in the prediction of epileptic seizures [3].

Structure of the paper is as follows. Section 1.1 represents introduction to why machine learning is to be used for medical applications. Section 2 describes significant work contributed by different authors in the area of epileptic seizure prediction which is based on EEG. Section 2.1 describes the basic flow derived from various literature studies following by Sect. 2.2, which describes the challenges posed by the seizure prediction models. Finally, Sect. 3 gives the conclusion of the analysis.

1.1 Machine Learning for Medical Applications

Now is the era of information processing. With evolving digitization in technology, data is flooded everywhere. Medical domain is also highly digitized which generate ample amount of data in hospitals and clinics. Specifically, medical data is growing in length (i.e. more number of samples is collected with time) and in width (i.e.

dimensionality of data is increasing). These data can generate accurate and innovative insights if processed and analyzed in a proper manner. This is one of the main reasons why machine learning is gaining popularity in the field of biomedical applications. Machine learning techniques can make use of latest processing capabilities and predict most accurate outcomes which are to be incorporated in clinical application. As a result, more intelligent healthcare services like personal medication recommendations, regular monitoring of various readings and decease tracking would be provided by the physicians. These became possible because machine learning techniques can relate complex relationships within large number of time evolving variables [5].

2 Related Work

Much work has been done in the area of seizure prediction based on EEG signals. Alotaiby et al. [2] developed a subject specific epileptic seizure prediction method. Scalp EEG signals have been used for the analysis. The features extracted using CSP are used for training a linear discriminant analysis classifier, which is then employed in the testing phase. Validation of the results generated by model is done by a leave-one-out cross-validation technique. The dataset considered for experiment is CHB-MIT of 24 patients. The performance evaluation parameters taken in to consideration are sensitivity, false prediction rate and prediction time. By considering the prediction horizon of 120 min, the obtained average sensitivity of 0.89, average false prediction rate of 0.39, and an average prediction time of 68.71 min has been achieved.

After the acquisition of EEG signals, preprocessing of the signal plays major role in artefact removal and feature extraction. These two key issues have direct implication on the performance of any seizure prediction system, i.e. anticipation time and true positive prediction rate. Considering this, Usman et al. [3] have proposed a model which focuses on both preprocessing and feature extraction techniques. Publically available dataset of CHB-MIT has been used by the authors to implement the model. After applying preprocessing methods, features were extracted in both time and frequency domain. Naïve Bayes classifiers, K-Nearest Neighbor and Support Vector Machine have been tested on the model, from which SVM has been selected by the authors for distinguishing testing data between pre-ictal and inter-ictal state since it performs better in terms of sensitivity. Seizure prediction was done sufficient time before the onset of the seizure by the model with better true positive rate. A higher true positive rate of 92.23%, maximum anticipation time of 33 min and average prediction time of 23.6 min was exhibited by the model. Authors suggested that, preprocessing of the EEG signal can be further improved by hybrid preprocessing methods with adaptive window size to get an increased sensitivity of seizure prediction. Also, online system for prediction of epileptic seizures can be implemented.

By incorporating the cloud computing and deep learning technology, Hosseini et al. [6] developed a BCI seizure prediction model based on both scalp EEG and intracranial EEG signal data. The method uses principle component analysis (PCA), independent component analysis (ICA) and differential search algorithm along with optimization approach on existing deep learning structures. For the storage of huge amount of data, real-time processing and automatic computing a cloud computing approach has been adapted.

It is very tough to acquire the scalp EEG signals at home for seizure detection and prediction system. To address this problem, Vandecasteele et al. [7] have developed a model which uses wearable devices based on electrocardiography (ECG) and photo-plethysmography (PPG). The model does seizure classification based on heart rate features extracted from the heart rate increase. Support Vector Machine with Gaussian kernel has been used for classification. Parameters such as Sensitivity, False Positive per Hour (FP/h) and Positive Predictive Value (PPV) have been used as the evaluation criteria. Authors suggested that in order to reduce the false alarm rate, a subject specific should be considered.

Development of seizure prediction system exhibits many challenges because of scalp/intracranial EEG signal. Interpretation of EEG signal may vary from patient to patient. So, features extracted from one subject's EEG signal might not scale well to other subject. Supervised feature extraction performs poor on new cases. Also, huge amount of data are being generated on regular bases for iEEG and sEEG which is a big data problem and requires special attention in terms of computational resources to store and process. To address these challenges, Hosseini et al. [8] have developed a cloud computing BCI system. Also, authors have proposed a seizure prediction technique which is deep learning based unsupervised feature extraction methods. To improve the performance in terms of increased accuracy and decreased computation time, dimensionality of EEG data has been reduced by principle component analysis. Stacked autoencoder has been used as a deep learning architecture for the analysis of EEG signals.

Fergus et al. [9] has developed seizure prediction model as a binary classification problem of seizure and non-seizure activity of EEG signals. K-nearest neighbour classifier has been used for the classification task which shows improved performance in terms of sensitivity, specificity and area under curve. Authors also suggested that first, there are many features which reported in literature are unexplored. Incorporation of those features might lead to an improved seizure prediction model. Second, use of evolved machine learning algorithm like convolutional neural network, recurrent neural network and support vector machine with various kernel functions for classification may improve classification accuracy. Third, consideration of variable window sizes to check the performance of seizure prediction system.

Numerous fields like speech recognition and computer vision have reported robust feature generation by using deep learning techniques as compared to manually crafted features. The architecture proposed by Pierre Thodoro et al. [10] is based on recurrent convolutional neural network (RCNN). The model is developed to simultaneously capture the spectral, temporal and spatial features while learning a general spatially-invariant representation of an epileptic seizure.

Most of the previous work of EEG based seizure classification has focused on detecting seizures using data from individual patients. Paul Fergus et al. [11] have focused on to generalise the seizure prediction model on various subjects of CHB-MIT dataset. An intense approach for preprocessing EEG signals was done by the authors and feature ranking techniques were used to extract features from pre-processed EEG signals. KNN and SVM classifiers have been used to evaluate the model in terms of sensitivity, specificity and area under curve. Authors directed that the evaluation of different parameter adjustment settings can be checked to see the performance of seizure prediction model. Also, advanced artificial neural network architectures like higher order and spiking neural network can be incorporated to improve the overall system.

Table 1 shows comparative analysis of various techniques adapted in literature for EEG based epileptic seizure prediction. Performance of any seizure prediction system is highly dependent on the type of dataset used (i.e. iEEG or sEEG), number of channels used, preprocessing methods used and classification techniques used. Also, the analysis suggests that various performance evaluation parameters have been considered with different system settings.

2.1   Basic Flow of Seizure Prediction System

From the available literature, generalized basic flow of seizure prediction system is detailed in Fig. 1. The sample EEG datasets used are CHB-MIT dataset, University of Freiburg dataset and Kaggle competition dataset.

Fig. 1. Basic flow of seizure prediction system

In data filtering, techniques such as independent component analysis (ICA), principle component analysis (PCA), singular value decomposition (SVD), factor analysis, nonnegative matrix factorization (NMF), and sparse component analysis (SCA) have been used for preprocessing and thus to improve the signal to noise ratio of the signal.

Features can be extracted from both the time and the frequency domains. Each of the domains has different feature selection criteria which influences the classification performance of the seizure prediction system. Frequency based methods such as fast Fourier transform, auto regression, moving average, combination of auto regression and moving average, short time Fourier transform and wavelet transform is widely used in the literature to extract features from EEG signals. The extracted high dimensional features impose burden to the classifiers. To reduce the numbers of features without affecting the classification accuracy, feature selection methods have been used. Majorly used method to reduce the dimensionality in terms of features is the statistics over extracted features in the EEG data signal. Work reported uses statistical features such as minimum, maximum, minimum power level, maximum power level, mean of the absolute values, standard deviation, average power and the ratio of the absolute mean. Finally, classifier models are used to distinguish different classes of EEG signals. Following are some of the majorly used classifiers: Linear Discriminant Analysis (LDA), Support Vector Machine (SVM), Neural Network (NN) and K-Nearest Neighbour (K-NN). Performance of classification are measured by different parameters like false negatives per seizure (FNPS), false positives per seizure (FPPS), detection delay, sensitivity, specificity and area under the RoC curve (AUC).

There are challenges involved on each stage in the system of seizure prediction. Descriptions of various challenges are listed in the following section.

2.2   Challenges

The methods adapted for EEG based seizure prediction poses a trade-off between sensitivity (being able to predict the seizure) and specificity (avoiding false alarms). No techniques have shown both a high sensitivity and zero false alarms per hour for clinical applicability of research. The main limitations of existing seizure prediction algorithms are unnecessary reduction of the number of features and the use of simplistic classification [12]. Preprocessing of the EEG data can be improved to get an increased sensitivity of seizure prediction. Other pre-processing methods can be tried, including those hybrid pre-processing methods and those that come with adaptive window sizes [3]. Another challenge in epileptic seizure prediction is to identify the proper metrics that characterize the performance on the imbalanced dataset [13]. Majority of the research focuses on the patient specific prediction models. A more generalized model can be proposed using domain adaptation or transfer learning approaches that would allow a model to be trained on one patient and still be applicable to another [13].

Table 1. Various techniques adapted in literature for EEG based epileptic seizure prediction

Sr. No.	Author	Dataset used	No. of subjects	EEG channels used	Preprocessing techniques	Classification methods	Evaluation Parameters
1	Alotaiby et al. [2]	CHB-MIT	24	18/23	CSP-based feature extraction stage	Linear Discriminant Analysis (LDA)	Sensitivity, Specificity, False Positive Rate (FPR)
2	Usman et al. [3]	CHB-MIT	24	23	Large Laplacian Spatial Filter, Common Averaging Filter, Common Spatial Pattern Filter (CSP)	Naïve Bayes, K-nearest neighbor, Support Vector Machine	Sensitivity, Specificity, Predction time
3	Larmuseau [4]	Kaggle Competition	4 (Dog - iEEG)	16	Standard Deviation	RNN	Average AUC
4	Hosseini et al. [6]	ECoG database - American Epilepsy Society	9 (iEEG)	15	Butterworth Bandpass filter, notch filter, deep learning	CNN, Stacked Autoencoder	Accuracy, Sensitivity, Specificity, FPR, FNR
5	Vandecasteele et al. [7] (Seizure Detection)	Hospital ECG, Wearable ECG, Wearable PPG	11	1	Filtering	SVM	Sensitivity, False Positive per hour, Positive Predictive Value
6	Hosseini et al. [8]	ECoG database - American Epilepsy Society	2	15	PCA	Stacked Autoencoder	Accuracy, Precision, Sensitivity, FPR, FNR
7	Fergus et al. [9] (Seizure Detection)	CHB-MIT	24	18/23	Root mean square, band pass filter, peak frequency, median frequency, variance, sample entropy, skewness, kurtosis	LDC, QDC, UDC, KNN, decision tree, SVM, Polynomial classifier, Logistic classifier	Sensitivity, Specificity, Receiver Operating Curve (ROC), Area Under the Curve (AUC)
8	Dadgar-Kiani et al. [13]	Kaggle Competition	1 (iEEG)	16	Fast Fourier Transform (FFT)	Logistic Regression, SVM, LSTM	True Positive Rate, FPR, AUROC, F1 score

3   Conclusion

Epilepsy is a complex public health problem that requires integrated multidisciplinary approach. A system which predicts the seizure before it occur, could highly improve the quality of life of person with epilepsy. To address the issue, epileptic seizure prediction models were studied. Though wide research has already been done in this area, no methods have achieved acceptable results which could be used in real-time applications for seizure prediction. With rapidly evolving computing power, machine learning paradigms, and big data technology, new methods can be proposed to predict the epileptic seizure with increased sensitivity and decreased false positive rate. Simulation and comparative analysis of existing techniques of epileptic seizure prediction can be performed to finalize the techniques that contribute best to improve the performance of epileptic seizure prediction.

References

1. Satishchandra, P., Sinha, S., Santhosh, N.S.: Epilepsy: Indian perspective. Ann. Indian Acad. Neurol. **17**(Suppl 1) (2014)
2. Alotaiby, T.N., Alshebeili, S.A., Alotaibi, F.M., Alrshoud, S.R.: Epileptic seizure prediction using CSP and LDA for scalp EEG signals. Comput. Intell. Neurosci. (2017)
3. Usman, S.M., Usman, M., Fong, S.: Epileptic seizures prediction using machine learning methods. Comput. Mathe. Methods Med. (2017)
4. Larmuseau, M.: Epileptic seizure prediction using deep learning. Master's dissertation, Department of Electronics and Information Systems, Universiteit Gent, Belgium (2015)
5. Esteban, C., Staeck, O., Baier, S., Yang, Y., Tresp, V.: Predicting clinical events by combining static and dynamic information using recurrent neural networks. In: International Conference on Healthcare Informatics (ICHI), pp. 93–101. IEEE (2016)
6. Hosseini, M.-P., Pompili, D., Elisevich, K., Soltanian-Zadeh, H.: Optimized deep learning for EEG big data and seizure prediction BCI via internet of things. IEEE Trans. Big Data **3**(4), 392–404 (2017)
7. Vandecasteele, K., De Cooman, T., Gu, Y., Cleeren, E., Claes, K., Van Paesschen, W., Van Huffel, S., Hunyadi, B.: Automated epileptic seizure detection based on wearable ECG and PPG in a hospital environment. Sensors **17**(10), 2338 (2017)
8. Hosseini, M.-P., Soltanian-Zadeh, H., Elisevich, K., Pompili, D.: Cloud-based deep learning of big EEG data for epileptic seizure prediction. In: Global Conference on Signal and Information Processing (GlobalSIP), pp. 1151–1155. IEEE (2016)
9. Fergus, P., Hussain, A., Hignett, D., Al-Jumeily, D., Abdel-Aziz, K., Hamdan, H.: A machine learning system for automated whole-brain seizure detection. Appl. Comput. Inf. **12**(1), 70–89 (2016)
10. Thodoroff, P., Pineau, J., Lim, A.: Learning robust features using deep learning for automatic seizure detection. In: Machine Learning for Healthcare Conference, pp. 178–190 (2016)
11. Fergus, P., Hignett, D., Hussain, A., Al-Jumeily, D., Abdel-Aziz, K.: Automatic epileptic seizure detection using scalp EEG and advanced artificial intelligence techniques. BioMed Res. Int. (2015)

12. Geethanjali, P.: Fundamentals of brain signals and its medical application using data analysis technique. In: Computational Intelligence for Big Data Analysis, Adaptation, Learning, and Optimization (2015). http://www.springer.com/gp/book/9783319165974. Accessed 26 Mar 2018

13. Dadgar-Kiani, E., Alkan, C., Shameli, A.: Applying machine learning for human seizure prediction (2017)

Simulation of a Self-Driving Car and Comparison of Various Training Methods

Bhavya Dube, Raef Kazi$^{(\boxtimes)}$, Akash Malya, and Manjusha Joshi

Mukesh Patel School of Technology Management and Engineering (Mumbai Campus), SVKM'S NMIMS University, Mumbai 400056, Maharashtra, India
1997bhavyadube@gmail.com, raefkazi27@gmail.com,
malyaakash97@gmail.com, manjusha.joshi@nmims.edu

Abstract. A simulation environment is presented in this paper for the autonomous driving of a car, along with its respective obstacles, tracks and tests. Unity 3D, which is a cross-platform game development platform and engine, powers this environment by providing basic navigation controls along with the functions for creating and recording the car's parameters as an input dataset. Using a convolutional neural network, the system predicts the future output, thus achieving complete autonomous navigation, compatible with any environment.

The paper also compares various pre-processing methods used on the input data, so as to find the most efficient model and to study which method contributes the most to effective autonomous driving.

Keywords: Simulation framework · Self-driving vehicle · Convolutional neural networks · Autonomous navigation

1 Introduction

With the recent technological advancements in the domains of computational power and software implementation, the field of vehicular navigation has seen a tremendous rise in its applications and is evolving into smarter and more intelligent vehicles. An intelligent vehicle can be defined as a system that can assist, or completely takeover from, the driver in handling the vehicle by sensing the nearby environment and conditions and provide information for vehicular control.

Intelligent vehicles have varying degrees of autonomy, depending on the need of the user. There are five basic types of vehicle navigation that are: no automation, driver assisted automation, partial automation, conditional automation, and full automation. While the first four conditions are already deployed into mainstream use, fully automated vehicles are still being researched and perfected, and are, as of the date of publication of the paper, unsuitable for widespread public usage.

As of today, there is an imperative need to improve road and vehicle safety. Each year, about 1.3 million people are killed in road crashes, with an average of 3287 deaths per day. Most of these numbers can be traced back to 4 major cause: distraction, speeding, drunk driving, and recklessness. Intelligent vehicles address this problem as they have the capacity to improve vehicle safety by assisting the driver or removing the driver completely; thereby completely eliminating the human error.

© Springer Nature Switzerland AG 2020
A. P. Pandian et al. (Eds.): ICICCS 2019, AISC 1039, pp. 392–403, 2020.
https://doi.org/10.1007/978-3-030-30465-2_44

Our paper uses three cameras (front, right, and left) to obtain data of the road and environment it is driving in. This data is run through a convolutional neural network which predicts steering angle as an output. The idea behind the project is to eliminate the need for hard-coding tons of features separately and coding each of the rules by hand, and instead to create a system that can learn to drive just by observing the human input. An important part of this end-to-end approach to a self-driving car is to predict the steering angle, and this in turn would allow us to explore the full power of convolutional neural networks. For example, only using steering angles as the training input, deep neural networks can extract features from the system automatically to help position the vehicle on the road and to make an accurate prediction.

With the help of different statistical tests, we aim to test various conditions and find the most efficient and cost-friendly method to help in autonomous driving.

2   Related Works

Pomerleau (1989) pioneered the system of using a neural network to aid autonomous vehicle navigation, by building the Autonomous Land Vehicle in a Neural Network (ALVINN) system. The model structure, though being relatively simple, was essential for predicting actions from pixel inputs affecting the motor outputs in daily scenarios. This system showed the potential of a neural network for implementing end to end autonomous navigation.

The NVIDIA Corporation recently released a paper showcasing a model based off of ALVINN. A basic Convolutional Neural Network (CNN) was employed to extract features from the given input images. By using a set of pre-processing techniques, such as positional shifts, rotations and image processing, the authors augmented the input data which was essential for the functioning of the network. Accurate steering angles were recorded using a set of cameras attached on the sides of the car. NVIDIA's model was moderately accurate in comparatively simple real-world scenarios, like driving in clear roads free of obstacles.

Recently, research on deep neural networks to take on various challenges of video classification, object detection, and parsing scenes is underway, which has led to designing more complicated CNN architectures in autonomous driving.

Udacity recently made their self-driving car simulation program free and open source. This simulation, designed on a cross-platform game engine, Unity 3D, uses NVIDIA's architecture as a backend for the car's autonomous navigation. The program has two modes available–a Training mode and an Autonomous mode. The training mode is used to collect the input data and store it as a sequence of images. The neural network is trained using these images, via a Python program. Autonomous mode is used to drive the car automatically, by acting as a server which collects the output data from the Python program.

3   System Overview

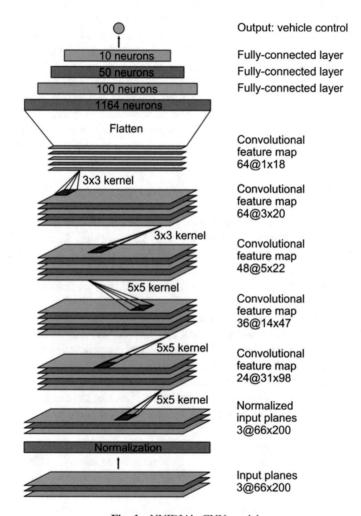

Output: vehicle control

Fully-connected layer
Fully-connected layer
Fully-connected layer

Convolutional
feature map
64@1x18

Convolutional
feature map
64@3x20

Convolutional
feature map
48@5x22

Convolutional
feature map
36@14x47

Convolutional
feature map
24@31x98

Normalized
input planes
3@66x200

Input planes
3@66x200

Fig. 1. NVIDIA's CNN model

The program acts as a simulator as well as server to the python program, which acts as a client. In training mode, the cameras mounted on the sides of the car collects images, which is recorded as input data. This input data is then pre-processed and augmented. The neural network acts on the augmented input data, giving an output in the form of steering angle and throttle values, which is collected back by the server program to drive the car autonomously (Fig. 2).

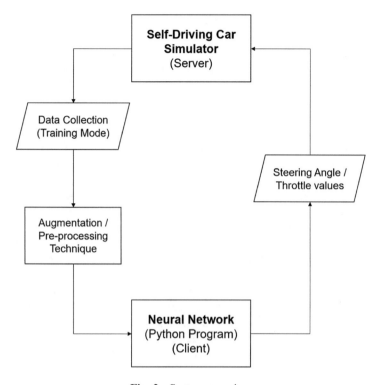

Fig. 2. System overview

The input is collected with the help of three cameras placed on the vehicle, present on the right, left and center. The pre-processing is applied to the image frames captured by the three cameras, which is then passed on to the convolutional neural network for training. This CNN uses the concept of backpropagation and weight adjustment to adjust the errors and increase the accuracy of the system. This model is then used to autonomously drive the car, with the cameras capturing the input frames in real time and the steering angle being displayed as output (Fig. 3).

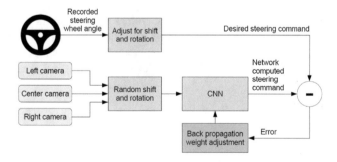

Fig. 3. Block diagram

The CNN architecture used is described by NVIDIA's model Fig. 1. Mean Square Error method is used to minimize the error between steering command output and the commands send by the human driver, which in turn trains the weights of the network.

9 layers are present in the network; these include 5 convolutional layers, 3 fully connected layers and a normalization layer. The input images are then split and passed on to the network.

The first layer of the network is specifically used to perform image normalization. This normalizer is not adjusted in the learning process and is hard-coded. The benefit of performing normalization in this network is that the network can be accelerated via GPU processing, and the normalization scheme is also allowed to be altered with the network architecture.

The next layers: the convolutional layers were built in such a wat so as to perform feature extraction and were chosen in an empirical fashion after performing a series of experiments and varying the layer configuration each time.

Strided convolutions were used in the first three convolutional layers with a 2×2 stride and a 5×5 kernel and a non-strided convolution with a 3×3 kernel size was used in the last two convolutional layers.

Three fully connected layers follow the five convolutional layers, leading to the inverse turning radius, which is an output control value. These layers that are fully connected are designed to function as a steering controller, but it is noted that upon performing end-to-end system training, it is impossible to make a clear break between those parts of the network that serve as a controller and those parts that function primarily as a feature extractor.

4   Dataset and Features

Udacity's open-source program (using Unity 3D engine) is used to collect the dataset for training. This includes pre-made and customizable tracks to train the car around.

To collect the data, three cameras are mounted onto the car; one collecting images from the center of the vehicle, one from the right of the vehicle and one from the left. Time-stamped video from the cameras mounted on the car, is captured at the same instant with the steering angle applied by the human driver. The vehicle has a maximum speed of 30 miles per hour and the steering angle has a maximum turn of 25° to the right and to the left side.

The training dataset consists of 6,108 individual frames recorded from the three cameras attached to the car during its training run. These frames have corresponding labels of torque, speed and steering angle. The data is further split randomly into a training and a validation set split in an 80/20 fashion. The original resolution of each frame is 320×160 pixels.

5   Data Augmentation Methods

Augmentation is the process of enhancing the input data so that the system can extract the maximum amount of information from it. By implementing various augmentation techniques, we can aid the system in extracting additional information from the input data. It is a method of utilizing the training data in such a manner that it increases the instances of training data.

Data augmentation is applied here so that the computational power is reduced for the system and the machine is able to make quicker decisions. Additionally, these augmentation methods help provide a much more generalized database by taking multiple parameters and scenarios into account, without having to manually train the machine for each individual scenario.

Some of the techniques of data augmentation applied in our model are:

5.1   Brightness Augmentation

This technique changes the brightness of the images to closely represent conditions like day, night, rainy, cloudy etc. It generates images with distinct brightness level by first converting the images to HSV, and then scaling them up or down the V channel and again converting back to the RGB channel. Thus, by training the vehicle on a single track in only one environment, we are able to simulate multiple environments that can be encountered (day, night, rainy, cloudy etc.) simply by applying brightness augmentation (Fig. 4).

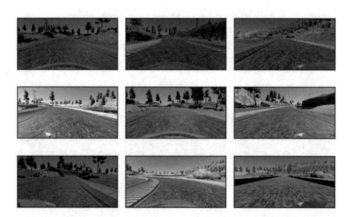

Fig. 4. Brightness augmentation

5.2   Shadow Augmentation

This technique of augmentation casts random shadow across the images. The aim is to have the model predict the correct steering angle even if the camera has been shadowed. This is achieved by shading on any side of the randomly chosen points all over the image. This helps the computer to make decisions for situations in which a random shadow may obstruct a part of the road by randomly applying shadows to different frames to generalize the data (Fig. 5).

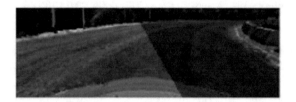

Fig. 5. Shadow augmentation

5.3 Using Left and Right Camera Images

Since three cameras are used to capture the data, we use this technique to simulate the result of car wandering off to the side, and recovering. In this method of augmentation, we add an angle of 0.26 to the left camera and subtract an angle of 0.26 from the right camera. The main idea is to move the left camera to right and the right camera to left to bring them in center.

5.4 Horizontal and Vertical Image Shifts

The images were moved horizontally to simulate the car being at different positions on the road, and according to it an offset was added to the steering angle. We added 0.0042 steering angle units per pixel shift to the right, and subtracted 0.0042 steering angle units per pixel shift to the left. To simulate the effect of the car driving up or down the slope, the images are shifted vertically by a random number (Fig. 6).

Fig. 6. Horizontal and vertical image shifts

5.5   Flipping

In addition to the augmentation above, we also flipped the images at random and inverted the sign of the predicted angle to simulate as it was driven in the opposite direction. Flipping helps solve the problem of overfitting the data. For instance, for a track that majorly veers off to the left, most of the data recorded is of a left turn as output due to the nature of the track. By flipping the data we are ensuring that the data is not skewed or favored towards one output, but rather is uniformly distributed and has equal instances of each output taking place.

5.6   Cropping

Some portion of the training frames are redundant for the learning process. The images are hence cropped from the top by 25% of image size and the bottom 30 pixels are removed. The images are then resized to 20 × 80 for faster computation. This helps the computer go through each frame of the video faster and analyze only the important and required sections of the image, rather than going through every pixel of the image which may prove to be redundant (Fig. 7).

Fig. 7. Cropping

6   Experiments and Results

In an attempt to find the optimal training parameters, along with the weights, we experimented with the different types of data augmentation and preprocessing to find which combination is the most accurate and computation efficient.

(i)   During the initial rounds of training, the variance of the data was low as the vehicle was being driven at a constant high speed and throttle, which resulted in a less generalized model; subsequently leading to poor autonomous performance.

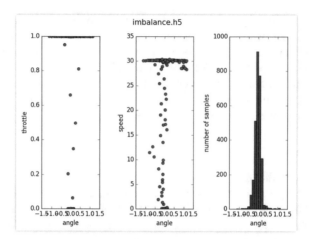

(ii) In the next iteration, we experimented with driving the vehicle at various speeds around the track. This had the effect of making the dataset more generalized and reduced errors, since the vehicle was now able to slow down while executing a sharp turn, whereas it used to crash previously due to less speed variation offered during training.

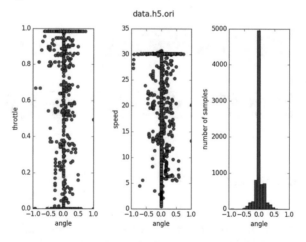

(iii) During autonomous driving, the predicted steering angles proved to be choppy as they varied abruptly from each other, resulting in the vehicle making abrupt turns instead of smoothly transitioning from side of the track to the other.

To counter this, we smoothed the steering data using method of moving averages. This helped in making the steering less choppy and resulted in better driving around the track and better, smoother turning.

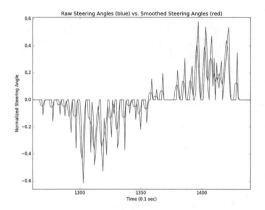

(iv) The final model consists of 4 convolution layers, 1 max pooling and 4 fully connected dense layers with relu activations. The optimizer used is 'adam' optimizer and MSE loss function is used.

Each different model was trained for 10 different epochs and epoch number 3 gave the best results with a loss of **0.022**.

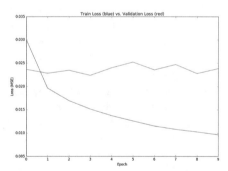

After training of the vehicle and applying pre-processing techniques is completed, the car is able to predict the output, i.e. the steering angle that needs to be applied to navigate around the track. The vehicle is able to navigate autonomously around the track multiple times without crashing. This trained model is applied to a previously unseen track and prediction about the steering angle to be applied is made successfully (Fig. 8).

Fig. 8. The blue line indicates the direction predicted by the steering angle depending on the track ahead

7   Conclusion and Future Work

Autonomous navigation has been achieved successfully by our trained model. The system can successfully predict the correct steering angle required for a smooth turn on the road, while different pre-processing and training methods give varying degrees of efficiency in the self-driving process. A basic hardware implementation of this model has also been done, where the wheels of the car turn either left, right, or remain stationary, depending on the steering angle prediction given by the neural network.

Thus, the model is able to make a decision based on the training set and does not need to be explicitly coded to extract features and predict output steering angle. It is able to detect the environment around it simply by providing data from the user, and this in turn helps improve the computational efficiency.

Due to this end-to-end approach, minimum training dataset is needed from human input for the network to learn how to drive in a traffic environment on local roads or highways that have or do not have lane markings present on them. It is also able to operate in places where there is unclear visual guidance such as unpaved roads and parking lots.

Currently, the trained model is able to perform autonomous navigation perfectly on the track it has been trained on. However, some errors occur when the track is a hitherto unseen one. More work is being done to improve the accuracy for such cases.

This concept of autonomous navigation can be further improved by implementing fuzzy logic while interfacing the hardware with the software. Particular ranges of the steering angle can be defined, each range sending a particular voltage value to the motors, making them turn accordingly. This can allow for accurate hardware steering. Various elements such as vehicle traffic and traffic signs can also be implemented in the simulation, which will make it closer to a real-life scenario. Further iterations can slowly lead to the creation of truly autonomous machines which can drive on their own without the need of human intervention.

This model is far from perfect and there is substantial research that can and needs to be done before this model can be widely deployed for public consumption.

References

1. Du, S., Guo, H., Simpson, A.: Self-driving car steering angle prediction based on image recognition. IEEE Trans. Consum. Electron. **57**(2) (2016)
2. Yang, Z., Zhang, Y., Yu, J., Cai, J., Luo, J.: End-to-end multi-modal multi-task vehicle control for self-driving cars with visual perceptions. In: 24th International Conference on Pattern Recognition (ICPR), August 2018
3. Navarro, A., Asher, Z.D.: Development of an autonomous vehicle control strategy using a single camera and deep neural networks. In: SAE International by Anthony Navarro, 11 April 2018
4. Grazioli, F., Kusmenko, E., Roth, A., Rumpe, B., von Wenckstern, M.: Simulation framework for executing component and connector models of self-driving vehicles (2017)
5. Marcelo Paulon, J.V.: AVCP: autonomous vehicle coordination protocol. Pontifícia Universidade Católica do Rio de Janeiro (PUC-Rio), December 2017
6. Bojarski, M., Testa, D.D., Dworakowski, D., Firner, B., et al.: End to end learning for self-driving cars, arXiv preprint arXiv:1604.07316 (2016)
7. Net-Scale Technologies, Inc.: Autonomous off-road vehicle control using end-to-end learning, Final Technical Report, July 2004
8. Bojarski, M., Yeres, P., Choromanska, A., Choromanski, K., Firner, B., Jackel, L., Muller, U.: Explaining how a deep neural network trained with end-to-end learning steers a car. arXiv preprint arXiv:1704.07911 (2017)
9. El Sallab, A., Abdou, M., Perot, E., Yogamani, S.: Deep reinforcement learning framework for autonomous driving. In: Autonomous Vehicles and Machines, Electronic Imaging (2017)
10. Farabet, C., Couprie, C., Najman, L., LeCun, Y.: Learning hierarchical features for scene labeling. IEEE Trans. Pattern Anal. Mach. Intell. **35**(8), 1915–1929 (2013)
11. Udacity: Self-driving car simulator. GitHub (2016)
12. Chen, C., Seff, A., Kornhauser, A., Xiao, J.: DeepDriving: learning affordance for direct perception in autonomous driving. In: Proceedings of the IEEE International Conference on Computer Vision, p. 2730 (2015)
13. Huval, B., Wang, T., Tandon, S., Kiske, J., Song, W., Pazhayampallil, J., Andriluka, M., Rajpurkar, P., Migimatsu, T., Cheng-Yue, R., et al.: An empirical evaluation of deep learning on highway driving. arXiv preprint arXiv:1504.01716 (2015)

Tri Layer Model for Color Image Ciphering Through DNA Assisted 2D Chaos

Nithya Chidambaram[✉], C. V. Sanjay Siddharth, P. Sherine, and Amirtharajan Rengarajan

Department of ECE, School of Electrical and Electronics Engineering, SASTRA Deemed to Be University, Thanjavur 613 401, India
cnithya@ece.sastra.edu

Abstract. In this modern digital world, there are many advancements that are made in the fields of networking. These days many multimedia messages, in the form of images are shared online for communicating with each other. The images thus send are vulnerable to attacks which are both internal (noise attack etc.) and external (hackers). An algorithm for securing these multimedia messages have been proposed here. The key employed in the proposed scheme is made from three different 2D chaotic maps. A colour image is split into RGB planes, it undergoes a 3-level confusion-double diffusion using Deoxyribo Nucleic Acid (DNA). A novel methodology has been approached to substantiate it over the currently existing encryption standards. The ability to withstand brute force attack, noise attack, chosen plain text attack of the method are to be tested.

Keywords: 2D chaotic map · Confusion · Diffusion · DNA encoding · Brute force attack · Chosen plain text attack

1 Introduction

Over the years, the development of technology and advancements in the fields of internet and cellular networks has led to the accumulation of volumes of data [1]. Going by the statistics of the National Security Agency, USA, around 70% of the total information exchanged between users are images [2]. Thus, securing the data from external cyber threats have become a necessity in the current era. All this information is managed efficiently using cloud computing, which provides on-demand IT services on computing and storage [3]. This information is not secure as they are viable to attacks made during the transformation of the information. Thus, prevention of unauthorized users from acquiring the image must be secured to provide confidentiality and integrity [4].

The existing encryption techniques such as Advanced Encryption Standard(AES), P-Fibonacci transform, wave transformation, random grids, Latin squares etc., are not strong enough to tackle the well-advanced cyber threats posed by the hackers to the information that is transmitted over the channel [5]. Pseudo randomness, high ergodicity and high sensitivity to initial values, non-convergence are just a few advantages of applying chaotic system on image encryption [6]. 1-D chaotic maps exhibit randomization behaviour even over higher order of iterations [7]. Non-linear combination of two different 1-D maps have been used as a method for the cryptosystem; this method

© Springer Nature Switzerland AG 2020
A. P. Pandian et al. (Eds.): ICICCS 2019, AISC 1039, pp. 404–416, 2020.
https://doi.org/10.1007/978-3-030-30465-2_45

posted advantages over the range of chaotic values, behaviours and distribution but compromises on the computational efficiency. Module based image scrambling is proposed to achieve a better shuffling effect and the inclusion of a 3rd map to enforce security [8]. Another scheme exists where the image is divided into 12 blocks and a combination map is used for encryption. The pitfall of this method is that each part suffers a loss during occlusion attack [9].

Over the past decade, DNA encryption and computing have become a viable field of study owing to its negligible power requirement, huge storage capacity and massive parallelism [10]. Many types of research in the field of modern biological cryptography have investigated, proposed and implemented encryption as well as decryption technologies based on DNA. One of the very prominent steganography methods involves protecting the message inside the DNA microdots [11]. An alternate method includes the use of the 3-D chaotic map hence performing a synchronous permutation and diffusion [12]. A technique involving DNA encoding along with the one-way coupled – map lattices have been proposed although it suffers from data loss while decrypting [13]. Another plausible approach uses key stream generation, DNA encoding and decoding along with pixel level diffusion – where only multiple rounds of the same algorithm produce efficient results [14]. One of the cryptosystems uses SHA-2 hash for acquiring the initial conditions used in the chaotic map [15]. The proposed algorithm has an efficient chaotic sequence generation and makes the system resistant to brute force attack. A multilayer chaining and modulo-2 addition make this cryptosystem resistant to hackers.

The paper is split like the following; Sect. 2 briefly explains the various chaotic, a clear vision of the encryption and decryption process. Section 3 contains the proposed methodology. Section 4 provides a study on security analysis via simulated attacks against the proposed method. Section 5 concludes the paper.

2 Preliminary

In this section, four main tools that are the origin of our proposed algorithm have are been elaborated. Three 2D chaotic maps are used for random sequence generation and DNA encoding adopted for data redundancy.

2.1 2D Tent Map

It is called so because of its bifurcation diagram being shaped like a tent. Equations (1) and (2) forms the Tent,

$$T_{1^n} = \begin{cases} b \times T_{1^n}; T < 0.5 \\ b \times (1 - T_{1n}); T_{1n} > 0.5 \end{cases} \tag{1}$$

$$T_{2^n} = \begin{cases} c \times T_{2n}; T_{2n} < 0.5 \\ c \times (1 - T_{2n}); T_{2n} > 0.5 \end{cases} \tag{2}$$

Where, b ∈ (0, 2) and c ∈ (0, 2). Use the sequence, to generate P-box, S-box for blue planes and DNA S-box for red plane.

2.2   2D Henon Map

The Henon map is an iterated discrete-time dynamical system that exhibits exhibit transitions to chaos through period doubling route. Equations (3) and (4) forms Henon map.

$$H_{n+1} = 1 - d \times H_n^2 + I_n \tag{3}$$

$$I_{n+1} = e \times H_n \tag{4}$$

Where, d ∈ (1, 2) and e ∈ (0, 1). Use the sequence, to generate P-box, S-box for red planes and DNA S-box for green planes.

2.3   2D Modified Piece Wise Linear Chaotic Map

PWCLM map has been employed to introduce complexity to the system where each chaotic system has its own characteristics. Sensitivity to initial condition and control parameter are both considered as 10^{-14}. The following Eq. (5) explains its complex chaotic behaviours,

$$P_{i+1} = \begin{cases} P_i/P_0; 0 < P_i < P_0 \\ (P_i - P_0)/(0.5 - P_0); P_0 < P_i < 0.5 \\ (1 - P_i); P_i \geq 0.5 \end{cases} \tag{5}$$

It was observed to show good ergodicity and determinacy. Some modifications were made to the above equations to obtain a complex random sequence. The following Eqs. (6) and (7) explains about its modified nature [16],

$$P_{i+1} = (P_i - \lfloor Q_i/Q_0 \rfloor * P_0)/P_0 \tag{6}$$

$$Q_{i+1} = (Q_i - \lfloor P_i/P_0 \rfloor * Q_0)/Q_0 \tag{7}$$

Where, $\lfloor x \rfloor$ is the maximum integral function, P ∈ (0, 0.5) and Q ∈ (0, 0.5)
Use the sequence, to generate P-box, S-box for blue planes and DNA S-box for red planes.

2.4   DNA Encoding

A DNA (Deoxyribonucleic Acid) is the hereditary material in living organisms. There are primarily 4 chemical bases that preserve information as a code; Adenine (A), Cytosine(C), Guanine (G), and Thymine (T). DNA pairs up with each other, that is (A with T) and (C with G). This type of pairing is called complementary pairs. Eight out of 24 available coding schemes meet the complementary rule of Watson-Crick. DNA encoding sequence is imposed on 2 bits at a time. The DNA encoding and exoring rule can be referred to in [10].

3   Proposed Method

The proposed methodology is given in Fig. 1 and explained in this section. Original image input, corresponding cipher image output and decryption output are shown in Fig. 2.

The following functions are illustrated for the red plane. It will be applied for each of the planes correspondingly.

3.1   Generation of P Box

Quantize the sequence H and sort the coefficients to create H_S. P box is generated using Eq. (8).

$$\left[H_S, H_{SP}\right] = \text{sort}\left(\text{mod}\left(H \times 10^{14}, M \times N\right)\right) \tag{8}$$

Where, H_S is the non-decreasing order sequence of H and H_{SP} is the respective index of H_S sequence.

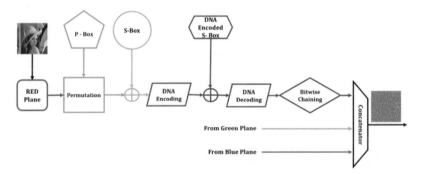

Fig. 1. Encryption scheme

3.2   Generation of S Box

Quantize a sequence I and reshape it to form S-box. S box is created using Eqs. (9) and (10).

$$I_{sub} = \text{mod}\left(I * 10^{14}, 256\right) \tag{9}$$

$$I_{sub} = \text{reshape}(I_{sub}, M, N) \tag{10}$$

Proposed Method

Original Image   Encrypted Image   Decrypted Image

Fig. 2. Sample input and output

3.3   Generation of DNA Encoded S-Box

Find the S box from the second equation form the chaotic maps using the Eq. (10). Perform DNA encoding for the obtained S box using rule 6 and Eq. (11).

$$T_1 sub = dnaencode\big((T_1 * 10^{14}, 256)\big) \tag{11}$$

3.4   Encryption Scheme

Step 1: Read the image and split into planes of Red, Green and Blue.

Step 2: Permutate each plane with its Corresponding P-Box. For example using Eq. (12),

$$Red_1[i] = Red[H_{SP}[i]] \text{ for } i = [1, M \times N] \tag{12}$$

Step 3: Diffuse each plane with its corresponding S-Box. For example using Eq. (13),

$$Red_2 = Red_1 \oplus I_{sub} \tag{13}$$

Step 4: Encode the diffused planes using DNA Rule.

Step 5: Perform Modulo 4 Addition among the resultant plane and its corresponding DNA encoded S-Box. For example using Eq. (14),

$$Red_3 = (Red_2 + T_{1 sub}) \bmod 4 \tag{14}$$

Step 6: Decode the DNA exored plane using DNA rule.

Step 7: Apply pixel-pixel Chaining on each plane to get Red_C, $Green_C$ and $Blue_C$ and merge them into an image.

3.5   Decryption Scheme

Step1: Read the encrypted image, split into planes of Red, Green and Blue, and apply pixel-pixel unchaining to arrive at corresponding cipher planes.

Step2: Encode each plane using DNA rule.

Step3: Perform Modulo 4 Addition among each plane and its corresponding DNA Encode S-box. For example using the Eq. (15),

$$Red_1 = (Red + T_{1\,sub}) \bmod 4 \tag{15}$$

Step 4: Decode the diffused planes using DNA Rule.

Step 5: Diffuse the resultant planes with its corresponding S-Box. For example using the Eq. (16),

$$Red_2 = Red_1 \oplus I_{sub} \tag{16}$$

Step 6: Permutate each resultant plane with its corresponding P-Box. For example using Eq. (17),

$$Red_3\left[H_{sp}[i]\right] = Red_2[i] \text{ for } i = [1, M \times N] \tag{17}$$

Step 7: Merge all the resultant planes to retrieve the original image.

4   Numerical Simulations and Discussions

4.1   Key Sensitivity Analysis

(a) *Key space:*

A good quality encryption scheme is dependent on secret keys rather than the algorithm. The key space must be vast, to make any brute force attack infeasible. In the proposed method, 14-bit precision is used and also consists of 12 parameters. The proposed method has a key space of 3×10^{56}. An efficient algorithm is one where key space is greater than in 2^{100}. Because recreating all possible keys available are out of reach with the present technology.

(b) *Key sensitivity:*

Brute force attack is a trial and error method used to decode encountered data through exhaustive effort rather than employing intellectual strategies. Even when one parameter of the key is slightly changed during the decryption process we cannot retrieve back the original image, therefore it withstands against brute force attack. This can be visualized in Fig. 3.

Original Image · Encrypted Image · Decrypted with original key MSE = 0, PSNR = INF · Decrypted with modified key MSE =8998.1869,PSNR = 8.5893

Fig. 3. Key sensitivity analysis

4.2 Frequency Analysis

(a) *Histogram analysis:*

The frequency distribution of the image is called histogram. It must conceal the notice of the original image and must not disclose the relationship between the original image and resultant encrypted image. The pixel values of all 3-colour channels are uniformly distributed over the range. Hence, it prevents the attackers from accessing info about the encrypted image. The histogram of the Lena is shown in Fig. 4.

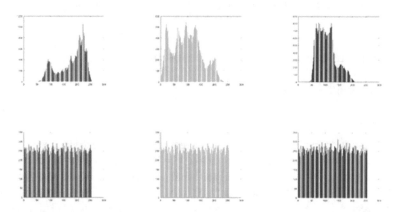

Fig. 4. Histogram of Lena: original and cipher

(b) *Information entropy:*

Information entropy is one of the standards to ensure the randomness of the distribution of the pixels. It is a measure of an average number of bits required to communicate or store one symbol of information. The values of entropy can be analytically analysed by the following equation:

$$IE(x) = - \sum_{i=1}^{N} K(x_i) \times \log_2 K(x_i) \qquad (18)$$

Where, X_i is the random variable with N outcomes,

$K(x_i)$ is the probability mass function of the outcome x_i

If N = 256, from Eq. (18), the ideally achievable value is IE(x) = 8. Practically, there exist a minute degree of predictability resulting in values less than but closer to eight. The values can be verified from Table 1.

(c) *Correlation of neighbouring pixels:*

Any image has higher pixel relativity in a horizontal, vertical and diagonal direction. To measure the correlation of adjacent pixels in original and cipher image, randomly selected 5000 pairs of pixels are subjected to pairwise correlation analysis.

The pixel correlation of an image is calculated as follows using Eqs. (19), (20) and (21):

$$r_{g,h} = \frac{\text{cov}(g,h)}{\sqrt{D(g)D(h)}} \tag{19}$$

Table 1. Correlation and information entropy

Test image		Horizontal	Vertical	Diagonal	Information entropy
Lena	Plane Red	0.9338	0.9641	0.9355	7.2796
	Plane Green	0.9049	0.9467	0.9079	7.6315
	Plane Blue	0.8597	0.9061	0.8606	6.9891
Lena cipher	Plane Red	0.0042	−0.0044	−0.0008	7.9972
	Plane Green	0.0058	0.0012	−0.0017	7.9971
	Plane Blue	0.0041	−0.0027	−0.0014	7.9971
Peppers	Plane Red	0.9424	0.9464	0.9178	7.3319
	Plane Green	0.9559	0.9617	0.9363	7.5242
	Plane Blue	0.9319	0.9406	0.9023	7.0793
Peppers cipher	Plane Red	−0.0009	−0.0017	0.002	7.9971
	Plane Green	0.0003	0.0076	−0.0014	7.997
	Plane Blue	−0.0018	0.0007	0.0042	7.9972
Couple	Plane Red	0.9493	0.9562	0.9163	6.2499
	Plane Green	0.9308	0.9534	0.8968	5.9642
	Plane Blue	0.9178	0.9442	0.887	5.9309
Couple cipher	Plane Red	0.0019	0.0015	−0.003	7.9975
	Plane Green	−0.0012	0.0033	−0.0067	7.9974
	Plane Blue	−0.0011	−0.0007	−0.0002	7.9972
Jellybean	Plane Red	0.9745	0.9763	0.9545	5.2626
	Plane Green	0.9757	0.9801	0.9598	5.6947
	Plane Blue	0.989	0.988	0.9815	6.5464
Jellybean cipher	Plane Red	0.0024	−0.003	−0.0039	7.9971
	Plane Green	−0.0049	−0.0054	0.0003	7.9974
	Plane Blue	−0.0006	0.0053	−0.0007	7.997

$$D(g) = \frac{1}{N} \sum_{i=1}^{N} (g_i - E(g))^2 \tag{20}$$

$$E(g) = \frac{\sum_{i=1}^{N} g_i}{N} \tag{21}$$

Where, g and h are adjacent pixels.

N, the total number of random pairs under study.

4.3   Image Quality Analysis

(a) *MSE (mean square error):*

Mean square error between the cipher image and the original image is an essential factor to evaluate the key sensitivity of an image encryption algorithm. Ideally, MSE = 0 for a lossless encryption algorithm. MSE values are low only when the main keys are correct. Mathematically MSE values are calculated as follows using Eq. (22):

$$MSE = \frac{1}{M \times N} \sum_{i=1}^{M} \sum_{i=1}^{N} (V(i,j) - W(i,j)) \tag{22}$$

Where, $V(i,j)$ and $W(i,j)$ are the mean pixel values at points (i, j)

M * N denotes the size of the colour image.

(b) *PSNR (peak signal to noise ratio):*

This analysis is used to test the image similarity after decryption. If the tested images are identical PSNR = infinity, which proves a lossless encryption algorithm. PSNR can be calculated numerically as follows using Eq. (23):

$$PSNR = 20 \times \log(\frac{L}{\sqrt{MSE}}) \tag{23}$$

Where, L is the maximum possible pixel value of the image. If and only the PSNR value is greater than 30 dB the image can't be perceived by the human eye.

4.4   Differential Cryptanalysis

(a) *NPCR (Number Of Pixels Change Rate)*

NPCR calculates the change rate of cipher image when only one pixel of the plain image is modified. Let OI1 undergoes a one bit change to form the image CI1, hence OI2 and CI2. The bipolar array D which is same in size of that CI1 and CI2. Mathematically, NPCR values are calculates using Eqs. (24) and (25).

$$NPCR = \frac{\sum_{i=1}^{M} \sum_{j=1}^{N} D(i,j)}{M \times N} \times 100\% \tag{24}$$

Where,

$$D(i,j) = \begin{cases} 0; & W_1 = W_2 \\ 1; & W_1 \neq W_2 \end{cases} \tag{25}$$

If all the pixels are the same there is no change. If D = 1, the corresponding values in CI1 and CI2 are different.

(b) *UACI (Unified Average Changes Intensity Test)*

UACI are used to evaluate the strength of image encryption algorithm in terms of resistance to differential attacks. The number of average changed intensity between cipher images is analysed through UACI parameters. The values of NPCR and UACI are tabulated in Table 2. Mathematically, UACI values are calculated using Eq. (26).

$$UACI = \frac{1}{M \times N} \left[\sum_{i=1}^{M} \sum_{j=1}^{N} \left| \frac{W_1(i,j) - W_2(i,j)}{L} \right| \right] \times 100\% \tag{26}$$

Where, $W_1(i, j)$ and $W_2(i, j)$ be the $(i, j)^{th}$ pixel

Table 2. Differential analysis of encrypted image

Test image	NPCR	UACI
Lena	99.6185	32.8877
	99.6109	33.5214
	99.7482	31.7384
Peppers	99.7650	21.9848
	99.7879	36.8425
	99.6475	32.0633

4.5   Resistance to Typical Attacks

There are four types of attacks, which are prevalent in the existing cyber world; the cipher text-only attack, chosen-cipher text attack, known-plain text attack and chosen-plain text attack. Among these four, the chosen plain text attack is the most powerful. Usage the same security keys to encrypt the images in existing algorithms poses as a potential loophole for attackers to predict the encryption algorithm. To solve this scenario, the proposed encryption algorithm imparts a pixel insertion process that is random. This generates an entirely different cipher image each time the same original message is encrypted even though security key remains unaltered. Let C1 be the encrypted image, C2 be the twice encrypted image, the pixel to pixel difference |C1-C2| is high. Therefore, the proposed method can withstand the chosen-plain text attack. The chosen plain attack can be visualised in Table 3.

Table 3. Chosen plain attack

Original image	Encrypted once	Encrypted twice	Difference

4.6   Robustness Attack

Noise attack

When the images are propagated through the transmission channel, they are vulnerable to various contaminations of noise, amplification and detection. For real time applications, the image must withstand all these contaminations. In our proposed method we have imposed three different noises such as salt and pepper noise, speckle noise and Gaussian noise. Figure 5 shows the decrypted image after the noise attack.

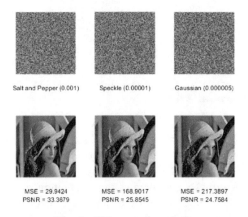

| Salt and Pepper (0.001) | Speckle (0.00001) | Gaussian (0.000005) |

| MSE = 29.9424 | MSE = 168.9017 | MSE = 217.3897 |
| PSNR = 33.3679 | PSNR = 25.8545 | PSNR = 24.7584 |

Fig. 5. Noise attack analysis

It is observed that the retrieved image is visually of higher quality.

Quality of proposed cryptosystem is compared with the existing algorithms where the information entropy is 7.9972 close to the idea value and better than the other algorithm. PSNR and MSE values after noise attack are acceptable. The neighbouring pixel correlation is also minimum in some planes, which implies that the cryptosystem has better statistical analysis. The comparison is made in Table 4 with the existing algorithm, which shows better statistical analysis and differential cryptanalysis.

Table 4. Performance analysis of the proposed method

Lena	Horizontal	Vertical	Diagonal	Entropy	NPCR	UACI
Proposed method	0.0042	−0.0044	−0.0008	7.9972	99.6185	32.8877
	0.0058	0.0012	−0.0017	7.9971	99.6109	33.5214
	0.0041	−0.0027	−0.0014	7.9971	99.7482	31.7384
Wu et al. [10]	−0.0112	−0.0026	0.0052	7.9895	99.6052	33.4280
	0.0050	0.0199	−0.0064	7.9894	99.6060	33.4966
	−0.0179	0.0120	−0.0161	7.9894	99.6113	33.3779
Wu et al. [12]	−0.0124	−0.0001	−0.0055	7.9903	99.6108	33.4525
	−0.0038	0.0059	−0.0086	7.9890	99.6058	33.4798
	0.0075	−0.0062	0.0006	7.9893	99.6057	33.4387
Rehman et al. [15]	−0.0073	0.0010	−0.0013	7.9966	99.6001	33.3575
	0.0011	−0.0020	0.0078	7.9972	99.5998	33.4287
	−0.0061	0.0058	−0.0003	7.9967	99.5997	33.3683

5 Conclusion

The proposed method stands at par on encryption and decryption of multimedia images based on a range of chaotic maps. The input color image was subjected to planar separation followed by pixel permutation process. The permuted image was diffused and encoded using DNA. After application of DNA Exor and DNA decoding, pixel wise chaining was performed to improve robustness. The existing literature were compared with the results arrived and the proposed method has better correlation values, entropy and NPCR values nearing the ideal value, thus ensures good encryption quality. The combination of three different chaotic maps ensures good resistance against statistical attacks, differential cryptanalysis and noise attack. This algorithm would be of much use for secure transmission of images over public, unprotected channels.

References

1. Wadi, S.M., Zainal, N.: Enhanced hybrid image security algorithms for high definition images in multiple applications. Multidimens. Syst. Signal Process. **29**, 1989–2012 (2018). https://doi.org/10.1007/s11045-017-0541-5

2. Huang, X., Sun, T., Li, Y., Liang, J.: A color image encryption algorithm based on a fractional-order hyperchaotic system. Entropy **17**, 28–38 (2015). https://doi.org/10.3390/e17010028

3. Balasubramanian, V., Mala, T.: A review on various data security issues in cloud computing environment and its solutions. ARPN J. Eng. Appl. Sci. **10**, 883–889 (2015)

4. Chidambaram, N., Thenmozhi, K.: DNA coupled chaos for unified color image encryption – a secure sharing approach (2018)

5. Bao, L., Zhou, Y.: Image encryption: generating visually meaningful encrypted images. Inf. Sci. (Ny) **324**, 197–207 (2015). https://doi.org/10.1016/j.ins.2015.06.049

6. Zhou, Y., Bao, L., Chen, C.L.P.: A new 1D chaotic system for image encryption. Sig. Process. **97**, 172–182 (2014). https://doi.org/10.1016/j.sigpro.2013.10.034

7. Huang, H., Yang, S.: Colour image encryption based on logistic mapping and double random-phase encoding. IET Image Process. **11**, 211–216 (2016). https://doi.org/10.1049/iet-ipr.2016.0552

8. Fu, C., Chen, Z.F., Zhao, W., Jiang, H.Y.: A new fast color image encryption scheme using chen chaotic system. In: Proceedings - 18th IEEE/ACIS Software Engineering, Artificial Intelligence, Networking, and Parallel/Distributed Computing SNPD 2017, pp. 121–126 (2017). https://doi.org/10.1109/SNPD.2017.8022710

9. Parvaz, R., Zarebnia, M.: A combination chaotic system and application in color image encryption. Opt. Laser Technol. **101**, 30–41 (2018). https://doi.org/10.1016/j.optlastec.2017.10.024

10. Wu, X., Kurths, J., Kan, H.: A robust and lossless DNA encryption scheme for color images. Multimed. Tools Appl. **77**, 12349–12376 (2018). https://doi.org/10.1007/s11042-017-4885-5

11. Enayatifar, R., Abdullah, A.H., Isnin, I.F., Altameem, A., Lee, M.: Image encryption using a synchronous permutation-diffusion technique. Opt. Lasers Eng. **90**, 146–154 (2017). https://doi.org/10.1016/j.optlaseng.2016.10.006

12. Wu, X., Wang, K., Wang, X., Kan, H.: Lossless chaotic color image cryptosystem based on DNA encryption and entropy. Nonlinear Dyn. **90**, 855–875 (2017). https://doi.org/10.1007/s11071-017-3698-4

13. Yap, W.S., Phan, R.C.W., Yau, W.C., Heng, S.H.: Cryptanalysis of a new image alternate encryption algorithm based on chaotic map. Nonlinear Dyn. **80**, 1483–1491 (2015). https://doi.org/10.1007/s11071-015-1956-x

14. Al-Mashhadi, H.M., Abduljaleel, I.Q.: Color image encryption using chaotic maps, triangular scrambling, with DNA sequences. In: International Conference on Current Research in Computer Science and Information Technology, ICCIT 2017, pp. 93–98 (2017). https://doi.org/10.1109/CRCSIT.2017.7965540

15. ur Rehman, A., Liao, X., Ashraf, R., Ullah, S., Wang, H.: A color image encryption technique using exclusive-OR with DNA complementary rules based on chaos theory and SHA-2. Optik (Stuttg) **159**, 348–367 (2018). https://doi.org/10.1016/j.ijleo.2018.01.064

16. Hu, Y., Zhu, C., Wang, Z.: An improved piecewise linear chaotic map based image encryption algorithm. Sci. World J. **2014**, 1–7 (2014). https://doi.org/10.1155/2014/275818

A Novel Linear Collaborative Discriminant Regression Classification and L1 Norm Based Algorithm for On-Chip Realization of Uncontrolled Face Recognition

J. N. Swaminathan[1]($\boxtimes$), A. Kavitha[2], R. Navaneethakrishnan[3],
S. Umamaheswari[3], and Ramalatha Marimuthu[3]

[1] Godavari Institute of Engineering and Technology, Rajahmundry,
Andhra Pradesh, India .
jnswaminathan@giet.ac.in
[2] Veltech Multitech Dr. Rangarajan Dr. Sakunthala Engineering College,
Chennai, Tamilnadu, India
kavivenkat99@gmail.com
[3] Kumaraguru College of Technology, Coimbatore, Tamilnadu, India
{navaneethakrishnan.ece,umamaheshwari.s.ece,
ramalatha.m.it}@kct.ac.in

Abstract. This paper proposes the novel algorithm that suits the on-chip realization of uncontrolled face recognition system. Face recognition has its own advantage and occupied all possible venues. It plays a key role in the identification or authentication of a person in wide range applications from unlocking the mobile phone to forensics. The biggest hindering factor is its speed when it is implemented along with other systems. The systems should be implemented as standalone devices. Design of Face recognition algorithms that fit for on-chip realization is far more important. With the existence of sophisticated System on Chip hardware and software platforms IP based system design is very much possible. In this work, Collaborative discriminates classification based face recognition is designed and implemented on Xilinx Zynq SoC platform. This proposed algorithm can be converted as an IP and the same can be reused for any design using VIVADO IP integrator. The proposed algorithm and architecture shows the improvements in speed and reduction of power consumption.

Keywords: VIVADO IP · L1 norm

1 Introduction

Face recognition through computing system was the great challenge and explored by number of researchers of Computer vision and artificial intelligent in the past decade. Face recognition is vital for both identification of a person and authentication. There are number of face recognitions were present in the literature. Every algorithm has its own advantage and limitations. In the initial stage Eigen Vectors were used broadly in face

© Springer Nature Switzerland AG 2020
A. P. Pandian et al. (Eds.): ICICCS 2019, AISC 1039, pp. 417–423, 2020.
https://doi.org/10.1007/978-3-030-30465-2_46

recognition algorithms [1]. Training face images be T1, T2, T3.... TM and the average of face set can be Ψ,

$$\Psi = \frac{1}{M}\sum_{n=1}^{M} T_n$$

Each differs from the average by $\Phi i = Ti - \Psi$. The set of very large vector here are subject to principal component analysis. Here the vectors Uk and scalar λk were considered as Eigen vector and Eigen values. The Covariance matrix is given by,

$$C = \frac{1}{M}\sum_{m=0}^{M} \Phi n\ \Phi n^T$$

$C = AAT$ The matrix $A = \{\Phi 1, \Phi 2 ...\Phi m\}$. The in this approach the face images were projected as face spaces. And the Eigen vectors and values were used to classify the faces. The angle of the face images should be same in order to recognize the faces using this method. PCA (principal Components Analysis) and Liner Discrimination analysis (LDA) are another possible ways in face recognition in which the Eigen values and vectors were used. Here the Eigen values are treated as mean Error is the sum of remaining Eigen values [2]. The within class metrics in this method are calculated as,

$$Sw = \frac{1}{M}\sum_{i=0}^{M} Pr(Ci) \sum i$$

$$Sb = \frac{1}{M}\sum_{i=1}^{M} Pr(Ci) \sum i\ (mi - m)(mi - m)^T$$

The within class matrix shows the average scatter $\sum i$ of the sample vectors x. In this work, the faces were tested with the faces which is not in the training set and the face images with different background. Sanjana et al., proposed an application where the face recognition algorithms has been used for Name – Face memory for the individuals who have memory disabilities [3]. Representation and classification of the images is all about face recognition. Some algorithms focus on different ways of representation of face images as a matrix, and some literatures concentrates the classification part [4, 5].

In PCA and LDA the squared euclidean distance emphasises the larger distances, this phenomenon is prone to outliers. Yang liu et al., proposed an algorithm which uses L2,1 norm. This is a discriminant manifold learning [6–8].

2 Clustering

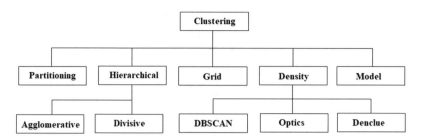

Fig. 1. Clustering

Clustering is an important aspect in any data processing. It groups the similar data in to sets. This eradicates the repetition or duplication of a same data in a set [9]. The clustering or representation of data should be done appropriately in order to perform the classification on it. Improper clustering will reduces the efficiency of classification [10, 11]. There are number of clustering algorithms were proposed in the literature. The large data set can be efficiently manipulated using DBSCAN [12]. The radius of nearest object can be found using density based clustering [13, 14]. K means, Expectation – Maximization, Partitioning around Medoids, Fuzzy C means are the commonly used clustering algorithms [15, 16] (Fig. 1).

3 Classification

The classification algorithm should be fast enough to continuously process the data. Most of the algorithms are good in processing controlled data [17]. i.e. Processing of defined or standard data. The combination LCDRC and L1 norm based face recognition systems are best suitable for un-controlled face recognition cases [18, 19]. i.e., the images need not to be in the predefined or standard data set. BCRE of an LDRC algorithm is given by,

$$BCRE = \frac{1}{n(c-1)} \sum_{\substack{i=1\,j=1}}^{n} \sum_{j=I(xi)}^{c} ||yi - yi\bar{j}inter||2$$

L1 norm based learning is said to be the robust in projection of vectors [20]. L1 and L2 norms can be implemented with PCA and LDA. The use of greedy algorithm will not maximize the criterian function. To overcome this issue non greedy algorithm also has been proposed by Luo et al [11]. It avoids the calculation of mean maximizing the varience. It is important to note that L2 norm and Suared L2 norm doesnot make much difference [21, 22].

Wopt of L1 is given by,

$$Wopt = \text{argwmax } Wopt = \text{argwmax } \frac{\text{tr}(WtXMXTW)}{(WtXLXTW)}$$

$$= \text{argwmar} \frac{\sum_{i,j=1}^{c} Bij\|wt(mi - mj)}{\sum_{i,j=1}^{nk} Aij\|wt(xi - xj)}$$

By solving the objective function independently, we can obtain the projection vector.

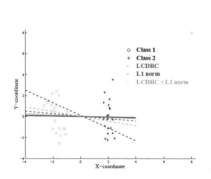

Fig. 2. Projection vectors.   **Fig. 3.** Classification accuracy

The Fig. 2 shows the graph of projection vectors for different classes on artificial database. In this case the outlier is to be considered. The following Table 1 shows the accuracy of the projection vector with or without outlier for different algorithms (Fig. 3).

Table 1. Accuracy of projection vector

Method	Without outlier	Outlier
LCDRC	96.79 ± 0.02	86.96 ± 0.05
L1 norm	96.73 ± 0.05	96.63 ± 0.02
LCDRC + L1	96.81 ± 0.08	96.69 ± 0.05

The implementation result shows the classification accuracy is high in the proposed method. It combines LCDRC and L1 norm.

4   System on Chip Realization

As the technology scales up, automation occupies its all possible avenues. As a result the intelligent systems become larger, power hungry and low speed. This is the perfect time where the paradigm shift is happening from System on Board Design to System on Chip design. The System on chip design provides the path to design the entire system on a single silicon die. This reduces the latency and increases the speed. In this work Xilinx Zynq 7000 has been used. This has ARM cortex A9 range Application processing unit which is capable of performing 2.5 DMIPS/MHz per CPU and wide range of sophisticated DSP block in programmable logic (Fig. 4).

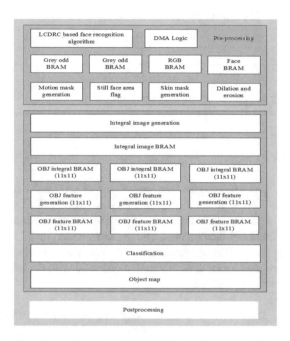

Fig. 4. Architecture of LCDRC based face recognition

The proposed architecture consist of 3 modules namely pre-processing unit, Classification unit and post processing unit. The pre-processing unit is responsible for feature extraction and clustering. The data sets will be processed through BRAMs and mask generation units. The prosed linear collaborative discriminant regression classification and L1 norm based classification algorithm is used in the classification unit. The post processing will be the representation of classified sets and producing the match results. Table 2 shows the resource utilization for the proposed architecture. Since the algorithm is all about data processing, large number of BRAM (Block RAM) is used.

Table 2. Resource utilization

Resource	Available	Used resource	Utilization (%)
Registers	82850	40300	49
LUTs	65700	47600	72
BRAM	260	222	85
DSP	350	150	43

Table 3 shows the comparison of face detection latency. It is noted that the SoC based implementation takes very less time when compared to software based systems. The SoC based implementation speedups the face recognition minimum of 32 times and maximum of 71 times in the 4 trails.

Table 3. Face detection latency comparison

Face	Software execution time	SoC execution time	Speedup
1	66	2	32
2	120	2.5	47
3	230	3.2	70
4	274	3.8	71

5 Conclusion

A novel linear collaborative discriminant regression classification and L1 norm based classification algorithm has been implemented on Xilinx Zynq 7000 SoC. The experimental result shows the proposed algorithm produces more accuracy when dealing with unconditional face images. The implementation of this algorithm on SoC platform speedups the system by minimum of 32 times and maximum of 72 times in the 4 tails. The face data sets has been handled carefully with the help of BRAM. The resources were efficiently utilized.

Acknowledgement. The Author like to acknowledge Shri. K. Sasi Kiran Varma, Vice Chairman of Godavari Institute of Technology, Rajahmundry, Andhra Pradesh, India.

References

1. Turk, M.A., Pentland, A.P.: Face recognition using eigenfaces. In: IEEE Computer Society Conference on Computer Vision and Pattern Recognition, pp. 586–591 (1991)
2. Zhao, W., Chellappa, R., Krishnaswamy, S.: Discriminant analysis of principal components for face recognition. In: Face Recognition. From Theory to Applications, pp. 73–85 (1998)
3. Nonavinakere, S., Aldana, J., Sisson, S., Cruz, E., George, K.: Memory aid device to improve face-name memory in individuals with Alzheimer's disease. In: IEEE International Conference on Healthcare Informatics, pp. 353–354 (2018)
4. Ram, A., Jalal, S., Jalal, A.S., Kumar, M.: A density based algorithm for discovering density varied clusters in large spatial databases. Int. J. Comput. Appl. **3**(6), 1–4 (2010)
5. Ankerst, M., Breunig, M.M., Kriegel, H.P., Sander, J.: OPTICS: ordering points to identify the clustering structure. In: Proceedings of the ACM SIGMOD International Conference on Management of Data, vol. 28, no. 2, pp. 49–60 (1999)
6. Vijayalakshmi, S., Punithavalli, M.: Improved varied density based spatial clustering algorithm with noise. In: Proceedings of the IEEE International Conference on Computational Intelligence and Computing Research (ICCIC), pp. 1–4 (2010)

7. Liu, Y., Gao, Q., Gao, X., Shao, L.: L2, 1-norm discriminant manifold learning. IEEE Access (2018)
8. Ke, Q., Kanade, T.: Robust l1 norm factorization in the presence of outliers and missing data by alternative convex programming. In: Proceedings of IEEE Conference on Computer Vision and Pattern Recognition, pp. 739–746. IEEE (2005)
9. Li, B., Yu, Q., Wang, R., Xiang, K., Wang, M., Li, X.: Block principal component analysis with non greedy 1-norm maximization. IEEE Trans. Cybern. **46**(11), 2543–2547 (2016)
10. Zhong, F., Zhang, J.: Linear discriminant analysis based on l1-norm maximization. IEEE Trans. Image Process. **22**(8), 3018–3027 (2013)
11. Luo, M., Nie, F., Chang, X., Yang, Y., Hauptmann, A., Zheng, Q.: Avoiding optimal mean robust PCA/2DPCA with non-greedy l1-norm maximization. In: Proceedings of International Joint Conference on Artificial Intelligence, pp. 1802–1808 (2016)
12. Rudin, W.: Functional analysis. In: International series in Pure and Applied Mathematics. Tata McGrawhill (1991)
13. Taylor, A.E., Lay, D.C.: Introduction to Functional Analysis. Wiley, New York (1958)
14. Setiono, R.: Extracting rules from pruned neural networks for breast cancer diagnosis. Artif. Intell. Med. **8**(1), 37–51 (1996)
15. Chou, S.-M., Lee, T.-S., Shao, Y.E., Chen, I.-F.: Mining the breast cancer pattern using artificial neural networks and multivariate adaptive regression splines. Expert Syst. Appl. **27**(1), 133–142 (2004)
16. Inan, O., Uzer, M.S., Yılmaz, N.: A new hybrid feature selection method based on association rules and PCA for detection of breast cancer. Int. J. Innovative Comput. Inf. Control **9**(2), 727–729 (2013)
17. Nauck, D., Kruse, R.: Obtaining interpretable fuzzy classification rules from medical data. Artif. İntell. Med. **16**(2), 149–169 (1999)
18. Larrain, T., Bernhard, J.S., Mery, D., Bowyer, K.W.: Face recognition using sparse fingerprint classification algorithm. IEEE Trans. Inf. Forensics Secur. **12**(7), 1646–1657 (2017)
19. Wei, C., Wang, Y.F.: Undersampled face recognition via robust auxiliary dictionary learning. IEEE Trans. Image Process. **24**(6), 1722–1734 (2015)
20. Peng, C., Gao, X., Wang, N., Li, J.: Graphical representation for heterogeneous face recognition. IEEE Trans. Pattern Anal. Mach. Intell. **39**(2), 301–312 (2017)
21. Roy, H., Bhattacharjee, D.: Local-gravity-face (LG-face) for illumination-invariant and heterogeneous face recognition. IEEE Trans. Inf. Forensics Secur. **11**(7), 1412–1424 (2016)
22. Moghaddam, B.: Principal manifolds and probabilistic subspaces for visual recognition. IEEE Trans. Pattern Anal. Mach. Intell. **24**(6), 780–788 (2002)

Simulation of Dual Polarization Radar for Rainfall Parameter and Drop Size Distribution Estimation

C. Pratibha, K. Manish Reddy, L. Bharathi, M. Manasa,
and R. Gandhiraj$^{(\boxtimes)}$

SIERS Research Lab, Department of Electronics and Communication
Engineering, Amrita School of Engineering, Amrita Vishwa Vidyapeetham,
Coimbatore, India
pratibhachennupatill03@gmail.com,
r_gandhiraj@cb.amrita.edu

Abstract. Disdrometers are devices that help in estimating DSD. The existing disdrometers in India are mostly hardware dependent are expensive and prone to wear and tear. An effective low cost RF based disdrometer can be implemented with the help of Dual-polarized radar. The complete working of S-band disdrometer is simulated and error is estimated. The rainfall estimation is done using the updated algorithm that uses polarimetric radar parameters to estimate the intensity of rainfall for the range of 2.7–2.9 GHz. The rainfall attenuation prediction is done using a power-law relation given by ITU-R recommendation.

Keywords: Disdrometer · Low cost · DSD · Rainfall estimation · Dual-polarization

1 Introduction

Disdrometer is a device that helps in estimating an important rainfall parameter called the drop size distribution (DSD). Drop size distribution is an important parameter for estimating soil erosion, climatic changes and infiltration of rainfall in to the soil. The existing disdrometer estimates this distribution by impact, optical, and acoustic based or laser based methods. Optical based devices, send light rays and a detector receives it. When a drop of rain crosses the light path casts a shadow. This varying intensity and time of variation provides the diameter and velocity of the rain drop. On the other hand impact dependent disdrometer uses piezo substrates to sense the drops by pressure change when the drops fall on the element [1]. Apart from the difference in technical terms, both of these disdrometers fall in the same cost range. All of these disdrometer use so much of hardware components which are of high cost and the power that is consumed by them is high. We prefer a cost efficient solution is a RF based disdrometer with the help of a SDR. Software Defined radio (SDR) is a system where most of the hardware components (which do the signal processing) are completely replaced by the software which is installed in a computer [2].

© Springer Nature Switzerland AG 2020
A. P. Pandian et al. (Eds.): ICICCS 2019, AISC 1039, pp. 424–433, 2020.
https://doi.org/10.1007/978-3-030-30465-2_47

Universal Applicable Hardware functions as a mode of the connection between the baseband (software) and RF end (antennas). The waveform is totally produced through programming and it is transmitted by the transmitting antenna and also on the receiving end, the signal is captured by a receiver antenna and. The simulation that is created in this project is a complete prototype of how a RF based disdrometer functions. Due to certain limitations in simulation environment, certain parameters like background noise, reflections from the surroundings, and the leakage from the antennas are not considered. And this project can be taken further by fabricating narrow beam antennas and experimenting in a real rainfall environment to validate and further can be made into a commercial product.

The upcoming topics in this report will give the thorough knowledge of the simulated model.

2 Block Diagram

See Fig. 1.

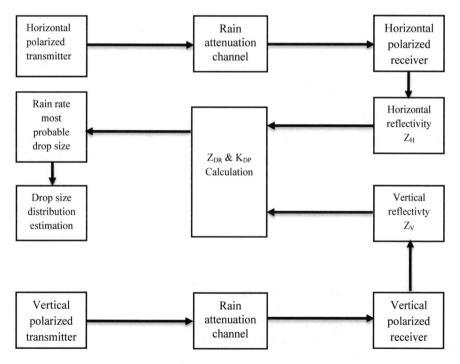

Fig. 1. Block diagram

2.1   Transmitter/Receiver Blocks

The transmission and reception is simulated using a software that supports radar simulation. As it was seen that the rain estimation algorithms work better at S-band among all the weather radar rages (S, X, K-band) available, the frequency band chosen for the experiment is S-band (2–4 GHz) with operating frequency 2.8 GHz [3]. A narrow beam width antenna helps in reducing the amount of transmitted power required, reduces the scattering of the power and helps reduces cross-polarization. The NEXRAD radar is a narrow beam horn antenna with a parabolic dish reflector. It is a doppler weather surveillance radar which has again of 53 dB and resolution of 250 m [4]. This pattern can be used with the required specifications.

The specifications given to the antenna in the experiment done:

- Radar type – Bistatic CW radar.
- Operating frequency – 2.8 GHz.
- Transmitter gain – 20 dB.
- Receiver gain – 20 dB.
- Transmitted power – 10 W.
- Distance between Transmitter and Receiver – 7.45 m.

2.2   Rain Attenuation Channel Block

The attenuation that a signal undergoes during transmission due to rainfall is simulated as mathematical model. The prediction model is discussed in Sect. 3.

2.3   Reflectivity Block

Horizontal reflectivity and Vertical reflectivity can be estimated from the transmitted power, received power (of the corresponding polarization) relation called Probert-Jones equation given below [5, 6].

$$P_r = \frac{P_t \cdot G^2 \cdot \theta^2 \cdot H \cdot \pi^3 \cdot K^2}{1024 \cdot \ln 2 \cdot \lambda^2 \cdot R^2} \times Z \qquad (1)$$

Where,

- P_r = Power received.
- P_t = Power transmitted from the radar.
- G = Gain of the radar antenna.
- θ = Beam width of the antenna.
- H = Pulse width of the transmitted RF energy.
- λ = Wavelength.
- Z = Reflectivity.
- R = Range of the target.
- K^2 = Constant (0.93 for rain).

2.4   Differential Reflectivity (Z_{dr})

It is the ratio of the horizontal reflectivity to the vertical reflectivity.

$$Z_{dr} = \frac{Z_H}{Z_V}$$

2.5   Specific Differential Phase (K_{dp})

The rate of change of difference in phase between the horizontal and vertical polarized signals with respect to distance. Using simulation this can be estimated using the relation based on horizontal reflectivity (Zh) and the differential reflectivity which is as follows for S-band frequencies.

$$K_{dp} = 0.000105 \cdot \zeta_H^{0.98} \cdot 10^{-0.26 \cdot Z_{dr}} \tag{4}$$

Where,
ζ_H – is horizontal reflectivity in $mm^6\ m^{-3}$.
Z_{dr} – is differential reflectivity in dB.

2.6   Rain Rate

Rain rate is estimated using the method mentioned in Sect. 4.

2.7   Drop Size Distribution

Drop size distribution plots the drop size distribution estimated at the corresponding rain rate. The plot is no of drop of the particular diameter vs the diameter of the drops. It gives the information on number of drops of a particular diameter present in a unit volume [7].

$$N(D) = \frac{\zeta_H \cdot D}{3.3987 \times 48 \times D_0^8} \times e^{-0.5 \cdot \left(\frac{D}{D_O}\right)^2}$$

Where,

ζ_H – is horizontal reflectivity in $mm^6\ m^{-3}$.
D – Diameter of drop.
D_O – Most probable drop size.

Most probable drop size is estimated by the relation given below.

$$D_O = \frac{45 \times \pi^{1.5} \times 10^{-3} \times \zeta_H}{48 \times 4 \times 3.3987 \times R}$$

Where,

ζ_H – Horizontal reflectivity in mm^6 m^{-3}.
R – Rain rate.

3 Rain Attenuation Prediction

The real-time effect of rain simulated in the software using a mathematical model which predicts the attenuation on the signal caused due to the rainfall. There are many models proposed for this purpose, but among these the prediction results were more accurate when the ITU-R proposed prediction model was used.

This model is derived by curve-fitting the real-time radar observations to get a mathematical model. The relation between rain rate and the amount of specific attenuation is given below [8].

$$\gamma_R = k.R^\alpha.L(dB)$$

Where,

- γ_R – The specific attenuation due to rain in dB.
- k, α – Frequency and polarization
- dependent coefficients.
- R – Rain rate.
- L – Path length.

The coefficients k, α can be estimated using the following relations [9]:

$$\log_{10} k = \sum_{j=1}^{4} a_j \, exp\left[-\left(\frac{\log_{10} f - b_j}{c_j}\right)^2\right] + m_k \log_{10} f + c_k$$

$$\alpha = \sum_{j=1}^{5} a_j \, exp\left[-\left(\frac{\log_{10} f - b_j}{c_j}\right)^2\right] + m_\alpha \log_{10} f + c_\alpha$$

Where,

f – Frequency in GHz.
k – Either k_H or k_V. α – Either α_H or α_V.

Value for the constants involved in the above relations are calculated at 2.8 GHz (S-band).

k_H – 0.0001425 k_V – 0.00017515
α_H – 1.1774 α_V – 1.0445

3.1 Path Length Correction

The attenuation that a signal undergoes during transmission due to rainfall is simulated as mathematical model. The prediction model is discussed in Sect. 3.

The transmitted signal doesn't travel in a straight line, due to the reflections and refractions the path length is not the same as the separation between the transmitter and receiver. The path correction is given in the updated model ITU-R P.530-14 [10–12].

The path correction factor 'r' is given by the following relation.

$$r' = 0.477 \cdot L^{0.633} \cdot R^{0.073 \cdot \alpha} \cdot f^{0.123} - 10.579 \cdot \left(1 - e^{-0.24 \cdot L}\right)$$

$$r = \frac{1}{r'}$$

Where,

L – Path length.
R – Rain rate.
f – Frequency in GHz.
α – Either α_H or α_V.

The updated ITU-R prediction model includes the path correction factor.

$$\gamma_R = k.R^{\alpha}.L.r (dB)$$

4 Rain Rate Estimation Algorithm

The rain rate can be estimated from the rainfall parameters like reflectivity (Z_H, Z_V), differential reflectivity (Z_{dr}) and differential phase (K_{dp}). There are many relations proposed for the estimation of rainfall. Among which the updated algorithm proved to be accurate for the tropical climates. Based on the threshold values of the rainfall parameters the algorithm for rain rate estimation is given as the following. All the coefficients involved in the algorithm below are frequency dependent and are calculated at 2.8 GHz (S-band) [13]. The relations are given below (Fig. 2).

The values of coefficients at S-band are calculated.

- $a_1 = 30.3$; $b_1 = 0.9298$
- $a_2 = 34.56$; $b_2 = 0.9496$
- $c = 0.0154$; $d = 0.7681$
- $p = 0.0084$; $q = 0.9284$; $r = -0.4055$
- $p_1 = 51.16$; $q_1 = 0.9311$; $r_1 = -0.0852$

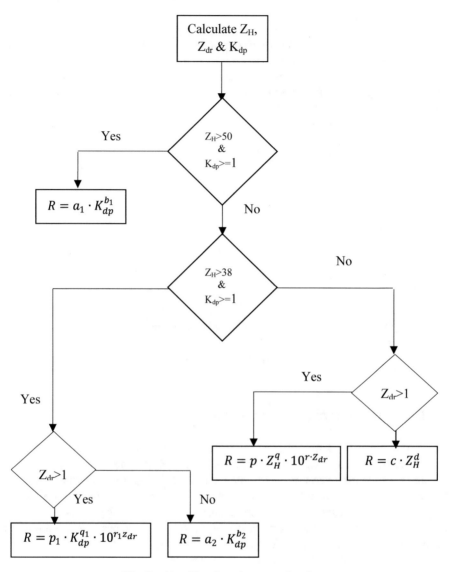

Fig. 2. Algorithm for rain rate estimation.

5 Results

See Fig. 3.

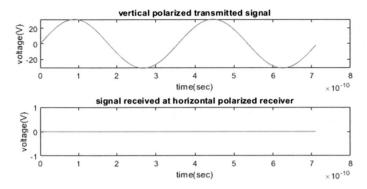

Fig. 3. Transmission and reception of a vertical polarized signal.

5.1 Rain Rate Calculation

The system serves the purpose only if elevation angles of both transmitter and receiver 0^0 (Fig. 4).

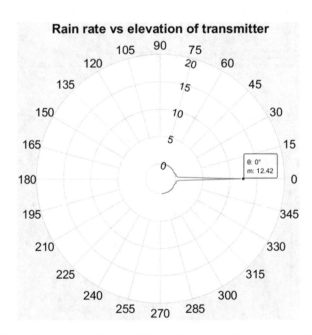

Fig. 4. Rainrate calculated (at 12 mm/hr) vs elevation of transmitter.

The rain rate estimation is also dependent on the separation between the receiver and the transmitter. For a transmitter peak power of 10Watts and antenna gain of 20 dB the distance should be fixed to be 7.45 m. As the distance is reduced the algorithm underestimates the rainfall rate. And as the separation increases the algorithm over-estimates the rainfall rates (Figs. 5 and 6).

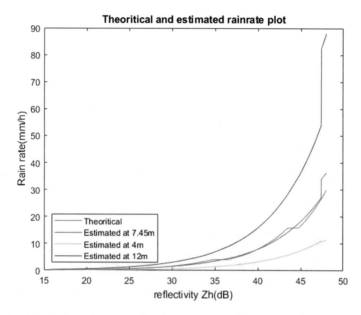

Fig. 5. Rainrate when separation between transmitter and receiver is 7.5 m.

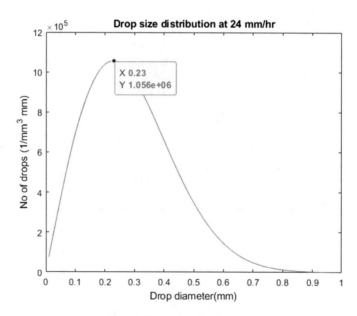

Fig. 6. Drop size distribution.

6 Conclusion

The results of the experiment done show that the dual polarization radar setup working at a frequency 2.8 GHz with antennas having a narrow beam width of 1^0, transmitter gain of 20 dB and receiver gain of 20 dB serves the purpose of disdrometer application only when the transmitting and receiving antennas are in LOS i.e. they are facing each other and the distance between them is fixed to be 7.45 m for a transmitted power of 10 Watts. From the simulation results obtained, it is understood that RF based application will work for the functioning of a disdrometer.

References

1. Kathiravelu, G., Lucke, T., Nichols, P.W.B.: Rain drop measurement techniques : a review (2016)
2. Hershberger, J., Pratt, T., Kossler, R.: A software-defined, dual-polarized radar system. In: 2016 IEEE Conference on Antenna Measurements Applications (CAMA), pp. 1–4 (2016)
3. Vulpiani, G., Giangrande, S., Marzano, F.S.: Rainfall estimation from polarimetric S-band radar measurements: validation of a neural network approach. J. Appl. Meteorol. Climatol **48**(10), 2022–2036 (2009)
4. NEXRAD-Wikipedia. https://en.wikipedia.org/wiki/NEXRAD
5. Rogers, P.J., Eccles, P.J.: The bistatic radar equation for randomly distributed targets. Proc. IEEE **59**(6), 1019–1021 (1971)
6. RADAR Reflectivity Measurement. https://training.weather.gov/nwstc/NEXRAD/RADAR/3-1.htm
7. Mali, P., Sarkar, S.K., Das, J.: Rain drop size distribution from radar reflectivity measurements **32**, 296–300 (2003)
8. Sundaresan, S., Anjana, C., Zacharia, T., Gandhiraj, R.: Real time implementation of FMCW radar for target detection using GNU radio and USRP. In: 4th IEEE International Conference on Communication and Signal Processing-ICCSP 2015, Department of Electronics and Communication Engineering, Adhiparasakthi Engineering College, Chennai, 2–4 April 2015
9. P.838: Specific attenuation model for rain for use in prediction methods. https://www.itu.int/rec/R-REC-P.838-3-200503-I/en
10. P.530: Propagation data and prediction methods required for the design of terrestrial line-of-sight systems
11. Gandhiraj, R., Soman, K.P., Sukesh, K., Kashyap, K.V.S., Yaswanth, K., Haswanth, K.: Design and parameters measurement of tin-can antenna using software defined radio. Commun. Comput. Inf. Sci. **837**, 558–568 (2018)
12. Semire, F.A., Mohd-Mokhtar, R., Akanbi, I.A.: Validation of New ITU-R rain attenuation prediction model over Malaysia equatorial region. MAPAN, November 2018
13. Zhang, Y., Liu, L., Wen, H., Wu, C., Zhang, Y.: Evaluation of the polarimetric-radar quantitative precipitation estimates of an extremely heavy rainfall event and nine common rainfall events in Guangzhou. Atmosphere **9**(9), 330 (2018)

Comprehensive Analysis of Different Modulation Techniques on a Multi-level Neutral Point Clamped Inverter in a Solar PV System

Mohsin Karim Ansari[1], Abdul Azeem[1], Adil Sarwar[1], Mohd Tariq[1], Md Reyaz Hussan[1(✉)], and Ahmed Riyaz[2]

[1] Department of Electrical Engineering, Aligarh Muslim University, Aligarh, India
{mkansari, adil.sarwar}@zhcet.ac.in,
azeem.abdul25@gmail.com, tariq.iitkgp@gmail.com,
mreyazamu@gmail.com
[2] Department of Electrical Engineering, BGSB University, Rajouri, Jammu and Kashmir, India
ahmedriyaz@bgsbu.ac.in

Abstract. In this paper, a performance comparison of three-level neutral point clamped (NPC) inverter using different modulation schemes are presented. Different level shifted carrier based PWM schemes and space vector modulation scheme for three-level NPC inverter is modeled and analyzed. The performance analysis is presented in order to understand its interaction with solar photo-voltaic (SPV) system. The power quality of the inverter output is compared using total harmonic distortion (THD). The whole system comprises a solar PV array (Waaree Energies WSM- 315), a boost converter (operated at MPPT using incremental conductance technique), a multi-level inverter (NPC), and a three-phase load (R and L). Then, the system is modeled in MATLAB®/ Simulink and the results obtained are presented and discussed in this paper.

Keywords: Neutral Point clamped converter (NPC) · Pulse Width Modulation (PWM) · Solar Photovoltaic Array (SPV) · Space Vector Modulation (SVM) · Incremental Conductance Maximum Power Point Tracking (INC-MPPT)

1 Introduction

Renewable energy sources have drawn a considerable interest from researchers across the globe to explore ways to tap this potential judiciously. The continuous increase in electricity demand and depleting conventional sources is calling for exploring alternative ways to generate electricity. Moreover, there are many remote areas in the developing countries which are geographically and technically quite difficult and financially not feasible to get connected to the grid. The advantages of renewable sources include clean, environment-friendly and inexhaustible. A detailed review on renewable energy resources is available in [1].

© Springer Nature Switzerland AG 2020
A. P. Pandian et al. (Eds.): ICICCS 2019, AISC 1039, pp. 434–441, 2020.
https://doi.org/10.1007/978-3-030-30465-2_48

The main advantage of solar Photovoltaic (PV) system is that it can be installed easily, and power availability is in plug and play mode. The PV cell generates dc electric power from solar energy, which is available free of cost in most of the parts of the world. Figure 1 illustrates the functional block diagram of an SPV system. PV arrays consist of several PV panels. Normally the input voltage of the PV panel is enhanced using a dc-dc boost converter to match the required voltage. A storage system is also needed if we want to use it as a backup.

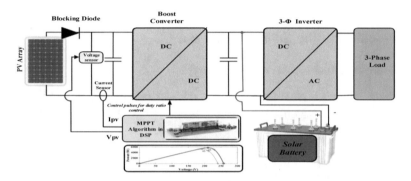

Fig. 1. A general block diagram of standalone alone PV with the storage system.

Researchers are continuously working to improve the performance of the power electronic converters for efficient utilization of solar energy. In line to the above, multi-level inverters have received importance in industries due to their higher efficiency and better power quality [2]. In literature, different topologies have been presented for multi-level operation such as Cascaded H-bridge(CHB) [3, 4], Flying Capacitor(FC) [5], Packed U-cell inverter (PUC) [6], etc. The problem with CHB is that it requires three different sources for three-phase power generations, whereas in the FC and PUC topologies the balancing of the high number of capacitor voltage is a complex process. In this work, a three-level Neutral Point clamped converter (NPC) inverter is used as it offers three phase power generation from a single dc source and relatively easier control in comparison to FC.

Modulation technique is used for controlling and regulating the output voltage in power electronic converters. Distortion, harmonics, switching losses are the important parameters to be considered before the selection of a modulation strategy for a particular family of converter topologies. Pulse Width Modulation (PWM) is also one of the most widely used modulation strategies. A targeted average output voltage and current in PWM is obtained by choosing a suitable value of duty ratio of the power converter switches. Some work has been reported on the comparison of carrier-based and space vector techniques in [7, 8, 9]. In [7] only carrier-based techniques are compared whereas in [9] only one carrier technique is compared with space vector modulation. Hence, there is a gap in the literature regarding the comparison of results with space vector technique and major carrier based PWM techniques in a solar PV environment. In this paper, three major carrier based PWM techniques, namely (IPD,

POD, and APOD) are compared with space vector technique. The objective is to compare and analyze the performance of three-level Neutral Point Clamped converter fed from solar PV panel with different modulation schemes.

The paper is structured as follows: The modelling of the PV array, boost converter, and NPC converter is presented in Sect. 2. The modulation techniques and the overall performance of three-level NPC inverter with different modulation schemes are discussed in Sect. 3. The comparative analysis for the above-mentioned modulation techniques is presented in Sect. 4. The conclusion is given in Sect. 5.

2 Modelling of the Solar PV System

2.1 Solar Photovoltaic Array (PV Array)

When light falls on a photovoltaic cell, it releases electrons, which is the cause of conversion of light energy into electricity. The equivalent circuit is shown in Fig. 2 and Eqs. (1) and (2) represents a photovoltaic cell [10].

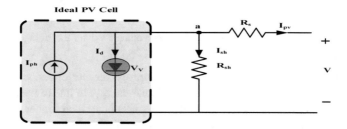

Fig. 2. Model of a PV cell

Applying KCL at node 'a' the photocurrent is given by,

$$I_{pv} = I_{ph} - I_d - I_{sh} \tag{1}$$

The current through the diode is given by,

$$I_d = I_o\left(e^{\frac{V_v \cdot q}{\eta kT}} - 1\right) \tag{2}$$

I_{sc} depends on weather conditions and ambient temperature.
Using Eqs. (1) and (2) becomes

$$I_{pv} = I_{ph} - I_o\left(e^{\frac{(I_{pv}R_s + V_{pv}) \cdot q}{\eta kT}} - 1\right) - \frac{I_{pv}R_s + V_{pv}}{R_{sh}} \tag{3}$$

Where; I_{pv}: photocurrent (A), I_{sc}: short circuit current (A), I_D: diode current (A), V_{pv}: terminal voltage of the cell (V), I_o: saturation current (A), η: ideality factor (manufacturing value between 1 and 2), q: electronic charge (1.6×10^{-19} Coulomb), T: junction temperature (K), k: Boltzmann's constant (1.38×10^{-23} Joules/K), N_p: Number of cells in parallel, N_s: Number of cells in series.

The simulation is done on the panel Waaree Energies WSM- 315, which is ranked among better solar PV panels in India for its high-power rating. Its model is given in the MATLAB®/ Simulink. The effect of variation in temperature is balanced by the incremental conductance MPPT controller.

2.2 Boost Converter

The output voltage of the boost converter can be expressed as:

$$\frac{V_o}{V_{in}} = \frac{1}{1 - D} \tag{4}$$

Where V_{in} is the average input voltage, V_o is the average output voltage. The boost converter has inductance 1.61 mH; capacitance is 652.3 μF and duty ratio D = 0.6523 shown in Fig. 3. Figure 4 shows the variation in the duty ratio. At t = 0.6 s a step change from 0.6 to 1 kW/ m^2 is shown in Fig. 5.

2.3 Three Level Neutral Point Clamped Inverter

This converter topology was introduced in 1981 [11]. The three-phase three-level neutral point clamped converter is shown in Fig. 6. At any given time two switches are on in a combined form. A k-level NPC converter produces k-level output phase voltage and a 2k − 1 level output line voltage [12]. The requirement of the blocking diode for each phase is given by (k − 1) × (k − 2) where k is the number of levels if its voltage rating and active switches voltage rating is same.

3 Modulation Techniques

In this paper, carrier-based PWM schemes and space vector modulation schemes are used. Their description are as follows:

3.1 Carrier-Based PWM Schemes

To synthesize k voltage levels, k − 1 triangular carrier signals are required. Modulating signal is a sinusoidal wave.

In level shifted modulation to get k voltage levels, k − 1 triangular carriers, all having the same peak to peak amplitude and same frequency is required. The k − 1 triangular waves are arranged vertically in contiguous bands. The frequency modulation index is given by $m_f = f_{cr}/f_m$ whereas the amplitude modulation index is defined as

$$m_a = \frac{\hat{V}_m}{\hat{V}_{cr}(m-1)} \quad \text{for} \quad 0 \le m_a \le 1 \tag{5}$$

where f_{cr} is carrier frequency, f_m is modulating frequency, $\hat{V}_m$ is the peak amplitude of the modulating wave, $\hat{V}_{cr}$ is the peak amplitude of each carrier wave and $f_{sw,\,dev}$ is device switching frequency. All the output voltages are taken at $m_a = 1$, $m_f = 40$, $f_m = 50$ Hz and $f_{cr} = 2$ kHz. Figures 7, 11, and 12 show the IPD modulation, the output voltage for IPD and THD (in %) in the output voltage (line-line) of the inverter at a load (40 Ω, 20 mH) respectively.

3.2   Space Vector Modulation Scheme

The space vector representation is also a form of PWM technique; the only difference is in converter switching duration. Figure 9 shows 27 vectors produced by a three-phase three-level converter: twelve internal (100, 211, 110, 221, 010, 121, 011, 122, 001, 112, 101, 212), three zero (000, 111, 222), six external (200, 220, 020, 022, 002, 202) and six middle (210, 120, 021, 012, 102, 201). The reference vector and inverter switching states are used for the calculation of switching times using method described in [9]. The six sectors are subdivided into four triangular regions shown in Fig. 10.

The three nearest stationary vectors give the reference vector V_{ref}. Figures 8, 13 and 14 shows the reference signals of three level space vector PWM, output voltage and THD(in %) respectively for the same load as in carrier-based schemes, i.e. 40 Ω, 20 mH.

4   Comparative Analysis of NPC Converter Under Different Modulation Schemes

Figure 15 shows a scatter plot of THD (in %) and m_a for each scheme for its line-to-line voltage v_{ab}. It could be seen that the harmonic profile of IPD-PWM is best among the carrier shifted schemes. The THD of SVM is comparable to IPD-PWM, but its control is somewhat more complex.

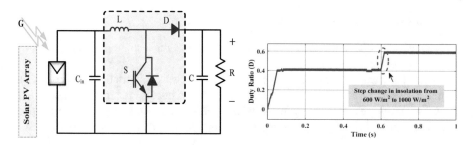

Fig. 3. Boost converter with PV array   **Fig. 4.** The duty ratio of the Boost converter

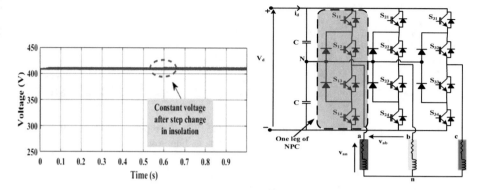

Fig. 5. The output voltage of the boost converter

Fig. 6. Three level Neutral Point Clamped (also known as diode clamped) converter.

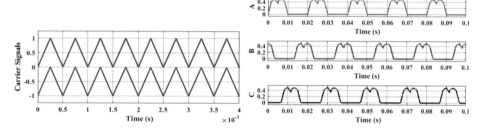

Fig. 7. In-phase disposition modulation (IPD) modulation

Fig. 8. A, B, and C are Space vector reference carrier signals

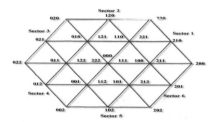

Fig. 9. Division of sectors and regions for three-level SVM

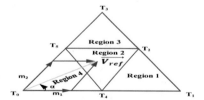

Fig. 10. Region selection with respect to modulation indexes

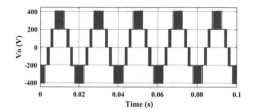

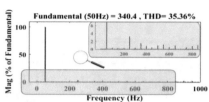

Fig. 11. IPD–PWM line-to-line voltage

Fig. 12. Harmonic profile of IPD-PWM line-to-line voltage

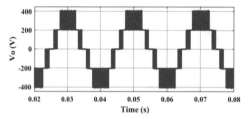

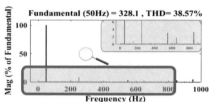

Fig. 13. SVM line-to-line voltage

Fig. 14. Harmonic profile of SVM line-to-line voltage

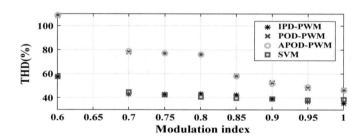

Fig. 15. Harmonic content comparison of v_{AB}

5 Conclusion

This paper has analyzed the performance of three levels NPC converter working under different modulation schemes with the solar PV application. A battery was also proposed to store a charge that can be used to drive the load when the supply from PV is not available. Among all the modulation schemes that were analyzed IPD-PWM gives the minimum total harmonic distortion (THD). The three-level NPC converter offers a good choice for the solar PV application and medium voltage drive. Simulation results show that SVM is comparative to IPD-PWM in terms of THD, but it involves more complex control and calculations.

References

1. Bose, B.K.: global energy scenario and impact of power electronics in 21st century. IEEE Trans. Ind. Electron. **60**(7), 2638–2651 (2013)
2. Sarwar, A., Asghar, M.S.J.: Simulation and analysis of a multilevel converter topology for solar PV basedgrid connected inverter. Smart Grid Renew. Energy **2**, 56–62 (2011)
3. Daher, S., Schmid, J., Antunes, F.L.M.: Multilevel inverter topologies for stand-alone PV systems. IEEE Trans. Ind. Electron. **55**(7), 2703–2712 (2008)
4. Siddique, M.D., et al.: Single phase symmetrical and asymmetrical design of multilevel inverter topology with reduced number of switches. In: 2018 IEEMA Engineer Infinite Conference (eTechNxT), pp. 1–6 (2018)
5. Trabelsi, M., Ben-Brahim, L.: Experimental photovoltaic power supply based on flying capacitors multilevel inverter. In: 3rd International Conference on Clean Electrical Power: Renewable Energy Resources Impact, ICCEP 2011, pp. 578–583 (2011)
6. Tariq, M., Meraj, M., Azeem, A., Maswood, A.I., Iqbal, A., Chokkalingam, B.: Evaluation of level-shifted and phase-shifted PWM schemes for seven level single-phase packed U cell inverter. CPSS Trans. Power Electron. Appl. **3**(3), 232–242 (2018)
7. McGrath, B.P., Holmes, D.G.: Comparison of multicarrier PWM strategies for cascaded and neutral point clamped multilevel inverters. PESC Record - IEEE Annual Power Electronics Specialists Conference, vol. 2, no. c, pp. 674–679 (2000)
8. Siddique, M.D., Sarwar, A.: Performance analysis of carrier based PWM technique for three level diode clamped multilevel inverter with different reference signals. In: 2016 IEEE 7th Power India International Conference (PIICON), vol. 6, pp. 1–6 (2016)
9. Yao, W., Hu, H., Lu, Z.: Comparisons of space-vector modulation and carrier-based modulation of multilevel inverter. IEEE Trans. Power Electron. **23**(1), 45–51 (2008)
10. Tan, Y.T., Kirschen, D.S., Jenkins, N.: A model of PV generation suitable for stability analysis. IEEE Trans. Energy Convers. **19**(4), 748–755 (2004)
11. Nabae, A., Takahashi, I., Akagi, H.: A New neutral-point-clamped PWM inverter. IEEE Trans. Ind. Appl. IA **17**(5), 518–523 (1981)
12. Rashid, M.H.: Power Electronics Handbook, 4th edn. Butterworth-Heinemann (2018)

Control Techniques of Packed U-Cell Multilevel Inverter: A Comprehensive Review

Mohsin Karim Ansari, Abdul Azeem, Adil Sarwar, Mohd Tariq,
and Md Reyaz Hussan$^{(\boxtimes)}$

Department of Electrical Engineering, Aligarh Muslim University, Aligarh, India
{mkansari, adil.sarwar}@zhcet.ac.in,
azeem.abdul25@gmail.com, tariq.iitkgp@gmail.com,
mreyazamu@gmail.com

Abstract. The importance of multilevel inverter is increasing continuously in recent decades from medium to high-level power conversion systems. Packed U-Cell MLI introduced recently is gaining increasing attention due to the reducing device count, lower THD and better power quality. Packed U-Cell has lesser number of devices to generate the same number of levels in the output voltages as compared to the other reduced device count MLI's reported in the literature. This paper focusses on seven level PUC inverter, which has been practically realised in many works. The basic operation, along with modulation strategies reported in the literature, has been presented. The later part of the paper deals with a detailed review regarding the control strategies, their implementation, and discussion on the future scope of the said inverter topology.

Keywords: Packed U-Cell (PUC) inverter · Multilevel Inverter (MLI) · Proportional Integral control (PI) · Proportional Resonant control (PR)

1 Introduction

The requirement of power is increasing exponentially in the modern world. DC to AC power converters are playing a very important role in various applications from generation to the utilization of electric power. Traditional 2-level and 3-level VSI's have high voltage harmonics that cause torque pulsations [1]. Higher input DC voltages are required while employing multilevel inverters for high voltage requirement. Different type of renewable energy sources such as PV panel, fuel cell, batteries and capacitors at the input for multiple dc sources can be used as voltage sources for achieving high voltages at the output.

Different topologies of MLI have been developed, each having their advantages and shortcomings. The diode-clamped converters introduced in [1], flying capacitor clamped [2] and cascaded H-bridge [3] MLI's are considered as classic topologies and are associated with many drawbacks when we go for a higher level in the voltage mainly due to a very high increase in the number of switching devices and capacitors. In the last decade, a series of reduced device count MLIs [4–6] have been proposed by many researchers, but most of the topologies were application specific. MLI's are needed to be very critically analyzed based on the active and passive component requirement,

© Springer Nature Switzerland AG 2020
A. P. Pandian et al. (Eds.): ICICCS 2019, AISC 1039, pp. 442–452, 2020.
https://doi.org/10.1007/978-3-030-30465-2_49

complexity and cost involved and their applicability. In this series PUC came out to be the most advantageous in terms of obtaining the highest number of levels with the use of least number of switches, the production cost is very low, and its component inter-connection is very easy [7–10]. It was introduced by Al-Haddad in 2008.

Figure 1(b) shows a single phase PUC with two input dc level; its switching pattern is given in Table 2. Initially, it was introduced as a 7-level PUC, and after that, 5-level PUC was introduced as a solution for the problem of capacitor voltage balancing. Being 2-level reduced, but the problem of capacitor voltage balancing is removed. The redundant switching states are to identified and used for better control of PUC.

The objective of this paper is to review the PUC topology, modulation, and control techniques. A detailed review is present on PUC in [11], but it does not focus on the control techniques and modulation schemes on PUC in detail. In this paper, Sect. 2 presents a review of the topology and its operating concept in detail. Section 3 presents the modelling of PUC MLI and the requirements of the controller for more than seven level operation. In Sect. 4, the different control techniques applied on PUC are reviewed, and their mathematical and technical aspects are evaluated, and Sect. 5 concludes this paper.

2 Topology and Modulation Strategy Implemented in PUC

2.1 Topology

The important part of MLI design includes its topology structure. A geometric pro-gression equation is formulated to obtain the number of voltage levels in PUC shown in Fig. 1.

$$x_n = ax_{n-1} + b$$
$$x_o = V. \tag{1}$$

The nth term is given by

$$x_n = x^n V + b \frac{1 - x^n}{1 - x} \tag{2}$$

Thus,

$$M_i = 2^{Mc_i + 1} - 1 \tag{3}$$

Where M_i is the number of voltage levels,
Mc_i is the number of capacitors used and i is an integer.

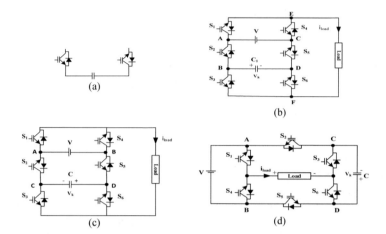

Fig. 1. (a) Basic U-Cell (b) and (c) are structures based on the polarities of the capacitor (d) more simplified form of PUC topology

Table 1. Comparison of components required to generate seven level

Topology	Components			
	Power semiconductor switches	Clamping diodes	Capacitors	Complexity of control
Diode clamped	12	10	6	Very high
Cascade H-bridge	12	0	3	Low
Flying capacitor	12	0	6	Very high
Hybrid cascade H-bridge	8	0	2	High
Packed U-Cell 7-level	6	0	2	Very high

Table 2 lists the number of level in voltage that can be obtained by different switching patterns in a general pattern as given in [12].

The number of levels produced in terms of the number of switches is given by the following equations

$$M_i = 2^{\frac{Msw_i}{2}} - 1 \tag{4}$$

Where M_{swi} is the number of power switches.

Table 1 gives the number of levels that can be generated based on the magnitude of the main and the auxiliary dc link. When $V = V_k$ three levels, $V = 2V_k$ five levels and when $V = 3V_k$ seven level output Is obtained.

2.2   Modulation Scheme and Operating Concept

Sinusoidal pulse width modulation (SPWM) is used for switching of the semiconductor devices. For n-levels (n-1) carrier waves are required. For five level operation, four carrier waves are required, and for seven levels, six carrier waves are required. The pulses obtained after the SPWM are then added to generate S waveform, as shown in Fig. 3. The S waveform is then used for obtaining pulses by applying conditions based on its amplitude. For example, if it is equal to two, then switches S1, S2 and S3 are turned on, as shown in Table 2. The carrier frequency should be a multiple of the fundamental frequency. For 50 Hz, 1000 Hz is taken as carrier frequency otherwise asymmetry will occur in the S signal obtained after summation.

Table 2. Switching table of three, five and seven level single phase PUC modulation scheme and operating concept

States	S1	S2	S3	Output voltage (VEF)	VEF (Three level)	VEF (Five level)	VEF (Seven level)	Magnitude of S waveform
1	1	0	0	V_1	$+E$	$+2E$	$+3E$	$+5$
2	1	0	1	$V_1 - V_k$	0	$+E$	$+2E$	$+4$
3	1	1	0	V_k	$+E$	$+E$	$+E$	$+3$
4	1	1	1	0	0	0	0	$+2$
5	0	0	0	0	0	0	0	-2
6	0	0	1	$-V_k$	$-E$	$-E$	$-E$	-3
7	0	1	0	$V_k - V_1$	0	$-E$	$-2E$	-4
8	0	1	1	$-V_k$	$-E$	$-2E$	$-3E$	-5

Table 3. THD obtained for two carrier based PWM schemes

Triangular PWM	THD in voltage (%)	THD in current (%)
Level-shifted PWM	18.01	7.78
Phase shifted PWM	17.96	2.79

Different modulation schemes have been applied in literature, and their performance is observed, which is given in Table 3 as discussed in [13]. Hybrid modulation is also applied on PUC using current vector modulation as discussed in detail in [14].

3   Modeling of Packed U-Cell Inverter

The switching function of the PUC inverter is defined as

$$S_i = \begin{cases} 0 \text{ if } S_i \text{ is off} \\ 1 \text{ if } S_i \text{ is on} \end{cases} \quad \text{where } i = 1, 2, 3 \tag{5}$$

From Eq. (2) and circuit diagram (Fig. 1(b)) output voltage of inverter can be written as

$$V_{EF} = V_{EA} + V_{AB} + V_{BF} \tag{6}$$

Where E, A, B and F nodes are illustrated in (Fig. 1(b)) and voltage across any two points can be obtained using switching function:

$$\left.\begin{array}{l} V_{EA} = (S_1 - 1)V \\ V_{AB} = (1 - S_2)(V - V_k) \\ V_{BF} = (1 - S_3)V_k \end{array}\right\} \tag{7}$$

Where V and V_k are the main DC link voltage and capacitor voltage, respectively. By substituting the Eq. (7) into Eq. (6), we get (Fig. 2)

$$\begin{aligned} V_{EF} &= (S_1 - 1)V + (1 - S_2)(V - V_k) + (1 - S_3)V_k \\ &= VS_1 - V_k + V - V_k - VS_2 + V_kS_2 + V_k - V_kS_3 \\ &= (S_1 - S_2)V + (S_2 - S_3)V_k \end{aligned} \tag{8}$$

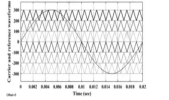 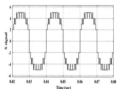 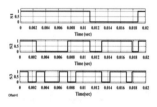

Fig. 2. IPD PWM scheme for 7-level   **Fig. 3.** S waveform obtained   **Fig. 4.** Switching pulses for 7-level

Since one of the switches in pairs of S_1 & S_4, S_2 & S_5, and S_3 & S_6 are turned ON, the current through switches is defined as the function of load current.

$$\left.\begin{array}{l} i_{AE} = S_1 i_{load} \\ i_{BA} = S_2 i_{load} \\ i_{FB} = S_3 i_{load} \end{array}\right\} \tag{9}$$

Applying KCL at node 'B.'

$$i_{FB} = i_{BD} + i_{BA} \tag{10}$$

and

$$i_{BD} = (S_3 - S_2)i_{load} \tag{11}$$

$$C_1 \frac{dV_k}{dt} = (S_3 - S_2) i_{load} \tag{12}$$

Rewriting the Eq. (12) as,

$$\frac{dV_k}{dt} = \frac{(S_3 - S_2) i_{load}}{C_1} - \frac{i_{load}}{C_1} S_2 + \frac{i_{load}}{C_1} S_3 \tag{13}$$

Calculating the Eq. (15) with the help of the Eq. (7), the following relation will be obtained:

$$\frac{di_{load}}{dt} = \frac{V}{L} S_1 + \frac{V - V_k}{L} S_2 - \frac{V_k}{L} S_3 - \frac{R}{L} i_{load} - \frac{V_{load}}{L} \tag{15}$$

Comparing the relations (15) and (16), the state space model can be obtained easily

$$\frac{dX}{dt} = A(x,t) + B(x,t)U + C(t) \tag{16}$$

By choosing the state variables as $x_1 = i_{AE}$ and $x_2 = V_k$ and using duty cycles (d_1, d_2, d_3) of switch (S_1, S_2, S_3) as the input matrix, the following state-space average model of the PUC inverter is derived [8].

$$A(x,t) = \begin{bmatrix} \frac{-R}{L} \\ 0 \end{bmatrix} \tag{17}$$

$$B(x,t) = \begin{bmatrix} \frac{-V}{L} & \frac{V-V_k}{L} & \frac{-V_k}{L} \\ 0 & \frac{-i_{load}}{C_1} & \frac{i_{load}}{C_1} \end{bmatrix} \tag{18}$$

$$u = \begin{bmatrix} d_1 \\ d_2 \\ d_3 \end{bmatrix} \quad \text{and} \quad C(t) = \begin{bmatrix} \frac{-V_{load}}{L} \\ 0 \end{bmatrix} \tag{19}$$

To regulate capacitor voltage (V_k) and obtain multilevel voltage waveform a well designed controller circuit is required that uses nonlinear mathematical state space model derived above. This state space model is the same for the five-level and seven level converter; the only difference is in the magnitude of capacitor voltage.

Again writing it in the form of switch combination as given in Eqs. (7) and (15) will give

$$\frac{di_{load}}{dt} = -\frac{V_{EF}}{L} - \frac{R}{L} i_{load} - \frac{V_{load}}{L} \tag{20}$$

and $V_{EF} = DV$, where $D = 0, \pm\frac{1}{3}, \pm\frac{2}{3}$ and ± 1 and D is the duty cycle which modulates the multicarrier PWM.

In another form, the average model can be written as

$$\left[\frac{di_{load}}{dt}\right] = \left[-\frac{R}{L}i_{load} - \frac{V_{load}}{L}\right] + \left[\frac{V}{L}\right][D] \tag{21}$$

4   Control Techniques

Different techniques have been applied on PUC MLI to control the voltage across the capacitor present in the dc link. In this paper, hysteresis current control method, proportional integral control, and proportional resonant control have been discussed.

4.1   Hysteresis Current Control Method

PUC MLI operation was introduced in [7] and in [15] having both the dc links as dc supply. The inverter operation with the help of simple switching techniques without any modulation methods appeared in [8] as shown in Fig. 4 for 7-level operation.

As it has been emphasized that PUC inverter is the optimization of H-bridge inverter and Flying capacitor clamped inverter (FCC) so to get the capacitor voltage maintained constant at $V_k = 1/2\,V$ or $V_k = 1/3\,V$ for five and seven level operation respectively, the need for capacitor voltage balance arrived. For this, a six-band hysteresis control was successfully implemented in [16] at the simulation level, and in [17], its experimental validation was performed. The mathematical formulation is as follows:

$$V_{EF} = k_{p1}\tilde{i}_{load} + k_{i1}\int \tilde{i}_{load}dt \tag{22}$$

Where $\tilde{i}_{load} = i^*_{load} - i_{load}$ and $\tilde{i}^*_{load}$ is the reference load current. For the capacitor voltage balance

$$i^*_{load} = k_{p2}\tilde{V}_k + k_{i2}\int \tilde{V}_k dt \tag{23}$$

Where $\tilde{V}_k = V^*_k - V_k$ and V^*_k is the reference voltage for the capacitor. Figure 5 shows the 7-level PUC controller using six band hysteresis current controller. The controller works in the following way, when the load current goes below the reference, then the positive voltages (sector 1 in Fig. 5) are applied this causes the rapprochement of the load current, and when the load current goes above the reference, then the negative voltages (sector 2) are applied. An important feature of the hysteresis control is that a sporadic switching frequency is obtained, unlike other methods that give a constant switching frequency of operation. A multiband hysteresis current controller for 7-level PUC-NPC is was discussed and presented in [18] and was extended to fourteen bands and a fifteen level PUC MLI in [19].

4.2 PI Control

The first simulation and experimental validation using a modulation scheme and PI control were given in [11]. Figure 7 shows the control logic. A mathematical formulation as in [20]

$$V_{EF} = i_{load}R + L_f \frac{di_{load}}{dt} + V_{load} \qquad (24)$$

Or,

$$\frac{di_{load}}{dt} = -\frac{R}{L}i_{load} - \frac{V_{EF}}{L} - \frac{V_{load}}{L} \qquad (25)$$

Since $3V_k = V$,
V_{EF} can be written as

$$V_{EF} = DV, D = 0, \pm\frac{1}{3}, \pm\frac{2}{3}, \pm 1 \qquad (26)$$

A 7-level output voltage is produced when the modulation is carried out using Eq. (26) where D is the duty cycle. Eqs. (27), (28) shows the average model as (Fig. 6):

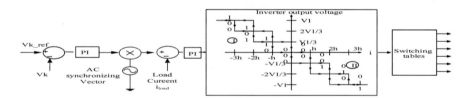

Fig. 5. Six band hysteresis controller of packed u-cell 7-level

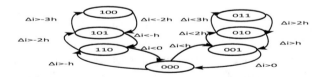

Fig. 6. Flow chart of the six band hysteresis current controller

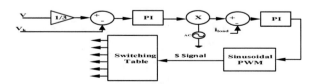

Fig. 7. PI Control logic for capacitor voltage balance control

$$\left[\frac{di_{\text{load}}}{dt}\right] = \left[-\frac{R}{L} \cdot i_{\text{load}} - \frac{V_{\text{load}}}{L}\right] + \left[\frac{V}{L}\right][D] \tag{27}$$

and

$$[D] = \left[\frac{V}{L}\right]^{-1}\left(F + \left[\frac{R}{L}\right][i_{\text{load}}] + \left[\frac{V_{\text{load}}}{L}\right]\right) \tag{28}$$

Where F is the Linearized signal. F can be obtained as

$$F = \left(i_{\text{load}} - \left(V_k - \frac{V}{3}\right)\left(k_{p1} + \frac{k_{i1}}{s}\right) \cdot \sin \omega t\right) \times \left(k_{p2} + \frac{k_{i2}}{s} + sk_{d2}\right) \tag{29}$$

4.3 Proportional Resonant (PR) Control

PR control in PUC 5-level was introduced in [21]. The advantages include good sinusoidal tracking when used with grid connection. The equation given below is the general form of the PR transfer function

$$G_{\text{PR}}(s) = K_p + \frac{K_1 s}{s^2 + \omega_o^2} \tag{30}$$

But at resonance PR control offers infinite gain which may result into large tracking for even a small change in reference frequency signal, so the equations have been modified as follows called quasi PR controller [22].

$$G_{\text{PR}}(s) = K_p + \frac{K_1 \omega_c s}{s^2 + 2\omega_c s + \omega_o^2} \tag{31}$$

$$G_F(s) = \frac{I_i(s)}{V_i(s)} = \frac{1}{L_i s} \frac{\left(s^2 + \frac{1}{L_g C_f}\right)}{\left(s^2 + \frac{(L_i + L_g)}{L_i L_g C_f}\right)} \tag{32}$$

Where ω_c is the cutoff frequency, ω_o is the angular frequency of the output signal, K_p is the proportional gain and K_1 is the integral gain constant. Due to the modification in the Eq. (30), sensitivity reduces. The control block diagram is shown in Fig. 8. The value of P is the magnitude of the sinusoidal reference signal. It will be four for five level and six for the seven-level operations.

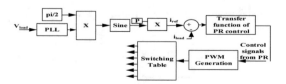

Fig. 8. PR control for PUC

5   Conclusion

This paper presents an overview of the control techniques that are implemented on PUC for the control of its capacitor voltage to generate seven levels in the output voltage waveform. But the PUC MLI which is an area of hot research has been successfully implemented and tested in the renewable energy system, as a static compensator, electric drives operation, incorporation of PUC in Modular Multilevel Converter, active buck-boost rectifier operation. Regarding various parameters like high power, better power quality, high reliability, and high-efficiency PUC topology are far better than all other reduced device count topologies. Continuous research on PUC is going on to extend it to three phase and its application in different areas.

References

1. Nabae, A., Takahashi, I., Akagi, H.: A new neutral-point-clamped PWM inverter. IEEE Trans. Ind. Appl. IA-**17**(5), 518–523 (1981)
2. Meynard, T.A., Foch, H.: Multi-level conversion: high voltage choppers and voltage-source inverters. In: PESC 92 Record, 23rd Annual IEEE Power Electronics Specialists Conference, pp. 397–403 (1992)
3. Peng, F.Z., Lai, J.-S., McKeever, J.W., Van Coevering, J.: A multilevel voltage-source inverter with separate DC sources for static VAr generation. IEEE Trans. Ind. Appl. **32**(5), 1130–1138 (1996)
4. Gupta, K.K., Ranjan, A., Bhatnagar, P., Sahu, L.K., Jain, S.: Multilevel inverter topologies with reduced device count: a review. IEEE Trans. Power Electron. **31**(1), 135–151 (2016)
5. Siddique, M.D., et al.: Single phase symmetrical and asymmetrical design of multilevel inverter topology with reduced number of switches. In: 2018 IEEMA Engineer Infinite Conference (eTechNxT), pp. 1–6 (2018)
6. Sarwar, A., Sarwar, M.I., Alam, M.S., Ahmad, S., Tariq, M.: A nine- level cascaded multilevel inverter with reduced count and lower harmoncs. In: Lecture Notes Electrical Engineering. Springer (2019, in press)
7. Ounejjar, Y., Al-Haddad, K.: A novel high energetic efficiency multilevel topology with reduced impact on supply network. In: 2008 34th Annual Conference of IEEE Industrial Electronics, pp. 489–494 (2008)
8. Ounejjar, Y., Al-Haddad, K.: A new high power efficiency cascaded U cells multilevel converter. In: 2009 IEEE International Symposium on Industrial Electronics, ISIE, pp. 483–488 (2009)
9. Siddique, M.D., Sarwar, A.: Performance analysis of carrier based PWM technique for three level diode clamped multilevel inverter with different reference signals. In: 2016 IEEE 7th Power India International Conference (PIICON), vol. 6, pp. 1–6 (2016)
10. Sarwar, A., Asghar, M.S.J.: Simulation and analysis of a multilevel converter topology for solar PV based grid connected inverter. Smart Grid Renew. Energy **2**, 56–62 (2011)
11. Vahedi, H., Kanaan, H.Y., Al-Haddad, K.: PUC converter review: topology, control and applications. In: Annual Conference of the IEEE Industrial Electronics Society, pp. 4334–4339 (2015)
12. Ounejjar, Y., Al-Haddad, K., Gregoire, L.-A.: Packed U cells multilevel converter topology: theoretical study and experimental validation. IEEE Trans. Ind. Electron. **58**(4), 1294–1306 (2011)

13. Tariq, M., Iqbal, M.T., Iqbal, A., Meraj, M., Roomi, M.M., Khan, M.S.U.: Comparative analysis of carrier schemes for PWM in multilevel PUC inverter for PV applications. In: 2016 4th International Conference on the Development in the in Renewable Energy Technology (ICDRET), pp. 1–6 (2016)
14. Onizuka, M.Y.V., Garcia, R.C., Pinto, J.O.P., da Silva, L.E.B.: Control of a 7-levels PUC based three phase inverter through vector current control and hybrid modulation. In: IECON 2016 - 42nd Annual Conference of the IEEE Industrial Electronics Society, pp. 6488–6493 (2016)
15. Ounejjar, Y., Al-Haddad, K.: Multilevel hysteresis controller of the novel seven-level packed U cells converter. In: SPEEDAM 2010, pp. 186–191 (2010)
16. Ounejjar, Y., Al-Haddad, K.: A novel six-band hysteresis control of the packed U cells seven-level converter. In: 2010 IEEE International Symposium on Industrial Electronics, pp. 3199–3204 (2010)
17. Ounejjar, Y., Al-Haddad, K., Dessaint, L.A.: A novel six-band hysteresis control for the packed u cells seven-level converter: experimental validation. IEEE Trans. Ind. Electron. **59**(10), 3808–3816 (2012)
18. Ounejjar, Y., Al-Haddad, K.: Multiband hysteresis controller of the novel three phase seven-level PUC-NPC converter. In: IECON Proceedings Industrial Electronics Conference, pp. 6257–6262 (2013)
19. Ounejjar, Y., Al-Haddad, K.: Fourteen-band hysteresis controller of the fifteen-level packed U cells converter. In: IECON 2010 - 36th Annual Conference on IEEE Industrial Electronics Society, pp. 475–480 (2010)
20. Vahedi, H., Al-Haddad, K., Kanaan, H.Y.: A new voltage balancing controller applied on 7-level PUC inverter. In: IECON 2014 - 40th Annual Conference of the IEEE Industrial Electronics Society, pp. 5082–5087 (2014)
21. Tariq, M., Iqbal, M.T., Meraj, M., Iqbal, A., Maswood, A.I. Bharatiraja, C.: Design of a proportional resonant controller for packed U cell 5 level inverter for grid-connected applications. In: 2016 IEEE International Conference on Power Electronics, Drives and Energy Systems (PEDES), vol. 2016, pp. 1–6 (2016)
22. Alves Pereira, L.F., Sanfelice Bazanella, A.: Tuning rules for proportional resonant controllers. IEEE Trans. Control Syst. Technol. **23**(5), 2010–2017 (2015)

Machine Learning Based User Interface Generation

Mihir Mistry, Ameya Apte, Varad Ghodake[✉], and S. B. Mane

Department of Computer and IT, College of Engineering, Pune, India
{mistrymu15.comp, apteaa15.comp, ghodakevv15.comp, sunilbmane.comp}@coep.ac.in

Abstract. Every product development starts with idea conceptualization and design. The ideators and designers first choice is sketching the idea to conceptualize the product. After a few iterations of product design, mockups/prototypes are passed to developers who have the job of converting these designs into a functional product by grasping the design concepts and using their coding skills. Utilizing the advancements in deep learning and computer vision technologies, gap between designers and developers is bridged by simplifying the product development process. This will also help people with little or no knowledge of web development to bring their design ideas into reality. The service will consist of two stages firstly accepting images of web mockups including hand-drawn sketches and identifying the various HTML components in the sketch using Convolutional Neural Networks and tag them appropriately, secondly, the identified components will be converted to structured JSON files which will be converted to code based on HTML5, Flexbox and Bootstrap.

Keywords: Deep learning · Computer vision · ROI detection · OCR · Front-end development · openCV · UI · Automated software development · Document layout

1 Introduction

Computer vision is a field of computer science that works on enabling computers to see, identify and process images in the same way that human vision does, and then provide appropriate output. It is like imparting human intelligence and instincts to a computer. In reality though, it is a difficult task to enable computers to recognize images of different objects. Computer vision is closely linked with artificial intelligence, as the computer must interpret what it sees, and then perform appropriate analysis or act accordingly. With the help of computer vision and deep learning the process of website development can be aided to make it more robust, flexible and fast.

© Springer Nature Switzerland AG 2020
A. P. Pandian et al. (Eds.): ICICCS 2019, AISC 1039, pp. 453–460, 2020.
https://doi.org/10.1007/978-3-030-30465-2_50

2   Existing System

User Interface Design process involves a lot a creativity that starts on a whiteboard where designers share ideas. Once a design is drawn, and after iterations among the design team, a design mockup is ready. Now, these mockups are passed on to the Development Team. The developers work on to convert these mockups into actual functional units. This is again verified by the design team and a final product is ready after many iterations. Work stops whenever one team finishes a portion of the project and passes responsibility to another team for feedback. This process takes several iterations and consumes a lot of effort and time. There is a big gap in how the ideas are exchanged between designers and developers. There are ways in which the designer designs an expected view with various drawing tools like Photoshop and CorelDraw. Certain modern products try and bridge the gap between the design and code by providing drag and drop user interfaces and exportable designs but in spite of such tools there is no service which allows conversion of simple design sketches to code. Developers still have to mimic the designs manually. The outputs of most of these drawing tools is of meagre help in terms of implementation process for the developer. There is a very striking gap in the process of designing and implementation and automation can greatly improve this process and have a great impact on the Web and Mobile app development.

3   Proposed Solution

Computer Vision is a discipline inside artificial intelligence that gives an application the capability to see and understand what it is seeing. Using Optical Character Recognition (OCR) the handwritten characters have be interpreted. Further leveraging the Convolution Neural Networks and its specific improvements like RCNN and YOLO algorithms, the Region of Interest (ROI) can be detected from an image. Further image processing can help to get the structure and positioning attributes of the various components detected.

The above can be combined with a robust Rule based Engine with a well defined set of rules observed after analysing many designs. This can work in hand with the machine learning models developed above and help in generating functional code of the detected structure.

We can use this technology to build a system that understands what a designer has drawn on a whiteboard and can translate that understanding to HTML code. This way we can generate HTML wireframes directly from a hand-drawn image giving an instant working design implementation thus streamlining the design process.

The steps to be performed by the user will be:

1. Draw a sketch of design on the whiteboard
2. Click an image of sketch
3. Upload it to our service
4. Get the functional code on which developer can further work

4   Overview of the Architecture

See Fig. 1.

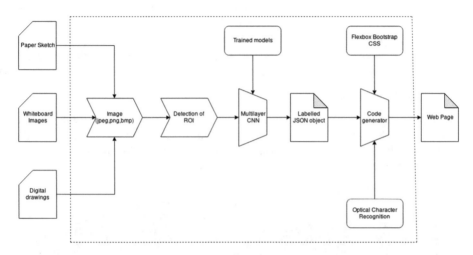

Fig. 1. Overview Architecture

5   Vision Model

An integral part of our solution is to detect the regions containing the layout elements from the input image the user provides. The difference between object detection algorithms and classification algorithms is that in detection algorithms, we try to draw a bounding box around the object of interest to locate it within the image. This problem cannot be solved by building a standard convolutional network followed by a fully connected layer. The reason being that the output layer here is not constant, because the image can have variable number of occurence of objects in an image. Therefore, algorithms like R-CNN [8], Faster-RCNN [9] and YOLO [10] are helpful in solving the problem of finding and classifying design layout elements [1].

With the experiments performed using these complex algorithms it was observed that more training data was required. Augmentation to dataset was done by adding variants like binarized images, changing the contrasts ratios of the already collected images. Creating more hand-drawn design layout was an option to train such complex models.

The results conspicuously hinted towards another approach by finding the Regions and Interests first, rather than Object Detection over the whole image, and then using modified deep learning algorithms to identify the DOM elements.

5.1 Detection of ROIs

The image captured by the user consists mainly of the white background of the paper and the black/blue sketch drawn on it. Before the various elements of the sketch can be detected it is necessary to figure out multiple regions of interest which can be then passed to a neural network for identification based on a confidence value. The above mentioned property of the input images can be extremely helpful for preprocessing the images. We have used the python interface for OpenCV for processing the images [6] and generating multiple bounding rectangle around the region of interest.

i. Grayscale conversion

The input image is a full color image, as the elements of interest have a binary color schema a simple linear transform is used to convert the RGB images to Grayscale (0–255). This helps reduce the dimensionality of the input image with almost no loss as all the information of the sketch is captured in the grayscale image (Fig. 2).

$$\text{RGB to Gray: } Y = 0.299 \cdot R + 0.587 \cdot G + 0.114 \cdot B$$

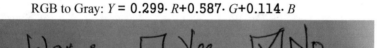

Fig. 2. Grayscale conversion

ii. Adaptive thresholding

Further the grayscale image can be reduced to binary image, this helps enhance the strokes of the characters and the figures drawn by the user and eliminates all the fine noise [5] in and around the image. A mean value of pixels in a kernel are calculated this is the threshold value of the kernel, a constant value is subtracted from this and the result is compared with the cutoff number, this generated a binary output which is used to binarize the image. The dimensions of the kernel are 7×7 and a very low cutoff value ~ 10 is used as the major component of the image is close to white i.e. the color of the paper. An inverted binary output is generated (Fig. 3).

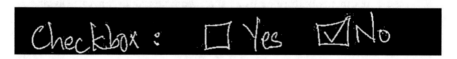

Fig. 3. Inverted Binary Output

iii. Dilation

The image is further dilated iteratively to group the nearby elements in order to avoid fragmented region detection. Dilation is a morphological operation which reduces noise and helps join disparate elements in the image. The primary goal of this step is to merge the nearby contours and hence generate bigger, less fragmented regions. A rectangular kernel is used for dilation as the natural structure of most sketches generated by humans are horizontal in nature. Eg. Words, Letters, Textboxes. Based on the size and the shape of the kernel we can generate regions which are horizontal or vertical merging of the fragmented contours. The iteration count, dimensions and shape of the kernel can be varied to generate different regions of interest (Fig. 4).

Fig. 4. Dilation

iv. Finding bounding rectangles

Multiple contours are detected in the dilated images. The contours are filtered based on the area and the dimensions of the contour to weed out possible regions which are very small or thin compared to the dimensions of the image. Bounding rectangles are created around the contours in order to mark the regions of interest. These bounding boxes can now be used to crop the regions of interest which can be passed to the Convolutional Neural Network for labelling. The predictions with the highest confidence score can be used for generating codes (Fig. 5).

Fig. 5. Bounding Rectangles Observation

5.2 Classification of ROIs

Convolutional Neural Networks are widely being used for image classification problems. We used CNN to learn a model by mapping input images to fixed size vector representing various DOM elements like textbox, radiobutton, checkboxes, labels etc. The Regions of Interests detected in the above step were needed to be classified. Some falsely identified regions needed to be discarded (Fig. 6a, b).

The above detected regions were resized into 150 × 150 images (not maintaining the aspect ratios) and were made grayscale as a part of preprocessing. The model consisted of two convolution layer and one fully connected layer as the dataset was simpler to learn. The model gave a validation accuracy of around 84% (due to confusion between radio buttons, check boxes, labels and headings). There is still scope to increase the accuracy of this model by further experimentation. This model also helped to discard the falsely identified regions as a confidence threshold was set for the classification.

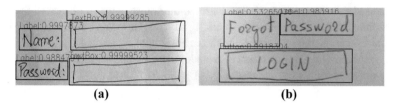

Fig. 6. a) & b): RoI classification**b**.

5.3   Handwriting Recognition

There are many cases where the text is written inside the design components. Some examples are hints in textbox, labels, checkboxes, paragraphs. These texts will be handwritten and need to be recognised accurately. Using various already available solutions on Optical Character Recognition like Tesseract [11] a robust model of text recognition was built.

Each detected design component from the above step was passed to this Text Recognition model to extract the handwritten content. This text will be then associated with the detected component for further steps.

6   Markup Generator

The results of the Neural networks are stored in a structured JSON file. Optical character recognition is performed for components containing text. The JSON object generated is a generic object which can be used for generating web elements based on different technologies. The Web engine converts the JSON formatted document layout file into Markup Language Code by detecting the DOM structure based on the coordinates of the regions and their height/width. To represent document layout in JSON, relative position of the element as well as its dimensions are calculated.

The format of an example DOM element in the JSON file is the following:

```
{
    "id":                           "b5e7d988cfdb78bc3be1a9c221a8f744",
    "type": "textbox",
    "height":"100",
    "width":"100",
    "x":"500",
    "y":"400",
    "value":"Type                                            here"
    "parent":                       "60c0b53095f81a7bf551b30c93fd20dd",
    "child":                                                    ""
}
```

Modern Browsers support 'flex-container' CSS class. This class has some useful built-in features to make web development easier. Vertical and horizontal stacking of elements can be done by just toggling the value of 'flex-direction' attribute of this class.

Besides all the dimension and type related information about the nodes, document metadata also stored in the JSON (e.g. Height, Width). Dependant information is calculated geometrically. Orientation and layout [7] of the tags of the HTML5 [12] document is decided from document height and width values available in the metadata and corresponding changes are reflected in the flexbox orientation. Web engine converts this document layout JSON into code.

7   Conclusions

The implemented solution provides an innovative and efficient way of designing front-end interfaces, helping to speed up development cycles in complex software development projects. Our solution leverages fundamental principles of computer vision and greedy approach to obtain better results. Thus the system can be a reliable and easy method to generate codes based on simple sketches by designers. Our solution is easily extensible as the JSON files describing the content of the sketches can be transformed to codes targeting variety of platforms. The results provided by the RCNN [8], Faster-RCNN [9] and YOLO [10] networks are promising but require an even larger datasets for improved results. Pre processing of images improved the detection of ROIs by openCV methods increased accuracy by a great extent for the dataset at hand compared with complex Object Detection Algorithms which are hard to train. A lot of this solution will be a step towards automated software development.

Generative Adversarial Networks GANs [3] have shown to be extremely powerful at generating images and sequences [3, 4]. Applying such techniques to the problem of generating computer code from an input image is so far an unexplored research area [1]. GANs could potentially be used as a standalone method to generate code or could be used in combination with our solution model to fine-tune results.

References

1. Beltramelli, T.: pix2code: Generating Code from a Graphical User Interface Screenshot. https://arxiv.org/pdf/1705.07962.pdf
2. Balog, M., Gaunt, A.L., Brockschmidt, M., Nowozin, S., Tarlow, D.: DeepCoder: learning to write programs. arXiv preprint arXiv:1611.01989 (2016)
3. Goodfellow, I., Pouget-Abadie, J., Mirza, M., Xu, B., Warde-Farley, D., Ozair, S., Courville, A., Bengio, Y.: Generative adversarial nets. In: Advances in Neural Information Processing Systems, pp. 2672–2680 (2014)
4. Reed, S., Akata, Z., Yan, X., Logeswaran, L., Schiele, B., Lee, H.: Generative adversarial text to image synthesis. In: Proceedings of the 33rd International Conference on Machine Learning, vol. 3 (2016)
5. Koike, S.: Adaptive threshold nonlinear algorithm for adaptive filters with robustness against impulse noise. IEEE Trans. Signal Process. **45**(9), 2391–2395 (1997)

6. Gurav, R.M. Kadbe, P.: Gesture recognition using OpenCV. In: 2015 International Conference on Industrial Instrumentation and Control (2015)
7. Bajwa, I.S., Siddique, I., Abbas Choudhary, M.: Web layout mining (WLM): a new paradigm for intelligent web layout design. In: 2006 ITI 4th International Conference on Information & Communications Technology (2006)
8. Rich feature hierarchies for accurate object detection and semantic segmentation. https://arxiv.org/abs/1311.2524
9. Faster R-CNN: towards real-time object detection with region proposal networks. https://arxiv.org/abs/1506.01497
10. You Only Look Once: Unified, Real-Time Object Detection. https://arxiv.org/abs/1506.02640
11. An Overview of the Tesseract OCR Engine. https://research.google.com/pubs/archive/33418.pdf
12. Rajesh, C.H.: Research on HTML5 in web development. Int. J. Comput. Sci. Inf. Technol. **5**(2) (2014)

Prediction Equations in Spirometry Using Demographic and Spirometric Values

Akash Patil[1](✉), Safna Hassan[1], Tejas Nayak[1], Vahida Attar[1],
Gajanan Sakhare[2], and Shardul Joshi[2]

[1] Department of Computer Engineering and Information Technology,
College of Engineering, Pune, India
akashmpatil111@gmail.com, safnahassan16@gmail.com,
tejasunayak@gmail.com, vahida.comp@coep.ac.in
[2] Research and Development Department, BRIOTA Technologies Private
Limited, Shivajinagar, Pune, India
gsakhare@gmail.com, shardul@briota.co

Abstract. Major research in prediction equations for spirometry in adults take only demographic data i.e. gender, age, height and weight into its consideration. Furthermore, the studies conducted in India on the prediction equations for spirometry in adults are inadequate and old, to make their validity unreliable. We have developed prediction equations for spirometry in the adults of Pune in Maharashtra, India by considering their measured spirometric data in addition to the demographic data. For doing so, a dataset composed of 2092 healthy subjects from Pune, who underwent spirometry tests and whose results were recorded in the database was used. Linear, Quadratic and Logarithmic prediction equations were developed using the dataset for forced vital capacity, and 13 other parameters, and the accuracy was compared with one another as well as the existent equations. The results have denoted a significant increase in accuracy when compared to previous studies. The correlation between spirometric parameters is convincing to take into account for developing more specific & accurate predictions.

Keywords: Data mining · Data science · Regression · Prediction equations ·
Accuracy · Pulmonary disease · Asthma · Spirometry

1 Introduction

Asthma and chronic obstructive pulmonary disease (COPD) have attracted research interest as a major public health problem of increasing concern to healthcare systems worldwide because of high prevalence and rising socioeconomic burden, especially in India [1–6].

According to the latest WHO estimates, released in December 2016, there were 3,83,000 deaths due to asthma in 2015 [7]. Asthma is one of the major noncommunicable diseases. It is a chronic disease of the air passages of the lungs which inflames and narrows them. Some 235 million people currently suffer from asthma. It is a common disease among children.

A. Patil, S. Hassan, T. Nayak, V. Attar, G. Sakhare and S. Joshi—Equal contribution.

© Springer Nature Switzerland AG 2020
A. P. Pandian et al. (Eds.): ICICCS 2019, AISC 1039, pp. 461–469, 2020.
https://doi.org/10.1007/978-3-030-30465-2_51

Spirometry [8, 9] is a highly informative and by far the most commonly performed investigation to evaluate pulmonary function in patients with asthma. The technical aspects of the equipment and test performance require a very meticulous attention to quality control and these have been well-standardised and revised from time-to-time by ATS/ERS.

1.1   Literature Study - Regression to Predict

Pulmonary function in healthcare is affected by an individual's ethnicity, gender, age and stature. For interpretation of the spirometric tests, a comparison is made with the values from the test results, and the expected values. These are called the "predicted" values and are developed by regression analysis of data collected from non-smoking and healthy individuals of the same population. Several prediction equations for spirometry parameters have been developed over the last few decades in ethnically diverse populations.

A few studies from different parts of India have reported prediction equations for spirometry over the last few years [1–6]. These studies have been carried out on more general and diverse population from Western India rather than specific populations, whose demographic details may vary from person to person.

Most of the studies that have been carried out for developing predicted values of spirometry have used linear regression analysis only and not logarithmic, quadratic etc.

Studies and surveys conducted for Indian population, moreover, deals with only three demographic parameters - age, height and weight - for each gender classification to predict a spirometric parameter.

1.2   Parameters Under Consideration

To develop and validate prediction equations for spirometry in adults of Pune, India, the following parameters were taken into consideration: demographic parameters - age, height, weight, gender, smoker/non-smoker and ethnicity; and measured spirometric parameters - FEV1[1], FEV3[2], FVC[3], FEV1/FVC[4], PEF[5], FEF25[6], FEF50[7], FEF75[8], FEF2575[9], FIVC[10], PIF[11], FIF25[12], FIF50[13], FIF75[14].

---

[1] FEV1 - Forced Expiratory Volume in the first second.

[2] FEV3 - Forced Expiratory Volume in the third second.

[3] FVC - Forced Vital Capacity.

[4] FEV1/FVC - Ratio of FEV1 and FVC.

[5] PEF - Peak Expiratory Flow.

[6] FEF25 - Forced Expiratory Flow with 25% interval.

[7] FEF50 - Forced Expiratory Flow with 50% interval.

[8] FEF75 - Forced Expiratory Flow with 75% interval.

[9] FEF2575 - Mean Forced Expiratory Flow in 25–75%.

[10] FIVC - Forced Inspiratory Vital Capacity.

[11] PIF - Peak Inspiratory Flow rate.

[12] FIF25 - Forced Inspiratory Flow with 25% interval.

[13] FIF50 - Forced Inspiratory Flow with 50% interval.

[14] FIF75 - Forced Inspiratory Flow with 75% interval.

2   Methods

2.1   Forming the Dataset

Subjects were drawn from spirometry dataset available from Private Hospitals from the city of Pune in Maharashtra collected over the last few years. They were chosen using criteria of normal spirometry result with non smoker & healthy BMI index criteria with no respiratory disease for last 3 years from the date of spirometry test We targeted the below age distribution (Table 1):

Table 1. Age distribution of the subjects

Age (in years)	Males (n = 1306)	Females (n = 786)
18 to 25	317	149
26 to 45	505	345
46 to 70	361	234
>70	123	58

The training datasets of males and females had 1044 and 628 subjects, respectively, while the test datasets had 262 and 158 subjects, respectively.

2.2   Sampling of Dataset

To confirm subject's test result as Healthy Person over any respiratory disease, GOLD/Hardie [11, 12] Interpretation is used for each patient record. If the maneuver quality is approved, the percentage of FEV1 that is constituted by FEV1/FVC is evaluated. This value is set according to the age of the subject:

- age < 70 years: 0.7
- 70 <= age < 80: 0.65
- age >= 80 years: 0.6

If the FEV1/FVC value is less than the FEV1 percentage limit, the potentiality of asthma is confirmed. Else, FVC and FEV1 values are compared with the respective Lower Limits of Normal values. If both the datapoint values are less than the Lower Limit of Normal values, the subject is treated as healthy [10].

2.3   Regression Analysis

To form prediction equations regression has been used. The independent variables (demographic and measured spirometric parameters) having a correlation factor > 0.5 (for the particular spirometric parameter) were entered, including quadratic terms and logarithmic terms (both separately), if it showed significant improvement in the regression analysis.

3   Statistical Analysis

The sampled dataset was used for performing the regression analysis. The analysis was done separately for the male and female subjects.

For each dependent variable, Pearson Correlation Analysis was performed to identify the significant independent variables, which included the demographic as well as the measured spirometric data values. The variables with the correlation coefficient greater than 0.5, i.e. those related with a moderate or a strong positive/negative linear relationship were considered for the regression analysis of the dependent variable.

The dataset was randomly apportioned into training (80%) and testing (20%) datasets. The training dataset was used to form the equations, which were later analysed on the testing dataset.

The equations were developed considering the linear as well as non-linear regression procedures. For the non-linear analysis, the dependent variables were taken in quadratic and logarithmic form. The accuracy for each model was calculated. If the accuracy increased from the linear model to the quadratic model or the quadratic model to the logarithmic model, the newer model was considered and vice versa. Other factors considered for choosing the model were complexity and clinical ease of application.

The analysis was repeated for the dataset in which the outliers were excluded. The results thus obtained were compared to the previous results giving an estimation of the fit of the model. If the change was insignificant, the model was accepted.

Moreover the equations obtained from the training dataset were applied on the testing dataset and thus obtaining the predicted values. These values were compared with the measured spirometric values to obtain the accuracy of the model.

To compare our model with the research done earlier in this domain, we referred to a previous paper [1] as this is the only model developed for the Western Indian Population. The accuracy of the models was compared by applying the respective prediction equations on the same dataset. As the former model considered only the demographic parameters in the equations, comparison of the actual equations was not possible.

4   Results

Approximately 2280 subjects were used for forming the train and test data. Out of 2800 subjects, 2092 were found eligible & were chosen as acceptable maneuvers shown by conventional spirometry device.

Results were obtained using methodologies discussed in the section above & compared within, for the most useful data. Final equations were chosen based on the following statistician criteria.

- Correlation >0.5
- Lowest Mean Squared Error

Expert's opinion was taken into consideration while selecting the best correlation for dependent spirometric parameters in between them. It was used to discard the least correlated parameters in the pulmonary function mechanism.

4.1   Comparison of Correlation of Dependent Parameters with Spirometric and Demographic Parameters

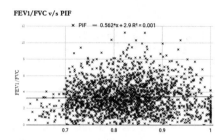

Fig. 1. Scatter plot of FEV1/FVC v/s PIF

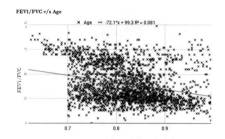

Fig. 2. Scatter plot of FEV1/FVC v/s Age

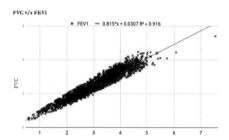

Fig. 3. Scatter plot of FVC v/s FEV1

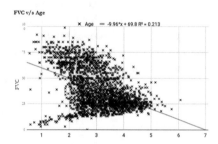

Fig. 4. Scatter plot of FVC v/s Age

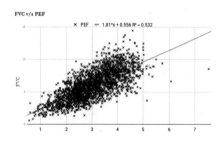

Fig. 5. Scatter plot of FVC v/s PEF

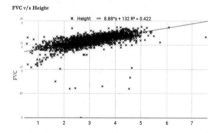

Fig. 6. Scatter plot of FVC v/s Height

The graphs in Figs. 1, 2, 3, 4, 5, 6, 7 and 8 represents a comparative study of scatter plots of a dependent parameter against a spirometric and a demographic parameter. Each scatter plot shows the correlational relationship between two variables and the deviation of actual points from the trend line.

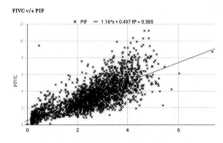

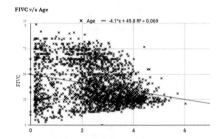

Fig. 7. Scatter plot of FIVC v/s PIF   **Fig. 8.** Scatter plot of FIVC v/s Age

For example, Fig. 3 represents a scatter plot of the dependent variable FVC with the spirometric parameter FEV1 and Fig. 4 with the demographic parameter age. A comparison of the graphs suggest a better correlation and a better fit for the variable FVC with FEV1 as compared to age.

However, it is not necessary that a spirometric parameter results in a better fit compared to a demographic one, as seen in Figs. 5 and 6, with FVC v/s height being closely fit to the trend line as compared to FVC v/s PEF.

Thus, a regression model should take into account both spirometric as well as demographic parameters while forming the prediction equations.

4.2   Prediction Equations

See Tables 2 and 3.

Table 2. Regression Equations of demographic & spirometric parameters in training dataset (Male)

Dependent parameters	Equation
FEV1	0.001 * Height + 0.003 * FEV3 + 0.799 * FVC + 0.032 * FEV1/FVC + 0.032
FVC	0.003 * FEV3 − 0.039 * FEV1/FVC + 1.21 * FEV1 − 0.039
FEV1/FVC	0.018 * Age + 0.006 * Weight + 0.007 * Height − 0.121 * FEV3 − 22.779 * FVC + 27.869 * FEV1 − 22.779
FEV3	1.2022 * FVC + 0.2792 * FIVC − 0.0047 * Age − 0.6105 * FEV1 − 0.0047
PEF	0.016 * Age + 0.006 * Weight + 0.014 * Height + 0.991 * FEF25 + 0.023 * FEF50 + 0.107 * FEF75 − 0.347 * FEF2575 − 0.347
FEF25	0.004 * Age − 0.002 * Weight − 0.002 * Height + 0.231 * FEF50 − 0.481 * FEF75 + 0.532 * FEF2575 + 0.626 * PEF + 0.532
FEF50	0.004 * Age + 0.002 * Weight − 0.003 * Height + 0.07 * FEF25 − 0.507 * FEF75 + 1.332 * FEF2575 − 0.014 * PEF + 1.332
FEF75	− 0.003 * Weight + 0.003 * Height − 0.109 * FEF25 − 0.252 * FEF50 + 0.949 * FEF2575 + 0.004 * PEF + 0.949
FEF2575	0.005 * Age + 0.0 * Weight + 0.0 * Height + 0.08 * FEF25 + 0.47 * FEF50 + 0.603 * FEF75 − 0.017 * PEF + 0.603
FIVC	0.009 * Age + 0.001 * Weight + 0.013 * Height + 0.431 * PIF + 0.169 * FIF25 − 0.138 * FIF50 − 0.037 * FIF75 − 0.037

(continued)

Table 2. (*continued*)

Dependent parameters	Equation
PIF	0.001 * Height + 0.4 * FIF25 + 0.459 * FIF50 + 0.173 * FIF75 + 0.071 * FIVC + 0.173
FIF25	0.002 * Age + 0.0 * Weight + 0.001 * Height + 0.949 * PIF + 0.09 * FIF50 − 0.252 * FIF75 + 0.062 * FIVC − 0.252
FIF50	0.001 * Age + 0.0 * Weight − 0.004 * Height + 0.674 * PIF + 0.048 * FIF25 + 0.331 * FIF75 − 0.026 * FIVC + 0.331
FIF75	0.003 * Age + 0.003 * Weight + 0.001 * Height + 0.545 * PIF − 0.352 * FIF25 + 0.616 * FIF50 − 0.03 * FIVC + 0.616

Table 3. Regression Equations of demographic & spirometric parameters in training dataset (Female)

Dependent Parameters	Equation
FEV1	0.002 * FEV3 + 0.833 * FVC + 0.022 * FEV1/FVC + 0.022
FVC	0.001 * Age + 0.0 * Weight + 0.001 * Height − 0.003 * FEV3 − 0.026 * FEV1/FVC + 1.162 * FEV1 − 0.026
FEV1/FVC	0.017 * Age + 0.011 * Weight + 0.023 * Height − 0.14 * FEV3 − 34.473 * FVC + 40.866 * FEV1 − 34.473
FEV3	0.038 * Age + 0.062 * Weight − 0.043 * Height − 19.727 * FVC − 0.931 * FEV1/FVC + 26.712 * FEV1 − 0.931
PEF	0.007 * Age + 0.005 * Weight + 0.008 * Height + 1.053 * FEF25 − 0.112 * FEF50 + 0.252 * FEF75 − 0.285 * FEF2575 − 0.285
FEF25	0.003 * Age − 0.002 * Weight − 0.002 * Height + 0.334 * FEF50 − 0.332 * FEF75 + 0.246 * FEF2575 + 0.712 * PEF + 0.246
FEF50	0.002 * Age + 0.001 * Weight − 0.003 * Height + 0.141 * FEF25 − 0.25 * FEF75 + 1.107 * FEF2575 − 0.05 * PEF + 1.107
FEF75	0.001 * Age − 0.005 * Weight + 0.0 * Height − 0.18 * FEF25 − 0.184 * FEF50 + 0.883 * ГЕГ2575 + 0.056 * РЕГ + 0.883
FEF2575	0.003 * Age + 0.001 * Weight + 0.004 * Height + 0.091 * FEF25 + 0.524 * FEF50 + 0.518 * FEF75 − 0.039 * PEF + 0.518
FIVC	0.009 * Age + 0.006 * Weight + 0.013 * Height + 0.215 * PIF + 0.182 * FIF25 + 0.187 * FIF50 − 0.132 * FIF75 − 0.132
PIF	0.001 * Age − 0.001 * Height + 0.323 * FIF25 + 0.417 * FIF50 + 0.333 * FIF75 + 0.045 * FIVC + 0.333
FIF25	0.001 * Age − 0.001 * Weight 0.728 * PIF + 0.378 * FIF50 − 0.359 * FIF75 + 0.086 * FIVC − 0.359
FIF50	0.001 * Weight − 0.001 * Height + 0.51 * PIF + 0.207 * FIF25 + 0.282 * FIF75 + 0.054 * FIVC + 0.282
FIF75	0.001 * Age − 0.001 * Weight + 0.002 * Height + 0.687 * PIF − 0.299 * FIF25 + 0.426 * FIF50 − 0.071 * FIVC + 0.426

4.3   Analysis of Pearson Coefficients and Mean Squared Error of Spirometry Parameters

See Tables 4 and 5.

Table 4. Regression Analysis of demographic & spirometric parameters on training dataset (Male)

Dependent variable	R^2	Adjusted R^2	Explained variance	Mean squared error
FEV1(L)	0.994	0.994	0.994	0.002
FVC(L)	0.993	0.993	0.993	0.004
FEV1/FVC	0.962	0.961	0.962	1.903
FEV3	0.685	0.684	0.685	0.337
PEF	0.89	0.889	0.893	0.596
FEF25	0.917	0.917	0.917	0.352
FEF50	0.948	0.948	0.949	0.115
FEF75	0.929	0.928	0.929	0.048
FEF2575	0.981	0.981	0.982	0.027
FIVC	0.639	0.637	0.639	0.546
PIF	0.984	0.984	0.984	0.067
FIF25	0.95	0.95	0.95	0.172
FIF50	0.981	0.981	0.981	0.091
FIF75	0.944	0.944	0.945	0.194

Table 5. Regression Analysis of demographic & spirometric parameters in training dataset (Female)

Dependent variable	R^2	Adjusted R^2	Explained variance	Mean squared error
FEV1(L)	0.995	0.995	0.995	0.001
FVC(L)	0.994	0.994	0.994	0.002
FEV1/FVC	0.965	0.964	0.965	1.945
FEV3	0.731	0.730	0.732	0.239
PEF	0.92	0.92	0.924	0.176
FEF25	0.947	0.947	0.948	0.125
FEF50	0.944	0.943	0.945	0.058
FEF75	0.904	0.903	0.906	0.035
FEF2575	0.978	0.978	0.979	0.017
FIVC	0.692	0.69	0.692	0.223
PIF	0.981	0.981	0.983	0.027
FIF25	0.942	0.941	0.943	0.087
FIF50	0.976	0.975	0.977	0.048
FIF75	0.934	0.934	0.939	0.101

5   Conclusion

The correlation analysis done using scatter plots defines the need to include both demographic as well as spirometric parameters in the prediction equations.

The spirometric prediction equations developed for all parameters show a relatively higher Pearson Coefficient value, as compared to the equations developed using regression, taking into consideration only demographic parameters (age, height and weight) as well as the existent equations from previous studies [1–6].

The correlation between the spirometric parameters was found to be conclusive enough to consider the addition of spirometric parameters' dependencies for the prediction equations.

Acknowledgement. We would like to thank BRIOTA Technologies Private Limited for their support and taking ownership of the research work presented in paper.

References

1. Desai, U., Joshi, J., Chhabra, S.K., Rahman, M.-U.: Prediction equations for spirometry in adults in western India. Indian J. Tuberc. (2016)
2. Chhabra, S.K., Kumar, R., Gupta, U., Rahman, M., Dash, D.J.: Prediction equations for spirometry in adults from northern India. Indian J. Chest Dis. Allied Sci. **57**(3), 204 (2015)
3. Biswas, M., Pranav, P.K., Nag, P.K.: Spirometry prediction equations for north-eastern indian population. Indian J. Physiol. Pharmacol. **62**(4), 431–438 (2018)
4. Chhabra, S.K., Kumar, R., Mittal, V.: Prediction equations for spirometry for children from Northern India. Indian Pediatr. **53**, 781–785 (2016)
5. Chhabra, S.K.: Regional variations in vital capacity in adult males in India: comparison of regression equations from four regions and impact on interpretation of spirometric data. Indian J. Chest Dis. Allied Sci. **51**, 7–13 (2009)
6. Marion, M.S., Leonardson, G.R., Rhoades, E.R., Welty, T.K., Enright, P.L.: Spirometry reference values for American Indian adults, results from the strong heart study. Chest **120**, 489–495 (2001)
7. World Health Organisation. https://www.who.int/news-room/fact sheets/detail/asthma
8. Spirometry. https://en.wikipedia.org/wiki/SpirometryParameters
9. Spirometry – NHS. https://www.nhs.uk/conditions/spirometry/
10. Culver, B.H.: How should the lower limit of the normal range be defined? Respir. Care. **57**(1), 136–145 (2012)
11. Vestbo, J., Hurd, S.S., Agusti, A.G., Jones, P.W., Vogelmeier, C., Anzueto, A., Barnes, P.J., Fabbri, L.M., Martinez, F.J., Nishimura, M., et al.: Global strategy for the diagnosis, management and prevention of chronic obstructive pulmonary disease, GOLD executive summary. Am. J. Respir. Crit. Care Med. **187**, 347–365 (2013)
12. Hardie, J.A., Buist, A.S., Vollmer, W.M., Ellingsen, I., Bakke, P.S., Morkve, O.: Risk of over-diagnosis of COPD in asymptomatic elderly never-smokers. Eur. Respir. J. **20**, 1117–1122 (2002)

Linguistic Feature-Based Praise or Complaint Classification from Customer Reviews

Sujata Khedkar[1][✉] and Subhash Shinde[2]

[1] Computer Engineering Department, VESIT, Mumbai, India
sujata.khedkar@ves.ac.in
[2] Computer Engineering Department, Lokmanya Tilak College of Engineering,
Koparkhairane, Navi Mumbai, India
skshinde@rediffmail.com

Abstract. Online reviews are very important in the customer's decision-making process in selecting the appropriate products in the online shopping portal. These reviews are then analyzed by business organizations to understand customer sentiment w.r.t. product/service. Traditional sentiment analysis techniques identify only positive, negative or neutral sentiment w.r.t. reviews and does not consider informativeness of reviews while analyzing sentiment. The extreme opinions like Praise and complaint sentences are considered as a subset of positive and negative sentences and becomes difficult to find. Praise sentences are more descriptive in nature. Praises contain more nouns, adjectives, intensifiers as compared to plain positive sentences and complaint sentences contain more connectives and adverbs rather than the plain negative sentences. This paper proposes a Linguistic feature-based approach for review sentences filtering and Hybrid feature selection method for classifying review sentence as Praise or Complaint.

These Praise and Complaint sentences can be further analyzed by business organizations to identify the reasons for customer satisfaction or dissatisfaction. It can also be used for creating automatic product description from online reviews in terms of pro and con of the product/service. The performance of the four different supervised Machine Learning classifiers, namely Random forest, SVC, KNeighbors, MLP with hybrid feature selection method is evaluated on three domains reviews using the parameters Accuracy, Precision, Recall, and F1-score. The proposed method showed excellent results as compared to the state of art classifiers.

Keywords: Text analytics · Sentiment analysis · Praises · Complaints · Big data · Machine learning · Hybrid feature selection

1 Introduction

In order to be successful and to survive in today's competitive business environment its essential for organizations to provide exceptional services/products to customers. With the various social media sites, customers can easily share their opinion on the social platform w.r.t product/service. It is important for business organizations to collect data from various channels, analyze it and measure customer satisfaction, understand

© Springer Nature Switzerland AG 2020
A. P. Pandian et al. (Eds.): ICICCS 2019, AISC 1039, pp. 470–481, 2020.
https://doi.org/10.1007/978-3-030-30465-2_52

customer behavior, recognize customer needs through reviews or social media channels. Many Business Intelligence applications across various domains use opinion mining as the key enabling technology for their Business Intelligence [1, 2] from big review data. The system is built by aggregating the opinions from various channels like blogs, comments, reviews or tweets. The major challenge is to find the extreme opinions in terms of praises and complaints to understand customer true opinions w.r.t product/service rather than just analyzing positive or negative sentiments [3, 4].

Sentiment analysis is a field that analyzes people's opinions, sentiments from unstructured text. Sentiment analysis works are three different levels of granularity namely Document level, Sentence level, and Aspect level [13–15]. Most early research on sentiment analysis was focused on document level and is used in domains like Hotels, Restaurants, E-commerce products, and Politics, etc. These online reviews have a significant impact on customer's purchase decisions [16, 18].

The sentiment classification classifies the text into three predefined categories as positive, negative or neutral based on a set of features using a classification method. However, it is possible to make other types of classifications by considering informativeness of reviews and to detect and classify reviews as praise or complaint. Praises are a subset of positive reviews and complaints are a subset of negative reviews These Praises and Complaints can be called extreme opinions.

Praises express the best views or judgments and complaints express worst views or judgments w.r.t product or service. The customer always wants to know about the best and worst aspects of product or service while making online purchase decisions. These very positive praise reviews and very negative complaint reviews have a very strong impact on product sales.

In this paper, we examine the effectiveness and limitations of different linguistic features based on linguistic properties of praise and complaint sentences and supervised sentiment classifiers to identify praise and complaint sentences from three domains reviews. Our main contribution is to analyze the effectiveness of linguistic features of praises and complaints in the supervised classification task. We performed various experiments and compare some supervised classifiers, namely Random Forest (RF), KNeighbors, SVC, MLP for binary classification tasks of identifying praise vs complaint sentences.

The rest of paper is organized as follows. In the following Sect. 2, we discuss the related work. Then Sect. 3, describes the proposed methodology. Section 4, describes the experiments, evaluations, and results. Conclusion and future work are described in Sect. 5.

2 Related Work

In related work, we found two main approaches to find the sentiment at the sentence level. First, machine learning technique's based on training corpus annotated with polarity information and, second strategies based on lexicons.

The classification performance depends on the identification of quality features and classification methods. Various types of features have been used in classification tasks such as lexical features, stylistic features, content features, sentiment features, semantic features [5]. The influence of these features for classification task has been evaluated and analyzed by some sentiment analysis studies [16, 18]. But some reviews are more informative (praise/complaints) as compare to plain positive/negative text. These informative reviews are very important from the Business Intelligence point of view for decision making. With an abundant amount of opinionated text, methods for automatic extraction of praises or complaints from a given text is not there in existence.

Customer reviews reflect the buyer's subjective perspective and include personal opinions, experiences, and complaints. Each review contains multiple sentences and its practically impossible to consume the whole corpus to know about customer's opinions. Customers often read only a few reviews and cannot get useful information from it. There are many noninformative sentences in the reviews. Most of the research previously was based on quantitative ratings given by the customer on online websites. The extreme opinions like Praises and Complaints are very challenging to identify from review corpus. Only 5%–10% reviews contains extreme opinions [10].

Ganesan et al. [1], studied the properties of praise and complaint sentences where Praise is a subset of Positive only reviews and Complaint is a subset of negative only reviews. The praise sentence contains more Adjectives, Intensifiers, Nouns, the Length of praise is longer than avg length of sentence. On the contrary Complaint, the sentence has fewer Adjectives and Nouns, more Past Tense and conjunctions. The existing sentiment analysis algorithms can be improved by only considering the analysis of praises and complaints sentences from big customer reviews data (Table 1).

Table 1. Shows examples of different types of sentences.

Sentence	Type
I don't like this hotel	Negative only
Unhappy with this hotel, price too high, rooms not clean	Complaint
I like this hotel because it's so cheap, nice food, good service, and big rooms	Praise
I like this hotel	Positive only

Krishnamoorthy [2], evaluated the features that contribute towards deciding the helpfulness of online reviews. They have used various features like review metadata, subjectivity, and readability related features for predicting helpfulness of reviews.

Othman, Hassan [3] has developed an opinion summarization system based on linguistic properties of review sentences and further classified as a direct, comparative or superlative opinion. The sentences are POS tagged and the sentences having POS tags as JJR, RBR are considered as comparative opinionated sentences, The sentences which contain JJS, RBS tags are considered as superlative opinionated sentence and sentences having JJ are considered as the direct opinioned statement. They have considered the overall review and extracted only opinionated sentences without further analysis of positivity or negativity within opinionated sentences.

Almatarneh and Gamallo [4, 5], have used Linguistic features to identify extreme opinions, they have used a bag of words, word embeddings, polarity lexicon and set of textual features to identify Complaint reviews from a set of reviews. They have constructed a lexicon of most positive and most negative words and used supervised algorithms like SVM, NB, etc.

Saumya [6], had developed Review Ranking system based on helpfulness score. The system classifies low quality and high-quality reviews based on hybrid features like nouns, adjectives, verb, difficult words, wrong words, entropy, review ranking, etc. and random-forest classifier. Low-quality reviews are ignored and only high-quality reviews are displayed.

Tripathy et al. [7], has compared the performance of Machine learning algorithms SVM, NV with tfidf, count vectorizer features to classify the reviews as either positive, negative or neutral. Alaei et al. [8–10], discussed the challenges of handling big review data and available approaches for sentiment analysis.

Abrahams et al. [11], discovered product defect using UGC in both the automotive and the consumer electronics domains. They have used principal component analysis and logistic regression to produce a multivariate explanatory analysis relating defects to quantitative measures derived from the text.

Hu and Zhang et al. [12], have used a novel structural topic model text analysis method for analyzing negative reviews of customers to understand customer's opinions about hotels in New York City.

Aurangzeb et al. [13], proposed domain independent rule based domain independent.

Customer reviews contents are spontaneous and insightful feedback of customers widely available free and easily accessible from anywhere, anytime. There does not exist a standard structured procedure that prescribes how organizations should acquire and analyze User-generated content in order to generate Managerial information for decision making. So, the proposed approach identifies the dimensions of customer satisfaction insights based on praises and complaints shared by users through big data from unstructured textual Reviews of three different domains Hotel, Amazon product reviews of Home Theater and Vacuum cleaner.

3   Proposed Approach

The proposed approach involves four different steps (Fig. 1):

- **Data Preprocessing** – the first step is to collect and extract user-generated contents i.e. customer reviews from review sites. Data Preprocessing cleans the extracted data before analysis. The reviews are converted into sentences. Data preprocessing techniques such as stop word removal, stemming, digits and special characters removal, etc are used to preprocess sentences. The Affin Dictionary is used to compute the sentiment score of each sentence The sentence's which are neutral sentences with score zero are eliminated from further analysis.
- **Linguistic Feature Based Sentence Filtering** – In this step the extracted sentences are linguistically evaluated. Sentences with subjective expression (complaints or Praises) are retained and sentences which are nonsubjective (facts, factual information) are discarded. The subjective sentences are divided into Praise_set and Complaint_set based on Affin sentiment score. Then the final Praise and Complaint Score of a sentence is calculated using formula 1 and 2.

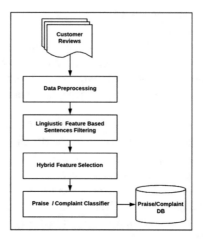

Fig. 1. A proposed approach for praise or complaint classification

$$P\_score = 0.1 * len(sentence) + 0.2 * len(nouns) \\ + 0.3 * len(adj) + 0.4 * len(intensifier) \tag{1}$$

$$Praise\_score = P\_score(sentence) * Sentiment\_score(sentence)/10 \tag{2}$$

$$C\_score = -0.1 * len(sentence) + 0.1 * len(nouns) \\ + 0.2 * len(conjection) + 0.3 * len(Adverb) + 0.3 * len(Verb) \tag{3}$$

$$Complaint\_score = C\_score(sentence) * Sentiment\_score(sentence)/10 \tag{4}$$

All Neutral sentences having low sentence score are eliminated from a corpus.

- **Hybrid Feature Selection** – This Module selects a subset of relevant features. There are two types of feature selection technique's Filter-based feature selection methods uses a statistical measure to provide a ranking of all features by the score and the best features are selected from the dataset Correlation-based Feature selection method extracts most important features from the set of sentiment words. Hybrid Feature selection Method uses various features like
  - Meta Features-the length of the sentence, number of non-stop words, average word length.
  - Synthetic features-the number of nouns, number of adjectives, number of verbs, number of intensifiers, number of adjectives, number of adverbs, etc.
  - Content Features-words which are Verbs, Nouns, Adjectives, Adverbs, Verb Phrases, etc.
  - Semantic Features-Positive or negative rating and Praise/Complaint score of the sentence from the Sentence filtering module.

Table 2 shows features that are considered for evaluation.

Table 2. Description of all linguistic features

Features	Descriptions
FS1	Unigram TF-IDF (1-gram)
FS2	Bigram TF-IDF (2-gram)
FS3	Trigram TF-IDF (3-gram)
FS4	FS1 (Unigram TF-IDF) + FS4 (Hybrid features-(Meta + Synthetic + content + Semantic))
FS5	FS2 (Bigram TF-IDF)+ FS4 (Hybrid Features)
FS6	FS3 (Trigram TF-IDF)+ FS4 (Hybrid Features)

Various experiments with a combination of the different feature set are performed on the selected dataset.

- **Praise/Complaint Classifier Module** – Many researchers have focused on the use of traditional classifiers such as Naïve Bayes, SVM, Maximum Entropy, etc. to solve classification Problems We have studied the usefulness of various classification algorithms such as SVC, Random Forest, MLP, KNeighbors algorithms for this problem.

4   Experiments

4.1   Data Preprocessing

In order to extract praises and complaints, we require to analyze customer reviews with opinion ratings and extract those documents associated with the lowest and highest scale. But as the numerical rating cannot be applied equally to all sentences that are part of the review, we have developed filtering method based on linguistic properties of reviews and Affin dictionary is used to compute the score of the sentence. We obtained the dataset of Amazon product reviews dataset (ACL15-Chan-Dataset) of and analyzed reviews of product's Vacuum Cleaner and Home Theater and Hotel reviews dataset (https://www.kaggle.com/harmanpreet93/hotelreviews/activity).

The datasets are preprocessed using the following steps:

(1) Neutral reviews (3-star rating) are deleted since they do not convey any views.
(2) Reviews are converted into sentences and stop words, digits, special characters are removed.
(3) Review sentences having short lengths i.e. Sentence length <30 characters are eliminated.
(4) Sentiment score of each sentence is calculated using Affin Dictionary.
(5) The neutral sentence's having Affin sentiment score zero are eliminated.

(6) Linguistic properties of each sentence are computed and Praise_score/Com-paint_score of each sentence is computed based on Eqs. 3 and 4. Table 3 shows the number of reviews in each class.

(7) Dataset is divided into Praise-set and Complaint-set based on their sentiment score as positive or negative

Table 3. Three dataset with a total number of review sentences in each class

Datasets	#of Review sentences	Negative only	Complaints	Positive only	Praises sentences
Hotel	**1,00,012**	14491	**23545**	26896	**35080**
Home Th.	**2031**	331	**708**	295	**697**
Vacuum cleaner	**4637**	579	**702**	1025	**2331**

Table 3 shows the number of Plain _Positive, Praise, Plain _Negative and Com-plaint sentences in Hotel and Amazon product datasets. There are 35.1% Praise and 23.5% Complaint sentences as extreme opinions in Hotel Reviews. The noninformative Plain_Positive (26.9%) and Plain_Negative (14.5%) sentences are eliminated from the analysis. From Amazon Product dataset, for Vacuum Cleaner there are 50.3% Praise and 15.1% Complaint and for Home Theater there are 49.3% Praise and 13.8% Complaint review sentences as shown in Fig. 2.

4.2 Training and Testing

We have used four supervised classification algorithms Random Forest, Kneighbours Classifier, SVC, and Multilayer Perceptron. We have compared and summarized and discuss the behavior of these classifiers with the linguistic features as discussed above. The dataset was randomly partitioned into training (80%) and testing (20%). We have used stratified sampling and used 5_fold_cross_validation. The output is evaluated using four parameters Precision, Recall, F1-Score, and Accuracy. Macro Avg of all parameters areca.

4.3 Results

Table 4 shows the classification results obtained by Random Forest, Kneighbours, SVC, Multilayer Perceptron classifiers for all datasets. These classifiers are evaluated based on base features and linguistic features and their combinations. The final scores were computed using Precision(P), Recall(R), F1 scores and Accuracy.

Tables 4, 5 and 6 shows Classification results obtained using Random forest, K neighbors, SVC, MLP with base features, linguistic features, and combinations.

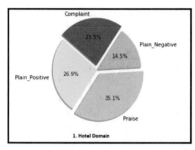

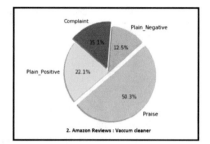

Fig. 2. Exploratory data analysis of customer reviews.

Table 4. Classification results for Hotel Reviews

Hotel domain reviews	Random forest			SVC			Kneighbour			MLP		
	P	R	F1	P	R	F1	P	R	F1	P	R	F1
FS1	0.877	0.820	0.844	0.420	0.50	0.456	0.809	0.705	0.740	0.948	0.928	0.983
FS2	0.870	0.839	0.925	0.420	0.50	0.456	0.788	0.677	0.710	0.924	0.898	0.910
FS3	0.864	0.838	0.851	0.420	0.50	0.456	0.793	0.652	0.687	0.923	0.897	0.909
FS4	0.998	0.997	**0.997**	0.841	1.0	**0.913**	0.946	0.990	**0.967**	0.997	0.998	**0.997**
FS5	0.996	0.999	0.997	0.841	1.0	0.913	0.945	0.989	0.967	0.995	0.998	0.994
FS6	0.998	0.994	0.996	0.841	1.0	0.913	0.945	0.987	0.966	0.995	0.998	0.994

Table 5. Classification results for Amazon product reviews

Home theater reviews	Random forest			SVC			Kneighbour			MLP		
	P	R	F1	P	R	F1	P	R	F1	P	R	F1
FS1	0.549	0.523	0.437	0.782	0.60	0.791	0.532	0.513	0.425	0.559	0.554	0.464
FS2	0.542	0.524	0.433	0.782	0.60	0.791	0.534	0.508	0.423	0.551	0.542	0.450
FS3	0.535	0.510	0.423	0.782	0.60	0.791	0.528	0.507	0.421	0.545	0.533	0.444
FS4	0.986	0.983	0.985	0.783	1.0	0.870	0.931	0.985	**0.957**	0.976	0.989	**0.983**
FS5	0.984	0.988	0.986	0.782	1.0	**0.877**	0.929	0.984	0.956	0.960	0.989	0.974
FS6	0.992	0.984	**0.988**	0.782	1.0	0.877	0.929	0.984	0.955	0.958	0.985	0.971

Table 6. Classification results for Amazon product reviews

Vacuum cleaner reviews	Random forest			SVC			Kneighbour			MLP		
	P	R	F1	P	R	F1	P	R	F1	P	R	F1
FS1	0.554	0.428	0.444	0.769	0.80	0.783	0.538	0.403	0.473	0.453	0.505	0.473
FS2	0.549	0.424	0.439	0.800	0.769	0.783	0.543	0.405	0.425	0.563	0.443	0.459
FS3	0.545	0.421	0.433	0.800	0.769	0.783	0.533	0.399	0.418	0.564	0.446	0.465
FS4	0.995	0.992	**0.994**	0.771	1.0	**0.870**	0.932	0.983	**0.957**	0.980	0.991	**0.985**
FS5	0.992	0.990	0.991	0.769	1.0	0.869	0.935	0.985	0.959	0.964	0.992	0.978
FS6	0.985	0.983	0.984	0.769	1.0	0.869	0.936	0.985	0.960	0.965	0.991	0.978

For Hotel domain reviews the Random forest, MLP algorithms with unigram tfidf and hybrid linguistic features gives F1-Measuere of 0.997 and outperforms other classifiers with Accuracy 99.6% (see Table 4). With Amazon Product Reviews for Home Theater product, Random forest with trigram tfidf and linguistic features outperform other classifier with F1-Measure of 0.988 and Accuracy 98.2% (see Table 5). For Vaccum cleaner reviews Randomforest with unigram tf-tfidf and hybrid features outperform other classifiers with F1-Measure of 0.994 and Accuracy 99.1% (see Table 6).

Base features combined with hybrid features have given better results as compared to simple base features as shown in Fig. 3.

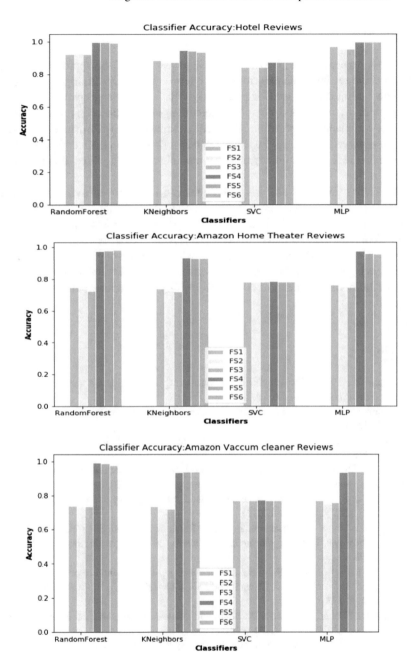

Fig. 3. Classifier accuracy for Hotel and Amazon product reviews

5   Conclusions

In this paper, to deal with big review data, we have proposed the approach for filtering noninformative review sentences from the review corpus based on linguistic properties of extreme opinions (Praise/Complaint). We have studied properties of extreme opinions such as praise and complaint sentences and proposed hybrid features for praise and complaint classification. We have examined the performance of these features with supervised learning classifiers (Random Forest, SVC, K neighbors, MLP) to identify extreme opinions like praises or complaints about three domains review dataset.

The existing sentiment analysis techniques can be improved by only considering informative sentences in the sentiment analysis process. The linguistic features have a high impact on increasing classification accuracy w.r.t extreme opinions.

The Random Forest classifier outperforms all other classifiers with (unigram + hybrid features). In the future, we will be evaluating other classifiers and compare it with Random Forest. We will evaluate the use of trained word embeddings and applicability of deep learning models in identifying extreme opinions from review corpus.

References

1. Ganesan, K., Zhou, G.: Linguistic understanding of complaints and praises in user reviews. In: Proceedings of NAACL-HLT (2016)
2. Krishnamoorthy, S.: Linguistic features for review helpfulness prediction. Expert Syst. Appl. **42**(7), 3751–3759 (2015)
3. Othman, M., Hassan, H.: A linguistic approach for opinionated document summary. Future Comput. Inf. J. (2017)
4. Almatarneh, S., Gamallo, P.: Linguistic features to identify extreme opinions: an empirical study. In: Lecture Notes in Computer Science, pp. 215–223 (2018)
5. Almatarneh, S., Gamallo, P.: A lexicon-based method to search for extreme opinions. PLoS ONE **13**(5), e0197816 (2018)
6. Saumya, S., Singh, J.P., Baabdullah, A.M., Rana, N.P., Dwivedi, Y.K.: Ranking online consumer reviews. Electron. Commer. Res. Appl. (2018). https://doi.org/10.1016/j.elerap. 2018.03.008
7. Tripathy, A., Agrawal, A., Rath, S.K.: Classification of sentimental reviews using machine learning techniques. Procedia Comput. Sci. **57**, 821–829 (2015). https://doi.org/10.1016/j. procs.2015.07.523
8. Alaei, A.R., Becken, S., Stantic, B.: Sentiment analysis in tourism: capitalizing on big data. J. Travel Res. (2017). 004728751774775
9. Liu, B.: Sentiment analysis and opinion mining. Synth. Lect. Hum. Lang. Technol. **5**, 1–167 (2012)
10. Turney, P.D.: Thumbs up or thumbs down?: Semantic orientation applied to unsupervised classification of reviews. In: Proceedings of the 40th Annual Meeting on Association for Computational Linguistics, Philadelphia, PA, USA, 7–12 July 2002
11. Abrahams, A.S., Fan, W., Wang, G.A., Zhang, Z.J., Jiao, J.: An integrated text analytic framework for product defect discovery. Prod. Oper. Manag. **24**, 975–990 (2015)

12. Hu, N., Zhang, T., Gao, B., Bose, I.: What do hotel customers complain about? Text analysis using structural topic model. Tour. Manag. **72**, 417–426 (2019)
13. Sarath Chandra Reddy, Ch., Uday Kumar, K., Dheeraj Keshav, J., Prasad, B.R., Agarwal, S.: Prediction of star ratings from online reviews. In: Region 10 Conference TENCON 2017 - 2017 IEEE, pp. 1857-1861 (2017)
14. Kharde, V.A., Sonawane, S.: Sentiment analysis of twitter data: a survey of techniques. Int. J. Comput. Appl. **139**(11), 5–15 (2016)
15. Pang, B., Lee, L.: Seeing stars: exploiting class relationships for sentiment categorization with respect to rating scales. In: Proceedings of the 43rd Annual Meeting on Association for Computational Linguistics, pp. 115–124. Association for Computational Linguistics (2005)
16. Zhao, Y., Xu, X., Wang, M.: Predicting overall customer satisfaction: Big data evidence from hotel online textual reviews. Int. J. Hospitality Manag. (2018)
17. Dataset: Hotel reviews dataset. https://www.kaggle.com/harmanpreet93/hotelreviews/activity
18. Aurangzeb, K., Baharum, B., Khairullah, K.: Sentiment classification from online customer reviews using lexical contextual sentence structure. Trends Appl. Sci. Res. (2011)

Implementation of European NCAP Standard Autonomous Emergency Braking Scenarios Using Two Leddar M16 Sensors

Ritesh Kapse[(✉)] and S. Adarsh

Department of Electronics and Communication Engineering, Amrita School of
Engineering, Amrita Vishwa Vidyapeetham, Coimbatore, India
riteshkapse.rk@gmail.com, s_adarsh@cb.amrita.edu

Abstract. In this paper, we re-generated and implemented EURO NCAP
standard scenarios for autonomous emergency braking using two Leddar M16
sensors. Out of total sixteen channels available in M16 module, we have used
middle channels since only data of longitudinal axis was required for selected
scenarios. The data is collected by mounting the sensors on hood of ego vehicle.
We have referred constant values of multiple parameters in the collision and
stopping time calculations from standard parameters defined in ADAS simulation toolbox of MATLAB 2018b. Data update and predicts of two datasets from
Leddar M16 sensor is carried out using Kalman Filter algorithm considering one
sensor data as ground truth data. Finally the final distance data output, from
updated datasets are used to display autonomous emergency braking and forward collision warning over the screen of python console. According to
decelerations at different stages, partial and full braking times are calculated.

Keywords: Light-emitting diode detection and ranging (Leddar) · Advanced
driver-assistance systems (ADAS) · European New Car Assessment Program
(Euro NCAP) · Autonomous emergency braking (AEB) · Full braking (FB) ·
Partial braking (PB) · Time-to-collision (TTC) · Forward collision warning
(FCW)

1 Introduction

In this 21st century, all you need is safety first and then leisure [1, 2]. Key studies
suggest that the safety system will reduce accidents by 27%, saving 8000 lives per year
on roads in Europe. USA road figures by the NHTSA, suggest that AEBS (autonomous
emergency braking systems) would lead to a reduction of 28,000 collisions, equating to
12,000 fewer injuries. The United Nations Economic Commission for Europe
(UNECE) have made such safety systems mandatory for new heavy goods vehicles in
2015. If driver is braking with insufficient force or late braking, it results into an
accident. There can be many reasons to do so such as listening music, talking over
phone call, poor visibility, when situation is difficult to predict, suddenly pedestrian
crosses the road etc [3]. Mostly people fails to deal the critical situations like not
applying the enough braking force to avoid an accident or no braking due to less time
to react. Hence it is required to have autonomous emergency braking implemented on

© Springer Nature Switzerland AG 2020
A. P. Pandian et al. (Eds.): ICICCS 2019, AISC 1039, pp. 482–490, 2020.
https://doi.org/10.1007/978-3-030-30465-2_53

vehicle [4, 9, 12]. Different sensors such as radar, (stereo) camera, leddar or lidar etc. are used to identify the collision threads ahead of the vehicle. Some of these systems can prevent collisions altogether by braking fully, but most are designed to simply reduce the speed of a car before it hits something.

[5] AEB city and inter-urban system is been included in the European New Car Assessment Program (Euro NCAP) safety rating from 2014. Road users such as pedestrians and cyclists are important to protect and hence Euro NCAP is continuously promoting AEB systems. For the accurate, reliable and robust detections we can use multiple sensors minimizing false positives. Data of two Leddar sensors is utilized to get the position and velocity of the most important object (MIO) using Kalman filter algorithm with respect to ego vehicle. The driver should react to the alert provided as forward collision warning and should apply the brakes with delay time. Time to collision (TTC) is calculated using mioVehicle distance and velocity. Forward collision warning alert is activated when the time-to-collision (TTC) is less than FCWStoppingTime time [6]. When there is scenario to apply emergency braking, it is applied as cascaded braking, that is multiple stage braking followed by full braking.

2 Proposed Method

Fig. 1. Block diagram for autonomous emergency braking using two Leddar M16 sensors

We have used two Leddar M16 sensors to get the distance data in respective 16 channels to check the position of Most Important Object (MIO) vehicle from Ego vehicle [7]. After data collection, updated distances of MIO are calculated using Kalman filter algorithm and is used in collision time calculations. AEB logic compares stopping times with collision time to pop up the warning to driver screen [11]. Below is the logic implemented while designing the autonomous emergency braking and forward collision warning (Figs. 1 and 2).

For autonomous emergency braking scenarios coverage we have shortlisted two standard AEB scenarios to regenerate and verify using two Leddar M16 sensors.

The NCAP scenarios which we have regenerated are explained below in details.

Scenario 1: AEB_Bicyclist_Longitudinal_50width
The ego vehicle collides with the bicyclist that is in front of it. Before the collision, the bicyclist and ego vehicle are traveling in the same direction along the longitudinal axis. At collision time, the bicycle is 50% of the way across the width of the ego vehicle. Ego vehicle moves at constant speed of 25 km/h (6.94 m/s) (Fig. 3).

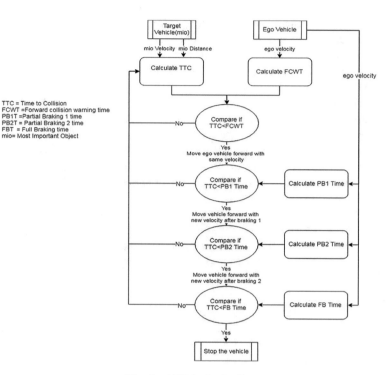

TTC = Time to Collision
FCWT =Forward collision warning time
PB1T =Partial Braking 1 time
PB2T = Partial Braking 2 time
FBT = Full Braking time
mio= Most Important Object

Fig. 2. AEB logic details

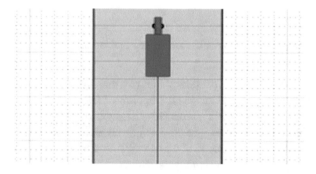

Fig. 3. Visualization of scenario 1 in driving scenario designer of Matlab simulation 2018b

Scenario 2: AEB_Pedestrian_Longitudinal_50width
The ego vehicle collides with the Pedestrian that is in front of it. Before the collision, the Pedestrian and ego vehicle are traveling in the same direction along the longitudinal axis. At collision time, the pedestrian is 50% of the way across the width of the ego vehicle. Ego vehicle moves at constant speed of 20 km/h (Fig. 4).

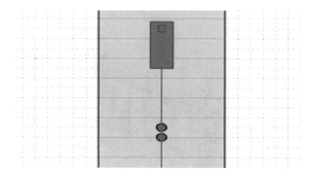

Fig. 4. Visualization of scenario 2 in driving scenario designer of Matlab simulation 2018b

3   Sensor Details

a. Leddar M16

The M16 modules has two main configuration types as the classic LED family and the new Laser family based on illumination methods. Both configurations are best suited for outdoor operation with different parameters as solid-state design with no motorized mechanisms, extensive range of temperature operation, performance in all weather and good immunity to lighting variations. The viewing angles of Leddar sensor in horizontal and vertical direction is 48° and 7.5° respectively. 16 independent active elements are combined by the Leddar® M16 Sensor module to give advanced sensing solution into a single sensor so that we can get continuous and accurate detection and ranging which includes lateral discrimination in complete beam width deprived of any moving parts. Infrared LED light source of the Leddar® M16 module gives wide-beam illumination up to 100 m range with six different field-of-view configurations (Fig. 5).

FOV SEGMENT DIMENSIONS (EXAMPLE OF A M16-LED 48°X8°)

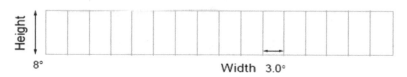

Fig. 5. Vertical and horizontal channel angular dimensions

The following sections present figures illustrating the sensitivity of the sensor across beam width (Segment amplitude efficiency) and height (amplitude vs tilt) (Fig. 6).

The emitter has a light source. Pulse rate of this light source is high (typically 100,000 pulses per second) and has very short pulse width. To shape the illumination beam, optical components may be used. The pulses of light start to travel via the detection area, receiver optics and a photodetector is used to capture the backscatter of

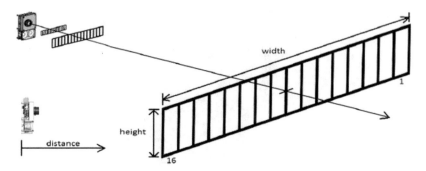

Fig. 6. Beam pattern width and height

reflected light. There can be single element or multiple elements components as photodetector, which provides several detection and ranging segments in the light beam. The receiver also contains signal amplification and digitization. The emitter and receiver functions are synchronized and driven by the Leddar Core. Also Leddar Core gives patented signal processing to increase the resolution and the signal-to-noise ratio of the digitized signal. The acquired signals are used for detections and distance measurements. Below figure explains the example of input signal and pulse received is reflection of the emitted pulse off an object. Location in the signal is used to calculate the distance of the object and pulse provides the detection function (Fig. 7).

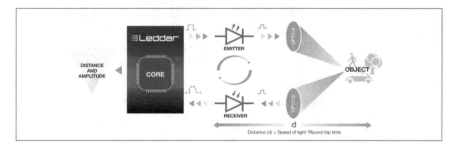

Fig. 7. Signal travelling through the main components of a Leddar sensing module

4 Experimental Setup

In our experiment, we have used two Leddar M16 sensor with same configurations. The first Leddar M16 data, we used to collect the datasets according to two scenarios defined above, and is used as ground truth data. The second sensor data of Leddar M16, we are considering as actual measurement by sensor and giving it to measurement function of Kalman Filter algorithm [8]. Hence by comparing the predicted and measured states, we can get updated results for inputs we are giving to Kalman filter algorithm. The updated measurements are nothing but the distance of most important object (MIO) from Ego vehicle. So change in distances with respect to change in time gets us MIO velocity, values of which is used in the expression of Time to Collision (TTC) calculations. Every time, calculated Time to Collision (TTC) is compared with different times such as forward collision warning time, partial braking times and full braking times to check the working of AEB logic.

There are certain standard constant values we are referring from simulation results of AEBTestBench module in MATLAB 2018b [10]. Following values are specified by EURO NCAP standard scenarios results generated in driving scenario generator of ADAS toolbox.

1. Driver reaction time is 1.2 s
2. Driver braking deceleration for forward collision warning is 4 m/s^2
3. 1st stage Partial Braking deceleration is 3.8 m/s^2
4. 2nd stage Partial Braking deceleration is 5.3 m/s^2
5. Full Braking deceleration is 9.8 m/s^2
6. Headway Offset is 3.7 m
7. Headway time margin is 0 s

Below are the equations necessary to derive different stopping time parameters (Figs. 8 and 9).

$$FCWStoppingTime = FCW.timeToReact + (egoVelocity/FCW.driver\_decel) \quad (1)$$

$$PB1StoppingTime = AEB.timeMargin + (egoVelocity/PB1decel) \quad (2)$$

$$PB2StoppingTime = AEB.timeMargin + (egoVelocity/PB2decel) \quad (3)$$

$$FBStoppingTime = AEB.timeMargin + (egoVelocity/FBdecel) \quad (4)$$

$$TTC = +/- [(mioDistance - AEB.headwayOffset)/(mioVelocity)] \quad (5)$$

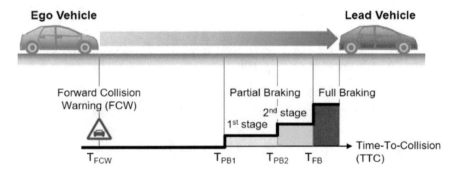

Fig. 8. Time calculations for multi-stage partial braking followed by full braking

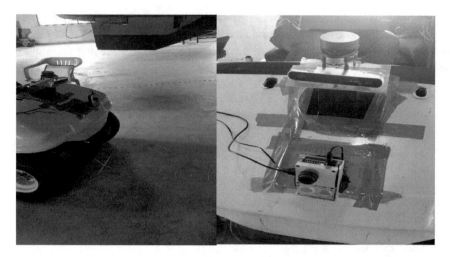

Fig. 9. Leddar M16 and other sensors setup for raw data collection as per AEB scenarios

5 Results

The data is processed as per Kalman Filter algorithm to get the updated sensor measurements. According to speed of ego vehicle, the different stopping times is calculated and compared with collision time (Table. 1).

Table 1. General calculations for stopping times with respect to ego velocity of car. (Refer formula's in experimental setup section)

Scenario	Vego (m/s)	FCW T(s)	PB1 T(s)	PB2 T(s)	FB T(s)
AEB_Bicyclist_Longitudinal_50width	6.94	2.93	1.82	1.30	0.70
AEB_Pedestrian_Longitudinal_50width	5.56	2.59	1.46	1.04	0.56

The result plots after execution of code in python for speed of ego vehicle as 25 km/h, 20 km/h respectively (Fig. 10).

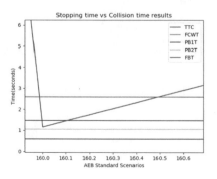
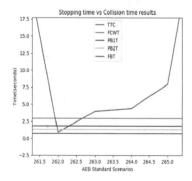

Fig. 10. Stopping time vs collision time results

Once the Time to Collision crosses the different stopping times, status of AEB and collision warnings changes and are displayed on drivers screen (Fig. 11).

```
IPython console
  Console 2/A ☒
Car is moving forward with set speed of 25(20)km/hr
Give forward collision warning only
AEBstatus = 0; FCWactivate = 1; decel = 0;
Apply partial braking stage one
AEBstatus =1;FCWactivate = 1; decel = 3.8;
Apply partial braking stage two
AEBstatus = 2; FCWactivate = 1; decel = 5.3;
Autonomous Emergency Brakes applied successfully
AEBstatus = 3; FCWactivate = 1;  decel = 9.8;
```

Fig. 11. Python console as warnings to driver screen

6   Conclusion and Future Scope

AEB scenarios are regenerated using Leddar M16 sensors. The python code which has AEB logic with Kalman filter algorithm for updated distances is compiled and ran successfully in Spyder (Python 3.7). Results of research has been verified with the standard simulation results of ADAS toolbox (AEBTestBenchModule) available in Matlab 2018b.

Since only few scenarios are verified for autonomous emergency braking according to EURO NCAP standards, next task would be to re-generate the remaining scenarios and implement them using combination of two Leddar M16 sensors.

References

1. Kumar, R.S., Stanley, P.K., Gandhi, A.S.: Raspberry Pi based vehicle collision avoidance system. In: Proceedings IEEE International Conference on Innovations in Electrical, Electronics, Instrumentation and Media Technology, ICIEEIMT 2017, pp. 211–215, January 2017. https://doi.org/10.1109/ICIEEIMT.2017.8116838
2. Schmitt, A.: Europe Will Use Vehicle Tech to Prevent Speeding, Save Thousands of Lives (2019). https://usa.streetsblog.org/2019/03/05/europe-will-use-vehicle-tech-to-prevent-speeding-save-thousands-of-lives/
3. Ariyanto, M., Haryadi, G.D., Munadi, M., et al.: Development of low-cost autonomous emergency braking system (AEBS) for an electric car. In: Proceeding - 2018 5th International Conference on Electric Vehicular Technology, ICEVT 2018, pp. 167–171 (2019). https://doi.org/10.1109/ICEVT.2018.8628442
4. Emani, S., Soman, K.P., Sajith, V.V., Adarsh, S.: Obstacle detection and distance estimation for autonomous electric vehicle using stereo vision and DNN. In: Proceedings of ICSCSP 2018, vol. 2 (2019). https://doi.org/10.1007/978-981-13-3393-4_65
5. Van Ratingen, M., Williams, A., Lie, A., et al.: The European new car assessment programme: a historical review. Chin. J. Traumatol. **19**, 63–69 (2016). https://doi.org/10.1016/j.cjtee.2015.11.016. English Ed

6. Kaempchen, N., Schiele, B., Dietmayer, K.: Situation assessment of an autonomous emergency brake for arbitrary vehicle-to-vehicle collision scenarios. IEEE Trans. Intell. Transp. Syst. **10**, 678–687 (2009). https://doi.org/10.1109/TITS.2009.2026452
7. Welch, G., Bishop, G.: An Introduction to the Kalman Filter, vol. 7, pp. 1–16. University of North Carolina, Chapel Hill (2006). https://doi.org/10.1.1.117.6808
8. Kapse, R.: Implementing an Autonomous Emergency Braking with Simulink using two Radar Sensors (2019)
9. Binoy, B.N., Keerthana, T.: A GSM-based versatile unmanned ground vehicle. In: International Conference on "Emerging Trends in Robotics and Communication Technologies", INTERACT-2010, Chennai, pp. 356–361 (2010)
10. Driving Scenario Generation and Sensor Models - MATLAB & Simulink - MathWorks India.   https://in.mathworks.com/help/driving/driving-scenario-generation-and-sensor-models.html. Accessed 18 Apr 2019
11. Hulshof, W., Knight, I., Edwards, A., et al.: Autonomous emergency braking test results. In: Proceedings 23rd International Technical Conference on Enhanced Safety of Vehicles, pp. 1–13 (2013). https://doi.org/13-0168
12. Wallner, J., Tang, T., Lienkamp, M.: Development of an emergency braking system for teleoperated vehicles based on lidar sensor data. In: 2014 11th International Conference on Informatics in Control, Automation and Robotics, vol. 02, pp. 569–576 (2014). https://doi.org/10.5220/0005114905690576

Tamper Detection of Medical Images Using Modified Hashing Algorithm

Nithya Chidambaram$^{(\boxtimes)}$, Kona Sai Harshavardhan Reddy,
Keertipati Vishal Varma, Kakamani Jagadeesh Sai Dheeraj,
Avija Sharan Reddy, and Amirtharajan Rengarajan

Department of ECE, School of Electrical and Electronics Engineering, SASTRA
Deemed to Be University, Thanjavur 613 401, India
cnithya@ece.sastra.edu

Abstract. In the present modern scenario, health remains as the top most priority of every human being. The identification of the problem and its diagnosis is complicated and remains as the most important part of it. Recently, many of our health related activities use data in the digital format. Digital data is easy and remains in a more comfortable form for transferring the data over long distances. Exchange of Medical Images must be done in a protective and secured manner due to security threats and the attacks of the unidentified intruders. If the privacy is questionable in medical data sharing, then the life of an individual concerned will remain under risk. This has led to the demand for integrity and authentication of medical images. In this paper, an algorithm based on the Secure Hashing Algorithm is proposed to maintain the medical image's integrity. The Region of Interest (RoI) is separated using spatial co-ordinates and the hash value of the RoI is generated and embedded into the Region of Non-Interest (RoNI) before transmission.

1 Introduction

One of the basic necessities of a human is to remain in a good health condition. The introduction of technology in the medical field has replaced all the inefficient paper records to its equivalent and most efficient digital format [1]. The clinical data such as scanned images of affected parts, textual prescriptions, reports etc. are present in a complete digital format [2]. The maintenance of this data is not a simple task and brings a sensitive issue on security and privacy of the data [3].

The availability of this digital data has led to an unimaginable growth in tools that are used to process this data [4]. It has become very easy to modify the data into other forms or to modify it within itself such that it cannot be detected [5]. Hence, the verification of the integrity of the data has become a requirement in the present world, especially with the medical images [6]. A perceptual image hashing is widely used now-a-days, taking its robustness and security into consideration. Hashing algorithms vary for different kinds of digital data. Based on the main content of the data hash value is generated through a hash function which is used for authentication. Hashing methods based on spatial domain [7] and frequency domain [8] are proposed in literature.

© Springer Nature Switzerland AG 2020
A. P. Pandian et al. (Eds.): ICICCS 2019, AISC 1039, pp. 491–498, 2020.
https://doi.org/10.1007/978-3-030-30465-2_54

As mentioned in the paper by Lahouari Ghouti [9], Cryptographic hashing is not suited for its one-way functions and change in the output for an insignificant change in the input values, and is not very suited for intended alterations made in the image. Many techniques had been put forward for image hashing like mentioned in paper by Sinari [10], proposing three different algorithms for hashing which does not include using spatial domain. For a medical image, the whole image is not important rather a part of the image contains the useful information is called Region of Interest (RoI) [11]. The RoI must be preserved and should not be allowed to change the data [12]. So RoI need to be verified for the tampering.

In this paper the proposed method is done in spatial domain discussed in Sect. 2. The finding of the proposed work is discussed under Sect. 3 which is results and discussions. The paper concludes in Sect. 4 with stated future scope of the proposed work.

2   Proposed Method

The RoI is separated from the original image using spatial co-ordinates. The hash value of the separated RoI is generated by using the proposed hash algorithm. The final binary hash values are embedded into the Region of Non-Interest and the same is transmitted. This algorithm is secured such that it is only known to the sender and the receiver. The receiver separates RoI from the received image and generates the hash value using the same algorithm proposed. The actual hash value is extracted from the Region of Non-Interest and is compared with the receiver generated hash value. With the result of the compared output, one can know whether the image is tampered or not and if the image is tampered, its position is also identified (Fig. 1).

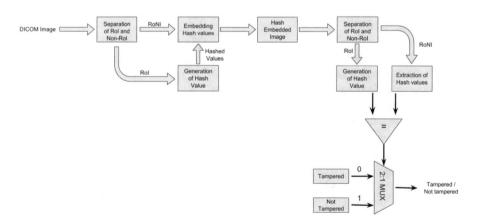

Fig. 1. Proposed method

2.1   Transmitter

2.1.1   Separation of RoI and RoNI

A 256 × 256 DICOM image is used as a sample test image with bit depth of 8 where indices are ranging between 1 and 256, from which the required part is separated as follows:

Step 1: Find r_j and c_j as final coordinates of the RoI by taking r_i and c_i as the initial coordinates, such that $j > i$

where i and j denotes indices ranging from 1 to 256.

Step 2: Calculate the size of RoI from the following expression,

$$[R\,C] = [(j - i)\,(j - i)] \text{ such that } R\%16 = 0 \text{ and } C\%16 = 0.$$

Based on r_i, c_i and r_j, c_j indices ranging from 1 to i and j to 256 both in rows and columns contribute to Region of Non-Interest.

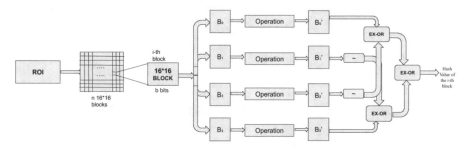

Fig. 2. Operation on each block

2.1.2   Generation of Hash Value for RoI

See Figs. 2 and 3.

Step 3: Divide the R × CRoI into 'n' number of 16 × 16 non-overlapping blocks.

For each block of size 16 × 16,

$$\text{No. of pixels} = a(\text{say})$$
$$\text{No. of bits} = b\,(\text{say})$$

Step 4: Split 'b' into 4 equal sub blocks. Let the blocks be B_0, B_1, B_2 and B_3.

Step 5: Divide each sub block into 16 words w_0 to w_{15}.

Step 6: Shift each word in the odd numbered positions by 7 times in the left direction i.e. left shift operation and each word in the even numbered positions by 5 times as done with the odd numbered words.

Step 7: The obtained words are named as H_0 to H_{15}, such that the even positioned words contain words shifted by 5 times and odd positioned words contain words shifted by 7 times.

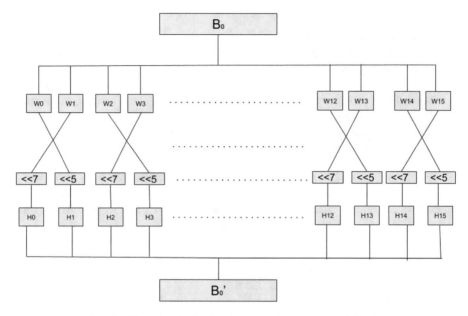

Fig. 3. Flowchart indicating the operation on each sub block

Step 8: Concatenate all the newly formed words to obtain a sub block with the same length of the previous sub block.

Step 9: Perform above operation on all the above sub blocks to obtain B_0', B_1', B_2' and B_3'.

Step 10: Perform EX-OR operation on the above sub blocks to obtain X, Y and Z from the expressions

$$X = (\sim B_2' \oplus B_0') \tag{1}$$

$$Y = (\sim B_1' \oplus B_3') \tag{2}$$

$$Z = X \oplus Y \tag{3}$$

Where, Z is the final hash value of 512 bits.

Step 11: Repeat the steps 3 to 10 for all the sub blocks of 'n' blocks in the RoI to obtain a hash values.

2.1.3   Embedding of Hash in RoNI

From the RoI of size R × C, embedding process is selected based on the P_{RoI} value.

$$P_{RoI} = \frac{(\text{size of DICOM image})}{(j - i) \times (j - i)} \times 100 \tag{4}$$

If $P_{RoI} <= 32$ then 1bit LSB embedding is performed.
If $32 < P_{RoI} < 50$ then 2bit LSB embedding is performed.
Once the embedding is done, the image is ready for transmission.

2.2 Receiver

2.2.1 Separation of RoI and RoNI

From the received 256×256 image, RoI of size $R \times C$ is extracted based on r_i, c_i and r_j, c_j.

So, RoNI is the remaining region apart from RoI.

2.2.2 Generation of Hash Value for RoI

A new hash value is generated for the RoI by the receiver with the proposed algorithm discussed in Sect. 2.1.2.

2.2.3 Extraction of Hash from RoNI

Using Eq. (4) in 2.1.3, P_{RoI} is calculated and the embedding part is found out accordingly.

16×16 block-wise generated hash and extracted hashes are compared. The image tampering results with the mismatched hash value through which the corresponding tampered position is identified.

3 Result and Discussions

In this section, the outcomes of the proposed algorithm and comparison of the Mean Squared Error (MSE) i.e., cumulative squared error between the processed image and the image being processed, the level of maintenance of the quality of the image after it is processed, Peak Signal Noise Ratio (PSNR), to check for minute loss of information from the image during transmission Structural Similarity Index (SSIM) and Normalized Correlation (NCC) values of the processed image and the original image are evaluated. A sample test image is considered for this purpose. The values of the above parameters are calculated for the different percentages of RoI and are tabulated.

$$MSE = \frac{\sum_{R,C} [I(r,c) - I'(r,c)]^2}{R \times C} \tag{5}$$

$$PSNR = 10 \log_{10} \frac{255^2}{MSE} \tag{6}$$

$$SSIM = \frac{(2\mu_o \mu_e)(2\sigma_{oe} + k)}{(\mu_o^2 + \mu_e^2 + k_1)(\sigma_o^2 + \sigma_e^2 + k)} \tag{7}$$

Where, μ_o & μ_e are mean intensity.
σ_o & σ_e are the standard deviation of original and embedded images respectively
k & k_1 are constants.

σ_{oe} is the covariance of both the images

$$NCC = \frac{\sum_{R,C}[I(r,c) \times I'(r,c)]}{\sum_{R,C}[I(r,c)]^2} \tag{8}$$

In the above Eqs. (5) and (8) I(r, c) represents the input DICOM image and I'(r, c) represents the final embedded DICOM image which is used for transmission.

From Table 1 the embedding size will be decided. The PSNR value for the various embedding size reaches the acceptable range from 39 dB to 49 dB. The NCC value for all the images with different RoIs obtained as close to one. SSIM arrived is also in acceptable range. If the size of the RoI is increased, then there is a need to use 3 bit LSB embedding which compromises the parameter values. Figures 4 and 5 shows the sample image with different RoIs and 1- bit as well as 2- bit embedded image. The proposed method is compared with the existing works and tabulated in Table 2.

Table 1. Details of the parameters of different images

Sample images	RoI (%) (P_{RoI})	MSE	PSNR	SSIM	NCC
Image I	25	0.1788	47.9380	0.9068	0.9996
	31.6406	0.1891	47.6944	0.9052	0.9996
	39.0625	1.8284	38.8374	0.8825	0.9991
	47.2656	1.7576	38.0090	0.9790	0.9991
Image II	25	0.1785	49.8750	0.9034	0.9999
	31.6406	0.1871	49.6750	0.8554	0.9999
	39.0625	1.8833	39.6405	0.8675	0.9996
	47.2656	1.7522	39.9534	0.8589	0.9996
Image III	25	0.1775	49.1553	0.9068	0.9998
	31.6406	0.1902	48.8556	0.8962	0.9999
	39.0625	1.8569	38.9561	0.8791	0.9995
	47.2656	1.7512	39.2122	0.8669	0.9997

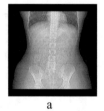
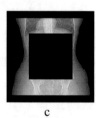
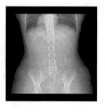

a   B   c   d

Fig. 4. Sample test images a. Original b. RoI c. RoNI d. 1-bit LSB embedded image

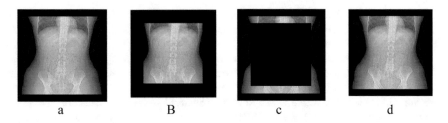

| a | B | c | d |

Fig. 5. Sample test images a. Original b. RoI c. RoNI d. 2-bit LSB embedded image

Table 2. Performance of proposed scheme in comparison to existing literatures

Methods	Embedding/ distortion inside RoI	Exact tampered blocks recognition inside RoI	Recognition of RoI in received image	Appropriate to medical images irrespective of size	Size of embedded data in RoNI
Das and Kundu [2]	Present	–	–	–	–
Liew et al., [11]	Not present	Possible	Not specified	No	Not needed
Eswaraiah and Sreenivasa [12]	Not present	Possible	Specified	Yes	Needed
Proposed method	Moderate	Possible	Specified	Yes	Not needed

4   Conclusion

This paper proposes the integrity verifiability. Here a unique and robust hashing technique for the detection of tampers inside RoI of a DICOM images has been proposed. Though the embedding capacity for each image and the RoI varies and is dependent on the input image, it is to be noted that the MSE, PSNR, SSIM and NCC are not compromised and the integrity of the image is maintained. Future works include the tamper detection in sonograms, satellite captured images and many other sensitive images. This algorithm can be further extended for retrieving the original DICOM image from the tampered image.

References

1. Kobayashi, L.O.M., Furuie, S.S., Barreto, P.S.L.M.: Providing integrity and authenticity in DICOM images: a novel approach. IEEE Trans. Inf. Technol. Biomed. **13**(4), 582–589 (2009)
2. Das, S., Kundu, M.K.: Effective management of medical information through ROI-lossless fragile image watermarking technique. Comput. Methods Programs Biomed. **111**(3), 662–675 (2013)

3. Brindha, M.: Confidentiality, integrity and authentication of DICOM medical images. In: Proceedings 2nd International Conference on Inventive Systems and Control ICISC 2018, ICISC, pp. 71–75 (2018)

4. Jamil, N., Aziz, A.: A unified approach to secure and robust hashing scheme for image and video authentication. In: Proceedings - 2010 3rd International Congress on Image and Signal Processing, CISP 2010, vol. 1, pp. 274–278 (2010)

5. Khor, H.L., Liew, S.C., Zain, J.M.: Region of interest-based tamper detection and lossless recovery watermarking scheme (ROI-DR) on ultrasound medical images. J. Digit. Imaging 30(3), 328–349 (2017)

6. Vasu, S., George, S.N., Deepthi, P.P.: An integrity verification system for images using hashing and watermarking. In: Proceedings - International Conference on Communication Systems and Network Technologies, CSNT 2012, pp. 85–89 (2012)

7. Mondal, S., Debnath, R., Mondal, B.K.: An improved color image steganography technique in spatial domain. In: Proceedings 9th International Conference on Electrical and Computer Engineering, ICECE 2016, pp. 582–585 (2017)

8. Basterrech, S., Janoušek, J., Snášel, V.: Intelligent Data Analysis and its Applications, Volume II. Advances in Intelligent Systems and Computing, vol. 298, pp. 459–468. Springer, Cham (2014)

9. Ghouti, L.: Robust perceptual color image hashing using randomized hypercomplex matrix factorizations. Multimed. Tools Appl. 77(15), 19895–19929 (2018)

10. Sinari, S., Aurora, A., Ruparel, D., Karamchandani, S.H.: Optimized techniques for image hashing. In: 2014 International Conference on Computing for Sustainable Global Development, INDIACom 2014, pp. 486–489 (2014)

11. Liew, S.C., Liew, S.W., Zain, J.M.: Tamper localization and lossless recovery watermarking scheme with ROI segmentation and multilevel authentication. J. Digit. Imaging 26(2), 316–325 (2013)

12. Eswaraiah, R., Sreenivasa Reddy, E.: Robust medical image watermarking technique for accurate detection of tampers inside region of interest and recovering original region of interest. IET Image Process. 9(8), 615–625 (2015)

Evaluation of Students Performance Using Neural Networks

B. Sai Kalyani, D. Harisha$^{(\boxtimes)}$, V. RamyaKrishna$^{(\boxtimes)}$, and Suneetha Manne$^{(\boxtimes)}$

Velagapudi Ramakrishna Siddhartha Engineering College,
Vijayawada, AP, India
kalyani.rosy@gmail.com, darisiharisha99@gmail.com,
ramyakrishnavadlamudil23@gmail.com,
hodit@vrsiddhartha.ac.in

Abstract. From Past decades, student performance is considered as an important factor for most of the educational institutions. The performance is evaluated based on various factors that plays a crucial part in the student career. In the recent years, students' performance prediction has become a significant challenge for all the institutions. Modern day educational institutions have adopted continuous evaluation to improve the performance. In the recent years, Neural Networks is used for predictions, which provides better results when compared to the classifiers. The data used for performance prediction will consists of the number of hours the student has spent for studying, his involvement in the academic activities and other contribution factors. These factors will play a crucial role in predicting the performance of the students. Thus, Neural Networks play a key role in predicting the students' performance. In particular, this paper uses Convolutional Neural Networks to predict the performance.

Keywords: Convolution neural networks · Academic performance · Predictions

1 Introduction

The major issue for the educational systems is to maintain the success rate of the students by helping the students to perform well in the examinations. There are many factors for the institutions to predict the performance of students. One important factor is to determine whether a particular student is eligible to get a scholarship from the institution. The second most important reason is to identify which students will fail in the semester end so that necessary steps can be taken to improve their performance. Normally the performance of the students is obtained manually by the teacher by observing the past scores or his performance in the classroom. This can be used even for the web based learning because the professor cannot get direct interaction with the student then it becomes difficult to measure the performance. But these assumptions may not work in all cases. As a result various algorithms are used to obtain accurate results.

© Springer Nature Switzerland AG 2020
A. P. Pandian et al. (Eds.): ICICCS 2019, AISC 1039, pp. 499–505, 2020.
https://doi.org/10.1007/978-3-030-30465-2_55

Initially various data mining classification models such as Naïve Bayes, Support Vector Machine, Decision Trees can be used for predictions but later it is found that the results are not perfect. Then machine learning algorithms have come into existence where the accuracy percentage is even better when compared to data mining algorithms. But with the introduction of Neural Networks the accuracy rate has increased rapidly. Neural Networks is a very powerful mechanism for predictions. Which predicts based on the loss calculated for every combination of the values and provides the best prediction. The most important problem is the large amount of data. These records must be transformed, calculate the percentage and provide the accurate predictions. But Neural Networks puts an end to this problem thereby enabling its usage for large amounts of data.

2 Literature Review

Ashraf, Anwer, Khan [1] compared various data mining techniques, classification algorithms to obtain the result. The main attributes that affect the students performance are also identified. Elbadrawyetal [2] used ANSI models by normalizing the results obtained in previous results. Three ANFIS models used models: ANFI GaussMF, ANFIS-TriMF, and ANFIS-GbellMF models and ANFIS-GbellMF model showed the best result. Kotsiantis, et al. [3] used six different machine learning algorithms to identify the poor performer students who are performing distance education. Prabha, Shanmuga Priyaa [4] used an evolutionary method that used Artificial Neural Networks and reducing the poor performance of the ANN using swarm optimization so that the weight values are properly updated thereby increasing the accuracy. Bendangnuksung, Prabu [5] used Deep Neural Networks to categorize the students based on their performance which observed a better accuracy than others. This paper will predict the percentage of each category the student belongs. This means that the percentages of the student becoming a low performer, high performer or a medium performer is predicted using Convolutional Neural Networks based on the given dataset.

3 Proposed Method

Convolutional Neural Networks (CNN) helps in designing the system to simulate the way, how the human brain analyzes and processes the given information. This technology helps in solving the problems that are impossible to solve by human brains and that beyond the imagination.

The detailed architecture is shown in Fig. 1.

Dataset

The dataset is a multivariate dataset consisting of 480 instances including categorical and numerical data. It consists of information of 305 males and 175 female students coming from different origins. The dataset is collected from two educational semesters which also consists of the attendance feature which is classified into two categories that is exceeding 7 absent days or under 7.

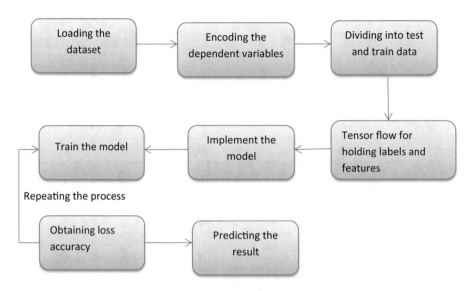

Fig. 1. Detailed architecture

Encoding
The dataset consists of categorical data which should be converted so that the performance can be predicted. Label encoding is used to convert the categorical values.

Tensor Flow
A symbolic math library called tensor flow is used so that it helps in building the model and helps in the creation of a large number of network layers.

The steps used are:

Step 1: Sequential is used to initialize the neural network.
Step 2: The first CNN layer uses ReLu activation function and output is given to next layer with same activation function but with dense value 32.
Step 3: These two layers are repeated. Sparse_categorical_entropy is calculated and the weight values are updated.
Step 4: The output is provided to the next layer with activation function sigmoid and accuracy value is obtained.

4   Experimental Results

Step 1: The dataset is in the form of .csv and is read to a variable.
Step 2: The data is extracted and is shown in Fig. 2.
Step 3: The labels are converted using Label Encoder shown in Fig. 3.
Step 4: The CNN model is implemented with 3 dense layers and activation function. Also the loss value is calculate and the values are updated as shown in Fig. 4.

gender	NationalITy	PlaceofBirth	StageID	GradeID	SectionID	Topic	Semester	Relation	raisedhands	VisITedResources	AnnouncementsView	Discussion	ParentAnsweringSchool
M	KW	KuwaIT	lowerlevel	G-04	A	IT	F	Father	15	16	2	20	Yes
M	KW	KuwaIT	lowerlevel	G-04	A	IT	F	Father	20	20	3	25	Yes
M	KW	KuwaIT	lowerlevel	G-04	A	IT	F	Father	10	7	0	30	No
M	KW	KuwaIT	lowerlevel	G-04	A	IT	F	Father	30	25	5	35	No
M	KW	KuwaIT	lowerlevel	G-04	A	IT	F	Father	40	50	12	50	No
F	KW	KuwaIT	lowerlevel	G-04	A	IT	F	Father	42	30	13	70	Yes
M	KW	KuwaIT	MiddleSchool	G-07	A	Math	F	Father	35	12	0	17	No

Fig. 2. Data extraction

```
le_gen = LabelEncoder()
le_nat = LabelEncoder()
le_pob = LabelEncoder()
le_top = LabelEncoder()
le_sem = LabelEncoder()
le_rel = LabelEncoder()
le_pas = LabelEncoder()
le_pss = LabelEncoder()
le_abs = LabelEncoder()
le_cls = LabelEncoder()
```

Fig. 3. Label conversion

```
def create_cnn():
    model = Sequential()
    model.add(Dense(64, activation='relu', input_dim=6))
    model.add(Dense(32, activation='relu'))
    #model.add(Dense(1, activation='softmax'))
    model.add(Dense(3, activation='sigmoid'))
    model.compile(optimizer="adam", loss="sparse_categorical_crossentropy", metrics=["accuracy"])
    return model
```

Fig. 4. Implementing the model

Step 5: The dataset is divided into train and test. The model is fit and 100 epochs are trained with batch_size 5 as in Fig. 5.

```
X = ndf.iloc[:,:-1]
y = ndf.iloc[:,-1]
print('Length of Features:{}'.format(X.shape))
print('Length of Labels:{}'.format(y.shape))
X_train,X_test,y_train,y_test = train_test_split(X,y,test_size=0.3,random_state=29)
print('Length of taining set={} & testing set={}'.format(len(X_train),len(X_test)))

model = create_cnn()
print(model.summary())
result = model.fit(X_train,y_train, epochs=100, batch_size=5,verbose=2,validation_data=(X_test,y_test))
y_pred = model.predict(X_test)
```

Fig. 5. Train test split

Test Case Analysis of Result

The details of a random student are provided as input by the user as shown in Fig. 6 and the percentage of performance is predicted.

```
while(True):
    print('Enter feature values of students you want to predict: ')
    """
    ['Relation','raisedhands',
     'VisITedResources','AnnouncementsView',
     'Discussion','StudentAbsenceDays','Class']
    """
    inp_rel = int(input('Relation[0=Father/1=Mother]: '))
    inp_rh = int(input('Raised hands count[Number]: '))
    inp_vr = int(input('Visited Resources[Number]: '))
    inp_av = int(input('Announcements View[Number]: '))
    inp_dis = int(input('Discussion[Number]: '))
    inp_abs = int(input('No. of Absent Days[0=Above-7/1=Under-7]: '))
    X_eval = [inp_rel,inp_rh,inp_vr,inp_av,inp_dis,inp_abs]
    X_eval = np.array(X_eval).reshape(1, -1)
    #print(X_eval,type(X_eval))
    y_eval = model.predict(X_eval)
    #print(y_eval[0],type(y_eval[0]),y_eval)
    res = ['High','Low','Medium']
    print('Our Neural Network Prediction of given student features is: ')
    for i in range(len(y_eval[0])):
        print('{} Performer: {}%'.format(res[i],round(y_eval[0][i]*100,2)))
    return
```

Fig. 6. Input given by the user

The input of a random student is given it is processed and the performance of a random student is as shown in Fig. 7.

```
Enter feature values of students you want to predict:
Relation[0=Father/1=Mother]: 0
Raised hands count[Number]: 44
Visited Resources[Number]: 59
Announcements View[Number]: 74
Discussion[Number]: 66
No. of Absent Days[0=Above-7/1=Under-7]: 0
Our Neural Network Prediction of given student features is:
High Performer: 76.78%
Low Performer: 2.63%
Medium Performer: 100.0%
Testing accuracy: 0.67599993657
```

Fig. 7. Test case of a student

5   Conclusion and Future Scope

The main purpose is to develop model that predicts the performance of students. The study shows that it is possible to predict the student graduation performance, which is based on the major impacting factors that affects the final output of the student. The experiment results are done in Python and convolutional neural networks is used for predicting the performance. It is concluded that the meta-analysis on predicting student's academic performance motivated us to do further research work in our own educational environment. With the usage of a large dataset the training is done better and more accurate results will be predicted. The scope of student performance prediction is very high in the world of neural network. With increase in the pressure for the educational institutions to increase the performance of the students, predicting it before their actual result will help in improving their performance.

References

1. Ashraf, A., Anwer, S., Khan, M.G.: A comparative study of predicting student's performance using data mining techniques. ASRJETS **44**, 122–136 (2018)
2. Elbadrawy, A., Polyzou, A., Ren, Z., Karypis, M.G., Rangwala, H.: Predicting student performance using personalized analytics. Computer **49**, 61–69 (2016). IEEE Computer Society
3. Kotsiantis, S., et al.: Preventing student dropout in distance learning systems using machine learning techniques. Appl. Artif. Intell. **18**(5), 411–426 (2003)
4. Prabha, T., Shanmugha Priyaa, D.: An evolutionary approach on students performance prediction and classification. Int. J. Pure Appl. Math. **119**(12), 15341–15361 (2018). http://www.ijpam.eu
5. Bendangnuksung, Prabu, P.: Students' performance prediction using deep neural network. Int. J. Appl. Eng. Res. **13**, 1171–1176 (2018)
6. Iyanda, A.R., Ninan, O.D., Ajayi, A.O., Anyabolu, O.G.: Predicting student academic performance in computer science courses: a comparison of neural network models. MECS **10**, 1–9 (2018)

7. Jayasree, C., Baseer, K.K.: Predicitng student performance to improve their employability by applying data mining and machine learning techniques. JCSE **6**, 1292–1308 (2018)
8. Surmenok, P.: Estimating an optimal learning rate for a deep neural network. Towards Data Science (2017)
9. Techniques. Procedia Computer Science (2015)
10. Sebastian, S., Puthiyidam, J.J.: Evaluating students performance by artificial neural network using weka. Int. J. Comput. Appl. **119**, 36–39 (2015)
11. Okubo, F., Yamshita, T., Shimada, A., Ogata, H.: A neural network approach for students' performance prediction. ACM (2017)
12. Manchandia, K., Khare, N.: Implementation of student performance evaluation through supervised learning using neural network. IJSERT (2017)

Disruption of Object Recognition Systems

Utsav Das[✉], Aman Gupta[✉], Onkar Singh Bagga[✉],
and Manoj Sabnis[✉]

Department of Information Technology, Vivekanand Education Society's
Institute of Technology, Mumbai, India
{2015utsav.das, 2015aman.a, 2015onkarsingh.bagga,
manoj.sabnis}@ves.ac.in

Abstract. In recent times, deep neural networks are being used in a wide variety of applications such as autonomous vehicles, medical imaging and surveillance. While they are becoming increasingly powerful, it is possible to disrupt their task by crafting adversarial inputs. These inputs are essentially perturbations added to the original inputs so that the application using the network, such as an object recognizer, is unable to classify the object in the image. Crafting such inputs to disrupt such a recognition task is termed an adversarial attack. Here, we implement two disruption strategies, Fast Gradient Sign Method (FGSM) and generating perturbations using a generator network. While FGSM requires access to the gradient calculated by the classifier with respect to the input image, the generator trains simultaneously with the classifier network to learn how to craft perturbations. Once the generator network is trained with a particular classifier (say, VGG16), it can disrupt other classifier networks in a black-box fashion as well. Using the same dataset, in this case CIFAR-10, it is possible to adversarially train the classifier to make it more robust to perturbed images. This involves training the classifier on the CIFAR-10 images with both the original images and the ones perturbed by the generator. In experiments, the attack using the generator achieves higher disruption accuracies than FGSM on very deep networks.

Keywords: Object recognition · Deep learning · Adversarial attacks

1 Introduction

Computer vision tasks using deep neural networks are being used in a wide variety of applications such as autonomous vehicles, surveillance, and manufacturing [13]. Considering their widespread use, it is necessary to understand how vulnerabilities in these systems can be exploited. It was found in [4] that by adding a hardly perceptible perturbation to the input images, it is possible to fool deep neural networks (DNNs) and force them to misclassify the input images. These perturbed images are termed adversarial samples and they can be used to conduct adversarial attacks on DNNs. There are several possible kinds of adversarial attacks based on the approach taken and the level of access they require to the model being attacked. In this paper, we provide the results obtained using two disruption strategies; crafting perturbations using the Fast Gradient Sign Method [1] (FGSM) and through a generator network [2].

© Springer Nature Switzerland AG 2020
A. P. Pandian et al. (Eds.): ICICCS 2019, AISC 1039, pp. 506–513, 2020.
https://doi.org/10.1007/978-3-030-30465-2_56

2 Related Work

One of the first and most popular adversarial attacks is the Fast Gradient Sign Method (FGSM) [1], where the attack is crafted by adjusting the input data to maximize the loss based on the gradients backpropagated by the neural network.

For an input image x, FGSM returns a modified input x':

$$x' = x - \in . \, sign(\Delta_x J(\theta, x, y)) \tag{1}$$

Here J is the loss function of the model under attack and $\in$ is the extent to which the modified image will be perturbed. This value should be chosen so that the perturbations are not easily apparent. The adversarial inputs crafted with FGSM are also highly transferable, in that the adversarial inputs can be used for different kinds of neural networks, as explained in [1]. Since the attack needs the gradients calculated by the model, it is of the white-box variety, which means it requires complete access to the network it is attacking. It is also possible to carry out attacks in a black-box fashion, as explained in [6]. Here, the attacker is first trained with a particular model in a white-box manner and then attacks another model to which it has no access (black-box).

With FGSM as a baseline, we run a second adversarial attack using a generator network [2]. The generator network is a neural network that is trained simultaneously with a DNN classifier. For an input image, the generator creates a perturbation and adds it to the image with the intention of fooling the classifier. The classifier has no clue as to the generator's existence and is trained only on unperturbed images. After a sufficient number of iterations, the generator network is able to craft perturbations that are effective enough to fool the classifier network it is trained with to a reasonable extent. Once the generator is trained, the perturbations are generated with a forward pass through the network. The approach is similar to that used in [9] except that the generator and discriminator are trained concurrently and the generator creates a perturbation instead of a completely modified image.

As seen above, adversarial attacks can be grouped into white-box and black-box attacks. They can also be grouped into targeted and untargeted attacks based on the goal of the attack. The goal of an untargeted attack is to get the attacked model to simply misclassify the output, without any concern for what the wrongly predicted class is. In a targeted attack, the goal of the attacker is to get the attacked model to misclassify the input as a particular/intended class. The attacks crafted by both FGSM and the generator network are untargeted attacks.

3 Approach

3.1 System Design

Architecture
See Fig. 1.

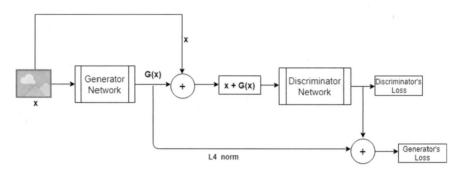

Fig. 1. System architecture

Steps:

- The input image is fed to the generator network.
- The generator network produces a minimum perturbation *(G(x))*. This gets added to the original image *(x)*, producing a perturbed image.
- The perturbed image is fed to the discriminator network.
- Based on the labels predicted, the discriminator's loss is computed using cross-entropy loss.
- The generator's loss is computed as *L4* norm plus its success on fooling the discriminator network.

3.2 Threat Model

Two networks – a discriminator D and a generator G – are trained in tandem with each other as in [2]. G is conditioned on the same input image x as seen by the classifier D. Specifically, G produces a small perturbation that can be added to x to produce an adversarial image x'. D remains oblivious to the presence of G while G's loss depends on how well it can fool D into misclassifying x'. Over time, G produces perturbations that can effectively fool the discriminator D it is trained with. Once fully trained, G can be used to generate image-conditional perturbations with a simple feed-forward operation. The loss functions for D and G are:

$$L_D(\theta_D, x, y) = -logD(x; \theta_D)_y \qquad (2)$$

$$L_G(x, y) = D(x + \epsilon . G(x); \theta_D)_y + c_g . max\left(||G(x)||_{p,} 1^{-3}\right) \qquad (3)$$

where $\in$ is the degree of perturbation, c_g is a weight on the L_G regularization term, and p denotes the p-norm. D has a regular cross-entropy loss while G's loss is the likelihood of the perturbed images and an L_p regularization term with a lower bound of 1^{-3}. The values used are $p = 4$, $c_g = 0.005$ and $\in= 1.0$ in (3) [2].

3.3   White-Box and Black-Box Attacks

The attack using both the generator and FGSM is transferable to other deep learning neural networks [6]. We use our trained G on a different D' that was not used to train G. Concretely, during training, G has white-box access to the discriminator D it is being trained with. Any subsequent adversarial attack by G on D is, therefore, a white-box attack. We can conduct black-box attacks on some unknown discriminator D' by training G with some known neural network classifier (D) and then using G, once fully trained, to generate adversarial images for D'.

3.4   Adversarial Training

We can train the discriminator D adversarially as shown in [3] and [2]. This encourages D to become more robust to perturbed images and encourages G to produce stronger perturbations that can fool the increasingly robust D. In this approach, the classifier under attack is trained on both the original and modified images, similar to a Generative Adversarial Network [5]. The loss function for the adversarially trained discriminator D is:

$$L_D\left(\theta_f, x, y\right) = -log\, D(x + \epsilon \cdot G(x); \theta_D)_y - log\, D(x; \theta_D)_y \qquad (4)$$

where the discriminator's loss is the cross-entropy over the regular inputs and the cross-entropy over the perturbed inputs. The generator loss remains unchanged from the original formulation. An adversarially trained generator can be used in black-box attacks.

4   Results

We test and compare these two disruption strategies on the CIFAR-10 dataset [11]. We use the generator network to attack two state-of-the-art models, VGG16 [7] and ResNet 18 [8]. Both attacks are untargeted. We perform both black-box and white-box attacks, with both non-adversarially trained and adversarially trained models. The architecture of the generator network is based on Lee et al. [3] and as shown in [2].

White-box attacks are only possible when both the discriminator and generator are trained together, either adversarially or non-adversarially. Along with the above two combinations, black-box attacks are also possible with only either the generator or discriminator being adversarially trained. However, in a black-box attack, the discriminator and generator have not been trained together.

As per [2], in the non-adversarial setting, we train the discriminator with loss (2). In the adversarial setting, we train the discriminator with loss (4). In both settings, we train the generator with loss (3). We use a batch size of 128, learning rate 0.001 and the Adam optimizer [10]. All experiments are done on the test set of the CIFAR-10 dataset which contains 10000 images.

In the tables below, column 1 indicates the classifier network and column 2 indicates the network with which the generator was trained. Column 3 indicates the classifier accuracy on the original CIFAR-10 dataset while columns 4 and 5 show the classifier accuracy after the FGSM and the generator network's attacks respectively.

4.1 White-Box Attacks

Table 1. Non-adversarially trained discriminator and generator

Classifier	Attacker	Unperturbed	FGSM	Generator
VGG16	VGG16	87.27	27.58	19.47

Table 2. Adversarially trained discriminator and generator

Classifier	Attacker	Unperturbed	FGSM	Generator
VGG16	VGG16	86.79	32.64	64.46

4.2 Black-Box Attacks

Table 3. Non-adversarially trained discriminator and generator

Classifier	Attacker	Unperturbed	FGSM	Generator
Res18	VGG16	83.29	23.11	19.48

Table 4. Non-adversarially trained discriminator and adversarially trained generator

Classifier	Attacker	Unperturbed	FGSM	Generator
Res18	VGG16	83.29	23.11	25.96

Table 5. Adversarially trained discriminator and Non-adversarially trained generator

Classifier	Attacker	Unperturbed	FGSM	Generator
Res18	VGG16	85.38	24.47	25.76

Table 6. Adversarially trained discriminator and generator

Classifier	Attacker	Unperturbed	FGSM	Generator
Res18	VGG16	85.38	29.2	26.09

4.3   Visualizations

FGSM Visualizations

The images below visualize the FGSM attack, with the original images followed by perturbed images and finally the perturbations that are added to the original images with epsilon value 0.15.

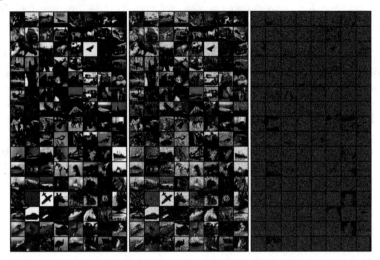

Generator Network Visualizations

The images below visualize the generator's attack with the original images followed by the perturbed images and finally the perturbations that are added to the original images with epsilon value 1.0.

5 Conclusion

We have implemented two disruption strategies on two state-of-the-art models, VGG16 and ResNet 18. The first strategy is FGSM and is used as a baseline to compare accuracies with a generator network's attack. The generator network's attack is crafted through jointly training a discriminator D and a generator G. G is responsible for learning to create image-conditional adversarial perturbations that can fool D. Once trained, G can be used to conduct white-box attacks against D and black-box attacks against other classifiers D'. For non-adversarially trained white box attacks, the generator network's attack outperforms FGSM (Table 1). The black-box attacks for both the adversarially and non-adversarially trained classifier and generator combinations are highly transferable to other DNNs as hypothesized by Goodfellow et al. [1] and shown in [6], resulting in significant drops in classification accuracy (Tables 3, 4, 5 and 6). After adversarially training the classifier (VGG16) and the generator, the classification accuracy rises from about 19% when non-adversarially trained to about 64% for a white box attack (Table 2), indicating that the classifier has become more robust to adversarial inputs. In the future, it is possible to study defense techniques to these adversarial attacks [12].

References

1. Goodfellow, I., Shlens, J., Szegedy, C.: Explaining and harnessing adversarial examples. arXiv preprint arXiv:1412.6572 (2014)
2. Crafting Adversarial Attacks with Adversarial Transformations. https://github.com/kawine/atgan
3. Lee, H., Han, S., Lee, J.: Generative adversarial trainer: defense to adversarial perturbations with GAN. arXiv preprint arXiv:1705.03387 (2017)
4. Szegedy, C., Zaremba, W., Sutskever, I., Bruna, J., Erhan, D., Goodfellow, I., Fergus, R.: Intriguing properties of neural networks. arXiv preprint arXiv:1312.6199 (2013)
5. Goodfellow, I., Pouget-Abadie, J., Mirza, M., Xu, B., Warde-Farley, D., Ozair, S., Courville, A., Bengio, Y.: Generative adversarial nets. In: Advances in Neural Information Processing Systems, pp. 2672–2680 (2014)
6. Papernot, N., McDaniel, P., Goodfellow, I., Jha, S., Celik, Z.B., Swami, A.: Practical black-box attacks against machine learning. In: Proceedings of the 2017 ACM on Asia Conference on Computer and Communications Security, pp. 506–519. ACM (2017)
7. Simonyan, K., Zisserman, A.: Very deep convolutional networks for large-scale image recognition. arXiv preprint arXiv:1409.1556 (2014)
8. He, K., Zhang, X., Ren, S., Sun, J.: Deep residual learning for image recognition. In: Proceedings of the IEEE Conference on Computer Vision and Pattern Recognition, pp. 770–778 (2016)
9. Baluja, S., Fischer, I.: Adversarial transformation networks: learning to generate adversarial examples. arXiv preprint arXiv:1703.09387 (2017)
10. Kingma, D., Ba, J.: Adam: a method for stochastic optimization. arXiv preprint arXiv:1412.6980 (2014)

11. Krizhevsky, A., Nair, V., Hinton, G.: The cifar-10 dataset (2014). http://www.cs.toronto.edu/kriz/cifar.html
12. Shen, S., Jin, G., Gao, K., Zhang, Y.: APE-GAN: adversarial perturbation elimination with GAN. arXiv:1707.05474 [cs.CV] (2017)
13. Computer vision. https://en.wikipedia.org/wiki/Computer_vision#Applications

An Efficient Hand Gesture Recognition System Using Deep Learning

R. Deepa[1(✉)] and M. K. Sandhya[2(✉)]

[1] Department of Information Technology, Loyola-ICAM College of Engineering
and Technology, Chennai, India
deepa@licet.ac.in
[2] Department of Computer Science and Engineering, Meenakshi Sundararajan
Engineering College, Chennai, India
mksans@gmail.com

Abstract. Most of the people perform various tasks by using a computer keyboard/mouse leading to repetitive wrist and hand motions, resulting in Carpal Tunnel Syndrome. This paper is geared towards developing a computer management system using hand gestures accomplishing virtual keyboard/mouse operations/commands to effectively eliminate the Carpel Tunnel Syndrome. Gesture Recognition provides an accurate estimation of hand gestures using deep learning algorithm. The complexity of hand structure in obtaining gestures and the rapidness of the movements of the hand or fingers are the problems of tracking algorithms. Thus, deep learning provides a rapid and precise estimate of hand gestures using Convolutional Neural Network (CNN) algorithm. This paper uses articulated CNN algorithm capturing possible gestures, accomplishing various keyboard/mouse operations/commands, thereby avoiding the syndrome. Compared to the conventional algorithm, the proposed work produces high accuracy, a good estimation of hand gestures and cost-effective.

Keywords: Gesture Recognition · Convolutional neural networks · Carpel
Tunnel Syndrome · Deep learning

1 Introduction

Computer technology has immensely grown up over the past decades and become an essential function of everyday lives. The two main accessories for Human-Computer Interaction are the keyboard and mouse. There are various kinds of keyboard and mouse that existed in cutting-edge innovation. The mechanical mouse was replaced by the optical mouse which used a LED sensor to detect table top movement. The laser mouse was then acquainted with improving the accuracy of the movement with the scarcest hand movement, it conquers the restrictions of the optical mouse which is the challenge to track high-gloss surface [1].

Due to the continuous and repetitive usage of keyboard and mouse to perform tasks, people may experience symptoms such as discomfort in the hand radiating to the elbow, hand burning, inflammation sensation, weakening of grip while picking up small items, problems with fine finger movements. These symptoms lead to the Carpal

© Springer Nature Switzerland AG 2020
A. P. Pandian et al. (Eds.): ICICCS 2019, AISC 1039, pp. 514–521, 2020.
https://doi.org/10.1007/978-3-030-30465-2_57

Tunnel Syndrome – a hand nerve entrapment disorder. It is because of the compression of the median nerve due to the swelling in the tunnel [2].

The proposed system avoids Carpal Tunnel Syndrome by identifying hand gestures using a convolutional neural network (CNN) to perform various commands/operations. CNN is one of the deep learning algorithms, which captures the possible variability of gestures performed by different people [3]. Deep learning models are analogous to the structure of the nervous system, where each neuron is connected and passing information to each other [4]. These models are at least three-layered model, where each layer receives the information from the preceding one and pass it on to the next layer. In this paper, the system is controlled using hand gestures by operating each and every function of the computer in an interactive way, thereby eliminating the use of keyboard/mouse operations. This system can also be useful for gaming, medical applications among others.

Section 2 presents the existing work. Section 3 explains the proposed Hand Gesture Recognition System. Section 4 discusses the performance analysis. Section 5 concludes with future work.

2   Existing Work

Human-computer interactions have become important in modern technology and programming, which allow unlimited ways to access computers. Hence, physical keyboards/mouse will no longer be needed and be replaced by video cameras that track gestures. In this section, we explore some of the recent methods for gesture recognition.

Zhengyou et al. proposed an interface framework named Visual Panel [5], which uses an arbitrary quadrangle-shaped panel, any tip-pointer tool to interact with the PC, but cannot acknowledge other shapes other than expressed shape. Kim et al. developed a sensor using ultra-wideband impulse signals for hand gesture recognition [6]. The waveforms reflected from the hand are determined by the reflection surface of a target in the time domain. Six hand gestures from American Sign Language are employed and presents a recognition accuracy of more than 90%. Deng et al. formulated an in-plane rotation model that explicitly solves hand detection and rotation estimation using CNN [7]. The network architecture employs faster R-CNN to generate rectangular region and extract local features.

Nuzzi et al. developed an intelligent hand gesture recognition system for Collaborative Robots to find the exact position of the hands in RGB images using a Faster R-CNN Object Detector [8]. The spatial requirements aid in filtering out the incorrect predictions made by the detector and improve the performance. But it identifies only four gestures. Varun et al. developed a model using gestures to help with people with difficulty in controlling or operating devices by applying CNN and Back Propagation methodologies [9]. Though the author in [9] claims high-level accuracy in gesture recognition, the gestures made by the disabled people may not be accurate and hence the system may not perform well in real time situations.

In this paper, the proposed system identifies hand gestures and are mapped to some of the keyboard/mouse operations to avoid Carpal Tunnel Syndrome.

3   Hand Gesture Recognition System (HGRS)

The architecture of the proposed system HGRS is shown in Fig. 1. The webcam captures the image, in which hand sequences are located by classifying the image into the foreground and background parts using skin color segmentation technique [10]. This technique uses RGB of skin color, which separates the hand, head or body from the background. The foreground contains the hand and background contains the rest of the image. Thus, the hand is located successfully within a noisy/cluttered background and identified the contour of the hand region from the image.

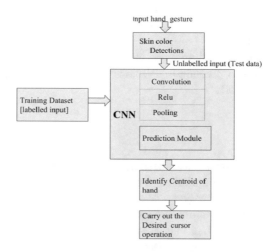

Fig. 1. Architecture of hand gesture recognition system

The steps to recognize hand gestures using the CNN model are

Step1: The processed image is input to the Convolutional layer to extract features
Step 1a: Increase the non-linearity in images using ReLU function
Step 2: Reduce the dimensionality of images using pooling layer
Step 3: Goto Step1 to perform the series of convolutions and pooling operations, repeatedly for eight iterations
Step 4: Convert the 3D data to one-dimensional data by process of flattening
Step 5: Predict gestures using a fully connected layer.
Step 6: Map these gestures to keyboard/mouse operations/commands to avoid Carpel Tunnel Syndrome.

3.1   Gesture Recognition Using Convolutional Neural Network

Deep learning networks are used to recognize a very complex pattern by breaking it into simpler patterns consisting of different layers. Generally, the deep networks take much longer time to train the network, but the recent technology advancements reduce the amount of time it takes to properly train a network. One of the deep learning algorithms is the Convolutional Neural Network. CNN is an arrangement of layers, where each layer changes one volume of activations to another through a differentiable function. In this paper, three major types of layers such as the Convolution, Pooling, and Fully-Connected are stacked to build CNN architecture as shown in Fig. 2.

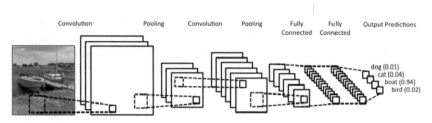

Fig. 2. General architecture of convolutional neural networks (source: http://www.wildml.com/2015/11/understanding-convolutional-neural-networks-for-nlp/)

The convolution layer extract features mainly from the image input. The input to this layer includes hand images with parameters such as batch size, timesteps, image height, width and depth and their values as 1, 1, 50, 50 and 5 respectively. This layer preserves the spatial relationship between pixels by using small squares of input data to learn image features. This layer convolves this input image with each of these eight filters individually, and apply zero paddings. Each filter will slide over each 5 × 5 set of pixels from the input itself until it slid over every 5 × 5 blocks of pixels from the entire image. This layer is applied with a zero padding of size 2. Zero padding pads the matrix input with zeros around the matrix border.

The rectified linear unit (ReLU) activation function is then applied to the matrix to detect sophisticated hand gestures gradually. ReLU increases the non-linearity in the output received from the convolution layer. This is done by setting all negative values in the matrix to zero and all other values are kept constant without affecting the receptive fields of the convolution layer. It is mathematically shown as

$$A(x) = \max(0, x)$$

If x is positive, it gives an output x. Otherwise, the output is 0.

ReLU is less computationally expensive with simpler mathematical operations. Only a few neurons are activated at a time, making the network sparse and efficient.

Next, pooling layer reduces image dimensionality by reducing the number of pixels in the previous convolution layer output. This layer has a filter of size 2 × 2 with a stride of 1 for max pooling operation. The filters are moved one pixel at a time when the stride is 1, as it slides across the image. This image extracts the most activated pixels/features and keeps these high values going forward while discarding the non-active pixels. Therefore, max pooling reduces the number of parameters in the network, thus reducing the computational load and overfitting.

Then, the network performs the series of convolutions and pooling layers with ReLU activations [11] between them repeatedly for eight iterations, before finally arriving at a traditionally fully connected network. After the convolution and pooling layers, the network would recognize high-level features of hand gestures named like an acknowledgement, fist, free hand, index finger, straight hand, palm, thumb and swing. Then, flatten the output from the pooling layer by converting the 3D-data to 1D data, which is fed to the fully connected layer.

The fully connected layer serves as a classifier on these extracted features and predicts class label based on the pre-trained dataset. This layer is a traditional multi-layer perceptron, where each neuron in the next layer connects every neuron in the previous layer. In a fully connected layer, Tangent Hyperbolic (Tanh) function is used in eight hidden layers. The tanh function is given as

$$f(x) = \tanh(x) = 2/(1 + e^{-2x}) - 1$$

This function is used by bringing mean close to 0 to center the data, which makes learning much easier for the next layer.

In a fully connected layer, the output layer uses Sigmoid function, which is plotted as a graph in the shape of 'S'. Compared to the Softmax function used in the model [9], the sigmoid activation function works well in the fully connected layer, the classifier, which leads to faster training process and convergence. Using this function, each identified gesture is independent of one another and belongs to exactly one class. The equation for the sigmoid function is given as

$$A = 1/(1 + e^{-x})$$

Therefore, based on the training data set, the fully connected layer classifies the extracted features into different classes. Then, the observed gesture is fed into the CNN model, which is already trained with 10000 images of different gestures. If the gesture is not matched, then it automatically takes the most relevant gesture and this gesture is added to the training dataset.

3.2   Cursor Operations

The gestures identified using the convolutional neural network are an acknowledge-ment, fist, free hand, index finger, straight hand, palm, thumb and swing. The gesture is recognized and the cursor point is mapped with the centroid of the hand region and desired cursor operation/commands are performed. These gestures are mapped to few keyboard/mouse operations/commands like select, cut, copy, paste, move and delete. The palm or freehand gesture is mapped to 'move' pointer operation, the fist to 'select' operation, the thumb/index to 'cut' operation, swing to 'copy' operation, acknowl-edgement to 'paste' operation, and straight hand to 'delete' operation. For example, these set of gestures can be used to copy and paste the file. These mapped operations avoid Carpel Tunnel Syndrome.

4 Performance Analysis

The proposed system uses Python and Java to implement gesture prediction and keyboard/mouse control operations respectively. Datasets for different gestures and a reasonable number of images are collected (in thousands) so that the output after training gives an accurate result. After collecting the datasets for different gestures, training is carried out on these datasets using Google Colaboratory (Colab) platform [12] equipped with TensorFlow [13] and Keras [14]. These integrate the deep learning framework and APIs. The hand gesture, as input is captured by the system webcam and is given to the prediction module. This image gets preprocessed, where it is shrunk down to 50 × 50 pixel and then fed to the CNN. Thus, the performance depends on the system, how it is able to train and run this network. The proposed system's performance is evaluated on a set of 10,000 hand images.

4.1 Prediction of Various Gestures

The proposed system predicts various gestures as described as follows. The palm or the freehand gesture is used for 'move' pointer operation, the fist for 'select' operation, the thumb/index for 'cut' operation, swing for 'copy' operation, acknowledgement for 'paste' operation, and straight hand for 'delete' operation. Figure 3 shows the prediction of various gestures such as acknowledgment gesture with the probability of 98%, freehand gesture with the probability of 99%, swing gesture with 99%, index finger with 99%, background (none) with 99%, straight hand with 99%, fist gesture with the 98%.

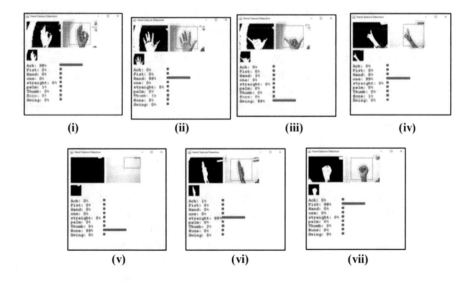

Fig. 3. Recognizing various gestures (i) ack gesture (ii) free hand gesture (Hand) (iii) Swing gesture (iv) index finger [One] (v) background (None) (vi) Straight (vii) Fist

4.2   Accuracy Evaluation

Figure 4 shows a graph, where the accuracy increases with an increase in the number of epochs. The accuracy is not a steep increase because the training model keeps learning batch wise at each iteration, where the batch size is 1000 and iterations is 10, for 1 complete epoch. This shows that different hand gestures are identified accurately and mapped to various commands as discussed in Sect. 3.2. Compared to the model explained in [9], which shows an accuracy of 1.0 theoretically, but this system improves the accuracy as the number of epoch increases.

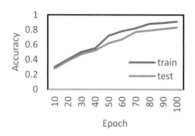

Fig. 4.   Accuracy

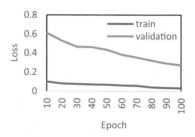

Fig. 5.   Model training & validation loss

4.3   Loss Evaluation

Figure 5 shows a graph with epoch against loss in training and validating the CNN model. The loss is inversely proportional to accuracy. The proposed system implements CNN to construct a model by looking at many examples and trying to find a model that minimizes loss. Loss is the penalty for a bad prediction, i.e. a number that shows how bad the prediction of the model was on a single example. If the prediction of the model is perfect, the loss is zero; if not, the loss is greater. The goal of training a model is to find, on average across all examples, a set of weights and biases that have a low loss. Therefore, in the proposed system, validation loss of hand gestures decreases steeply, when the epoch increases.

5   Conclusion

In this paper, hand gestures are recognized using the convolutional neural network model to perform different keyboard/mouse operations to avoid Carpel Tunnel Syndrome. The proposed system improves accuracy and minimizes training and validation loss. The response time heavily relies on the hardware of the machine including the processing speed, available RAM size, and the available the webcam features. This system may have better performance when it is running on a standard machine with a webcam that performs better in different types of lightings. This system can be further enhanced by using sensors to be more user-friendly, accurate, and flexible for various devices. This system can be extended to use in Augmented and Virtual reality environments to position, scale, and animate 3D virtual objects.

References

1. Pereira, C., Neto, R., Reynaldo, A., Luzo, M., Oliveira, R.: Development and evaluation of a head-controlled human-computer interface with mouse-like functions for physically disabled users. Clin. Sci. **64**, 975–981 (2009)
2. Thomsen, J.F., Gerr, F., Atroshi, I.: Carpal tunnel syndrome and the use of computer mouse and keyboard: a systematic review. BMC Musculoskelet Disord. **9**, 134 (2008). https://doi.org/10.1186/1471-2474-9-134
3. Wu, D., Pigou, L., Kindermans, P., Le, N.D., Shao, L., Dambre, J., Odobez, J.: Deep dynamic neural networks for multimodal gesture segmentation and recognition. IEEE Trans. Pattern Anal. Mach. Intell. **38**(8), 1583–1597 (2016)
4. Marblestone, A.H., Wayne, G., Kording, K.P.: Toward an integration of deep learning and neuroscience. Front. Comput. Neurosci. **10**, 94 (2016). https://doi.org/10.3389/fncom.2016.00094
5. Zhang, Z., Wu, Y., Shan, Y., Shafer, S.: Visual panel. In: International Journal of Advanced Research in Computer Engineering & Technology (IJARCET), November 2001
6. Kim, S.Y., Han, H.G., Kim, J.W., Lee, S., Kim, T.W.: A hand gesture recognition sensor using reflected impulses. IEEE Sens. J. **17**(10), 2975–2976 (2017). https://doi.org/10.1109/jsen.2017.2679220
7. Deng, X., et al.: Joint hand detection and rotation estimation using CNN. IEEE Trans. Image Process. **27**(4), 1888–1900 (2018). https://doi.org/10.1109/TIP.2017.2779600
8. Nuzzi, C., Pasinetti, S., Lancini, M., Docchio, F., Sansoni, G.: Deep learning-based hand gesture recognition for collaborative robots. IEEE Instrum. Meas. Mag. **22**(2), 44–51 (2019). https://doi.org/10.1109/MIM.2019.8674634
9. Varun, K.S., Puneeth, I., Jacob, T.P.: Hand gesture recognition and implementation for disables using CNN'S. In: 2019 International Conference on Communication and Signal Processing (ICCSP), Chennai, India, pp. 0592–0595 (2019). https://doi.org/10.1109/iccsp.2019.8697980
10. Brancati, N., De Pietro, G., Frucci, M., Gallo, L.: Human skin detection through correlation rules between the YCb and Ycr subspaces based on dynamic color clustering. Comput. Vis. Image Underst. **155**, 33–42 (2017)
11. Maas, A.L., Hannun, A.Y., Ng, A.Y.: Rectifier nonlinearities improve neural network acoustic models. In: Proceedings ICML, vol. 30 (2013)
12. Carneiro, T., Nóbrega, R.V.M.D., Nepomuceno, T., Bian, G., Albuquerque, V.H.C.D., Filho, P.P.R.: Performance analysis of google colaboratory as a tool for accelerating deep learning applications. IEEE Access **6**, 61 (2018)
13. Abadi, M., Agarwal, A., Barham, P., Brevdo, E., Chen, Z., Citro, C., Ghemawat, S.: TensorFlow: large scale machine learning on heterogeneous distributed systems. arXiv preprint arXiv:1603.04467 (2016)
14. Bastien, F., Lamblin, P., Pascanu, R., Bergstra, J., Goodfellow, I., Bergeron, A., Bengio, Y.: Theano: new features and speed improvements. arXiv preprint arXiv:1211.5590 (2012)

A Blockchain Framework for Proptech: Success Model Through Disintermediation and Self-regulation

Vibhor Arya[1], Praneeth Naganathahalli[2], and Shekhar Shukla[3(✉)]

[1] Deloitte India, Bengaluru, India
pgp17.vibhor@spjimr.org
[2] Paytm, Noida, India
pgp17.praneeth@spjimr.org
[3] Information Management Area,
S. P. Jain Institute of Management and Research, Mumbai, India
shekhar.shukla@spjimr.org

Abstract. Land records are traditionally recorded via physical documentation method, where the documents are isolated with different stakeholders in order to ensure their integrity. Of late, governments across the world are ushering exponential changes to transform Property Technology (PropTech) i.e. a facet of digital transformation in the property industry. Can government itself be a part of the blockchain ecosystem by paving the way for disintermediation? Is self-regulation a way forward for blockchain in PropTech. This question remains unanswered in the present-day scenario. Therefore, this paper explores the possibility of regulating PropTech through Blockchain, and presents a framework for Blockchain adoption in PropTech & the role, the Government can play within this ecosystem.

Keywords: Blockchain · PropTech · Disintermediation · Self-regulation · Governance

1 Introduction

Blockchain or Distributed Ledger Technology has been regarded as one of the most significant innovations that have the potential to disrupt a wide spectrum of industries. Many organisations have joined the bandwagon of this technology characterised as immutable, irrevocable & transparent [3, 16].

Blockchain's grand promise is to do for transparency of the internet's contribution towards communication [4, 6]. This transparency is particularly necessary for contexts where the counterparty trust is sceptical. Trustworthiness of registry records such as births, deaths, marriages, lands, financial transactions are few examples of such contexts [2].

Land records have traditionally been recorded in physical documents that have been isolated with different stakeholders to ensure their integrity. Of late, governments across the world are ushering exponential changes to transform Property Technology (PropTech) i.e. a facet of digital transformation in the property industry.

© Springer Nature Switzerland AG 2020
A. P. Pandian et al. (Eds.): ICICCS 2019, AISC 1039, pp. 522–528, 2020.
https://doi.org/10.1007/978-3-030-30465-2_58

The government of Andhra Pradesh has piloted the key initiative of adopting blockchain for managing land records & vehicle registrations. While the government agencies in the United Kingdom, Brazil, Canada & 20 other countries are examining blockchain architecture for the provision of government services.

However, so far the implementation of blockchain has been piloted through third-party organisations, while the government itself played the role of an external regulator. Can government itself be a part of the blockchain ecosystem paving the way for disintermediation? Is self-regulation a way forward for blockchain in PropTech. This question remains unanswered in the present-day scenario [7]. This paper, therefore, explores the possibility of regulating PropTech through Blockchain & the role, the Government can play within this ecosystem.

The remaining of the paper is organized as mentioned. Section 2 contains a brief literature review on PropTeach and Blockahins. Sections 3 presents the research problem. Section 4 demonstrates the proposed research methodology which is a framework. Section 5 discusses the results and conclusion and Sect. 6 discusses the limitation and future work for the proposed research.

2 Literature Review

Many articles/papers have been published to date on Blockchain and its applications in a variety of fields. With digital disruption affecting the real estate industry [11, 12], quite a few research activities have been published on the potential impact that Blockchain can create in the PropTech space [13]. While most of them give us a general sense of the advantages of Blockchain applications like disintermediation, decentralization, security etc., *there are a few that focus on the challenges in implementing one*. Over the course of our research, we have managed to use some of the existing works to understand the various use cases of blockchain in the PropTech industry and analyze the shortcomings of the technology implementation.

Lemieux [9, 10], discusses about the use of Blockchain technology for property registration system in developing countries. The paper presents a risk-based approach to evaluating the use of Blockchain in real estate. The findings show that Blockchain technology can be effectively used to address the most important issue of data integrity and avoid the inefficiencies of intermediaries. It, however, argues that there could be several limitations in the long-term because of the issues related to the reliability of information and ever-changing real estate market in terms of price and the associated regulations.

Veuger [14], also conducted research on the application of Blockchain in the real estate industry and discusses about simplification of the existing transaction process in the real estate industry through the use of smart contracts that could be executed based on a predetermined set of rules. It also points out to another significant social impact, the manner in which the principle responsibilities of the stakeholders can change, leading to disintermediation.

Some of the most important elements of a blockchain transaction in the real estate industry are – Registration, ownership, possession, characteristics of the property and transfer of ownership. The paper questions the role of different stakeholders in the real estate value chain which primarily includes the owner, the notary and the banks to suggest that complete disintermediation is not possible because of the inconsistent contracts and documents that form an important part of establishing the ownership. Human intervention will be required to comprehend the documents and establish the rightful ownership based on the rules governing the real estate industry. The paper, however, recommends the use of Artificial Intelligence to validate documents and execute contracts by themselves without having the need of intermediaries like the notary or any other third-party agents.

Deloitte [1] consulting has came up with its report on highlighting key aspects on why Blockchain based technology solution can be really helpful in business arenas like PropTech. The key highlights of the report present aspects of Distributed Database, Immutability and Authenticated Modification, Trust, Decentralized Network and Unique Incentivization procedure [5, 8, 15].

A review of the above studies helps us understand that the application of blockchain in the real estate industry brings in the advantage of transparency, security, speed and efficiency. Though the challenges of doing away with the stakeholders like notary, third-party agencies, brokers, banks etc. are of multiple dimensions, the idea of disintermediation in PropTech through blockchain is widely talked about. It was observed during the review of some of the existing studies that the impact of regulatory authority which is one of the most important stakeholders in the real estate industry in any developing economy is under-explored. The regulatory authority makes sure that the existing rules that govern the real estate market are consistently followed and necessary reforms are introduced for consumer protection and standard business practices. Regulations include the rules of the registration process, liabilities of buyers and sellers in real estate transactions, public law permits and obligations etc. Any blockchain implementation that secures the real estate transactions should either be regulated by the existing authority in the region or allowed to operate in a self-regulated environment. However, a regulatory authority that sets rules to govern blockchain implementation may defeat the purpose of disintermediation and on the other hand; a self-regulated real estate environment seems too risky a choice to make for any government.

3   Research Problem

The paper questions the role of different stakeholders in the real estate value chain which primarily includes the owner, the notary and the banks to suggest that complete disintermediation is not possible because of the inconsistent contracts and documents that form an important part of establishing the ownership. It was observed during the review of some of the existing studies that the impact of regulatory authority which is one of the most important stakeholders in the real estate industry in any developing economy is under-explored. The regulatory authority makes sure that the existing rules that govern the real estate market are consistently followed and necessary reforms are introduced for consumer protection and standard business practices. The aim of this research is to analyse the implications of blockchain technology in the real estate industry in the regulatory environment. A review of some of the above mentioned studies bring us to the following research questions.

- Can complete disintermediation and self-regulation through blockchain be a sustainable way forward in the real estate industry?
- The possibility of regulating PropTech through Blockchain & the role the Government can play within this ecosystem.

4 Methodology

Though blockchain in real estate promises huge potential, the implementation is either under contemplation or in the initial stages of a pilot run. The approach, therefore, involves studying various blockchain use cases in the real estate industry and understanding the regulations around it. These will be required to answer some of the critical questions on how various governments are planning to regulate the real estate market through blockchain and the changes required at various levels. The approach also includes understanding the nuances between different consensus mechanisms of blockchain and the possibilities of facilitating a regulated, yet, a fairly disintermediate environment.

While there are multiple projects being initiated in the real estate industry using blockchain, governments are still trying to understand the implications of using the technology and its interference with the existing regulations in the industry.

Most of the current models include the concerned authorities planning to regulate the blockchain platform from outside the network by setting guidelines and rules for the technology partners.

The authorities are contemplating new legislation to address issues or disputes related to the blockchain implementation and the associated smart contracts. It is absolutely required that new laws, legislation, and regulations need to be brought in for a controlled implementation of the technology in a sector, which, if left to the people to self-regulate, can cause high volatility in the market. However, governments could also consider regulating the blockchain implementation by being part of the network itself.

We propose a framework for Blockchain implementation in Real Estate/PropTech Industry using a involved or incorporated state of governance which provides an optimum balance of disintermediation.

The key challenges that query this proposition of government being an outside agent to the Blockchain are immense and usually make the advent of Blockchain in the scene futile. The government not being a direct onlooker or node in the Blockchain can lead consortium of selling stakeholders who might influence the sanctity of transparent transactions in Blockckhain. There can also be a problem of price fluctuations as the prices might be controlled by influential agents and might lead to eventual loss of naive public.

The presence of governance authority also reassures the collaborative functioning of Blockchain network without formation of cartels and help in seamless and trustworthy transactions saving the time and economy of transaction execution. Further, any imposition of corrupt practices in terms of introducing additional crypto currency charges etc. cannot be imposed without the nod of the government.

5   Conclusions

Any transaction in a blockchain network is based on a consensus mechanism. The security, robustness, and trustworthiness of a blockchain system are highly dependent on its consensus model. In this case, the regulatory authority of a particular region could oversee and regulate a blockchain based network by building a strong consensus mechanism and being part of the model.

Figure 1 represents the current approach in which governments are planning to implement regulations in the blockchain system for real estate. The authority is not part of the network but would set rules and regulations for the contracts and transactions that are executed. Regulatory reporting becomes essential for the platform provider to ensure the credibility of the network.

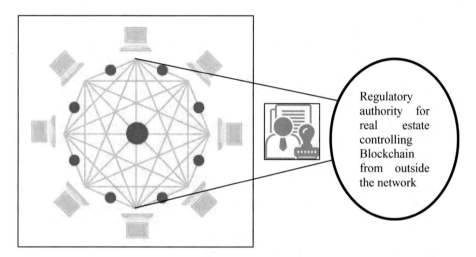

Fig. 1. Present state of research for blockchain based real estate [17]

The proposed approach of regulating blockchain implementation in real estate where regulatory authority is part of the blockchain system and monitors the network efficiently

In Fig. 2, the regulatory authority forms a node of the blockchain so that it can approve transactions, execute contracts and at the same time monitor the network for any regulatory issues. One of the consensus mechanisms adopted in blockchain is the use of majority votes to decide whether a record/transaction is valid and if it can enter the blockchain. The authority regulating the blockchain could be provided with a majority vote on approving blocks in the network so that all the applicable rules and regulations set are followed, and any discrepancy is reported immediately in real time. We see a clear scope of following improvements in the PropTech space with the proposed framework:

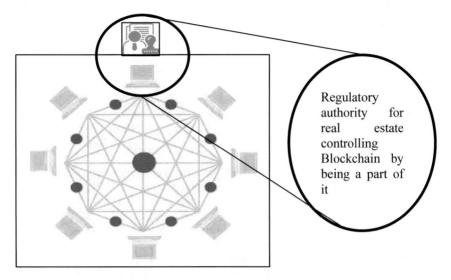

Fig. 2. Proposed state of research for blockchain based real estate [17]

- Transparency
- Trust
- Price Control
- Elimination of Fraudulent Practices
- Seamless Transaction

6 Limitations and Future Work

The proposed approach to regulate the blockchain technology in the real estate industry requires the network to provide a majority vote share to the node that represents the regulating authority. The easiest way forward would be to assign a 51% vote share to the regulating node so that any invalid transaction could be withheld and not allowed to enter the blockchain. However, this would defeat the purpose of attaining disintermediation and decentralization through blockchain. Also, any vote share less than 50% for the regulating node may lead to a possibility of collusion between all the other parties in the network which can be used to bypass any regulatory requirements.

It, therefore, becomes important to analyze the participating stakeholders and come up with a model that could recommend the most optimal distribution of the percentage of vote share among all the nodes, including the regulatory authority so as to reap the core benefits of the blockchain and also maintain the required regulations in the network.

References

1. Andavarapu, V.: Blockchain in commercial real estate: the future is here
2. Beck, R., et al.: Blockchain–the gateway to trust-free cryptographic transactions (2016)
3. Drescher, D.: Blockchain Basics: A Non-technical Introduction in 25 Steps, 1st edn. Apress, Frankfurt am Main (2017)
4. Haridas, S.: This Indian City Is Embracing BlockChain Technology - Google Search
5. Hawlitschek, F., et al.: The limits of trust-free systems: a literature review on blockchain technology and trust in the sharing economy. Electron. Commer. Res. Appl. **29**, 50–63 (2018)
6. Hughes, L., et al.: Blockchain research, practice and policy: applications, benefits, limitations, emerging research themes and research agenda. Int. J. Inf. Manag. **49**, 114–129 (2019)
7. Iansiti, M., Lakhani, K.R.: The truth about blockchain. Harv. Bus. Rev. **95**, 118–127 (2017)
8. Lacity, M.C.: Addressing key challenges to making enterprise blockchain applications a reality. MIS Q. Exec. **17**(3), 201–222 (2018)
9. Lemieux, V.L.: Evaluating the use of blockchain in land transactions: an archival science perspective. Eur. Prop. Law J. **6**(3), 392–440 (2017). https://doi.org/10.1515/eplj-2017-0019
10. Lemieux, V.L.: Trusting records: is blockchain technology the answer? Rec. Manag. J. **26**(2), 110–139 (2016)
11. Marchand, D.A., Wade, M.R.: Digital business transformation: where is your company on the journey. http://www.research/perspectives-for-managers/digital-business-transformation-where–is-your-company-on-the-journey/
12. Marchand, D.A., Wade, M.R.: Digital business transformation. http://www.research/insightsimd/digital-business-transformation/
13. Spielman, A.: Blockchain : digitally rebuilding the real estate industry. Massachusetts Institute of Technology (2016)
14. Veuger, J.: Trust in a viable real estate economy with disruption and blockchain. Facilities **36**(1/2), 103–120 (2018)
15. Wood, G.: Ethereum: a secure decentralised generalised transaction ledger. Ethereum Proj. Yellow Pap. **151**, 1–32 (2014)
16. Yli-Huumo, J., et al.: Where is current research on blockchain technology?—a systematic review. PLoS ONE **11**(10), e0163477 (2016)
17. What is Blockchain Technology? A Step-by-Step Guide For Beginners. https://blockgeeks.com/guides/what-is-blockchain-technology/

Agentless Automation Model for Post Exploitation Penetration Testing

Saraswati Maddala$^{(\boxtimes)}$ and Sonali Patil

Department of Information Technology, KJ Somaiya College of Engineering,
Mumbai, India
{saraswati.m, Sonali.patil}@somaiya.edu

Abstract. In a world where even the most mundane of tasks requires the intervention of the cyberspace, it is particularly important to be aware of the risks and vulnerabilities that come with cyberspace. This has led to the development of a field better referred to as vulnerability assessment and penetration testing (VAPT). There are multiple parts to perform penetration testing, in this paper we will be focusing on the Post Exploitation part. Post exploitation outlines the activities to perform in order to breach the security of a target system by leveraging a discovered exploit. In this paper we propose an agentless model for the automation of post exploitation activities, where agentless means that no third party software is required to be installed on the target machine.

Keywords: Post exploitation · Agentless · Penetration testing · Exploit

1 Introduction

Post exploitation, as the name suggests, in literal terms means the activities to be performed post or after an exploit has been found. It is made up of the phases of operation once a victim's system has been compromised by the attacker. The value of the compromised system is determined by the value of the actual data stored in it and how an attacker may make use of it for malicious purposes. The whole post exploitation concept rose from this fact only as to how you can use the victim's compromised system's information. This phase actually deals with collecting sensitive information, documenting it, and having an idea of the configuration settings, network interfaces, and other communication channels. These may be used to maintain persistent access to the system as per the attacker's needs. For many penetration testers, gaining access to a system or finding an exploit is said to be the easier part. The post exploitation activities and the actual fetching of data is what is the real challenging and time consuming part which requires a great deal of patience, experience and knowledge in the field of penetration testing. In the next few sections, we will include an in depth study of the research related to penetration testing and its automation done thus far, this will be done as part of the literature survey. The literature survey also discusses the problems phased with current penetration testing methods. Thereafter, we will explain our proposed methodology to solve the aforementioned problems phased by current methods of penetration testing. We will provide and in depth explanation of the proposed methodology, including all its parts and how they will function individually and

© Springer Nature Switzerland AG 2020
A. P. Pandian et al. (Eds.): ICICCS 2019, AISC 1039, pp. 529–539, 2020.
https://doi.org/10.1007/978-3-030-30465-2_59

also as a part of the integrated penetration testing system. We will then proceed forward to discuss the results obtained from the implementation of the proposed methodology following which we will conclude the paper and discuss the future scope too.

2   Literature Survey

The number of cyber-attacks are increasing almost exponentially on a day-to-day basis. The stronger and more intelligent attackers get, the bigger threats and vulnerabilities we face. In order to make sure our systems are protected, we have to ensure that there are no vulnerabilities present [9]. Vulnerabilities are basically a fault in the system which could be in the form of a software or implementation error or both. These could be exploited by the attacker in order to breach the system and perform unauthorized activities [11]. Vulnerability Assessment and Penetration Testing (VAPT) is a cyber-defense activity which involves finding any flaws or vulnerabilities in the system and network architecture. There are various tools which can be used to perform VAPT, these tools perform various tests using different techniques in order to identify vulnerabilities. Hence, having identified the vulnerabilities, protective action can be taken. These actions can safe guard the system against any future cyber-attacks [1]. The primary aim of penetration testing is to enhance data security. Security information and weaknesses that are specified in penetration testing are considered confidential and shall not be disclose until complete resolution of defects [3, 10].

Manual penetration testing or post exploitation activities are known to have numerous problems. Initially, the security of a the target system is checked by the penetration testing team based on their expertise and skill. However they may fail to cover all aspect of threat handling and being 100% sure is almost mythical. An overlooked threat however could result in the compromise of the entire system which usually implies the breach of confidential data (e.g. identity theft, stealing of personal data, bank data leakage, etc.). This means that although penetration testers might have the required qualification on paper, it is extremely tough to find candidates with professionalism of the required level. In addition, with manual penetration testing it is possible to leave out a substantial ration of the attack space uninvestigated. Moreover, when it comes to manual penetration testing, identifying all possible attack vectors can be an extremely tiresome and time consuming endeavor. Hence, the process of manual penetration testing is considered to be costly and labour-intensive [2]. Until recently, penetration testing was a very much complex and manual process that was carried out by only security specialists with vast experience in the field [8]. Testers typically use their judgement to understand the exploit needed for a given situation and then write their own custom exploits [12]. Most of these exploits and tasks are time-consuming and tedious to perform. Comprehensive, manual penetration testing typically demands and extensive team of specialists who have diverse skill sets, which most organizations cannot afford to uphold in-house even on contract basis [7].

3   Proposed Solution

In this report we are dealing with the post exploitation part of penetration testing, we are working with an assumed breach scenario. This basically means that we have already got a hold on the authentication details of the target machine, these many include details like username, password, administrator's password, stored authentication hashes, etc. We will suggest an algorithm to systematically generate a penetration testing model which can be used for post exploitation in this research. To achieve this goal, number of methodologies has been followed. The proposed methodology is an agentless method, which basically means that there is no third party software that will be installed on the target machine. Agent based exploitation involves the installation of a third party software on the compromised machine, this software will be used to continually monitor the machine. For convenience, the process of post exploitation can be better explained using Fig. 1 below.

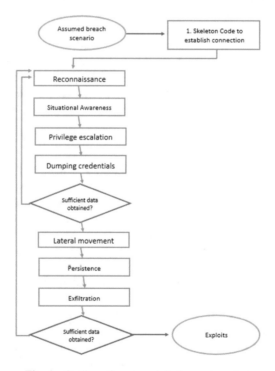

Fig. 1. Outline of post exploitation automation

As we start off the model by assuming a breach, this means we have access to the credentials of at least one of the machines on the network, we use the credentials to establish a connection with the target machine via the skeleton code. The skeleton code, if it manages to establish a connection with the target machine then performs the reconnaissance, situational awareness, privilege escalation and finally the dumping credentials phases. These are the initial four phases and the outputs of these initial phases decide whether we can proceed forward to the phases. If sufficient data is not obtained from the four initial phases, the first phase we will go back to the first phase. If sufficient data has been obtained we will proceed forward by using data obtained from initial phases to perform the next three phases, i.e. lateral movement, persistence and exfiltration respectively. After this we once again check if we have actually obtained enough data, if not, we once again revert back to the first phase. If sufficient data has been obtained, we proceed forward using data obtained in all the preceding phases in order to achieve the end goal of any penetration test, that is the exploit phase. The aforementioned algorithm is written as a python script and each of the modules in the penetration test as PowerShell modules embedded in said python script This is how automation is achieved in each module, as using the python script, each PowerShell module is executed one after the other with no human intervention needed. The skeleton code is used to establish a connection to the target machine and gain access to a remote shell via which each module will be implemented. The entire model is agentless, which means that there will be now files dropped on the target. Hence the PowerShell scripts corresponding to each module are stored and fetched for execution from a server hosted by us. This aids us in successfully achieving the goal of making the model agentless as no files touch the target machine and no trace is left on the target machine proving that we were ever there. We will briefly explain the function of each module or phase in post exploitation along with their respective sub-phases or sub-modules:

3.1 Skeleton Code

The creation of the skeleton code is the primary phase which is of the highest value in the process of post exploitation. In this phase, the actual connection between the attacking machine and the victim machine is established. After getting the authentication details, it is important to know what to do with them and how to use them in the stealthiest way possible i.e. use the authentication details without leaving any trace of entry on the victim's machine. The skeleton code can be used to achieve such a connection. It is the base of all the further exploits, failure to establish a connection will mean the failure of the whole activity that is post exploitation.In the proposed methodology, we have outlined and nominated two protocols, either of which will be used in the skeleton code to establish a seamless connection between the attacking machine and the target machine. This is the most crucial phase since we aim to design an agentless module and this phase is what gives us that attribute. Figure 2 below shows the outline of the skeleton code with the selected authentication methods:

Fig. 2. Authentication method

The Server Message Block (SMB) protocol is a communication protocol which works on client-server basis. It is used for the sharing of network resources such as access to files, printers, serial ports, etc. While WMI (short for Windows Management Instrumentation), is an API designed by Microsoft that allows the control of systems and devices in a network.

3.2 Reconnaissance

The second phase is reconnaissance which is intel gathering or information gathering and getting to know as much as possible about the victim's. Reconnaissance can be described as a set of techniques used to covertly discover and assemble information about a quantified target. The foremost objective of this phase is to assemble as much information about the specified target as possible. Figure 3 shows the modules to be included in the phase of reconnaissance in the proposed method.

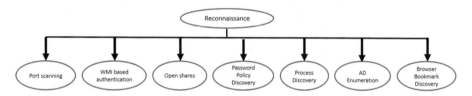

Fig. 3. Reconnaissance

The port scan can be used to identify which ports in the network are open and receiving information. It can also be useful in identifying any security devices such as firewalls that are present in the target network. Open shares can be used to identify documents related to any particular user and password policy discovery process, as the name suggests, can be used to identify any stored passwords on the target machine. AD (active directory) enumeration involves fetching data from the network's active directory, this is done using a PowerShell script known as SharpHound. Finally, the browser history and well as recovery of security patched can be done, the latter of the two will leave the system exposed to previously identified vulnerabilities again.

3.3 Situational Awareness

Situational awareness can be defined as the comprehension of the cyber threat environment present in the target machine. This means identifying any risks and how they could impact the system and also the identification of any measures to mitigate these risks. A keen appreciation of the threat landscape having entered a system and carried out reconnaissance on the said system, the next step is to identify all the probable attack vectors of the system. The various modules entailed in situational awareness in this methodology are shown in Fig. 4 below.

These modules include the identification of the Anti-Virus software, Endpoint detection and response as well as the Host Intrusion Prevention System agent present on the target machine. It also includes capturing the network traffic or packets going to and from the target machine and also looking for credentials stored in the target system registry.

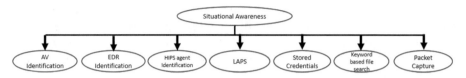

Fig. 4. Situational awareness

3.4   Privilege Escalation

Privilege escalation is a technique used by adversaries to use the low-level foothold
they have gained on a given target in order to gain access to a high level account (e.g.
system administrators, domain administrators, domain controllers, etc.) which has
access to critical resources. Many times, the account that has been compromised could
be one which is low in the organization hierarchy, meaning it might not contain any
data or confidential resources appealing to an adversary. Hence, the adversary will try
to make his way through the organization's architecture to get to the top level sensitive
data. Using elevated privileges, attackers can perform various malicious activities such
as mess with security settings, configurations and data. Attackers usually get access to
lower privilege accounts first and then utilize them to acquire high-level privileges and
eventually attain full access to the organization's IT infrastructure. Figure 5 below
shows the sub-modules that will be included in the module of privilege escalation in
our proposed solution.

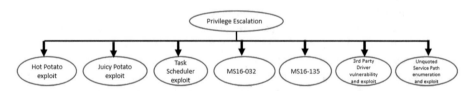

Fig. 5. Privilege escalation

The hot potato exploit is a culmination of two Windows vulnerabilities that is
NBNS spoofing and NTLM relay with a locally running fake WPAD proxy on the
target machine. Similarly, Juicy potato is basically exploits the way Microsoft handles
tokens. M16-032 and M16-135 are security updates for secondary logon to address
elevation of privilege. In the reconnaissance phase, we can identify whether these
updates have been done on the target machine, if so we need to find a way to bypass
them and if not, the entire phase of privilege escalation has become easier. In addition
to this, we can also exploit the task scheduler to gain system privileges. There is a
vulnerability in the Microsoft Windows task scheduler which deals with handling
ALPC. This is a provision which allows an authenticated user alter the data in a file
which was supposed to be protected by the file system ACLs. Hence, attackers may
abuse this in order to obtain higher level privileges. The unquoted service path exploit
abuses the fact that every Windows Service has an executable which is stored under

some known definite path. This can be used for privilege escalation because if an adversary gains access to a certain folder in the same path containing the whitespace or separator then he can insert any malicious in the parent path before the whitespace. The third party driver vulnerability involves the exploitation of graphics drivers and low quality third party code. This is a good exploit for discovering sandbox escapes and EoP vulnerabilities.

3.5 Dumping Credentials

Credential dumping, as the name quite explicitly states, is the process of finding login information (such as passwords, usernames, hashes, etc.) stored on the target machine or the network that machine is on. These credentials can then be used in the next phase, which is Lateral Movement, in order to traverse through the network looking for juicy or confidential data. In order to move through a network, cybercriminals must gather necessary login credentials. Figure 6 below shows the modules to be used for credential dumping.

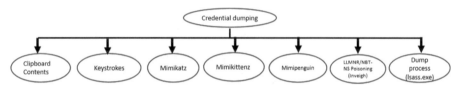

Fig. 6. Credential dumping

Credential dumping as the name suggests, is a pretty self-explanatory phase. In this method, we will be using techniques such as viewing the target's clipboard contents and monitoring keystrokes. We will also inculcate predefined credential dumping modules such as mimikatz, mimikittenz, mimipenguin and LLMNR poisoning scripts.

3.6 Lateral Movement

Lateral movement is made up of techniques which aid the attacker in traversing across the network in which the target machine resides. In this phase the adversary tries to use the credentials obtained in the Credential Dumping phase in order to gain access to other machines on the network and hence try to find confidential data on their machines too. Many times, no additional tools are even required to perform lateral movement. Figure 7 below shows the modules to be included in lateral movement.

Password spraying using SMB or Kerberos protocols can be used, this in an attempt of using the data obtained from credential dumping in a hope to move through the system. In addition to password spraying a script known as PSExec can be used to run PowerShell scripts on the target machine, hence aid us in traversing across the network.

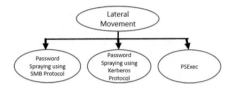

Fig. 7. Lateral movement

3.7   Persistence

Once the adversary has a foothold on the target network, it is very essential for that foothold to be maintained. This means that once the adversary has access to a system on the network, he must make sure that he will not lose that access due to any activities on the victim machine. For example if the point of compromise was using the user credentials, if they are changed by the victim then the adversary will lose access to the system. Hence the adversary has to come up with ways to prevent losing the foothold on the target system. Figure 8 below shows the persistence modules we will be considering.

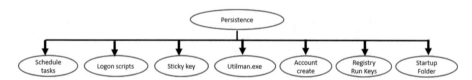

Fig. 8. Persistence

Tasks as simple as creating or adding a user and running registry keys can be used to maintain access to a system. In addition to this, logon scripts can be put in the start-up folder such that every time the target machine reboots, it will automatically connect to the attacker's machine.

3.8   Exploits

This is the phase in which actual finding of useful data or carrying out of the exploit occurs. An exploit is a malicious task used to gain access to data on the target machine. There are hundreds of exploits that can be used, they have to be chosen based on the results of all the preceding phases. In our model we have opted to use four exploits as can be seen in Fig. 9 below.

The ms17-010 SMB Exploit is a Remote Code Execution exploit (RCE). It is used to describe an attacker's ability to remotely execute any command of choice from one computer to another. An attacker can exploit and possibly take full control of a computer which is vulnerable to RCE. Cisco SMI switches come with an inherent vulnerability which allows any potential customer to not only have a switch shipped to any location but also start it without any authentication or extra configuration. Dirty CoW exploit is a Linux Kernel exploit which leaves the system vulnerable to privilege

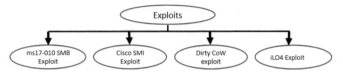

Fig. 9. Exploits

escalation. Using this exploit a user who is on a lower privilege level in the system architecture can attain access to memory mappings which should be read-only. Lastly, iLO4 Exploit is an authentication bypass in HP iLO 4 1.00 to 2.50, triggered by a buffer overflow in the Connection HTTP header handling by the web server. Exploiting this vulnerability gives full access to the REST API, allowing arbitrary accounts creation.

3.9   Exfiltration

Data exfiltration is the illicit transmission of confidential information from the victim's network to the adversary. It is transferred stealthily in such a way that it is tough to differentiate between the legitimate network traffic and malicious data exfiltration. Data customarily moves in and out of any organization's networks, exfiltration of data can be expertly done such that it appears to be legitimate network traffic. Figure 10 shows the exfiltration modules that will be covered in this method.

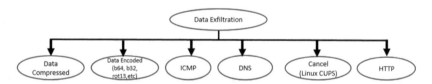

Fig. 10. Data exfiltration

Exfiltration techniques for seamless data transfer include compressing or encoding the data. Having done this it can then be transferred by obfuscating it in ICMP, HTTP or DNS packets. This will help prevent the network traffic from appearing anomalous and hence avert tracing.

4   Results

Having discussed the working of the modules in the proposed architecture, we can now proceed forward to see the implementation of said architecture. Since we are dealing with an assumed breach scenario, we can use the details that were required to establish the foothold as an input for the fields which are required to establish a connection with the target machine i.e. the IP Address, Username and Password fields. Having entered these, the penetration testing of the entire network takes place on it's own without the requirement of any human intervention of any sort. All the modules cascade and implement one after the other, in many cases the output of preceding modules is automatically being fed as the input to the proceeding modules. Figure 11 below gives a little limpse of how the output of the automation tool looks.

```
root@kali:~/RTA/examples# ./rta_sc.py
Please enter the address of target machine:192.168.1.5
Please enter the username:Administrator
Please enter the password:password
Scanning all the ports in the network...
Hostname       : 192.168.1.0
alive          : False
openPorts      : {}
closedPorts    : {}
filteredPorts  : {}
finishTime     : 6/5/2019 3:16:43 AM
Hostname       : 192.168.1.1
alive          : True
openPorts      : {443, 3389, 445, 139...}
closedPorts    : {80, 23, 21, 110...}
filteredPorts  : {}
finishTime     : 6/5/2019 3:16:43 AM
Hostname       : 192.168.1.2
alive          : True
openPorts      : {}
closedPorts    : {}
filteredPorts  : {80, 23, 443, 21...}
finishTime     : 6/5/2019 3:16:43 AM
```

Fig. 11. Output of proposed methodology

The system analyses which module's output to use based on preset conditions and feeds them as input to the modules that require it. Since the output of the entire tool, i.e. concatenation of outputs of all modules, is several pages long it is not feasible to include the entire output as part of this paper.

5 Conclusion

This paper discusses the importance of penetration testing in the increasingly cyber dependent world. Post exploitation in penetration testing is explained in detail. This paper also discusses the challenges faced by the current methods of post exploitation in penetration testing. Finally, we have proposed a method for post exploitation automation. The proposed method is agentless, meaning that automation will be done without the intervention of any third party software. There are nine stages to post exploitation in penetration testing and the automation of each of the modules to be included in the automation of each of these stages is described. As for the future scope of the proposed methodology, the skeleton code is flexible such that it can accommodate the addition of new modules and the removal of older ones. New exploits can be added to the model as and when they are discovered.

References

1. Goel, J.N., Asghar, M.H., Kumar, V., Pandey, S.K.: Ensemble based approach to increase vulnerability assessment and penetration testing accuracy. In: 2016 International Conference on Innovation and Challenges in Cyber Security (ICICCS-INBUSH) (2016)

2. Almubairik, N.A., Wills, G.: Automated penetration testing based on a threat model. In: 2016 11th International Conference for Internet Technology and Secured Transactions (ICITST) (2016)
3. Stefinko, Y., Piskozub, A., Banakh, R.: Manual and automated penetration testing. Benefits and drawbacks. Modern tendency. In: 2016 13th International Conference on Modern Problems of Radio Engineering, Telecommunications and Computer Science (TCSET), Lviv, pp. 488–491 (2016)
4. (ISC)2 Government Advisory Council Executive Writers Bureau: Penetration testing: Pros and cons of attacking your own network (2013). https://llgcn.comlarticles/2013/02/04/pros-cons-penetration-testing.aspx
5. Almubairik, N.A., Wills, G.: Automated penetration testing based on a threat model. In: 2016 11th International Conference for Internet Technology and Secured Transactions (ICITST), Barcelona, pp. 413–414 (2016)
6. Chen, C., Zhang, Z., Lee, S., Shieh, S.: Penetration testing in the IoT age. Computer **51**(4), 82–85 (2018)
7. Shebli, H.M.Z.A., Beheshti, B.D.: A study on penetration testing process and tools. In: 2018 IEEE Long Island Systems, Applications and Technology Conference (LISAT), Farmingdale, NY, pp. 1–7 (2018)
8. Chu, G., Lisitsa, A.: Poster: agent-based (BDI) modeling for automation of penetration testing. In: 2018 16th Annual Conference on Privacy, Security and Trust (PST), Belfast, pp. 1–2 (2018)
9. Tetskyi, A., Kharchenko, V., Uzun, D.: Neural networks based choice of tools for penetration testing of web applications. In: 2018 IEEE 9th International Conference on Dependable Systems, Services and Technologies (DESSERT), Kiev, pp. 402–405 (2018)
10. Zitta, T., et al.: Penetration testing of intrusion detection and prevention system in low-performance embedded IoT device. In: 2018 18th International Conference on Mechatronics - Mechatronika (ME), Brno, Czech Republic, pp. 1–5 (2018)
11. Yevdokymenko, M., Mohamed, E., Onwuakpa, P.: Ethical hacking and penetration testing using raspberry PI. In: 2017 4th International Scientific-Practical Conference Problems of Infocommunications. Science and Technology (PIC S&T), Kharkov, pp. 179–181 (2017)
12. Shinde, P.S., Ardhapurkar, S.B.: Cyber security analysis using vulnerability assessment and penetration testing. In: 2016 World Conference on Futuristic Trends in Research and Innovation for Social Welfare (Startup Conclave), Coimbatore, pp. 1–5 (2016)

The Effect of Global Political Risk on Stock Returns: A Cross-Sectional and a Time-Series Analysis

Karen Vargas[1], Angelica Gonzalez[2], and Jesus Silva[3(✉)]

[1] Universidad de la Costa, Barranquilla, Colombia
kvargas17@cuc.edu.co
[2] Universidad de Edimburgo, Edimburgo, Scotland
angelica.gonzalez@ed.ac.uk
[3] Universidad Peruana de Ciencias Aplicadas, Lima, Peru
jesussilvaUPC@gmail.com

Abstract. Given the rise of political uncertainty, it is important to develop an understanding of their effect on financial markets. We use a political risk measure to calculate their effect on stock markets based on a political risk measure. The political risk proxy is related to cross-country returns and two portfolios: one with upside and other with downside political risk. Time-series and cross-sectional analysis are conducted to measure the effectiveness of this measure on global markets. The results evidence that an increase in global political risk is negatively correlated with an upside portfolio containing global stock returns.

Keywords: Policy uncertainty · Asset pricing · Political risk · Stock markets

1 Introduction

Political risk can be defined as laws, regulations, or contracts which governments unreasonably or reasonably change or fail to enforce and affect investor returns. Global stock markets tend to respond to political events. However, different countries are affected by a common political risk but in different magnitudes. For instance, the cumulative effect of US tariff policies announcements has had a different impact on financial markets. Let us consider the response of the US announcement about new tariffs on China in May of 2019 on global stock returns. After a failure of the long trade deal between Washington and Beijing, the United States has decided to raise its tariffs from 10 to 25 per cent for Chinese goods. The trade war between US and China lowered stock prices in the US in 700 basis points, in Japan in 70 points and the European stock indexes fell by more than 1000 points [1]. We can evidence that political uncertainty in the US has a negative impact on global financial markets. There are two significant current discussions in the literature about political risk. First, political risk is an entry barrier in developed and emerging markets and it can be accounted as a systematic risk to explain cross-country differences [2]. Second, changes in government policies affects negatively global markets. This paper evaluates the

© Springer Nature Switzerland AG 2020
A. P. Pandian et al. (Eds.): ICICCS 2019, AISC 1039, pp. 540–548, 2020.
https://doi.org/10.1007/978-3-030-30465-2_60

impact of a global political risk measure on global stock markets. Research to date focuses on measure the impact of political risk on local and international markets [3] in developed or in emerging markets [2, 4]. The three questions addressed in this document are: 1. Is the global political risk proxy a source of systematic risk on asset pricing models? 2. What is the impact of the global political risk measure in the cross-section of excess of returns? 3. What is the impact of the global political risk measure in the time-series of two portfolios that represent countries with upside and downside political risk?

2   Literature Review

2.1   Asset Pricing

The Capital Asset Pricing Model (CAPM), is the classic asset pricing model in the literature. The CAPM was developed in 1964, 1965 and 1966 by Sharpe, Lintner and Mossing [5]. The CAPM theory establishes that expected returns are a linear function of its systematic risk: the market beta. [5] evidence that the CAPM was theoretically correct but not empirically. Consequently, most important asset pricing models are only a simply modifications of the CAPM assumptions [6]. [7, 8] modified the classic CAPM grouping stocks in portfolios instead of using one asset. In effect, they obtained better results than latter studies. CAPM studies suggest that portfolios with high returns present high betas. An important issue in asset pricing is to identify sources of risks. Then, if these sources of risks are identified in an economy, pricing factors would be easy recognized to value assets. However, macroeconomic risks are not stable in economies. Although, in the literature there are some factor risks which are explained as empirically as theoretically such as the political risk measure presented by the International Country Risk Guide (ICRG) institution. [9] study theoretically the impact of political uncertainty on asset pricing. They claim that political risk is orthogonal to the investment set of opportunities. Therefore, political risk can be considered as a plausible source risk factor on an asset pricing model. Others studies, like [2, 10] evidence that emerging and developed asset returns are explained by the political risk (PR) measure developed by the ICRG. In consequence, the political risk measure presented by the ICRG used in the literature is considered on this paper but as a source of risk of global linear asset pricing models. The most common techniques in the literature to test linear asset pricing models are the time-series approach (SA) and the cross-sectional approach (CSA). Both techniques complement each other. However, each one of these techniques offer advantages and disadvantages depending of the data. TSA is more pertinent for balanced panels and small number of assets and CSA for a large number of assets. Both techniques are used to test asset pricing models in this paper. The Error-in-Variables (EIV) problem is presented in the CSA. EIV occurs due to factor loadings are estimated in the first step using the TSA and then these results are used as factors to evaluate cross-section returns [6]. Therefore, their final parameters will need a correction in their standard errors. Consequently, [7], presents the Fama-Macbeth methodology to correct the EIV problem by adjusting standard errors in their estimations. Then, the Fama-Macbeth procedure is selected to measure the cross-section in global returns for this study. Besides, recent global asset pricing models are evaluated through this two-stage procedure. The Fama-Macbeth approach is an extension of the CSA. However, The Fama-Macbeth estimates the market price and the pricing errors through time-series averages. Consequently, the cross-section of 46 countries stock market returns and the time series of two global portfolios that represent upside and downside PR countries are evaluated to measure the impact of a global political risk.

2.2   Political Risk

The first research about political risk took place after the Second World War as a result of the increase of flow from US to Europe. How ever, most of these indices were qualitative. As a result, in the literature there are few political risk indices to do empirical studies. The Political Risk Service (PR) measure is commonly used in the literature. This proxy is a combination of 12 qualitative characteristics. The ICRG converted the qualitative to quantitative through point conversion: Government Stability (12 points); Socioeconomic Conditions (12 points); Investment Profile (12 Points); Internal Conflict (12 Points); External Conflict (12 Points); Corruption (6 Points); Military in Politics (6 Points); Religious Tensions (6 Points); Law and Order (6 Points); Ethnic Tensions (6 Points); Democratic Accountability (6 Points); Bureaucracy Quality (4 Points). The PR measure is selected for these reasons: First, there were many institutions that measure political risk by a country offering qualitative information. However, qualitative information is not suitable for empirical analysis then the quantitative one provided by the Political Risk Service (PR) is selected. Second, this measure is able to capture past and future trends of political risk. Third, this database is available for most developed and emerging markets which make a representative sample for our research. Fourth, existing literature uses this measure to evaluate impact on volatilty returns [10], and asset returns [4, 11].

3   Methodology

3.1   The Global Political Risk Measure

The proxy to construct the global political risk measure using the ICRG measure is defined following the methodology presented by [4, 12, 13]. This measure is calculated as the annual average of change of political risk across countries in the sample. This measure is called global political risk and is presented in Table 1. This proxy captures the idea that an increase in political risk produces a reduction in global returns and a decrease in political risk produces an increase in global returns.

3.2   Stock Market Data

There are 47 country excess of returns and two downside and upside political risk portfolios used to measure the impact of the global political risk index. Data-Stream and Bloomberg are the main source of data to calculate excess returns and to construct

Table 1. The global political risk proxy (Panel A) and Summary Statistics (Panel B)

Panel A: Notation

$Ch.\,PR_{it}$.   Change in political risk for country i

$(PR_{it+1}-PR_{it})/PR_{it}$ where PR_{it} is country's i political risk in year t

$f^{GPR}t$ The global political risk average change factor in year t

$$= \sum_{i=1}^{N} Ch.\,Pr_{it}/N$$

Panel B: Summary Statistics	
	f^{GPR}
Mean	0.026
Std.dev	0.04
Auto Correlation	0.19
P-Value	0.6397
N	11
Sample	1998-2008

these risk portfolios. These measures are in US dollars, and its frequency is yearly. First, the excess market returns are calculated using price indices. Then, the logs of the returns are estimated in an annual basis and the US risk-free rate is subtracted of the log of returns. Second, following [4] procedure the downside and upside portfolios are formed. There is evidence that countries that present an upside in political risk impact negatively stock returns. On the other hand, countries that evidence a downside in political risk impact positive stock returns, The first portfolio includes countries that experimented downgrades (decrease in political risk). The second portfolio includes countries that experimented upgrades (increase in political risk). These portfolios are constructed weighting each country's return with the absolute value of their average risk change. Those portfolios are rebalanced each year [4].

3.3 Cross-Country Model: Exposition of Global Stock Returns to the Global Political Risk Measure

The relation between country i's asset return and its political risk measure (f^{GPR}).

$$E(R_i^e) = \beta_i\gamma \tag{1}$$

R_i^e is the country i's equity risk premium. β_i is the country i's risk loading. γ is the market price associated with the political risk measure. These parameters are obtained applying Fama-Macbeth regressions. Recent studies use this approach to test global asset pricing models. [13, 14] apply the two-step Fama Macbeth procedure to evaluate the cross-section of global stock returns. [13] evaluate the impact of international instability in the cross-country of stock returns. Similarly, [1] apply this approach to evaluate the three Fama and French factors in the cross-section of global portfolios. *First Stage:* The time series regressions is performed by regressing each one of the country's asset returns to the global risk factor.

$$R_{i,t}^e = a_i + \beta_i f_t + \in_{i,t} \tag{2}$$

f_t is the global political risk factor f^{GPR} observed in year t. and $\in_{i,t}$ is the yearly error term. Betas (β) are estimated in the first stage to be used as parameters in the second stage. *Second Stage:* The market exposure γ_t is calculated each period t by regressing cross- countries returns with the factor loadings obtained in the first section.

$$R_{i,t}^e = \gamma_{o,t} + \beta_i \gamma_t + \alpha_{i,t} \tag{3}$$

γ_t is the estimated market risk at time t, α_t the pricing errors and β_i the factor loadings estimated in the first stage. Then, the factor risk premium γ and pricing errors α_i are calculated through a time-series averages using the γ_t and the α_t. The risk premium γ is the time series average of γ_t the pricing error α is the time series average of α_{it}

$$\gamma = \frac{1}{T} \sum_{t=1}^{T} \gamma_t \tag{4}$$

$$\alpha_i = \frac{1}{T} \sum_{t=1}^{T} \alpha_{it} \tag{5}$$

The advantage of the Fama-Macbeth procedure over the cross-section approach is that there is no cross-sectional correlation in pricing errors. Then, variances are not estimated each period. Instead, we calculate variances in the time series average after γ and α. Then, pricing errors and risk premium variances are estimated as follow:

$$Var(\gamma) = \frac{1}{T^2} \sum_{t=1}^{T} (\gamma_t - \gamma)(\gamma_t - \gamma) \tag{6}$$

Autocorrelation of excess of returns may result in an autocorrelation in estimations (4) and (5). This issue is not important in monthly frequency. We solved this problem applying Newey-West standard errors corrections in formulas (6) and (7) [6].

$$Var(\alpha) = \frac{1}{T^2} \sum_{t=1}^{T} (\alpha_{it} - \alpha_i)(\alpha_{jt} - \alpha_j)' \tag{7}$$

Test: Using sampling theory, pricing errors are tested whether are jointly zero:

$$\alpha' cov(\alpha)^{-1} \alpha \sim X^2(n-1) \tag{8}$$

3.4   Time-Series Approach: Exposition of Upgrade and Downgrade Risk Portfolios to the Average Change of Political Risk in Global Markets

The time-series approach is applied to evaluate the impact of the average change of downside and upside political risk in two portfolios. This approach is applied for two reasons: First, [4] and [2] used this approach to measure the impact of changes on PR on developed and emerging countries. Second, these portfolios are balanced panels and are a small number of assets [6]. The following equation shows an unconstrained time series regression:

$$R_t = \alpha + \beta f_t + \in_t \tag{9}$$

$$\gamma = \frac{1}{T}\sum_{t=1}^{T} f_t \tag{10}$$

$$\sigma(\gamma) = \frac{\sigma(f_t)}{\sqrt{T}} \tag{11}$$

R_t is either the downgrade or the upgrade global portfolio at year t. f_t is the global average change return either downgrade or upgrade at year t. $\in_t$ is the residual at time t. We estimated in Eq. 9: Pricing errors, Factor sensitivities and Residuals. The Factor Premium: The factor premium in the time-series approach is the mean of the factor. Standard errors $(\in_t)$ are iid over time. Test: The t-test is used to evaluate if pricing errors are zero. This test is used because only one asset is evaluated per regression. The t-test test is valid when the regression errors are no correlated and homoscedastic.

4   Results

4.1   Cross-Country Asset Pricing Model

Two parameters are required to explain global political risk changes: β Country i's risk loading on the global political risk measure f^{GPR} and the γ market price. The standard Fama-Macbeth procedure is applied to obtain β_i and γ. β_i is calculated in the first regression (Eq. 1) [7]. The excess of returns of 47 country markets indices and the factor f^{GPR} are the dependent and the independent variables, respectively. The market price risk γ is estimated in the second regression of the Fama-MacBeth procedure (Eq. 2). This parameter is presented in Panel B: Table 3. Regression results of stage I are reported on Panel A: Table 3. Panel A shows a statistically significant positive coefficient for the United States and for Greece at 10%. This Panel also presents a statistically significant negative coefficient for Malaysia at the 10% level. These results are corrected by heteroskedasticity and autocorrelation applying the Newey-West standard errors [6]. Regression results of the stage II are reported in Panel B: Table 3. Panel B shows that the risk premium γ is statistically insignificant. Then, there is no evidence that the global political risk proxy impact cross-country returns.

4.2   Effect of Upside and Downside Political Risk in the Time-Series of Global Returns

[2, 4] evaluate the impact of changes of political risk in the time-series of emerging and developed stock markets. [2, 4] form portfolios that represent countries with upgrade or downgrade in political risk. Following these findings two portfolios that represent upside and downside in political risk are formed using global markets. Then, the impact of the global political risk proxies are evaluated in the time-series of two portfolios using the Eq. 9. Panel A shows that average returns are higher if there is a downgrade in political risk and that average returns are lower when there is an upgrade in political risk. Approximately, the downgrade portfolio returns exceed the upgrade portfolio in 5.5% yearly. Panel B: Table 2 reports that the upgrade political risk factor is strongly significant and negative in global returns at 1%. The heteroskedasticity and the auto-correlation test was applied in both regressions in the residuals. This test showed a positive autocorrelation of one lag in the downgrade portfolio. Then, the standard errors were corrected through the Newey Regression. This result supports the hypotheses that an increase of global political risk affects world markets negatively.

Table 2. Time-Series of Upgrade and Downgrade Portfolios of the Global Political Risk Measure

Panel A: Description of the sample

Sample period: 1998-2008

Portfolio	Average returns (%)	Average Change in Risk (%)
Downgrade (D)	-2.97	-10.16
Upgrade (U)	-8.50	14.26

Panel B: time-series regression

	Portfolio D	Portfolio U
Beta	5.33	(-3.827**)
t-statistics	1.19	-2.17
Alpha (pricing error)	0.51	0.46
t-statistics	1.29	-1.74
Price of risk	(-0.10***)	0.14***
t-statistics	-3.0	2.8
R2	0.3	0.34
N	11	11

*** p<0.01, ** p<0.05, * p<0.1

Table 3. Cross-country of the Global Political Risk Measure, This table shows the factor risk loadings and the market risk price applying Fama-Macbeth regressions. Panel A provides the results of the first stage of the Fama-Macbeth procedure. Panel B provides the market price estimated in the second stage. * significance at 10% level. ** significance at 5% level. *** significance at 1% level.

Panel A: Factor Loadings

Developed Countries	β	Std.Errors	α	Std.Errors	R²	Emerging Countries	β	Std.Errors	α	Std.Errors	R²
US	1.214*	0.576	-0.042	-0.03	0.33	India	0.571	-1.364	0.015	-0.063	0.019
UK	0.836	-0.656	-0.033	-0.03	0.153	Brazil	-1.023	-1.666	0.045	-0.077	0.04
Japan	0.218	-0.996	-0.033	-0.05	0.005	Indonesia	1.095	-1.909	-0.044	-0.089	0.035
Australia	0.416	-0.625	-0.004	-0.03	0.047	Thailand	-0.326	-1.424	-0.025	-0.066	0.006
Hong_Kong	-0.381	-0.961	0.000	-0.04	0.017	Turkey	1.906	-2.062	-0.050	-0.096	0.087
Italy	1.248	-0.816	-0.028	-0.04	0.206	Argentina	-0.229	-1.56	-0.036	-0.073	0.002
Austria	0.271	0.653	0.009	0.05	0.01	Chile	-0.325	-0.933	0.007	-0.043	0.013
Belgium	0.692	-0.825	-0.020	-0.04	0.115	Colombia	-0.0316	-1.594	-0.007	-0.074	0
France	0.518	-0.803	-0.008	-0.04	0.044	Hungary	1.055	-1.312	-0.013	-0.061	0.067
Portugal	1.234	-0.904	-0.031	-0.04	0.171	Korea	0.438	1.953	0.003	0.050	0.008
Sweden	0.59	-1.338	-0.012	-0.06	0.021	Mexico	0.0635	-1.107	0.024	-0.051	0
Switz.	0.832	-0.667	-0.020	-0.03	0.147	Pakistan	0.0322	-1.505	0.002	-0.070	0
Finland	1.552	-1.432	-0.013	-0.07	0.115	Poland	-0.31	-1.154	0.017	-0.054	0.008
Greece	2.015*	1.031	-0.037	-0.05	0.298	SriLanka	0.618	-1.032	-0.029	-0.048	0.038
Israel	0.313	-0.735	0.002	-0.03	0.02	Malaysia	-3.066*	1.672	0.040	-0.078	0.272
Netherland	0.614	-0.749	-0.022	-0.03	0.069	South Africa	-0.195	-1.052	0.008	-0.049	0.004
NewZ	-0.149	-0.800	-0.012	-0.04	0.004	Czech republic	-0.385	-1.365	0.058	-0.063	0.009
Germany	0.672	-0.973	-0.020	-0.05	0.05	Egypt	0.948	-2.097	-0.029	-0.097	0.022
Canada	0.446	-0.810	0.006	-0.04	0.033	Morocco	0.835	-0.996	0.002	-0.046	0.073
Denmark	0.159	-0.803	0.046	-0.04	0.004	Peru	-0.389	-1.037	0.015	-0.048	0.015
Spain	1.18	-0.789	-0.034	-0.04	0.199	Philippine	-1.08	-1.203	-0.026	-0.056	0.082
Ireland	0.992	-0.734	-0.022	-0.03	0.169	Taiwan	0.298	-1.089	-0.041	-0.051	0.169
Singapore	-0.598	-0.980	0.004	-0.05	0.04	China	-0.633	0.999	0.007	0.057	0.014
						Venezuela	-2.928	-1.71	0.0233	-0.081	0.268

Panel B: The Market Price Risk

Price of risk	0.01
Std.Errors	0.045

Newey West Regressions Standard errors

Korea-China-Austria

5 Conclusions

We analyze the relationship between a global political risk measure and asset pricing models. First, using cross-country regressions we find that there were no significant correlatón between the global political risk proxy and the cross-country returns. The global political risk is not a valid source of systematic risk in the international stock markets. Following the findings of [2, 4] we construct two portfolios that represent countries with upside and downside in political risk. The results provides evidence that positives changes in political risk affect negative stock returns and that a decrease affect positive stock returns. Also, that an upside in political risk have a significant negative effect on the time-series of global returns. Therefore, it is evidenced that the upside global political risk factor is a valid source of systematic risk in the international stock markets.

References

1. Ward, A.: Why China's new tariffs could make the US trade war even worse (2019). https://www.vox.com/2019/5/13/18617818/china-usa-trade-war-trump-tariff
2. Erb, C., Campbell, H., Viskanta, T.: Political risk, economic risk and finantial risk. Financ. Anal. J. **52**, 29–46 (1996)
3. Brogaard, J., Detzel, A.: The asset-pricing implications of government economic policy uncertainty. Manage. Sci. **61**, 3–18 (2015)
4. Diamonte, R., Liew, J., Stevens, R.: Political risk in emerging and developed markets. Financ. Anal. J. **52**, 71–76 (1996)
5. Miller, M., Scholes, M.: Rate of return in relation to risk: a reexamination of some recent findings. In: Jensen, M.C. (ed.) Studies in the Theory of Capital Markets, pp. 47–78. Praeger, New York (1972)
6. Goyal, A.: Empirical cross-sectional asset pricing: a survey. Fin. Markets. Portfolio Mgmt. **26**(1), 3–38 (2012)
7. Black, F., Jensen, M., Scholes, M.: The capital asset pricing model: some empirical tests. In: Jensen, M.C. (ed.) Studies in the Theory of Capital Markets. Praeger, New York (1972)
8. Fama, E., McBeth, J.: Risk, return and equilibrium: empirical tests. J. Polit. Econ. **71**, 607–636 (1973)
9. Pastor, L., Veronesi, P.: Uncertainty about government policy and stock prices. J. Finance **67**(4), 1219–1264 (2012)
10. Boutchkova, M., Hitesh, D., Durnev, A., Molchanov, A.: Precarious politics and return volatility. Rev. Financ. Stud. **25**(4), 1111–1154 (2012)
11. Ferson, W.E., Harvey, C.R.: Fundamental Determinants of National Equity Market Returns: A Perspective on Conditional Asset Pricing, Working Papers (1996)
12. Erb, C., Campbell, H., Viskanta, T.: Political risk, economic risk and finantial risk. Financ. Anal. J. **7**, 29–46 (1996)
13. Gaitán-Angulo, M., et al.: Company family, innovation and colombian graphic industry: a bayesian estimation of a logistical model. In: Tan, Y., Shi, Y., Tang, Q. (eds.) Data Mining and Big Data. DMBD 2018. Lecture Notes in Computer Science, vol. 10943. Springer, Cham (2018)

MRAS Based Speed Sensorless Vector Controlled PMSM Drive

Badini Sai Shiva and Vimlesh Verma[⊠]

Department of Electrical Engineering,
National Institute of Technology Patna, Patna, India
{badinisai.eepg16,vimlesh.verma}@nitp.ac.in

Abstract. A Fictitious Quantity (Y) based Model–Reference–Adaptive–System (MRAS) is proposed for speed–estimation of Vector Control Permanent Magnet (PM) Synchronous Motor (PMSM) drive. Online speed/position estimation techniques are essential to make the drive mechanically robust, and to increase reliability. The proposed Y-MRAS speed/position estimation technique is quite robust, independent of integrator and differential-terms and stable in all four-quadrant operation. But, sensitive to stator-resistance (R_s) at very low-speeds; that can be compensated with online resistance estimation technique. The model is developed completely in the rotor reference frame (d- and q-axes) by using reference-voltages and actual-currents. The proposed Y-MRAS based rotor-speed (ω_r) estimation-technique is simulated and verified in MATLAB/SIMULINK platform for various speed and load.

Keywords: PMSM · Speed-estimation · Vector-control · Speed-sensorless · MRAS

1 Introduction

The Permanent Magnet Synchronous Machine received its demand from few decades in application of Electrical-Vehicles (EV), robots, industries, because of its high efficiency, large torque/volume ratio, simple structure, and consistency in operation. The presence of PM in rotor make the advantage over induction and wound rotor type synchronous motors in EV. The position is essential for the PMSM drive. The mechanical/electronic speed/position sensors make drive expense and reduce its consistency, and reliability. By considering signal-transmission, and environmental-issues; sensorless operation is preferable. From all the control techniques available in the literature; the vector-controlled drive has a great advantage over other control techniques with its dynamic performance [1, 2].

Many speed sensorless techniques are presented in the literature for the PMSM drive [2–11]. These are mainly classified into Back EMF based, Signal Injection (SI) based, State-Observer based, Model-based techniques and others like Artificial intelligence. The Back-EMF based estimation techniques use Back-EMF to estimate the position/ speed and this performs satisfactorily at high and moderate-speeds; but, under zero/low-speed operation; back-emf becomes very-low and it's unable to track it. That makes speed estimation very difficult and also, sensitive to machine-parameters.

© Springer Nature Switzerland AG 2020
A. P. Pandian et al. (Eds.): ICICCS 2019, AISC 1039, pp. 549–556, 2020.
https://doi.org/10.1007/978-3-030-30465-2_61

SI-based techniques estimate the speed based on the saliency of the machine. This is the only technique which performs high accuracy at zero speed. But it requires an external-hardware for High Frequency (HF) SI and to extract the position. The combination of Back-EMF and SI-based method make superior to all techniques with its advantages at high- and low-speed operation respectively. Observer-based like Extended Luenburger Observer, Kalman Filter, Sliding Model, etc. techniques are more sensitive to machine parameters and huge complexity. The primary advantage in this method is it treats the parameter's as state variables and can be estimated along with shaft speed. But some methods require filters and initial condition, this degrades the advantage of the techniques.

MRAS computes functional candidate which expressed in different forms with equal–quantities and are used in adjustable and reference-model [6, 8, 12, 13]. The Reference Model is independent of the shaft–speed (ω_r) (i.e., unknown–quantity) and the Adjustable Model is dependent on ω_r. The difference between these two quantities will form error (ε) and passed through PI Controller (i.e., Adaption-Mechanism). The output of PI (i.e., speed) is feed to the Adjustable Model; and this continues till the ε is zero. The MRAS based techniques are more attractive with its advantages: stability, simplicity, less computation-complexity, no external-Hardware, many are free from integrator and differentiation terms, and less dependent on machine parameters, The other techniques like Artificial Intelligence (ANN, fuzzy, etc.) are a recent approach which requires huge-memory to train the system and more complex to build.

The Q-MRAS [6] speed estimation technique facing problem in zero-crossing; that is overcome by the β-MRAS [8] and Y-MRAS technique's. The proposed method depends on stator-resistance and mutual-flux linkage. Mutual-flux varies with the aging of magnets/flux weakening operation methods. The R_s varies with machine stator-windings temperature. This will deteriorate the performance of the position/ speed estimation technique at low–speeds and at high–speed vector control ("$i_{ds} = 0$") is failed but machine tends to rotate at reference speed. So, we require online R_s estimation technique for R_s compensation [8, 14, 15].

Section 2 presents the Modeling of PMSM. Section 3 presents the proposed rotor speed estimation technique (i.e., *Y-MRAS*) for vector–control PMSM drive. Section 4 MATLAB/Simulation results are shown and Sect. 5 concludes the work.

2 Modeling of PMSM

The PMSM machine modeling is taken from [2]. Equation 1 shows the stator–voltage in rotor reference frame (i.e.,"*d*" and "*q*"–*axes*) for PMSM.

$$\begin{pmatrix} v_{ds} \\ v_{qs} \end{pmatrix} = \begin{bmatrix} R_s + L_d P & -\omega_s L_q \\ \omega_s L_d & R_s + L_q P \end{bmatrix} \begin{pmatrix} i_{ds} \\ i_{qs} \end{pmatrix} + \begin{pmatrix} 0 \\ \omega_s \lambda_{af} \end{pmatrix} \tag{1}$$

$$T_e = \left(\frac{3}{2}\right) P\{i_{qs}\lambda_{af} + (L_d - L_q)i_{ds}i_{qs}\} \qquad (2)$$

$$T_e - T_L = J\frac{d\omega_r}{dt} + B\omega_r \qquad (3)$$

The Eq. 2 shows the torque developed. The Eq. 3 shows the dynamics equation. Where T_e and T_L are electric and load-torque respectively. $Þ$ = derivative term $\left(\frac{d}{dt}\right)$, $\omega_s = P\omega_r$, the PMSM machine is nonsaliency type with a sinusoidal Back–EMF waveform. Parameters are shown in Table 1 [6].

3 Y-MRAS Based Rotor Speel D Estimation Technique

The expression for reference and adjustable model for MRAS [6, 13, 16].
   The Fictitious quantity (Y) is expressed as:

$$Y_1 = v_{qs}^* i_{qs} - v_{ds}^* i_{ds} \qquad (4)$$

Substituting (1) in (4)

$$Y_2 = R_s\left(i_{qs}^2 - i_{ds}^2\right) + \left(L_q i_{qs}\frac{di_{qs}}{dt} - L_d i_{ds}\frac{di_{ds}}{dt}\right) + \omega_s i_{qs}i_{ds}(L_d + L_q) + \omega_s i_{qs}\lambda_{af} \quad (5)$$

Under steady–state condition: $\frac{d}{dt} = 0$, Y_2: becomes:

$$Y_3 = R_s\left(i_{qs}^2 - i_{ds}^2\right) + \omega_s i_{qs}i_{ds}(L_d + L_q) + \omega_s i_{qs}\lambda_{af} \qquad (6)$$

The condition for vector control in PMSM is "$i_{ds} = 0$", Y_3 becomes:

$$Y_4 = R_s i_{qs}^2 + \omega_s i_{qs}\lambda_{af} \qquad (7)$$

Among Eqs. 4 to 7, Y_2 and Y_3 are reliant on all machine parameters and the noise is increased due to the presence of "$\frac{d}{dt}$". So, Y_4 and Y_1 are considered as Adjustable and Reference–Model respectively. Y_1 is independent of ω_r and Y_4 is dependent on it. The error (ε_p) developed from Eq. 8. ε_p is passed through a PI–Controller. The output (i.e., ω_{r-est}) is used to tune the adjustable–quantity, this continues till $\varepsilon_p = 0$. Equation 9 shows the rotor–position (ρ_{ms}) electrical angle from speed.

$$\varepsilon_p = (v_{qs}^* i_{qs} - v_{ds}^* i_{ds}) - (R_s i_{qs}^2 + \omega_r(\varepsilon, t)P i_{qs}\lambda_{af}) \qquad (8)$$

$$\rho_{ms} = \int \omega_s dt = P\rho_{mr} = P\int \omega_r dt \qquad (9)$$

4   Simulation Results

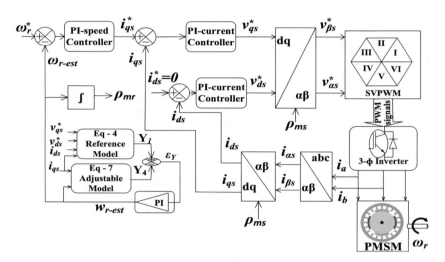

Fig. 1. Block diagram of Y-MRAS based speed sensorless vector-controlled PMSM drive.

The proposed Fictitious–quantity $(Y)MRAS$ based speed-estimation technique for speed sensorless vector–controlled PMSM drive (shown in Fig. 1) is verified in *MATLAB/SIMULINK* under various operation speed/load and is shown below. Two current sensors are used in the drive to measure three phase stator currents. The speed loop is closed with the estimated speed. Position is extracted from estimated speed (Eq. 9) and used for frame transformation by using Clarke and Park transformation. The machine parameters are taken from and are shown in Table 1. The proposed technique performs satisfactorily over $Q - MRAS$ [6] at zero crossing. The estimated and the actual-speed/position overlap each-other and that shows how robust is the $Y - MRAS$ speed-estimation technique. The machine currents are shown in d and $q - axes$ rotor-reference frame; as $i_{ds} = 0$, shows the sucessfull operation of vector-control and i_{qs} shows the torque-component current. Estimated-Position (ρ_{ms}) information is ploted with Actual-Position (ρ_{ms-act}) in electrical angle (i.e., in rad). The position error is shown between the actual and estimated position. Proper tuning of PI-Controller gains will improve the performance of the system. The ω_r is plotted with ω_{r-est} and ω_{ref} (reference speed) on the same scale, to show the perfection of $Y - MRAS$ speed estimation technique.

4.1   Drive Response for Forward and Reverse-Motoring

Figure 2(a) and (b); shows the forward and reverse–motoring of PMSM with a generator type load. In Fig. 2(a), the ω_{ref} is changed in step command from 10 rad/sec to -10 rad/sec and back to 10 rad/sec at t = 2 s and t = 4 s respectively. İn Fig. 2(b), the ω_{ref} is changed in a ramp-type from 4 rad/sec to −4 rad/sec. Figure 2 presents the load on the machine, ω_r ploted with ω_{r_est} and ω_{ref}. The stator currents are presented in d and q-axes (rotor reference frame). ρ_{ms} and ρ_{ms-act} are presented on same sale. The position error $\left(\rho_{ms-act} - \rho_{ms}\right)$ gives the accuracy of the proposed algorithm.

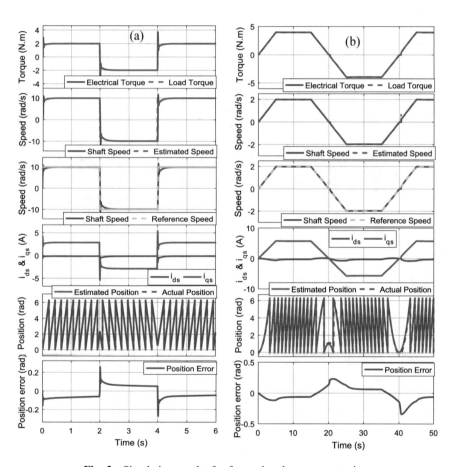

Fig. 2. Simulation results for forward and reverse-motoring.

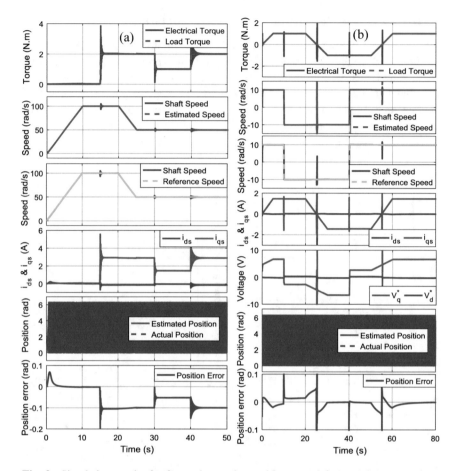

Fig. 3. Simulation results for forward motoring and low-speed four-quadrant operation.

4.2 Drive Response for Step Type Load Command

İn Fig. 3(a), the ω_{ref} is changed in ramp form, from 0 rad/sec to 100 rad/sec in 10 s and maintained for the next 10 s. At t = 20 s to t = 25 s, ω_{ref} is changed from 100 rad/sec to 50 rad/sec in the slow ramp form and maintained. Initially, there is no load on the machine. a step load on the machine with 2 N.m, 1 N.m and 2 N.m respectively at t = 15 s, 30 s and 40 s.

4.3 Drive Response for Four-Quadrant Operation

The proposed method is verified for four quadrant operation and simulation results are plotted in Fig. 3(b). The ω_{ref} is changed in the form of step command between 10 rad/sec to -10 rad/sec and the smooth variation of the load is seen. Figure 3(b) presents the load on the machine, ω_r is ploted with ω_{r-est} and ω_{ref}. The stator reference voltage and actual currents are presented in d and q-axes (rotor–reference frame). ρ_{ms-act}, ρ_{ms} and position error are presented in electrical radiance.

Table 1. PMSM machine parameters [6]

"Pole pair (P)"	3
"Nominal speed (ωn)"	$1000\,rpm.$
"d − axis inductance (Ld)"	$6.66\,mH$
"q − axis inductance (Lq)"	$5.80\,mH$
"Mutual flux linkage (λaf)"	$0.1546\,Wb - turn$
"Stator phase winding resistance (R_s)"	$1.4\,\Omega$
"Machine inertia (J)"	$0.00176\,Kg - m^2$
"Viscous co − efficient (B)"	0.00038818

5   Conclusion

This paper proposes, the new Fictitious quantity $Y - MRAS$ based shaft speed estimation technique for $PMSM$ drive and performs satisfactorily in all four-quadrant operation of speed sensorless $PMSM$ drive. The $MRAS$ uses instantaneous and steady–state–quantities of Y to estimate the shaft speed. The proposed technique is free from the differential, integrating terms. But this depends on stator resistance. We require an online stator-resistance estimation technique for stator resistance compensation; especially for low-speed operation.

Acknowledgment. "This work was supported by the Science & Engineering Research Board (FILE NO. ECR/2016/000900), under Early Career Research Award".

References

1. Bose, B.K.: Modern Power Electronics and AC Drives. Prentice Hall PTR, Upper Saddle River (2002). 711

2. Krishnan, R.: Magnet Synchronous and Brushless DC Motor Drives. CRC Press/Taylor & Francis, Boca Raton (2017). https://doi.org/10.1201/9781420014235
3. Yan, Z., Utkin, V.: Sliding mode observers for electric machines-an overview. In: IEEE 2002 28th Annual Conference of the Industrial Electronics Society, IECON 2002, vol. 3, pp. 1842–1847 (2003). https://doi.org/10.1109/iecon.2002.1185251
4. Zhang, G., Wang, G., Xu, D.: Saliency-based position sensorless control methods for PMSM drives - a review. Chin. J. Electr. Eng. 3, 14–23 (2019). https://doi.org/10.23919/cjee.2017.8048408
5. Borsje, P., Chan, T.F., Wong, Y.K., Ho, S.L.: A comparative study of Kalman filtering for sensorless control of a permanent-magnet synchronous motor drive. In: 2005 International Conference on Electric Machines and Drives, pp. 815–822 (2008). https://doi.org/10.1109/iemdc.2005.195816
6. Maiti, S., Chakraborty, C., Sengupta, S.: Simulation studies on model reference adaptive controller based speed estimation technique for the vector controlled permanent magnet synchronous motor drive. Simul. Model. Pract. Theory 17, 585–596 (2009). https://doi.org/10.1016/j.simpat.2008.08.017
7. Sai Shiva, B., Verma, V., Khan, Y.A.: Q-MRAS-based speed sensorless permanent magnet synchronous motor drive with adaptive neural network for performance enhancement at low speeds. In: Innovations in Soft Computing and Information Technology, pp.pp. 103–116. Springer, Singapore (2019). https://doi.org/10.1007/978-981-13-3185-5_10
8. Shiva, B.S., Verma, V.: Speed and parameter estimation of vector controlled permanent magnet synchronous motor drive. In: 2nd International Conference on Power, Energy and Environment: Towards Smart Technology, ICEPE 2018, pp. 1–6. IEEE (2019). https://doi.org/10.1109/epetsg.2018.8658882
9. Yongdong, L.,Hao, Z.: Sensorless control of permanent magnet synchronous motor – a survey. In: 2008 IEEE Vehicle Power and Propulsion Conference, pp. 1–8 (2008). https://doi.org/10.1109/VPPC.2008.4677421
10. Batzel, T.D., Lee, K.Y.: An approach to sensorless operation of the permanent-magnet synchronous motor using diagonally recurrent neural networks. IEEE Trans. Energy Convers. 18, 100–106 (2003). https://doi.org/10.1109/TEC.2002.808386
11. Xu, D., Wang, B., Zhang, G., Wang, G., Yu, Y.: A review of sensorless control methods for AC motor drives. Trans. Electr. Mach. Syst. 2, 104–115 (2019). https://doi.org/10.23919/TEMS.2018.8326456
12. Landau, Y.D.: Adaptive control: the model reference approach. IEEE Trans. Syst. Man. Cybern. SMC-14, 169–170 (1984). https://doi.org/10.1109/tsmc.1984.6313284
13. Verma, V., Chakraborty, C.: New series of MRAS for speed estimation of vector controlled induction motor drive. In: IECON 2014 - 40th Annual Conference of the IEEE Industrial Electronics Society, Dallas, TX, pp. 755–761 (2014). https://doi.org/10.1109/iecon.2014.7048585
14. Badini, S.S., Verma, V.: A new stator resistance estimation technique for vector controlled PMSM drive. In: 2018 IEEE International Conference on Power Electronics, Drives and Energy Systems, pp. 1–6 (2019). https://doi.org/10.1109/pedes.2018.8707533
15. Maiti, S., Chakraborty, C., Hori, Y., Ta, M.C.: Model reference adaptive controller-based rotor resistance and speed estimation techniques for vector controlled induction motor drive utilizing reactive power. IEEE Trans. Ind. Electron. 55, 594–601 (2008). https://doi.org/10.1109/TIE.2007.911952
16. Verma, V., Chakraborty, C., Maiti, S., Hori, Y.: Speed Sensorless vector controlled induction motor drive using single current sensor. IEEE Trans. Energy Convers. 28, 938–950 (2013). https://doi.org/10.1109/TEC.2013.2273935

Multiple Model Filtering for Vehicle Trajectory Tracking with Adaptive Noise Covariances

M. Nithin and Manoj Panda$^{(\boxtimes)}$

Department of Electronics and Communication Engineering,
Amrita School of Engineering, Amrita Vishwa Vidyapeetham, Coimbatore, India
mk_panda@cb.amrita.edu

Abstract. As the automotive world is moving towards its ultimate aim of fully autonomous vehicles, predicting the trajectories itself and that of the neighboring vehicles is essential for each vehicle. This paper proposes a novel filtering method for predicting the path of an ego vehicle (i.e., a vehicle of interest) using measurements by a Global Positioning System (GPS) device and combining such measurements with multiple candidate kinematic motion models. The proposed Multiple Model filtering method adapts to the noise conditions as inferred by measurements. It is shown that the proposed adaptive method provides a considerable level of improvement compared to the existing non-adaptive multiple model filtering.

1 Introduction

As Intelligent Transport Systems (ITS) are becoming closer to realization in the true sense, accurate tracking of the trajectories of vehicles has become very important. In order to execute the envisaged ITS applications for safety and infotainment, the vehicles will require an accurate knowledge of their own trajectories and that of their neighbors on the same and/or nearby road(s). Each vehicle can track its own trajectory by using on-board sensors, e.g., a Global Positioning System (GPS) device and an Inertial Measurement Unit (IMU), and applying a *data-driven* approach [1] or a *model-based* approach [2] of estimation. The tracking of neighbors' trajectories can be performed in two ways. If the vehicles cooperate, then each vehicle can periodically broadcast its trajectory-related information to its neighbors using Vehicle-to-Vehicle (V2V) communication. If the vehicles do not cooperate (e.g., because of trust-deficit), then there are two options available. Firstly, a vehicle can make use of additional sensors such as LiDAR, radar, image processing, etc., if available, to track its neighbors' trajectories. Secondly, vehicles can send their trajectory-related information to a trusted cloud server using Vehicle-to-Infrastructure (V2I) communication, and the server can share the processed information back to the vehicles.

This paper is concerned with the problem of tracking of a vehicle's own trajectory (which, as discussed above, is also the prerequisite for tracking the neighbors' trajectories using V2V and V2I communication). To that end, each vehicle

© Springer Nature Switzerland AG 2020
A. P. Pandian et al. (Eds.): ICICCS 2019, AISC 1039, pp. 557–565, 2020.
https://doi.org/10.1007/978-3-030-30465-2_62

uses a GPS device and an IMU and applies a model-based approach, as in [3]. As argued in [3], the model-based approach is preferred over the data-driven approach due to the reliability concerns of sensors. The work [3] proposed a multiple model Unscented Kalman Filter (UKF) approach. Four kinematic models, viz., Constant Velocity (CV), Constant Acceleration (CA), Constant Turn rate (CT) and Constant Turn Rate and Acceleration (CTRA) [4–6], were used to make four independent UKFs run in parallel. Multiple model filtering was introduced because, at any moment the vehicle should obey any of the above kinematic models. Suppose only CV model was used, then when the vehicle takes a turn the model fails miserably. To avoid such scenarios multiple models are used. The Maximum Likelihood (ML) approach was used to select the most appropriate kinematic model at any time step, and thereby predict the most likely path of vehicles. This so-called Maximum Likelihood Multiple Model (MLMM) filtering turned out to be better than the UKF using any one kinematic model. Note that the UKF approach has found applications in other areas as well [7].

In [3], however, the noise covariance matrices Q and R, representing the state noise and measurement noise, respectively, are assumed to be constants and are not updated during the execution of filtering operations, even though the noise levels may actually change over time. In a model-based approach, the accuracy of filtering is critically dependent on how accurately the noise covariance matrices are modelled. The use of inaccurate Q and/or R matrices can degrade the filtering performance and can even make the filter unstable [8]. Therefore, the matrices Q and R should be continually adjusted to minimize the error.

In this paper, we extend the MLMM filtering approach in [3] to Adaptive MLMM (AMLMM), by adaptively adjusting the state noise covariance matrix Q at each time step. The measurement noise covariance matrix R is kept to be a constant assuming the sensor quality to be consistent. An innovation-based approach [9–11] (to be described soon) is used to adaptively adjust the Q matrix for the UKFs corresponding to each kinematic model. With the AMLMM approach, we could achieve a reduction of $\approx$30% in the Root Mean Square (RMS) error in position estimates compared to MLMM.

The remainder of the paper is organized as follows. In Sect. 2 we provide a survey of the related work. In Sect. 3 we develop the methodology adopted to track the trajectory of an ego vehicle. The results are discussed in Sect. 4. The paper concludes in Sect. 5.

2   Literature Survey

The estimation of noise covariance matrices has been an active area of research. There are different methods available to estimate the noise covariance matrices, which can be generally divided into *on-line* and *off-line* methods [12]. The adaptive filtering methods form the on-line methods, and can be classified as: the Bayesian methods, the Correlation methods, the Maximum-Likelihood methods, and the covariance-Matching methods [13]. The innovation-based approach adopted in this paper is a covariance-Matching method.

Several studies have been conducted to predict the path of the vehicles using adaptive Kalman filtering techniques. The work [14] applies an adaptive Extended Kalman Filter (EKF) with multiple model filtering approach. Here, the authors assign weights to each model output and the overall estimate is a weighted sum. The weights are then adaptively updated at each time-step according to measurements from the sensors. However, in [14], the Q and R matrices do not adapt with time. In this paper, we adapt the Q matrix.

The work [9] proposed a covariance-matching method known as the *sage-husa* estimation for linear time variant (LTV) systems. However, the system equations involved in vehicle path prediction are often nonlinear. The work [15] extends the sage-husa estimator concept to a modified sage-husa estimator which makes it suitable to use with nonlinear systems too. The work [16] applies Kalman filter techniques along with the covariance-matching based adaptive update of the Q and R matrices for predicting the path of autonomous vehicles. However, in [15] and [16], only a single kinematic model is applied. In this paper, we apply a multiple model approach to deal with the model uncertainty.

3   Methodology

This section first discusses the adaptive method for estimation of state noise covariance matrix Q and then explains how this new estimation method can be incorporated into the MLMM filter for improved results. As in [3], we have four UKFs running in parallel, each pertaining to one of the models CV, CA, CT and CTRA, respectively. However, unlike in [3], when modeling the vehicle motion the state noises are assumed to be unknown and time-varying, and an adaptive UKF is established with recursively estimating the state noise covariance matrix Q. As mentioned before, the measurement noise covariance matrix R is assumed to be constant.

Let the *state equation* and the *measurement equation* for the ith model are given by

$$\mathbf{x}_i(k{+}1) = \mathbf{f}_i(\mathbf{x}_i(k)) + \boldsymbol{\eta}_i(k), \text{ and} \tag{1}$$
$$\mathbf{z}_i(k) = \mathbf{h}_i(\mathbf{x}_i(k)) + \boldsymbol{\nu}_i(k), \tag{2}$$

where $\mathbf{x}_i(k)$ and $\mathbf{z}_i(k)$ denote the 'state of the vehicle' and the 'measurement obtained from the sensors', respectively, and $\boldsymbol{\eta}_i(k)$ and $\boldsymbol{\nu}_i(k)$ denote the additive 'process noise' and the 'measurement noise', respectively, at discrete time k. The vector functions $\mathbf{f}_i$ and $\mathbf{h}_i$ denote the ith 'motion model' and its corresponding 'measurement model', respectively. The ith model is characterized by the motion model $\mathbf{f}_i$ and measurement model $\mathbf{h}_i$.

From (1), the state noise can be estimated as

$$\hat{\boldsymbol{\eta}}_i(k-1) = \hat{\mathbf{x}}_i^+(k) - \mathbf{f}_i(\hat{\mathbf{x}}^+(k-1)) \tag{3}$$
$$= \hat{\mathbf{x}}_i^+(k) - \hat{\mathbf{x}}_i^-(k)$$
$$= K_i(k)\boldsymbol{d}_i(k),$$

where $\hat{\mathbf{x}}_i^-(k)$ and $\hat{\mathbf{x}}_i^+(k)$ denote the estimates, immediately before and after incorporating the measurement, respectively, $K_i(k)$ denote the Kalman gain and $\boldsymbol{d}_i(k) = \boldsymbol{z}_i(k) - \mathbf{f}_i(\hat{\mathbf{x}}^-(k))$ denotes the *innovation*, all pertaining to the ith model-dependent UKF at time slot k. Therefore, the state noise covariance matrix $Q_i(k)$ pertaining to model i at time-step k can be estimated as

$$
\begin{aligned}
\hat{Q}_i(k-1) &= E[\hat{\boldsymbol{\eta}}_i(k-1)\hat{\boldsymbol{\eta}}_i^T(k-1)] \qquad (4)\\
&= E[K_i(k)(\boldsymbol{d}_i(k)\boldsymbol{d}_i^T(k))K_i^T(k)]\\
&= K_i(k)E[(\boldsymbol{d}_i(k)\boldsymbol{d}_i^T(k))]K_i^T(k)\\
&= K_i(k)S_i(k)K_i^T(k).
\end{aligned}
$$

To keep the estimate immune to short-lived high fluctuations, one can obtain a smoothened estimate, or the so called Sage-Husa estimate, as [9,17]

$$
\hat{Q}_i(k) = (1 - w_i(k-1))\hat{Q}(k-1) + w_i(k-1)[K_i(k)S_i(k)K_i^T(k)], \qquad (5)
$$

where the weight factor $w_i(k-1)$ is given by

$$
w_i(k-1) = (1-n)/(1-n^k) \qquad (6)
$$

where n is a factor between (0.95, 0.995).

The entire Adaptive UKF process is summarized in Fig. 1. As shown in (3) and (4) the innovation can be used to estimate the state noise covariance matrix. Introducing (5) to existing UKF- MLMM algorithm [3] results in adaptive update of matrix Q and reduces the estimation error in the state noise covariance matrix Q. Thus the filter performance is expected to improve. All other steps involved in Fig. 1 are the same steps in a normal UKF algorithm. So these steps are not explained here, as it is already available in many well written papers including [18]. The same process is to be followed for each motion model, and hence a generalized diagram is shown.

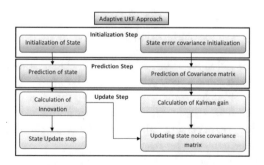

Fig. 1. Implementation of Adaptive UKF, Flowchart

4   Results

The main objective of the experiment was to compare the performance of the AMLMM filter with MLMM filter algorithm and analyze whether adaptive algorithm improves the performance of MLMM filter. The Warrigal Dataset [19] has been used for testing the algorithm.

Consider the two scenarios shown in Figs. 2 and 3 from the Warrigal Dataset. The scenarios are marked with the most likelihood models selected by the AMLMM filtering algorithm at any particular instant.

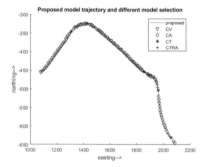

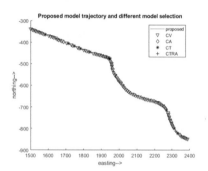

Fig. 2. Trajectory of Scenario 1 with motion models marked.

Fig. 3. Trajectory of Scenario 2 with motion models marked.

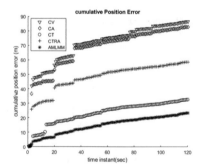

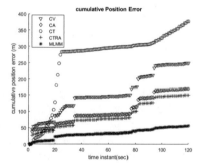

Fig. 4. Scenario 1: model-wise cumulative error with adaptive state noise covariance matrix.

Fig. 5. Scenario 1: model-wise cumulative error with non-adaptive state noise covariance matrix.

Figures 4 and 5 show the cumulative position errors of scenario 1 with and without adaptive MLMM. It's clear from the figures that cumulative error of all models has been reduced to a considerable level.

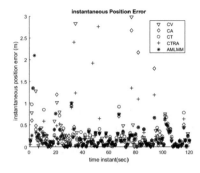

Fig. 6. Scenario 1: model-wise instantaneous error with adaptive state noise covariance matrix.

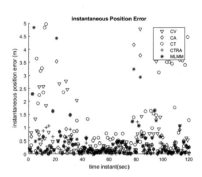

Fig. 7. Scenario 1: model-wise instantaneous error with non-adaptive state noise covariance matrix.

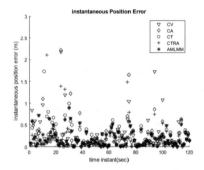

Fig. 8. Scenario 2: model-wise instantaneous error with adaptive state noise covariance matrix.

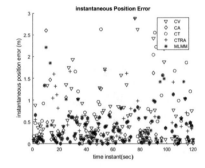

Fig. 9. Scenario 2: model-wise instantaneous error with non-adaptive state noise covariance matrix.

Figures 6 and 7 shows the instantaneous error plot of the same scenario. From the figures it can be inferred that, AMLMM has smaller instantaneous error than individual model-dependent adaptive UKFs. Similar conclusions are derived from Figs. 8 and Fig. 9 for Scenario 2.

To make it clearer, in Figs. 10 and 11 we show the instantaneous errors pertaining to the overall maximum likelihood (ML) model for Scenario 1 and 2, respectively. From the Figs. 10 and 11 it can be seen that the darker circles (AMLMM points) are more settled compared to hollow ones (MLMM points). Clearly, the Adaptive MLMM gives better results for Scenario 2 also.

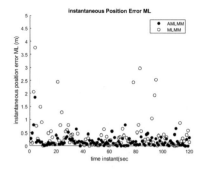 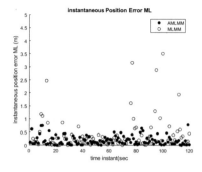

Fig. 10. Scenario 1: overall instanta-neous error with the ML model adaptive and non-adaptive state noise covariance matrices.

Fig. 11. Scenario 2: overall instanta-neous error with the ML model adaptive and non-adaptive state noise covariance matrices.

Table 1 shows the comparison of the Root Mean Square (RMS) position errors of AMLMM and MLMM. The percentage reduction in error (with reference to the value without adaptive MLMM) is calculated to show the improvement while using AMLMM.

Table 1. Comparison of RMS position error

Model	Scenario1			Scenario2		
	MLMM	AMLMM	% reduction	MLMM	AMLMM	% reduction
CV	1.4446	0.8515	41.055	1.4606	0.6162	57.8081
CA	1.1736	0.8269	29.5466	1.2999	0.5913	54.5075
CT	1.7950	0.5130	71.4204	3.3561	0.5355	84.0422
CTRA	1.0882	0.7222	33.6358	1.3268	0.5894	55.5773
MLMM	0.6904	0.4660	32.4969	0.9184	0.4238	53.8508

5   Conclusion

In this paper, an adaptive MLMM filtering method was developed and the com-parison with MLMM filter using real trajectory data showed that the adaptive MLMM filter's performance is significantly better compared to MLMM filter. Moreover, the error in position estimation has been found to be reducing for every model used for the experiment as well. proposed Future scope of this work is to predict the trajectories of neighboring vehicles using V2V and V2I commu-nication and study the effect of packet losses on AMLMM filter.

References

1. Chen, C.P., Zhang, C.-Y.: Data-intensive applications, challenges, techniques and technologies: a survey on big data. Inf. Sci. **275**, 314–347 (2014)
2. Dellaert, F., Pomerlau, D., Thorpe, C.: Model-based car tracking integrated with a road-follower. In: Proceedings of the 1998 IEEE International Conference on Robotics and Automation, vol. 3, pp. 1889–1894. IEEE (1998)
3. Vashishtha, D., Panda, M.: Maximum likelihood multiple model filtering for path prediction in intelligent transportation systems. Proc. Comput. Sci. **143**(1), 635–644 (2018)
4. Li, X.R., Jilkov, V.P.: Survey of maneuvering target tracking. Part I. Dynamic models. IEEE Trans. Aerosp. Electron. Syst. **39**(4), 1333–1364 (2003)
5. Lefèvre, S., Vasquez, D., Laugier, C.: A survey on motion prediction and risk assessment for intelligent vehicles. Robomech. J. **1**(1), 1 (2014)
6. Schubert, R., Richter, E., Wanielik, G.: Comparison and evaluation of advanced motion models for vehicle tracking. In: 2008 11th International Conference on Information Fusion, pp. 1–6. IEEE (2008)
7. Ranganathan, V., Prabha, G., Narayanankutty, K.: Constant modulus hybrid recursive and least mean squared algorithm performance comparable to unscented Kalman filter for blind beamforming. In: 2016 IEEE Annual India Conference (INDICON), pp. 1–4. IEEE (2016)
8. Almagbile, A., Wang, J., Ding, W.: Evaluating the performances of adaptive Kalman filter methods in GPS/INS integration. J. Glob. Position. Syst. **9**(1), 33–40 (2010)
9. Sage, A., Husa, G.W.: Adaptive filtering with unknown prior statistics. In: Proceedings of Joint Automatic Control Conference, pp. 760–769 (1969)
10. Akhlaghi, S., Zhou, N., Huang, Z.: Adaptive adjustment of noise covariance in Kalman filter for dynamic state estimation. In: 2017 IEEE Power Energy Society General Meeting, pp. 1–5 (2017)
11. Wang, J.: Stochastic modeling for real-time kinematic GPS/GLONASS positioning. Navigation **46**(4), 297–305 (1999). https://doi.org/10.1002/j.2161-4296.1999.tb02416.x
12. Duník, J., Šimandl, M.: Estimation of state and measurement noise covariance matrices by multi-step prediction. In: Proceedings of the 17th IFAC World Congress, pp. 3689–3694 (2008)
13. Dunik, J., Straka, O., Kost, O., Havlík, J.: Noise covariance matrices in state-space models: a survey and comparison of estimation methods - part 1. Int. J. Adapt. Control Signal Process. **31**(11), 1505–1543 (2017). https://doi.org/10.1002/acs.2783
14. Barrios, C., Himberg, H., Motai, Y., Sad, A.: Multiple model framework of adaptive extended Kalman filtering for predicting vehicle location. In: 2006 IEEE Intelligent Transportation Systems Conference, pp. 1053–1059 (2006). https://doi.org/10.1109/ITSC.2006.1707361
15. Shi, Y., Han, C., Liang, Y.: Adaptive UKF for target tracking with unknown process noise statistics. In: 12th International Conference on Information Fusion, pp. 1815–1820 (2009)
16. Liu, Y., Fan, X., Lv, C., Wu, J., Li, L., Ding, D.: An innovative information fusion method with adaptive Kalman filter for integrated INS/GPS navigation of autonomous vehicles. Mech. Syst. Signal Process. **100**, 605–616 (2018)

17. Myers, K., Tapley, B.: Adaptive sequential estimation with unknown noise statistics. IEEE Trans. Autom. Control **21**(4), 520–523 (1976). https://doi.org/10.1109/TAC.1976.1101260
18. Wan, E.A. Van Der Merwe, R.: The unscented Kalman filter for nonlinear estimation, pp. 153–158 (2000). https://doi.org/10.1109/ASSPCC.2000.882463
19. Ward, J., Worrall, S., Agamennoni, G., Nebot, E.: The warrigal dataset: multi-vehicle trajectories and V2V communications. IEEE Intell. Transp. Syst. **6**(3), 109–117 (2014)

Automated Criminal Identification System Using Face Generation

Amogh Karve, M. Balasubramanian, Kartik Chaudhari[✉],
and S. B. Mane

College of Engineering Pune, Pune, Maharashtra, India
{karve15.it, balasubramanianml5.it, chaudhariks15.it,
sunilbmane.comp}@coep.ac.in

Abstract. The goal of the project is to aid the police department to ease the process of criminal recognition. We first initiate the recognition by the generation of face images based on image generation by using their sketch and visual attributes. The images we are generating are hyper-realistic in nature. This is followed by facial identification based on the image generated in the previous phase. The Model we have chosen here is a Generative Adversarial Network. The model learns the association by the use of a generator and a discriminator. We are using a context based Generative Adversarial Network to do the same.

Then, we are running the process of facial identification for recognizing the person from the image generated with the help of facial recognition. For this we use the HOG Face recognition algorithm. We work towards aiding the police department in reducing their sketching time and criminal recognition time.

Keywords: Machine learning · Generative Adversarial Network · Face generation · Face recognition

1 Introduction

Machine Learning

Machine Learning is an application of Artificial Intelligence (AI) that gives the computer the ability to become more accurate in predicting outcomes without being explicitly programmed. Machine learning has been affecting our lives to a great extent in the past few years with its applications ranging from speech recognition, effective web-search to even self-driving cars. Machine learning algorithms generate models that help to unravel patterns in observed data, thus helping to understand the world better and predict things.

Generative Adversarial Networks

The basic idea of these networks is that you have 2 models, a generative model and a discriminative model. The discriminative model has the task of determining whether a given image looks natural i.e. if it is an image from the dataset or looks like it has been artificially generated by the generator for fooling the discriminator. The task of the generator is to create natural looking images that are similar to the original data

© Springer Nature Switzerland AG 2020
A. P. Pandian et al. (Eds.): ICICCS 2019, AISC 1039, pp. 566–573, 2020.
https://doi.org/10.1007/978-3-030-30465-2_63

distribution but are actually fake, so that it can fool the discriminator into believing that it is an image from the dataset.

Our proposed solution is an end-to-end system that mainly comprises of 2 stages. The first stage involves face generation, while the latter takes care of face recognition. Thus, we build a system that takes visual attributes and a sketch as input and identify the criminal as output.

2 Related Work

The rapid development of deep learning has accounted for recent exciting progress in image generation, especially the introduction of generative adversarial networks (GAN) [1]. Simply using a Vanilla GAN didn't generate hyper-realistic images from merely specifying the visual attributes. So, based on a straightforward extension of the GAN model, a conditional GAN model was designed by Jon Gauthier in his research paper [1]. In Conditional GANs, he added the capability for each network to condition on some arbitrary external data which describes the image being generated or discriminated. The faces with specific attributes can be generated by merely using random noise and the variation in the conditional information provided to the extended GAN. Conditional data consists of facial attributes that one should be able to tweak corresponding to particular attributes and see reasonable changes in the output faces. This model ran on the LFW dataset [2] and the process stops when the discriminator is maximally confused.

The main drawback in this approach is that there wasn't a clear control over the image generated and moreover the likelihood of test set under the conditional GAN and Vanilla GAN was roughly the same. So, merely using the attributes to generate an image did not provide great results.

Hence, only attributes are not enough to generate the image one is seeking for. One possible model according to [3], could be to bring in a roughly drawn freehand sketch of the criminal, along with which the visual attributes will be specified. In the training phase, the input to the Context Based Generative Adversarial Network is an image which consists of two parts, the first one being the sketch of the image and the second one being the actual image. So, the dataset is preprocessed in order to generate such joint representation of sketch and images. The sketch is generated by edge detection techniques. Apart from these the visual attribute vector is fed to the GAN. In training, the generator learns the natural appearance of faces such that any noise vector in the latent space would be able to generate a visually plausible image. This approach doesn't have potential drawbacks and hence, we have adopted a similar approach.

Now, coming to the second stage, the face identification part, which according to recent works employ deep neural networks [6]. Our approach on the other hand is based on [4], which is a purely data driven method which learns its representation directly from the pixels of the face. The accuracy of this method as tested on the Labeled Faces in the Wild (LFW) dataset, achieves an accuracy of 99.63% and on the YouTube Faces DB it achieves 95.12%. Hence, we have incorporated a similar approach for identification of faces once the image is generated.

3 Proposed Solution

We intend to provide a solution which is an end-to-end application that would take a sketch and visual attributes as inputs and try to identify a particular criminal suspect as output. This application is intended in aiding the work done by the police dept.

In particular, our proposed solution consists of two stages, the face generation stage and the face identification stage. The former deals with generation of hyper-realistic facial images when a roughly drawn freehand sketch and the visual attributes are provided. We initially started with an attribute only based image generation technique [7, 8] but, it was observed that using sketches in a real life scenario makes the process more realistic in police work use. Whereas the latter deals with identification of the facial image generated in stage one and mapping it to an actual person present in the criminal database. Solving the problems individually and pipelining them seems to be an ideal solution to this problem (Fig. 1).

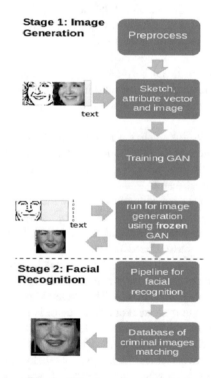

Fig. 1. Overview of methodology

4 Overview of Architecture

We have chosen an architecture that is useful for criminal recognition in a very efficient pipelined fashion. The architecture is divided in to the User interface, Face generator and face identifier. The face generator consists of the context based GAN which in turn consists of a generator and a discriminator. The face identifier consists of an HOG algorithm. The face generator and identifier are both based on a python backend.

The free hand sketch along with a marked attribute list needs to be uploaded in the face generation module. This outputs realistic facial image which is sent to the identification module. The identification module then identifies the relevant criminal (Fig. 2).

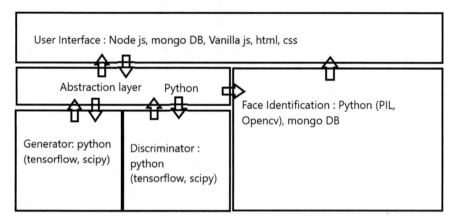

Fig. 2. Architecture of proposed solution

The face generation module consists of scipy and tensorflow [11].

The face identification module consists of face_recognition module from PIL.

The graphical user interface has been created using html, css and javascript, Node.js [10] and express has been used for writing apis in backend which call the python ode written for face generation and face identification modules.

5 Working

5.1 Face Generation

We begin with the training of the context based GAN. The Adam optimizer has been used for the generator and discriminator network. During the training period, the GAN uses the correspondence between the image and sketch. The generator and discriminator make up the generative adversarial network. The generative network consists of a 5 up-convolutional layers layers and has 4 layered convolutional network. The Leaky RelU is used in all the layers for the generative network with tanh at the final layer. The network is trained with a batch size of 64 and epochs of 200. During the back propagation, the updation is being done on the basis of contextual and perceptual loss. The update is done over 500 iterations. We then freeze the network and use it to test some of our data. This number of iterations we used for testing were also 500. We observed that the generated images become clear enough to be passed on to the face generation module at around 500 iterations. The images that are sent into the network are sketch images attached to the masked images as the network works by the correction of an image. The Generative network works on it by considering the problem to be an image completion problem over the masked region. The masked region is where the generated image occurs (Fig. 3).

Fig. 3. Generated image

5.2 Face Recognition

To implement this stage one can avail the face_recognition package [9] present in python. Using this library, one can recognize the face in the image and then identify whose image it is from the given database. So, this module facilitates easier identification of faces given in the pictures with an accuracy of 99.38%. Essentially, the steps followed in this algorithm are:

- Encode the picture using the HOG algorithm to create a simplified version of the image. Using the simplified image, the part of the image that looks like a generic HOG encoding of a face is found out.
- The pose of the face is figured out by finding the main landmarks in the face. Once we find those landmarks, use them to wrap the image so that the eyes and mouth are centered.
- Pass the centered face image through a neural network that knows how to measure features of the face and save those measurements.
- Looking at all the faces we've measured in the past, see which person from the database has the closest measurements to the face's measurements. That's the match required! (Figs. 4 and 5).

ATTRIBTUE AND SKETCH TO FACE GENERATOR

5 o Clock Shadow	Big Lips	Big Nose	Chubby	Double Chin
○ Yes ◉ No	○ Yes ◉ No	○ Yes ◉ No	○ Yes ◉ No	○ Yes ◉ No

Eyeglasses	Goatee	Heavy Makeup	High Cheekbones	Male
○ Yes ◉ No	○ Yes ◉ No	◉ Yes ○ No	◉ Yes ○ No	○ Yes ◉ No

Mouth_Open	Mustache	Narrow Eyes	Pale Skin	Pointy Nose
◉ Yes ○ No	○ Yes ◉ No	○ Yes ◉ No	○ Yes ◉ No	◉ Yes ○ No

Rosy Cheeks	Smiling	Wearing Lipstick
○ Yes ◉ No	◉ Yes ○ No	◉ Yes ○ No

Browse... No file selected.

submit

Fig. 4. Graphical interface for providing input attributes and the roughly drawn sketch image

Name : Natasha Romanoff
Age   : 32

Fig. 5. The output image generated along with the details of the criminal after face identification

6   Results

The results clearly show us that by using our proposed solution, the time required for the identification process of each criminal has come down remarkably. In addition to being sped-up, manual effort required for the process has been reduced. Automation of the face generation has resulted in about 10 times increase in speed on average. Here the total time for our proposed solution indicates the time required for attribute feeding, rough sketch, image generation and face recognition.

The dataset used for training our model was the CelebrityAttribute (CelebA) [5] dataset that contains over 200 thousand images of celebrities labeled to 40 facial attributes. From this dataset, we used 140K and 18 attributes were used for training and generating the frozen GAN model. This model was then tested on the remaining 60K images (Fig. 6 and Table 1).

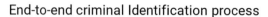

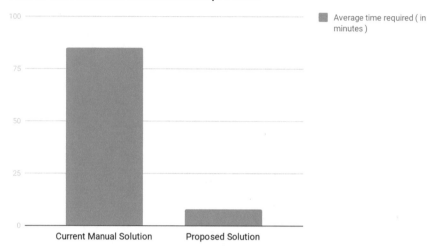

Fig. 6. Graph of average time required for current manual solution and our proposed solution

Table 1. Comparison of time required using current manual solution and proposed solution

Approach	Average time required (in minutes)
Current manual solution	85
Proposed solution	8

7 Conclusion

In this paper we have proposed a solution in the form of an application that is an end-to-end system to be used for criminal recognition and identification. Our system will aid the Police department in identifying suspected criminals. Using a generative model for face-images automates the process and reduces the dependency on specialized sketchers for doing the same. Our application will bring down the time of this entire process by a great extent. The Police department can also store the mapping between visual attributes and suspected criminals for future references. In this way our proposed solution can save both time and skilled resources by replacing the current manual task by an automated system based of generative models and face recognition.

References

1. Goodfellow, I., Pouget-Abadie, J., Mirza, M., Xu, B., Warde-Farley, D., Ozair, S., Courville, A., Bengio, Y.: Generative adversarial nets. In: NIPS, pp. 2672–2680 (2014). http://papers.nips.cc/paper/5423-generative-adversarial-nets.pdf

2. LFW Dataset - Consists of about 13,000 color images and 73 attributes per image. Each image has confidence values for a large number of facial expression attributes and related features, which include: race (Asian, Indian, black, white), age (baby, child, senior), and emotion (frowning, smiling). http://vis-www.cs.umass.edu/lfw/
3. Lu, Y., Wu, S., Tai, Y.-W., Tang, C.-K.: Image generation from sketch constraint using contextual GAN. https://arxiv.org/abs/1711.08972
4. Schroff, F., Kalenichenko, D., Philbin, J.: FaceNet: a unified embedding for face recognition and clustering. https://www.cv-foundation.org/openaccess/content_cvpr_2015/app/1A_089.pdf
5. CelebA dataset. https://www.kaggle.com/jessicali9530/celeba-dataset
6. Wang, M., Deng, W.: Deep face recognition: a survey. https://arxiv.org/abs/1804.06655
7. Reed, S., Akata, Z., Yan, X., Logeswaran, L., Schiele, B., Lee, H.: Generative adversarial text to image synthesis. www.jmlr.org/proceedings/papers/v48/reed16.pdf
8. Yan, X., Yang, J., Sohn, K., Lee, H.: Attribute2Image: conditional image generation from visual attributes. https://arxiv.org/abs/1512.00570
9. face_recognition. https://pypi.org/project/face_recognition/
10. Node.js. https://nodejs.org/en/docs/
11. TensorFlow. https://www.tensorflow.org/overview

Automated UVM Based Verification of Device Life Cycle Management IP

Srivasthav Kotha[1]([✉]), Rajin Ravimony[2]([✉]),
and N. Mohankumar[1]([✉])

[1] Department of Electronics and Communication Engineering, Amrita School
of Engineering, Coimbatore, Amrita Vishwa Vidyapeetham, Coimbatore, India
srivasthavkotha@gmail.com, n_mohankumar@cb.amrita.edu
[2] Ignitarium Technology Solutions Private Limited, Bangalore, India
rajin.ravi@ignitarium.com

Abstract. Device lifecycle management intellectual property (IP) is used to configure the generic register transfer level (RTL) based on the customer requirement instead of designing multiple designs of different specifications. In this work, a generic universal verification methodology (UVM) testbench is scripted for verifying the device lifecycle management IP and the script can generate automatically a specific verification environment for the provided specification through an excel sheet instead of creating different verification environments for different variant IP's. An UVM based verification environment is created for verifying the intellectual property because of its rich base class libraries, phasing, UVM factory, reporting mechanism, metaprogramming etc. from all other verifying methodologies. In this work, an UVM based verification testbench for all the blocks of an IP is scripted and simulation is verified using cadence ncsim, code coverage is analyzed using cadence integrated metrics center.

Keywords: UVM · Intellectual Property (IP) · Device lifecycle management

1 Introduction

In the market, generally the same functional Intellectual properties (IP's) are released with different variants based on customer requirement. Theses IP vendors design and fabricate a generic register transfer level (RTL) and customize the IP based on the customer requirement instead of designing different variant IP's. This customizing of an IP is done after the fabrication of the generic IP and before reaching the final customer, during this phase the IP has to undergo different lifecycle states and needs a different set of permissions to customize the IP and these set of processes are done by reading or writing to a device lifecycle management IP. This device lifecycle management IP can be further protected by adding a watermarking during the design phase or fabrication phase in order to prove the ownership [1].

The design of the device lifecycle management IP is complex and considering every aspect of the design is very difficult for the design engineer. So, as the complexity of the design grows higher, there can be a chance of making functional errors in the design. So, this leads to the need for functional verification of an IP.

© Springer Nature Switzerland AG 2020
A. P. Pandian et al. (Eds.): ICICCS 2019, AISC 1039, pp. 574–583, 2020.
https://doi.org/10.1007/978-3-030-30465-2_64

1.1 Existing Methods of Verification

For the verification of an IP, there are many existing verification methods available like starting with basic Verilog, system Verilog, open verification methodology (OVM), verification methodology manual (VMM) and universal verification methodology (UVM) [2] testbenches etc.

The basic Verilog test benches cannot be used for the complex designs because it cannot consider all the test scenarios for verifying the IP. System Verilog was the first verification testbench which was built on the object-oriented programming language and it is very efficient for verifying the design but there is no proper framing of the language and sufficient base class libraries [3, 4] which makes the designing of test-bench complex. Finally open verification methodology (OVM), verification methodology manual (VMM) are tool-dependent based testbenches which cannot be used by all the tool vendors. So, there is a need of standardization in the test bench methodology in order to make feasible usage of test code on all the tools and UVM is first verification methodology which is standardized in 2011 by IEEE standards association and IEEE 1800.2 stands for Universal Verification Methodology.

UVM is also built on the object-oriented programming language similar to system verilog but there are many advantages of UVM [5] over system verilog like rich base class libraries, reporting mechanism, configuration class, phasing, UVM factory, transaction level modeling and reusability [6]. These features make the design of testbench and debugging of errors easier.

1.2 Verification and Automation

Moreover in order to make a generic testbench which is feasible for verifying different intellectual property having the same set of features and also to reduce the design time of testbench. We can make use of a scripting language which can automate the whole testbench based on the given set of features. In this work, testbench all the blocks are scripted using python and automated based on the given set of specification through an excel sheet.

2 Device Lifecycle Management IP

A newly fabricated generic IP has to undergo many lifecycle states before reaching the end customers specification and as we go from one state to other, the set of access controllability's are reduced based on the customer requirement and loop back to the previous state is restricted in order to avoid re-customizing of an IP. This set of access permissions and lifecycle states are controlled by using the device lifecycle management IP. Figure 1 shows the block diagram of device lifecycle management IP.

OTP Fuse Module: OTP fuse module is a onetime programmable device which is initially set to zero and the set of permissions, data are stored along with the marker and parity bits in the permitted lifecycle state. The OTP Fuse module is programmed using the OTP programming module and read using the OTP read module.

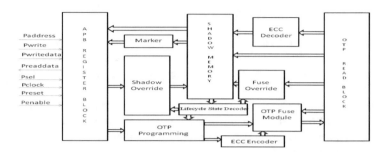

Fig. 1. Device lifecycle management IP

Lifecycle Decoder: The lifecycle state decoder module is a combinational block and used to the check the current lifecycle state based on the priority from high lifecycle state to low and used to avoid re-customization of an IP.

ECC Encoder Block: Whenever the data is programmed into the OTP fuse module, the corresponding ECC code is also calculated and stored along with the data in corresponding data and ecc_data locations. This block is designed to prevent the errors during the transmission of data.

OTP Programming Module: The OTP programming module provides the OTP write protocol for enabling the OTP Fuse module as write operation and the corresponding address, data is sent to OTP fuse block. The write operation of data is done bit by bit to the OTP fuse and write enable is made high only when a particular bit is one.

OTP Read Block: The OTP read module is used to clone the data from the OTP fuse module to the shadow memory block and to avoid the more power consumption by the OTP fuse module. During the cloning, data is passed on to the ECC decoder, fuse override blocks along with the shadow memory in order to verify the correctness of data and remapped address location, data.

ECC Decoder Block: During the transfer of data from the OTP fuse array to shadow memory. The error controlling code (ECC) block samples the data from data location, parity bits from ecc location and checks for the corrupted data. If it is a single bit correction a flag on one_bit_error signal is raised and corrected data is sent out along with the corresponding address and enable out signal or if it is a two-bit correction a flag on two_bit_error signal is raised and no correction of the data is done because the ECC block is a single bit error correction and double bit error detection block.

Fuse Override Block: The fuse override block contains the address of all the remapped address in the OTP module with corresponding correct data. Whenever the data is transferring from the OTP module to shadow memory the fuse override block checks for the remapped address and if available, the correct data is written to the shadow memory block by enabling the enable_out signal high and placing the corresponding data and address at the data_out and address_out correspondingly. The byte signal is used to represent the correction in the LSB or MSB location.

Shadow Override Block: The Shadow override block is used to modify or change the data in the shadow memory only if the particular lifecycle state as access permission to override the data. Whenever the particular lifecycle state is having access permission and enable_in signal is high then the corresponding address_in and data_in is put on to the address_out and data_out in the next clock cycle by making an enable_out signal high. During this two clock cycles, the busy signal is set high which indicates the transmission of address and data from the input side to the output side and error signal is set high whenever there is no access permission to override the data.

Shadow Memory: The shadow memory is the clone of the OTP efuse block. At the boot time, data from the OTP efuse is copied into the shadow memory and the OTP efuse module is turned off. This is done in order to avoid the high power consumption by the OTP fuse module. During the boot time, the complete data of every address location is made available by the otp_read_address, otp_read_data signals to the shadow memory and read when enable_OTP_read is set high.

APB Register Block: The APB register block acts as an interface between the device lifecycle management IP and the APB protocol bus architecture to communicate with the other modules either to read data from device lifecycle management IP or to write data to device lifecycle management IP from processor based on the APB protocol [7].

The APB register block also contains the four registers to know the status of the modules and to read and write address, data into the device lifecycle management IP or from device lifecycle management IP.

Finite State Machine: The OTP fuse module requires a timing operation either to write to the OTP fuse block or to read from the OTP fuse block and this timing operation is provided by the finite state machine. The clock cycles and interrupts define the state of the FSM.

2.1 Verification of Device Lifecycle Management IP

ECC Encoder Block: The error controlling code is also a combinational block where the data_out is expected immediately after the data_in is provided. So, consideration of delay parameters in the scoreboard can be neglected (Fig. 2a).

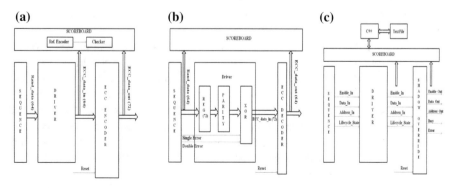

Fig. 2. (a) ECC encoder (b) ECC decoder (c) Shadow overide block

In the ECC encoder, block data is driven randomly and the driven data is sampled, written to a scoreboard by a monitor class. A reference encoder block is created as a function in the scoreboard and whenever the data is written into the scoreboard by monitor, the function is called and an expected data is returned which is compared with the actual response of DUT.

OTP Read Block: During the boot time the data is copied from the OTP fuse module to the shadow memory with a fixed timing protocol. Therefore, the OTP_read is verified using an assertion [8, 9] based check. Assertions are triggered at every positive edge of a clock and verified based on the properties provided. If any property is failed then the error message is displayed.

Shadow Override Block: For verification of this block, the memory block is stored as a text file and read using C++ program [10] and whenever it is called, the corresponding permission signal is returned to the UVM based verification test bench.

In the verification environment, an enable signal is driven randomly along with the address, lifecycle state and data by the driver and this signal are sampled at every positive edge of the clock and driven to the scoreboard by a monitor.

In the scoreboard, the sampled address and lifecycle state is feed as an input to the direct programming interface which calls the C++ program and provides the address and lifecycle to read the corresponding permission bit from a text file and return to the UVM scoreboard. If the returned permission bit is 1 then the output address, data and compared with corresponding input address, data.

Fuse Override Block: Fuse override block is similar to the shadow override block, the only extension is the additional address check. If the provided address matches with the address in the fuse override locations, active bit is high and if permission is available then the enable out is made logic 1, address out is equal to address in and based on the select bit the data out and byte bits are assigned.

Lifecycle State Decoder: This Lifecycle state decoder block is verified for two test cases (i) random lifecycle state data (ii) Incremental life cycle state data

A shaped sequence [11] is generated in the random lifecycle data test case and prioritization of the block is verified. In the incremental lifecycle state test case, the main focus is to check for the normal functionality of the lifecycle state decoder block.

ECC Decoder Block: The decoder block is driven with a random data along with calculated parity bits and verified for mainly three test scenarios (i) one-bit error case (ii) two-bit error case and (iii) without any error scenario.

To drive the data along with parity bits, first in the sequence a random data of 64 bits is stimulated and sent to the driver for padding with parity bits. In the driver class first, 64 bits is converted to 72 bits and padding with zeros in the parity locations, next the 72-bit data is driven as an input to encoder functionality block which stimulates the data along with corresponding parity bits and finally driven as an input to the decoder block.

During the one bit error check and two-bit error check a signal from sequence class to the driver is set high and the 72 bit data is xored with the 72 bit error data where only zero, one or two bits are made logic one randomly based on the no error check or one bit error check or two bit error check.

If the decoder is tested for no error check or one-bit error check then the same data stimulated by the sequence is expected at the output. Therefore in the scoreboard, the DUT response signals and driver input signals are compared.

APB Register Block: APB register block is the interface between the APB protocol bus and the device life cycle management block, it also contains the four internal registers for storing the address, data and status of the device lifecycle management IP and finally for enabling the IP block. So, the verification test environment must check read, write operation of the internal registers based on the APB protocol.

Write Check: The data is driven based on the APB write protocol and verified by simultaneous checking at the register output lines.

Read Check: The read operation is verified by first driving a write data to the registers and held to expect the same data during the read operation.

Internal Register Check: The sequence class is modified in order to generate a successive write and read operation on the same address and drive to the register in order to check same data is written and read during successive write and read operations.

Finite State Machine: In the IP, the finite state machine is used for the timing requirements of the intermediate blocks and to be specific, the output of a FSM is more like a timing protocol for read or write operations in between the blocks. Therefore the verification test environment is built on the assertion-based verification.

3 Automation of Testbench

Blocks which are having the same functionality but different specification need not be designed or verified separately. A generic design or testbench script can create an automated design or testbench based on the requirement.

In this work, every testbench is written in the python programming language and the user inputs for creating a specific testbench are handled using the file operations in script. Such that whenever a specific type of testbench as to be generated, the user as to define the values in the excel sheet and compile the python file, which takes the generic information from excel sheet using the file handling and generates a user defined teestbench for the data information provided through excel sheet by user.

Basically in the excel sheet, the number of input, output ports, labels of ports, size of registers, format of registers, memory elements, number of states in the finite state machines, external triggers, transition from one state to other etc. are declared as specifications.

This generic script helps in creating or automating a multiple testbench by just passing a generic information through a file handling operation and efficiency of this testbench can be analyzed by using a coverage report [12].

4   Results

ECC Encoder Block: In the Fig. 3a, rst indicates the reset signal, data_in indicates the input data-driven and data_out indicates the corresponding output calculated and Fig. 3b shows the code coverage report of the testbench which is given as 100% by Integrated Metric Coverage (IMC) tool.

(a)   **(b)**

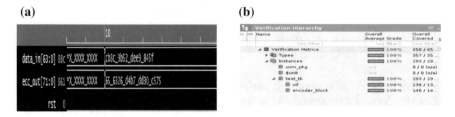

Fig. 3. (a) ECC encoder waveform (b) ECC Encoder Coverage report

ECC Decoder Block: In Fig. 4a, the ecc_in is the random data along with the parity bits, data_out is the obtained output and error, bit_position indicates the error type and error position. Figure 4b shows the code coverage report by IMC tool for the testbench which is given as 72.49% hit for decoder design, 94.64% hit for interface block.

(a)   **(b)**

Fig. 4. (a) ECC decoder waveform (b) ECC Decoder Coverage report

APB Register Block: In Fig. 5a, Psel, Pwrite, Paddr, Penable, Pwdata, Prdata are the signals driven from the APB block to the device lifecycle state IP in order to access the IP and otp_pg_addr, otp_pg_data, otp_pg_bc, otp_pg_en, otp_pg_er are the signals driven from the APB internal registers to the IP and from IP to APB registers.

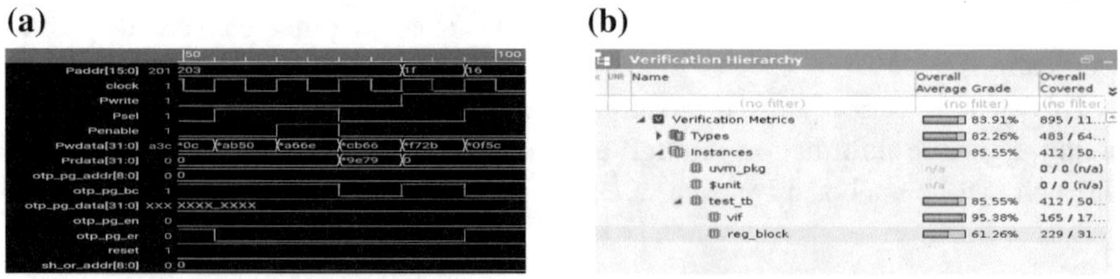

(a)   **(b)**

Fig. 5. (a) APB register waveform (b) APB register coverage report

In the waveform, Paddr indicates the address register and whenever the Pwrite and Penable are high, data is written into the internal address register and during the read operation in the next cycle, the address is placed on Pread bus as shown in a waveform. Figure 5b shows the code coverage report by IMC tool for the testbench which is given as 61.26% for register block and 95% hit for interface block.

Finite State Machine: The waveform consists of the input clock(clk), reset(rst), in1, in2, in3, triggers from one state to other (t_ext_S1_S2, t_ext_S2_S0) and output preset state(ps), O1, O2, O3, O4. Figure 6b left most part shows the code coverage report by IMC tool for the testbench which is given as 67.86% hit by the fsm block and right most part shows the states and transitions covered from one state to other state.

(a)   **(b)**

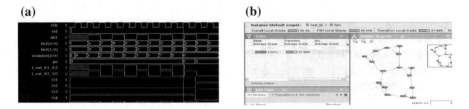

Fig. 6. (a) FSM waveform (b) FSM coverage report

OTP Read Block: The signals *CS, READEN, RST, read_done* undergoes a timing operation after the reset(rst) is set high and this timing initialization indicates the read operation from the OTP fuse to the shadow memory. At every 6th state the data is sampled, enable signals of fuse override (*en_fo*), shadow memory(*en_or*) is set high and in the 5th state ecc enable is set high. In Fig. 7b, the testbench coverage is 77.78% (Table 1).

(a)   **(b)**

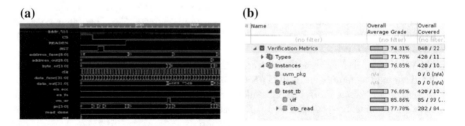

Fig. 7. (a) OTP read waveform (b) OTP read coverage report

Table 1. Coverage metrics.

Block	Code coverage	Functional coverage
ECC encoder block	100%	100%
OTP read block	77.78%	60.61%
ECC decoder block	72.49%	100%
APB register block	61.26%	87.34%
Finite state machine	67.86%	86.72%

5   Conclusion and Future Scope

A generic verification environment for the device lifecycle management is designed, scripted using the universal verification methodology and python. Moreover, for a particular design specification, the scripted testbench is compiled, verified using the cadence simvision and functional coverage, code coverage is calculated using cadence incisive metrics center. The work proposed can be extended in the future to achieve the 100% code coverage.

References

1. Mohan Kumar, N., Devi, M.N., Nath, D.B, Scaria, A.: VLSI architecture for compressed domain video watermarking. In: Communications in Computer and Information Science, vol. 205. Springer, Heidelberg (2011)
2. Accellera: Universal verification methodology (UVM) 1.1 class reference (2011)
3. Spear, C., Tumbush, G.: SystemVerilog for Verification: A Guide to Learning the Testbench Language Features. Springer, Heidelberg (2012)
4. Bromley, J.: If SystemVerilog is so good, why do we need the UVM? Sharing responsibilities between libraries and the core language. In: Proceedings of the 2013 Forum on specification and Design Languages (FDL), 24 September 2013, pp. 1–7. IEEE (2013)
5. Salah, K.: A UVM-based smart functional verification platform: Concepts, pros, cons, and opportunities. In: 2014 9th International Design and Test Symposium (IDT), 16 December 2014, pp. 94–99. IEEE (2014)
6. Ni, W., Zhang, J.: Research of reusability based on UVM verification. In: 2015 IEEE 11th International Conference on ASIC (ASICON), 3 November 2015, pp. 1–4. IEEE (2015)
7. Mefenza, M., Yonga, F., Bobda, C.: Automatic UVM environment generation for assertion-based and functional verification of systemc designs. In: 2014 15th International Microprocessor Test and Verification Workshop 15 December 2014, pp. 16–21. IEEE (2014)
8. Sohofi, H., Navabi, Z.: Assertion-based verification for system-level designs. In: ISQED (2014)
9. Konale, S., Rao, N.B.: C-based predictor for scoreboard in Universal Verification Methodology. In: 2014 International Conference on Advances in Engineering and Technology Research (ICAETR-2014), 1 August 2014, pp. 1–5. IEEE (2014)

10. Ma, C., Liu, Z., Ma, X.: Design and implementation of APB bridge based on AMBA 4.0. In: 2011 International Conference on Consumer Electronics, Communications and Networks (CECNet), 16 April 2011, pp. 193–196. IEEE (2011)
11. Peryer, M.: Seven separate sequence styles speed stimulus scenarios. In: DVCON (2013)
12. Elakkiya, C., Murty, N.S., Babu, C., Jalan, G.: Functional coverage - driven UVM based JTAG verification. In: 2017 IEEE International Conference on Computational Intelligence and Computing Research, ICCIC, 14 December 2017 pp. 1–7. IEEE (2017)

Survey on Soft Computing Methods for Accident Condition and Severity Predictions

Ashwani Kumar Kannojiya[(✉)], Richa Maurya, and B. Rajitha

Motilal Nehru National Institute of Technology, Allahabad, India
ashwanikumar.mnnit@gmail.com,
richamaurya.mnnit@gmail.com, rajitha@mnnit.ac.in

Abstract. Roadways are a boon for both the agricultural and industrial sectors and are referred to as lifelines for any nation. The ever-increasing volume of transport and motor vehicles has literally choked these road networks resulting in traffic jams and road accidents. Road accidents are a combined result of the negligence of the vehicle driver, weather conditions, road condition and lighting conditions which results in a huge loss of life and property every year. Hence, there is a very strong need to develop an automated system to predict the severity of accidents and the possible reasons behind it. So, that preventive actions can be taken to avert these accidents in the future. This paper deals with the description of the prediction model and compares the model with the other soft computing methods to find out the most efficient method to predict the accident conditions and severity. SVM is found to have good average accuracy value among the available methods.

Keywords: Decision tree · k-nearest neighbor · Logistic regression · Naïve Bayes · Support vector machine · Linear Classifier (LC) · Severity of accidents

1 Introduction

Roadways act like arteries of the social, economic and cultural health of any country. Road transport provides the basic infrastructural facilities and serves as a support system for all sectors of any developing country (like India) and are one of the means which keeps a nation running. India had seen a surge augment in its road networks since independence and is having one of the largest road networks in the world. This extensive network of roadways is accompanied by the ever-increasing number of transport and motor vehicles hence, increasing the burden on these roadways. According to data, in 2015, there were about five lakh road accidents in India, which killed about 1.5 lakh people and injured about five lakh people [1]. Factors behind road accidents include rash driving, negligence of traffic rules, weather conditions, road conditions, lighting condition, etc. These road accidents cause a huge loss to life and property every year. The paper proposes an efficient model by comparing the various soft computing methods by their average accuracy value. The model can be used for efficient prediction of severity of accidents as well as finding the causes behind the

A. P. Pandian et al. (Eds.): ICICCS 2019, AISC 1039, pp. 584–591, 2020.
https://doi.org/10.1007/978-3-030-30465-2_65

accidents. The model aims to get an efficient prediction of the severity of accidents that may happen at a given route at a particular road condition, lighting condition, and weather condition, so that we can efficiently formulate preventive ways for that particular set of conditions based on the severity of accidents that may happen. Secondly, the system will help us to predict the possible weather condition and road condition that may lead to the accident at a given set of conditions. Once we get the basic idea about the severity of accidents that may happen and the possible condition that led to the accidents, we can develop ways to prevent these situations like blocking the road during particular weather condition, enhancing the road and lighting conditions, restricting entries for a set of vehicles that may be vulnerable to accidents. These preventive measures will help us to decrease the chances of loss of life and property due to these accidents. India, as a signatory to the Brasilia declaration, intends to reduce road accidents and traffic fatalities by 50% by 2022. We do not have any control over weather conditions nor on the negligent attitude of any driver. We can only develop preventive measures to avert that scenario and this model will hopefully prove to be a help for the same.

The organization schema of the paper is as follows: Sect. 2 deals with a thorough literature review and earlier works related to this area. Section 3 has a thorough description of our proposed model as well as it contains a description of the dataset used in this paper. Section 4 is the result and observation section, showing the results after applying the classification methods on the accident dataset. The last section i.e. Sect. 5 contains the conclusion of the paper and future scope.

2 Related Work

In our study about, we found a lot of literature related to the prediction of severity and probability of accidents using classifiers. Many of these works have compared the different known classifiers with each other using different measurement metrics. For example Yisheng [2] used KNN and c-means clustering Real-time Highway Traffic Accident Prediction. The result showed that it is promising to use the k-nearest neighbor method in real-time traffic accident prediction. In his other paper [3] he used Support vector machine for in real-time traffic accident prediction. Lu [4] in his work using the logistic regression method for the traffic accident hotspot prediction. His results showed that his model could provide good predictions against traffic accident with 86.67% correct rate. In his paper [5] Miao Chong1 analyzed the GES automobile accident data from 1995 to 2000 and investigated the performance of the neural network, decision tree, support vector machines and a hybrid decision tree – neural network based approaches to predicting drivers' injury severity in head-on front impact point collisions. Krishnaveni [6] in his work have compared Naive Bayes Bayesian classifier, AdaBoostM1 Meta classifier, PART Rule classifier, J48 Decision Tree classifier and Random Forest Tree classifier for classifying the type of injury severity of various traffic accidents.

A Perspective Analysis of Traffic Accident using Data Mining Techniques for comparing classifiers Efficiency criterion as a measurement metrics was used Kim in [7] for comparing Decision Tree, ANN, and Linear Regression. In this paper, it has taken into account an accident dataset for evaluation of our classifier methods. So that it can predict the severity of accidents based on the weather and lighting factors. Kumar in [8] used efficiency evaluation for comparing ANN and regression classifiers method. In his work Regression and ANN was used on real and simulated data and the end results have been reported. The results showed that if the data include errors with the absence of real values of attributes, the statistical method of regression could act better than the ANN method and its performance is much superior than the other one Using AUC (area under curve) as a metric measure Huang et al. [9] have compared NB, DT, and SVM with each other. His work applied classifiers on the real data, and the result showed that the AUC criterion is better than accuracy for comparing the classification methods. Abdel-Aty et al. [10] used a probabilistic neural network as the classification algorithm to predict crashes. Luo and Garber [11] found that the three different pattern recognition techniques - the C-means clustering method, the Naïve-Bayes method, and the Discriminant Analysis - was only able to identify the patterns leading to crash with an overall classification error rate at about 50%, and they analyzed reasons for the unsuccessful identification of the patterns leading to crash. Most of the works which have been done in this field have ignored the accuracy factor for comparing different classifiers.

This paper takes a relatively simple approach for evaluation. It uses accuracy as a metric for comparison (Accuracy is calculated as the ratio of correctly predicted observations to the total number of observations).

3   Proposed Methodology

3.1   Dataset Description

The dataset used in this paper training our model is Road traffic accidents dataset of Leeds City Council [12]. This Road traffic accidents dataset contains data from different years of Leeds City Council. The data contains the following parameters: Type of vehicles involved, road surface, lighting conditions, weather conditions, victim category (Driver/passenger) and most importantly the severity of the accidents. The dataset comprises of 5000 accident records on an average.

This section deals with a brief description of the proposed methodology for prediction of severity of the accidents and the possible cause behind a given accident scenario. The success of the proposed model is governed by the type of soft computing method used. For this, a thorough comparative study of different available soft computing methods is done and the method with the highest and reasonable amount of accuracy is chosen.

MODEL I: For predicting the severity of the accidents. Prediction of the severity of the accidents involves training the model with the accident dataset. For this, the dataset is preprocessed and three classes of severities are formulated. Firstly, the High severity (Fatal) class include accidents which resulted in the loss of lives and extreme damage to the involved vehicle. Medium Severity (Serious) accidents class include accident which resulted in serious injuries and damage to the vehicle and the last one is the Low Severity (Slight) accidents class which resulted in slight injuries and little or no damage to the vehicle. Once the proposed model is trained, it is ready to predict the severity of the accidents on the test datasets. The trained system classifies the test dataset into three aforementioned categories. If the model predicts high severity outcomes for given weather, road, and lighting condition then this is the matter of the concern for the authorities to take preventive measures regarding the same. Either they can warn and forbid people or block those routes until those conditions pass or they can take necessary steps to enhance the road and lighting condition (Fig. 1).

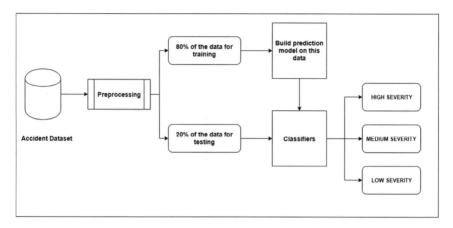

Fig. 1. MODEL I – Accident severity prediction

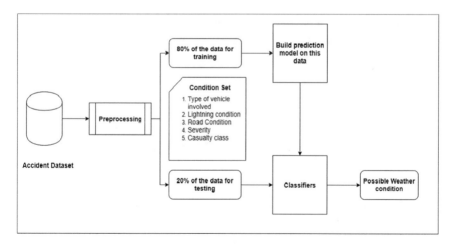

Fig. 2. MODEL II – Vulnerable weather condition prediction based condition set

MODEL II: For predicting the possible condition (weather, lighting or road) on which the accident may occur at a given set of conditions. The second model deals with prediction of the possible cause of the accident under a given set of conditions. This can be understood as follows: Suppose we are given with a particular route with a set of conditions as follows: Route condition={Vehicle type, Road type, Lighting condition, Severity….}. The model can predict particular weather condition (based upon these set of conditions) in which the accidents may happen. Once the weather condition is known, preventive actions can be taken to avert accidents in that weather condition. The predicted severity and the cause of the accident will govern the intensity of preventive measures needed to avert the danger of mishaps. The same approach can be applied for predicting other conditions like vulnerable vehicle type, Road type, etc (Fig. 2).

4   Performance Analysis

For thorough comparative analysis of the different available soft computing available methods, the model was trained and tested with the accident dataset to check which method provides us with the most accurate results. For training and testing the accuracy, the dataset is split in an 80:20 ratio i.e. into two categories, train, and test. The training part is used for training the model and the test class is used for testing the accuracy of the method. The model was compared with the available soft computing methods to evaluate the accuracy of each method, the most accurate method was chosen at the end for the severity prediction model. The paper involves a comparison of some well-known classifiers-Decision Tree, KNN, Logistic Regression, Naïve Bayes, Support Vector Machine, and Random forest.

   Calculation of accuracy is done as follows:

$$Accuracy = (Tp + Tn)/Total\ data\ (here\ total\ number\ of\ rows)$$

Tp (True Positive): The number of correct predictions that the occurrence is positive, Tn (True Negative): The number of correct predictions that the occurrence is negative. The results are tabulated in Tables 1, 2 and 3.

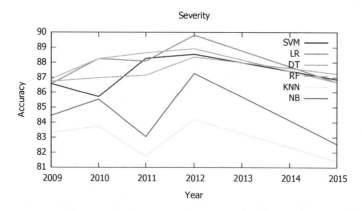

Fig. 3. Graphical representation of accuracy levels obtained from MODEL I

Table 1. Accuracy values for MODEL I by different soft computing methods based on Severity of accidents

Year	SVM	Logistic regression	Decision tree	Random forest	KNN	Naive Bayes
2009	86.57	86.58	86.90	86.74	83.30	84.45
2010	85.71	88.25	88.24	86.98	83.73	85.53
2011	88.27	88.08	88.64	87.15	81.75	83.05
2012	88.54	89.81	88.91	88.36	84.18	87.27
2015	86.87	86.68	86.68	87.24	81.43	82.55
Average	87.19	87.88	87.87	87.29	82.88	84.57

Table 2. Accuracy values for MODEL II on different soft computing methods based on the possible weather condition

Year	SVM	Logistic regression	Decision tree	Random forest	KNN	Naive Bayes
2009	89.69	83.96	85.76	91.65	85.43	85.76
2010	91.14	83.54	85.35	91.86	87.16	86.44
2011	90.50	86.78	86.96	93.48	89.01	86.78
2012	93.64	89.64	89.82	95.45	92	91.64
2015	89.68	79.55	78.99	88.18	87.05	84.24
Average	90.93	84.69	85.38	92.12	88.13	86.97

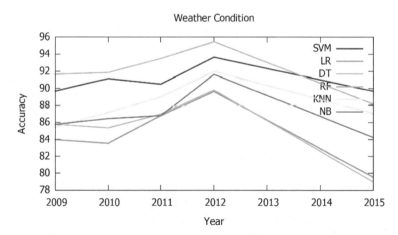

Fig. 4. Graphical representation of accuracy levels obtained from MODEL II taking weather condition as the unknown variable

Table 3. Accuracy values for MODEL II on different soft computing methods based on the possible road condition

Year	SVM	Logistic regression	Decision tree	Random forest	KNN	Naive Bayes
2009	83.63	81.99	82.32	86.91	81.83	83.31
2010	88.97	81.01	80.83	88.07	83.91	86.62
2011	87.34	85.85	85.66	91.25	82.50	84.92
2012	89.27	89.27	89.45	91.45	88.36	90.90
2015	91.37	82.93	81.61	85.55	85.74	86.87
Average	88.12	84.21	83.97	88.65	84.47	86.52

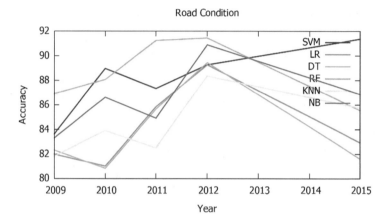

Fig. 5. Graphical representation of accuracy levels obtained from MODEL II taking road condition as the unknown variable

Figures 3, 4 and 5 present the graphical representation of results tabulated in Tables 1, 2 and 3 respectively. As it can be observed that Random forest performs better than the other methods for any condition.

5 Conclusion and Future Scope

Road accidents destroy thousands of lives and families every year and there is a strong need for a model to predict the causes which possibly lead to an accident so that, it can be averted. This paper dealt with proposing an efficient model for predicting the severity of accidents and the possible cause behind it. To make the system highly efficient, the paper involves a comparison based analysis to find the best among the available soft computing methods. After the analysis, it was found that SVM shows the good accuracy level among all the methods and hence can be used for building an efficient prediction model. The model can be extended to predict the age group which is most vulnerable to accidents, time of the day at which most of the accidents happens, etc.

References

1. Yang, Y., Liu, X.: A re-examination of text categorization methods. In: Annual ACM Conference on Research and Development in Information Retrieval, pp. 42–49, USA (1999)
2. Yisheng, L., Tang, S., Zhao, H.: Real-time highway traffic accident prediction based on the k-nearest neighbor method. In: 2009 International Conference on Measuring Technology and Mechatronics Automation, vol. 3. IEEE (2009)
3. Yisheng, L., et al.: Real-time highway accident prediction based on support vector machines. In: 2009 Chinese Control and Decision Conference. IEEE (2009)
4. Tao, L., et al.: The traffic accident hotspot prediction: based on the logistic regression method. In: 2015 International Conference on Transportation Information and Safety (ICTIS). IEEE (2015)
5. Miao, C., Abraham, A., Paprzycki, M.: Traffic accident analysis using machine learning paradigms. Informatica **29**(1) (2005)
6. Krishnaveni, S., Hemalatha, M.: A perspective analysis of traffic accident using data mining techniques. Int. J. Comput. Appl. **23**(7), 40–48 (2011)
7. Kim, Y.S.: Comparison of the decision tree, artificial neural network, and linear regression methods based on the number and types of independent variables and sample size. J. Expert Syst. Appl., pp. 1227–1234 (2008)
8. Kumar, U.A.: Comparison of neural networks and regression analysis: a new insight. J. Expert Syst. Appl. **29**, 424–430 (2005)
9. Huang, J., Jingjing, L., Ling, C.X.: Comparing Naive Bayes, decision trees, and SVM with AUC and accuracy. In: Third IEEE International Conference on Data Mining. IEEE (2003)
10. Abdel-Aty, M., Pande, A.: Identifying crash propensity using specific traffic speed conditions. J. Saf. Res. **36**, 97–108 (2005)
11. Luo, L., Garber, N.J.: Freeway crash predictions based on real-time pattern changes in traffic flow characteristics, Office of University Programs, Research Innovation and Technology Administration, U.S. Department of Transportation UVACTS-15-0-101 (2006)
12. Dataset source. https://data.gov.uk/dataset/6efe5505-941f-45bf-b576-4c1e09b579a1/road-traffic-accidents

FPGA Based Acceleration of Security Algorithm Using Co-design Approach for WSN Applications

Vilabha S. Patil[1][(✉)], Shraddha S. Deshpande[2],
and Yashwant B. Mane[2]

[1] Department of E & TC, Rajarambapu Institute of Technology,
Rajaramnagar, Islampur, India
vilabha.mane@ritindia.edu
[2] Department of Electronics, Walchand College of Engineering,
Sangli, India
{shraddha.deshpande,
yashwant.mane}@walchandsangli.ac.in

Abstract. The optimum security in wireless sensor network (WSN) is an emerged research area. A cryptographic algorithm is a technique for securing the data collected and transmitted in wireless sensor network. Number of cryptographic algorithms are available but they differ from each other in terms of key size, data size, security complexity and processing time. Tiny Encryption Algorithm (TEA) known as a lightweight and competent security algorithm, which will be appropriate for WSN applications. This paper presents TEA encryption or decryption algorithm implementation as a custom component based on ALTERA FPGA with cyclone IV E device. With this customization, the design of TEA is further accelerated with parallelism, while adding as custom instruction. The paper aims to compare the custom codesign approach i.e. custom instruction approach and full software approach with NIOS II processor. The obtained result shows that the performance of proposed custom codesign aspect outperforms the software only implementation in terms of execution clock cycles and power consumption.

Keywords: TEA · Security · Custom instruction · NIOSII processor

1 Introduction

WSN is self-organized with remotely placed sensor node devices. The WSN node should comply with critical design objectives like security, low power consumption and low cost. Hence sensor nodes must be designed with power efficient security algorithm with prime concern. Therefore many security algorithms like TEA, DES, AES, SEA, ECC, NTUREncryp have been emerged. These security algorithms are

V. S. Patil—Research Scholar, Walchand College of Engineering, Sangli, India.
S. S. Deshpande and Y. B. Mane—Walchand College of Engineering, Sangli, India.

© Springer Nature Switzerland AG 2020
A. P. Pandian et al. (Eds.): ICICCS 2019, AISC 1039, pp. 592–603, 2020.
https://doi.org/10.1007/978-3-030-30465-2_66

computationally intensive, thus requires more power of sensor node processor for data processing. This leads research development in hardware implementation of security algorithm on hardware during recent years.

Field programmable gate arrays (FPGA) is become attractive solution for hardware implementation due to its parallel processing and reconfigurable features. Cryptography calculations are positioned by their speed of encryption, unscrambling and their power to confront the assaults [1].

TEA is one of the simple, fast security algorithm leads to power saving and well appropriate for WSN applications. Massive writing is available for implementation and execution of TEA and its different variations. Many software implementations of TEA for the sensor nodes have been explored in [2–4]. Due to less size of TEA code it was mainly introduced for software implementation as implemented by researchers in [2–4]. These software implementations with microcontrollers and processors make them functionally fixed after deployment. Thus Field programmable gate array(FPGA) is attractive option for implementation due to its features like re-programmability, re-configurability [5]. FPGA is the platform which makes it possible to update the security algorithm after deployment in sensor nodes. FPGA based TEA algorithm implementation is explored in [6]. Additionally hardware implementation of TEA algorithm is introduced in [7, 8]. These are specially application specific i.e. targeting power constraint application such as radio frequency identification (RFID) tags.

Though many researchers have been explored full hardware or full software design and implementation, but nobody has addressed custom codesign approach with effect of parallelism. This approach leads to improve the performance with power saving.

In this study, TEA algorithm is synthesized in FPGA, where parallel computing is possible with concurrent execution. Thus it is proposed hardware software codesign approach for TEA custom component with NIOS II processor. Furthermore, this algorithm is tailored with effect of parallelism for acceleration.

This proposed design aims to determine significance of TEA implementation as a custom component with hardware software codesign approach with effect of parallelism for encryption/decryption computation. Hence the codesign encryption and acceleration using parallelism effect computations are addressed in the this work to show how it can affect the performance and power consumption.

2   TEA Security Algorithm

TEA was introduced by Roger Needham and David Wheeler in 1994 [9]. This is lightweight, simple algorithm which produces short and simple cipher does not rely on large tables or precomputations. TEA customs simple addition, shift operations and XOR operations. It has very small code size. It makes TEA perfect security algorithm to provide security for WSN nodes which have the inadequate size of memory, power, and cost. TEA works on 64bit data and utilizes 128 bit key divided into four subkeys like k[0], k[1], k[2], k[3]. It has the Fiestel structure with 32 rounds.

The pseudo code for TEA is as mentioned below with following considerations.

N: Number of iterations
v0 = Least significant 32 bits of data block
v1 = most significant 32 bits of data block

k[0], k[1], k[2], k[3] sub keys of 128 bit key
Delta = 0x9E3779B9 Sum = 0
TEA encryption/decryption
for i = 0 to N
do
sum = sum +/− delta;
v0 = v0 +/− ((v1 < < 4) + k [0]) ^ (v1 + sum) ^ ((v1 > > 5) + k [1]);
v1 = v1 +/− ((v0 < < 4) + k [2]) ^ (v0 + sum) ^ ((v0 > > 5) + k [3]);

2.1   Algorithm Flow

Figure 1 illustrates the TEA security algorithm flow of encrypting 64 bit input data
using given keys. In this paper 64 bit data is focussed with 32 rounds. It explores
Verilog hardware implementation model of the TEA function. The TEA architecture as
presented below is implemented as Verilog model. This module is written in Verilog HDL language and compiled using Xilinx ISE 14.1 environment. The RTL design

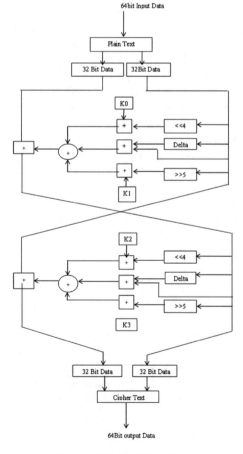

Fig. 1. TEA algorithm flow

generated from Verilog was modeled primarily using simulator directing verification and functionally checking.

3   Implementation

The development board utilized for this design is DE2 115 board with of ALTERA FPGA platform integrated with Cyclone IV E device the directed FPGA platform embedded with NIOS II soft core processor. This processor can be used to synthesize the entire security system on a system on a programmable chip.

NIOS II Processor [10]:
NIOSII is 32 bit configurable embedded soft-core processor specially designed for ALTERA family of FPGA. The NIOS II's basic functionality can be extended by adding custom peripherals or custom instructions. By the addition of custom instructions time, critical software operations can be accelerated as a custom logic block. It has three different member families like NIOS II/f(fast), NIOS II/s(standard) and NIOS II/e(economy). Each of these is having a specific price and performance range. It has pipelined RISC architecture with separate 16 bit instruction bus and 32 bit databus. The register file is configurable with 128, 256 or 512 registers. At a time only 32 registers are accessible as general purpose registers.

The NIOSII processor and many other components such as custom peripherals and standard peripherals are integrated into ALTERA DE2 [12] board to form the total system as shown in Fig. 2. CYCLONE IVE FPGA device enables the process of interfacing the NIOS II processor and the peripherals to the DE2 board. Avalon bus is used to form the interconnection network by connecting these components.

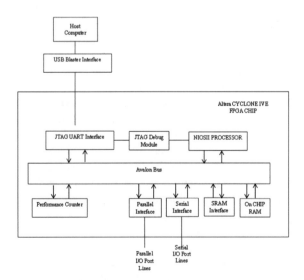

Fig. 2. NIOS II implemented on FPGA

3.1   System Design

The system design architecture implementation explored in three different types:

1. TEA implementation in hardware using Verilog
2. Software only implementation with NIOS II processor
3. Hardware-software co-design implementation in combination with the NIOS

II soft core processor

QSYS is the system integration tool used for NIOS II hardware system design. This is a part of Quartus II software tool that can be called immediately for configuration and generation of whole NIOS II system design. Quartus II 13.1 onwards software contains the integration tool QSYS. This integration tool is used to develop and integrate processor, Memory controllers, peripherals, different IP cores, along with custom components. Later the Quartus II tool is used accomplish synthesis, placement and routing. Then the system is generated for selected FPGA device with system interconnect generation.

In this system design the components required for implementation of TEA encryption using QSYS tool consists of following tools as shown in Fig. 4.

Clk: This is clk source which provides the clock to other components connected in system.

NIOS II Processor: It is considered as processing unit of the sensor device. NIOS II has three versions of processor NIOS II Economic/(E), NIOS II Fast/(F), NIOS II Standard/(S). In this paper implementation is carried out with two versions of NIOS II processor i.e. E and F.

Memory module: On chip memory RAM module is selected with 48000 bytes of total memory size. When a program from NIOSII eclipse IDE is executed, it will be automatically downloaded and executed into On-Chip Memory.

JTAG UART: This is IP of QSYS component library used for communication between host PC and QSYS system design. Communication is done through serial data streams. It is used for debugging purpose of system design.

TEA custom hardware block: This custom block is designed to improve the performance of security algorithm for WSN applications. This custom block of TEA as security algorithm is implemented as custom component to NIOS II processor. This custom component takes two 32 bit data input as plain text through dataa and datab registers of NIOS II processor and cipher text is displayed as encrypted output on NIOS II console window [11]. The custom hardware block is further modified with parallelism leads to improve the performance of security algorithm.

Performance counter: This is used for to track the clock cycles and timing of multiple sections of software program. It can be started and stopped with the help of single instruction used at the beginning and end of the section. It provides the performance of TEA security custom component with NIOS II processor i.e. codesign approach and TEA security algorithm of software only implementation with NIOS II processor.

Input, Output GPIO'S: These are integrated for testing and synchronization.

In this work, the Cyclone IV E device is used with FPGA based DE2-115 development board.

The proposed SOPC contains the above-mentioned components. All these components are connected through Avalon bus as shown in Fig. 4. As shown in fig. TEA hardware custom module is connected to NIOSII processor through GPIO and Avalon Interface bus. The encryption or decryption of data collected from sensor is performed by TEA hardware module in cooperation with software written in NIOS processor.

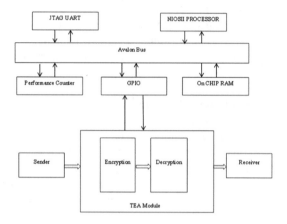

Fig. 3. Connection between TEA module and SOPC system

The QSYS integration tool generates the interconnect logic to integrate the above mentioned basic components of hardware system. Figure 3. shows the components used in system design and the system generation for system implementation through NIOS IDE. In this work, the Cyclone IV E device is used with FPGA based DE2-115 development board.

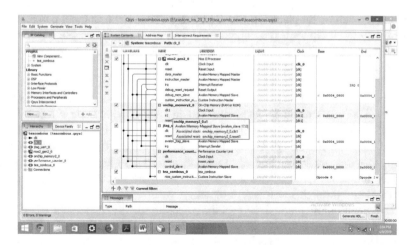

Fig. 4. System design using QSYS

As shown in Fig. 5 tea_combcus_0 is custom component generated through hardware software codesign. It is connected to NIOS II processor through JTAG UART and on chip memory.

3.2   System Implementation

The combination of system generated using QSYS SOPC tool with Quartus II takes place. Later the pin assignments have been done by seeing the pin assignment of the Cyclone IV E of DE2 115. Subsequently system is programmed inside the hardware, with the DE2115 Cyclone IV E of ALTER FPGA is connected to host PC through USB blaster cable.

3.2.1   TEA Implementation in Hardware using Verilog

The basic algorithm flow of TEA is fully presented in previous section. This basic algorithm is written in Verilog language. This model is implemented as FPGA model and was compiled in Xilinx ISE 14.2. The verification and functionality checking of code was done through simulation. In the last step the design was compiled and synthesized using Quartus II. Further the TEA algorithm implementation in Verilog is accelerated through applying parallelism. The respective code is also verified for proper functionality.

3.2.2   Software only Implementation

The TEA algorithm is realized through C code. After the generation of system in QSYS without integration of custom instruction, it is compiled in Quartus II soft-ware. Next the .sopc file is used by NIOS II IDE to create the project. In this NIOSII IDE project consists of C/C++ code is compiled and generated as execut-able code using GNU compiler. Later the new project is produced within NIOSII IDE. The TEA algorithm C code is added to this project. The project is compiled with build project command. Then the project is implemented on FPGA cyclone IV Device.

Finally the results for the TEA security software only implementtaion using the NIOS II EDS with 64 bit plaintext, four 32 bit keys, sum, delta as input parameters are obtained with 64bit cipher text as output dispayed on NIOS II console window.

3.2.3   Custom instruction Co-Design Implementation

Basic TEA module and TEA block accelerated with parallelism is synthesized in Verilog as a hardware blocks as mentioned in previous section. These blocks are implemented as customized logic blocks and integrated with basic NIOSII System using QSYS tool. The NIOSII hardware system is integrated with the custom logic block and recompiled with QUARTUSII software. There are two important blocks of this design as custom instruction (customized logic block) and the software macro. During the implementation of custom logic, the software macro is used to access the accelerated logic through a software program.

After generation of QuartusII project the .sopcinfo file is called in NIOSII IDE as a hardware platform. Subsequently the software macro is modified with custom instruction and compiled with NIOS II built activity. With the command run as NIOS II hardware the output is displayed on NIOS II console window as shown in Fig. 5 i.e. cipher text with custom instruction acceleration with the performance parameters like clock cycles and time of execution.

Fig. 5. NIOS console window

4 Results and Discussions

In this section, the details of the results and experimental analysis are presented. The comparison and analysis is done with the help of performance parameters like power consumption, clock cycle count and execution time. The custom configuration as basic approach, with parallelism and software implementation has been tested for the two versions of NIOS II processors (E and F). The implementation results of TEA (encryption/decryption) and comparison in terms of clock cycles and time of execution is as shown in Table 1.

Table 1. Comparison of clock cycles and time of execution

Item	With software only implementation		With custom instruction		Custom instruction with parallelism	
	NIOS II E	NIOS II F	NIOS II E	NIOS II F	NIOS II E	NIOS II F
Clock cycles	4817	847	63	33	30	16
Time of execution (μSec)	96	16	1.26	0.66	0.6	0.32

From the results in Table 1, it is observed that the clock cycles and time required for implementation of TEA encryption algorithm in software only implementation i.e. without custom instruction are significantly higher as compared to the clock cycles required for implementation of security algorithm with custom instruction co design approach.

This custom instruction based implementation has comparable cycle count performance and 96% less cycle count as compared to software only implementation. Additionally, the parallelism based custom instruction approach of this work has a comparable improvement in execution time and clock cycle count than the serial aspect of custom instruction implementation. Thus acceleration of custom block with parallelism reduces the cycle count by 50% as compared to custom block implementation without parallelism.

Thus, implementation of security algorithm with custom codesign approach drastically reduces the clock cycle count by 60 to 70 times as compared to typical software implementation with NIOS II processor. Such improvement is possible due to acceleration of security algorithm as custom instruction approach, additionally it reduces the overhead of processing unit.

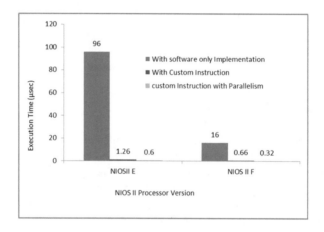

Fig. 6. Comparison of execution time w.r.t. software only implementation

Figure 6 shows significant improvement in terms of execution time of custom implementation as compared to software only implementation.

The power consumption results of TEA security are as shown in Table 2. The Power consumption is measured with the power play analyzer tool of Quartus II.

Table 2. Comparison of power consumption

Item	With software only implementation		With custom instruction		Custom instruction with parallelism	
	NIOS II E	NIOS II F	NIOS II E	NIOS II F	NIOS II E	NIOS II F
Power consumption (mw)	0.95	1.15	0.84	0.99	0.80	1.0

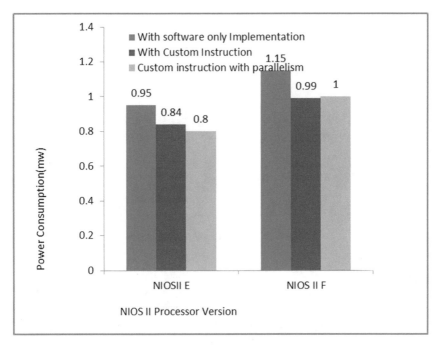

Fig. 7. Comparison of power consumption

With this comparison it is observed that acceleration of security algorithm as a custom instruction can save power dissipation of soft core processor in comparison to software only implementation.

Regarding analysis with both versions of the processor, it is noticed from Fig. 8 that greatest improvement i.e. by 47.61% is seen by F version of NIOS II processor in terms of execution time w.r.t. E version. Conversely it is noticed from Fig. 7 and Table 2 that NIOS II F processor although provides significant performance improvement in terms of clock cycle but on the other hand consume high power for the execution of task. It consume 15% more power w.r.t NIOS II E processor. It depends on the operating frequency of processor version i.e. completing given task in fewer clock cycles at higher operating frequency can increase dynamic power consumption.

Performance Evaluation

The clock cycle count can be used as performance evaluation metric in order to identify the efficiency enhancement of implementation using the respective clock rate. The performance of design implementation approaches in terms of clock cycles is calculated as,

$$\text{Performance} = \frac{\text{NIOSII Processor clock rate (50 MHz)}}{\text{Number of clock cycles required for given implementation}}$$

Figure 8 shows that performance of custom codesign implementation is higher w.r.t. software only implementation.

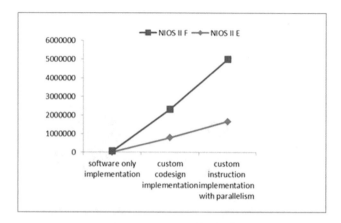

Fig. 8. Comparison of performance

5 Conclusion

In this paper, the problem of security and power dissipation in WSN node has been resolved with acceleration of TEA security algorithm. The acceleration is addressed by custom codesign approach of NIOSII soft core processor. The custom block of TEA have been implemented using basic approach and designed as a custom instruction to NIOSII processor. This implementation reduces the number of clock cycles required to complete the execution by 96%. Additionally, it leads to a reduction in power consumption by 12%. Further the custom implementation is accelerated with parallelism in TEA Verilog code hardware block. In comparison with an initial custom implementation, this acceleration led to 35.48% significant performance improvement along with 14% power saving. Thus this design offers a flexible architecture which improves overall performance in terms of fewer execution cycles. This design provides effective cryptosystem by reducing the overhead of processing unit of the sensor node.

References

1. Dener, M.: Security analysis in wireless sensor networks. Int. J. Distrib. Sens. Netw. **10**(10), 303501 (2014)
2. Pavlin, M.: Encription using low-cost microcontrollers. In: 42nd International Conference on Microelectronics, Devices and Materials and the Workshop on MEMS and NEMS, Society for Microelectronics Electronic, pp. 189–194 (2006)
3. Liu, S., Gavrylyako, O.V., Bradford, P.G.: Implementing the TEA algorithm on sensors. In: ACM-SE 42: Proceedings of the 42nd annual Southeast regional conference, pp. 64–69. ACM Press, New York (2004)
4. Niati, R., Yazdani, N.: A more energy efficient network setup method for wireless sensor networks. In: Asia-Pacific Conference on Communications, pp. 640–643 (2005)

5. Patil, V. S., Mane, Y. B., Deshpande, S.: FPGA based power saving technique for sensor node in wireless sensor network (WSN). In: Computational Intelligence in Sensor Networks, pp. 385–404. Springer, Heidelberg (2019)
6. Hussain, M.A., Badar, R.: FPGA based implementation scenarios of TEA block. In: 13th International Conference on Frontiers of Information Technology (FIT). IEEE (2015)
7. Israsena, P.: Securing ubiquitous and low-cost RFID using tiny encryption algorithm. In: Symposium on Wireless Pervasive Computing, 4 p. IEEE, Los Alamitos (2006)
8. Israsena, P.: Design and implementation of low power hardware encryption for low cost secure RFID using TEA. In: Information, Communications and Signal Processing, pp. 1402–1406, December 2005
9. Wheeler, D.J., Needham, R.M.: TEA, a tiny encryption algorithm. In: International Workshop on Fast Software Encryption. Springer, Heidelberg (1994)
10. Altera, Nios II processor: the world's most versatile embedded processor (2013). http://www.altera.com/devices/processor/nios2/ni2-index.html. Accessed 10 June 2013
11. Altera, Nios II custom instruction user guide (2013). http://www.altera.com/literature/ug/ug_nios2_custom_instruction.pdf. Accessed 10 June 2013
12. Altera, DE2 development and education board (2013): http://www.altera.com/education/univ/materials/boards/de2/unv-de2-board.html. Accessed 10 Feb 2013

Blind Forgery Detection in Digital Images: An Analysis

Nagaveni Hebbar[(⊠)] and Ashwini Kunte[(⊠)]

Department of Electronics and Telecommunication, Thadomal Shahani
Engineering College, Mumbai University, Mumbai, Maharastra, India
{nagaveni.hebbar,askunte}@gmail.com

Abstract. In today's world, the usage of images in digital forms have become common in various fields for information exchange because of the development in handy digital devices like cameras and mobiles. Checking the authenticity of the image has become necessary as there could be unethical modifications possible with the images using several image editing tools and software that are easily available over the Internet, which leaves no clues of modification to the viewer. Detection of blind or passive forgery becomes more difficult as we do not have any prior knowledge of the forged image. This paper gives a detailed analysis of various blind forgery techniques and detailed classification of several forgery detection methods. It also gives a comparative analysis of different blind forgery detection methods with respect to different parameters that will give clear idea to the researcher about the performance of forgery detection methods used currently and give directions to future scope of improvement over the existing methods.

Keywords: Digital image authentication · Copy move forgery ·
Image splicing detection

1 Introduction

Images are one of the easiest ways to convey information in this digital world. It is said that 'an image worth thousand words' and hence they are used in our daily life in many applications, namely digital television, as evidence in court, medical, science and technology, military, education, social media and most of the contents on Internet. Due to the advancement in the handy digital devices like cameras and mobile phones, it has become easier to acquire digital images and share with anyone. Seeing is not believing [1], as manipulating digital images is easy, but detecting it could be difficult. The images are changed to improve its appearance, to attract the viewers, to get the publicity of a product, to conceal some meaningful information, to hide or remove the evidence of certain illegal activity captured in the image, etc. With the advancement in technology, image manipulation has become easy as there are several image editing tools or software available, such as GNU Image Manipulation Program (GIMP), Pixlr, Photoshop, etc. Hence many times, the images conveyed may not be same as it is acquired.

© Springer Nature Switzerland AG 2020
A. P. Pandian et al. (Eds.): ICICCS 2019, AISC 1039, pp. 604–614, 2020.
https://doi.org/10.1007/978-3-030-30465-2_67

The act of modifying the digital image fraudulently to cover certain information is called as Image forgery. Hence authentication and integrity checking of a digital image is necessary before using it in any critical application. Detection of such manipulation is difficult just by seeing the image. Photo manipulation was first recorded in the 1860s [2] when the face of Abraham Lincoln was pasted to the body of Southern politician John Calhoun. Before the invention of digital cameras and scanners, the photo manipulation was done with some traditional art method. The images were edited during the printing process. With the advent of digital equipment and technology, analog image editing became obsolete and many digital image editing software and tools emerged with time. Hence, there is a need for efficient and robust method to identify the forgery in digital image and to locate the forged region in the image. There are several methods existing to detect the manipulations in an image, but every method has its own weaknesses and strengths.

In this paper, Sect. 2 gives the classifications of digital image forgeries. Section 3 gives in detail about different types of forgery detection techniques and the analysis of recent methods used for forgery detection. The challenges in the existing forgery detection methods are discussed in Sect. 4 and finally the Sect. 5 gives the conclusion and future directions based on analysis made in previous section.

2 Digital Image Forgery Types

Digital images could be easily manipulated in many ways using the different image manipulation software or tools which are easily and many times freely available. Image editing tools are used by many people for different purpose. If the images are edited for unethical reason, then it is called forgery. Different ways of forging [3] digital images are given in "Fig. 1".

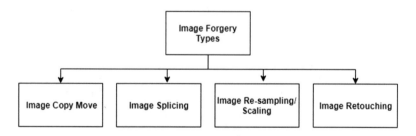

Fig. 1. Digital image forgery types

In Copy move forgery [4], copy of a small part of an image is taken and moved to another region in the same image to conceal certain information. Such manipulation may completely change the interpretation of the image.

Splicing is another common method used for image tampering. Image Splicing is the act of copying a small part of one image and pasted [5] in another image. This manipulation type is usually done to create an image evidence illegally or for hiding certain information by inserting a new object in the image.

In image re-sampling [6], the forged region is made to undergo certain geometrical transformation such as up-sampling or down-sampling which is done to resize the forged portion to hide from detection. Other geometrical transformations like rotation, scaling, skewing, Flipping or Mirroring are also done to the forged portion to hide the forgery from viewers. A certain post processing mechanism like JPEG compression, blurring or noise addition methods are used on the forged image to conceal the traces of manipulation from detection.

Image retouching [7] is usually done after acquiring the image to enhance the appearance of the image. There are several image enhancing mechanisms to make the image look better which can even remove certain real features of the image like hiding aging factor, remove scar or pigmentation from the face, conceal gray hair etc. It becomes forgery if retouched image is used for unethical purposes.

3   Classifications of Manipulation Detection Methods in Images

Digital image manipulation detection [8] has become a necessity in present world as the usage of digital images have increased and forgery cases also have increased with years. Forgery detection in digital images can be broadly categorized as active and passive/ blind methods. The classifications of different techniques used in digital image manipulation detection are given in "Fig. 2".

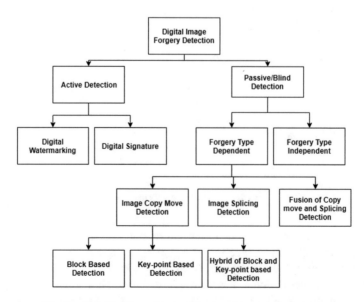

Fig. 2. Classification of image forgery detection techniques

3.1 Active Detection

In Active detection methods, the authenticity is checked by using certain embedded information stored in the image either while acquiring the image or before using it for any application. There are two types of active methods namely Digital watermarking detection [9] and Digital signature detection [10]. In watermarking technique, information about the image like logo or symbol is inserted in the image during acquisition or before distributing the image. To check the authenticity, the embedded watermark is extracted and verified with the obtained watermark. In digital signature method, using a hash algorithm the message digest of an image is calculated, which is also called as signature. This signature is sent along with the image. During authentication, signature of the image is recalculated for the received image and verified with the attached signature. The signature will not be same if the image is forged. The problems in using watermark method is that, all digital cameras are not capable of inserting watermark and cameras having this watermark embedding feature are expensive. A legitimate modification of an image for quality improvement purpose may also be detected as forgery.

3.2 Passive/Blind Detection

Blind method [11, 3] works when knowledge about the digital image like watermark or signature is completely absent. This method makes use of traces left out during image processing steps in different phases of acquisition and storage of digital images. Passive detection works on the principle that the modification of an image changes certain statistical property in the image, which may help in detection of the modified region. Some passive techniques use the analysis of the image in transformed domain to detect the traces of forgery. In some methods, changes in consistent characteristics like lighting, shadow and noise variation is analyzed to detect the forgeries in the image. Blind detection is the only option, if we do not have the watermark or signature embedded in the image. Passive methods are again classified into three types. They are based on forgery type, source camera features and statistical characteristics of the image.

Forgery Type Dependent. These forgery detection methods in images [12] are mainly used in copy move forgery and splicing. These types of detection algorithms use few common steps as shown in "Fig. 3".

Copy Move Detection. In this method, copy of a small region is moved on to the same image. Copied and moved parts are identified by comparing the similar regions in the image. Copy move forgery [13] is very easy but very hard to detect as the pasted region has almost same characteristics as copied region. Copy move tampering detection methods are categorized as Block, Key-point and Hybrid of block and key-point based methods.

In block-based techniques [14, 15], the image is initially segmented into overlapping or non-overlapping blocks. The feature extraction is done for all the segmented blocks. Then matching of extracted feature vectors with each other is done using any block based matching algorithm. Some examples of feature extraction algorithms [12] in block based methods are Histogram of oriented gradients HOG, Histogram of Oriented Gabor Magnitude (HOGM), DCT, DWT, Dyadic Wavelet Transform (DyWT) [16], Local Binary Pattern (LBP), etc. The matching algorithms used are K-Dimensional Tree, Lexicographic sorting, Radix sort, Counting blooms filter, Euclidean distance, Phase correlation etc. Then some post processing is done to exactly locate the copied and pasted region. Block base methods give better accuracy in detection in simple forgery but are not invariant to rotation and scaling of forged region.

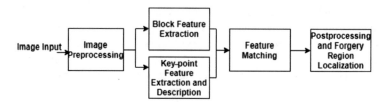

Fig. 3. Basic steps involved in forgery type dependent detection techniques

In key-point based techniques [17], the image key-points are extracted by feature extraction algorithm which extracts high frequency components such as corners, sharp edges etc., from the image. Key-point feature matching algorithm is used to find the tampered region in the image. Then some post processing is done to exactly locate the copied and pasted part. In Key-point based methods, computation speed is high compared to block-based methods and variant to rotation and scaling, but it may skip detection when tampering is done to smooth region. Some examples of key-point based feature extraction and local feature description algorithms are Scale Invariant Feature Transform (SIFT) [18], Speed Up Robust Features (SURF) [19, 20], Harris corner points, etc. Examples of key-point based matching algorithm [12] are 2 Nearest Neighbor (2NN), generalized 2NN, best bin first, clustering based etc.

To minimize the weakness of block and key-point based techniques, hybrid of these two methods [21] are used.

Image Splicing Detection. This is another commonly used image tampering method in which a small part of one image is copied and pasted in another image. In this method since the copied portion is from another image, the characteristics of pasted part and rest of the image are different. In Splicing detection [22, 23] methods several features like Bi-coherence, DCT or DWT coefficients, invariant image moments, camera response function, etc. are extracted. Several methods also use machine learning algorithm like Support Vector Machine (SVM) for forged image classification and sometimes to automate the selection of threshold for feature matching.

Hybrid of copy move and splicing methods [5] are used to detect any type of forgery in a given image. Many existing techniques find out if forgery is there or not and its type in first phase and finds the copy move or spliced region in second phase. Few methods also use SVM for image classification and matching features for detecting forgery. Recent methods use Convolutional Neural Network (CNN) to improve the detection performance and for the automation of tasks like feature extraction and feature matching. From the survey of existing methods, it is understood that certain features can detect a specific type of forgery. Better features must be formulated to detect any type of forgery.

Forgery type Independent methods are not dependent on how manipulation is done on the image. They are based on some internal changes in the characteristics of image due to the forgery process or few post processing transformations made on the tampered image to hide detection. Forgery type independent methods are based on various image processing operations like compression format, filtering process and re-sampling. Another category of forgery type independent method is based on lighting inconsistency and various camera features.

Comparative analysis of different recent blind techniques of forgery detection is explained in Table 1. Major problem with key-point based copy move forgery detection technique is to detect the forgery in small and smooth region. In [24] authors have used Swarm intelligent algorithm along with SIFT to overcome this problem. Another problem with copy move detection techniques is to find out manipulations in case of post processing operations on forged images. A method was proposed in [25] to overcome this problem using multi-radius Polar Complex Exponential Transform (PCET) but this method requires more execution time. This method uses Lexico-graphical matching which could be replaced by other method to reduce execution time. In the copy move detection method in [26], authors have used stationary wavelet transform based features. This method gives better performance with increasing block size, but it can't detect the forged region with post processing operations like scaling and rotation. A fusion based technique was introduced by the authors in [27], which uses KAZE interest point detector and SIFT feature extraction to overcome the weaknesses of block and key-point based methods. Another fusion based approach was proposed in [28] outperforms block-based techniques in efficiency and key-point based techniques in accuracy.

[29] is based on illumination feature for image splicing detection as illumination of light may vary for forged part and remaining part of the image. Compared to Method based on Dichromatic Line (MDL) and Method based on Illuminating Map (MIM), this gives better accuracy but slow computation speed. In this method, none of the post processing attacks like blurring, adding noise, etc. are considered. Another splicing method which was proposed in [30] using wavelet domain which performs well with gray scale images, but performance should be improved for color images. One more Image splicing detection method was proposed in [31], which is based on correlation of neighborhood pixel by measuring statistical parameters. According to authors it performs well, but some optimization could be applied to automate the process of feature matching. Method proposed in [32] is a block-based method in which the noise level is evaluated in each block and relation between noise levels with other blocks are calculated to detect forgery. This method doesn't perform well with larger block size and bigger forged region. A hybrid method to detect copy move and splicing detection using combination of DWT and DCT is used in [33], which is a simple method, but the accuracy is not good. Authors in [34] have proposed a better approach to detect image splicing, copy move forgery and combination of splicing and copy move. This method detects if the given image is forged or not in first phase and detects forgery type in second phase. It gives good performance but some optimization algorithm like SVM could be

Table 1. Comparison table for developments in blind/passive forgery detection

Author/Year	Forgery type	Detection method	Performance	Remark/Future scope
Zhao, et al./ (2017) [24]	Copy move	Image segmentation is done to get non overlapping image blocks and then Swarm intelligent algorithm is used to get independent layer according to smooth degree, then SIFT feature extraction is used	Effectively detects forgery in small and smooth regions	Higher false positive rate needs to be addressed
Wo Yan, et al./(2017) [25]	Copy move	This technique uses PCET. The multi-radius PCET with graphic processing is used to extract the features. Then lexicographic matching is used to detect forgery	Performs well with rotation and scaling, JPEG compression, noise Degradation and smoothing	Execution time needs to be reduced
Mahmood, et al./(2018) [26]	Copy move	This method uses Stationary wavelet transform based features to detect forgery	Reduced false detection and execution time. Accuracy is 95% in 4×4 image block and 98% in 8×8 image block	Different post processing attacks to be addressed
Yang Fan, et al./(2017) [27]	Copy move (Hybrid of block and key- point based)	A robust interest point detector KAZE is used along with SIFT to extract more features. Better matching algorithm is used, and improved post processing is used to localize the forged region	Performs well with rotation and scaling Also works well for smooth region and small region of forgery	Accuracy could be improved
Zheng, Jiangbin, et al./(2016) [28]	Copy move (Hybrid of block and key -point based)	In this technique, block-based approach (Zernike moments) and key-point based approach (SIFT) are used to identify modifications in smooth and non-smooth regions	Outperforms key-point based methods in accuracy of detection and block-based methods in efficiency	Optimization of Threshold value for feature matching needs to be done
Fan, et al./ 2015 [29]	Image Splicing	Image is divided into horizontal and vertical patches, illuminance is estimated for each patch, and fake patches are produced by intersection of forged horizontal and vertical band and represented in the map of forgery detection. Finally, by combining all the maps, spliced part of the image is identified	Compared to MDL and MIM, this gives better accuracy but slow computation speed	Detection of forgery with Attack on illumination feature to be tested

(continued)

Table 1. (*continued*)

Author/Year	Forgery type	Detection method	Performance	Remark/Future scope
Park, et al./ 2016 [30]	Image Splicing	Image Splicing forgery is detected by calculating the characteristic function moments for the inter-scale co-occurrence matrix using wavelet transform	More than 95% Accuracy tested for four different datasets for splicing detection	To improve performance for color image splicing
Zhu, et al./ (2018) [32]	Image Splicing	Image is divided into non-overlapping blocks. Noise variance and the sharpness of each block are estimated. Then, the relationship between sharpness of blocks and noise variances is estimated to detect forgery	Works well with small spliced region and forgery is detected with smaller block size. Detects if same camera is used for spliced region and main image	Improvement in detection with bigger block size and bigger spliced region could be done
Sundaram, et al./(2017) [31]	Image Splicing	Correlation among neighborhood pixel by measuring statistical parameters	Performance and response time are better than existing methods	Optimization could be applied
Hayat, et al./(2017) [33]	Copy move/Splicing	DWT is initially used to select lowest frequency sub-band and then DCT is used to get feature vectors	Accuracy is 73.6%	Detection accuracy should be improved
Huynh-Kha, Tu, et al./2016 [34]	Copy move/Splicing	DWT is used in first phase to identify the suspicious forged region. Second phase uses matching method to identify copy move or spliced forgery	Accuracy is 92%	Use of canny edge detector to detect forged region accurately
Rao, et al./ 2016 [35]	Copy move/Splicing	A Deep learning approach, which uses CNN to extract best features from the test images, and then fusion of different features is used to obtain the final features	Accuracy is 97%	Not robust to various attacks

used to automate the process between two phases. A deep learning approach is proposed in [35] for better performance to detect forgeries, which gives an accuracy of 97%.

Given an image, we can't say that which type of forgery is done just by seeing it. A better and easy way is required to identify if the image is authentic or not, to identify the forgery type and to localize the forged region accurately.

4 Challenges in Blind Forgery Detection Methods

All the passive detection techniques analyzed in this paper has one or the other weakness as listed below.

- Some methods are good in accuracy but computationally complex. Reduction of computational complexity with increased accuracy is a big challenge.
- Detection methods should be robust to various transformation or post processing operations performed on the tampered image like rotation, noise addition, blurring, flipping, edge sharpening, scaling, compression, change in intensity etc.
- Another challenge is to choose the threshold value in feature matching. Most of the time the choice of threshold value is not automated. It varies with some characteristic of image like image size, block size, number of features, etc. It is important for each technique to define the selection procedure of optimal threshold value by identifying the characteristics. Optimization algorithm should be used to automate this process.
- Devices used to acquire the image and image forgery methods continue to emerge with time. It is necessary to identify the invalid detection methods.
- There are several attacks made on forged images to hide detection. There are several known attacks include, removing pattern noise, removal of JPEG artifacts, extracting pattern noise from another camera and adding it to forged image to confuse identification, etc.
- Selection of dataset is also important as new tampering tools and software are emerging. Dataset selected should include new ways of tampering techniques.
- Most of the blind forgery detection methods are based on statistical features of an image. If small part of the image is tampered, the statistical features may not change and hence detection becomes difficult.
- Few methods just detect the forged region but cannot precisely localize the forged region.

5 Conclusion

A complete overview of blind (passive) digital image forgeries and its detection techniques is given in this paper. In this digital world, due to the growth in technology, several digital tools, software and mobile applications are emerging, with different ways to edit digital images and are used for some information exchange. Editing the image using easily available software to convey wrong information is called forgery. This paper gives the classification of different image forgeries and its detection techniques. A detailed analysis of blind/passive forgery detection techniques are done in this paper. From the analysis, a conclusion can be drawn that the existing blind detection methods are not robust enough to identify and locate the forged region accurately in case of different attacks on forged images. Better feature extraction and matching algorithms should be used to minimize the detection time and to improve the detection accuracy. Few steps in the forgery detection process must be automated using optimization algorithms to improve the performance. Few researchers have used hybrid method and machine learning approach to improve the results. In the recent methods the deep learning approaches are used for better performance. The weaknesses of existing methods listed gives better direction to the researchers to provide better solution to this problem.

References

1. Farid, H.: Seeing is not believing. IEEE Spectr. **46**(8), 44–51 (2009)
2. Walia, S., Kumar, K.: An eagle-eye view of recent digital image forgery detection methods. In: International Conference on Next Generation Computing Technologies, pp. 469–487. Springer, Singapore (2017)
3. Birajdar, G.K., Mankar, V.H.: Digital image forgery detection using passive techniques: a survey. Digital Invest. **10**(3), 226–245 (2013)
4. Shivakumar, B.L., Baboo, L.D.S.S.: Detecting copy-move forgery in digital images: a survey and analysis of current methods. Global J. Comput. Sci. Tech. (2010)
5. Asghar, K., Habib, Z., Hussain, M.: Copy-move and splicing image forgery detection and localization techniques: a review. Aust. J. Forensic Sci. **49**(3), 281–307 (2017)
6. Peng, A., Wu, Y., Kang, X.: Revealing traces of image resampling and resampling antiforensics. Adv. Multimedia (2017)
7. Shah, H., Shinde, P., Kukreja, J.: Retouching detection and steganalysis. Int. J. Eng. Innovations Res **2**(6), 487(2013)
8. Haouzia, A., Noumeir, R.: Methods for image authentication: a survey. Multimedia Tools Appl. **39**(1), 1–46 (2008)
9. Rey, C., Dugelay, J.L.: A survey of watermarking algorithms for image authentication. EURASIP J. Appl. Sig. Process, 613–621 (2002). Special issue on Image Anal. Multimed. Interact. Serv
10. Singh, M., Kaur, H., Kakkar, A.: Digital signature verification scheme for image authentication. In: 2nd International Conference on Recent Advances in Engineering & Computational Sciences (RAECS), pp. 1–5. IEEE (2015)
11. Mahdian, B., Saic, S.: A bibliography on blind methods for identifying image forgery. Sig. Proc. Image Commun. **25**(6), 389–399 (2010)
12. Walia, S., Kumar, K.: Digital image forgery detection: a systematic scrutiny. Aust. J. Forensic Sci. pp. 1–39 (2018)
13. Warif, N.B.A., Wahab, A.W.A., Idris, M.Y.I., Ramli, R., Salleh, R., Shamshirband, S., Choo, K.K.R.: Copy-move forgery detection: survey, challenges and future directions. J. Netw. Comput. Appl. **75**, 259–278 (2016)
14. Mahmood, T., Nawaz, T., Ashraf, R., Shah, M., Khan, Z., Irtaza, A., Mehmood, Z.: A survey on block based copy move image forgery detection techniques. In: International Conference on Emerging Technologies (ICET), pp. 1–6. IEEE (2015)
15. Warbhe, A.D., Dharaskar, R.V., Thakare, V.M.: Computationally efficient digital image forensic method for image authentication. Procedia Comput. Sci. **78**, 464–470 (2016)
16. Muhammad, G., Hussain, M., Bebis, G.: Passive copy move image forgery detection using undecimated dyadic wavelet transform. Digital Invest. **9**(1), 49–57 (2012)
17. Warbhe, A.D., Dharaskar, R., Thakare, V.: A survey on keypoint based copy-paste forgery detection techniques. Procedia Comput. Sci. **78**, 61–67 (2016)
18. Amerini, I., Ballan, L., Caldelli, R., Del Bimbo, A., Serra, G.: A SIFT-based forensic method for copy-move attack detection and transformation recovery. IEEE Trans. Inf. Forensics Secur. **6**(3), 1099–1110 (2011)

19. Bay, H., Ess, A., Tuytelaars, T., Van Gool, L.: Speeded up robust features (SURF). Comput. Vis. Image Underst. **110**(3), 346–359 (2010)
20. Pandey, R.C., Singh, S.K., Shukla, K.K., Agrawal, R.: Fast and robust passive copy-move forgery detection using SURF and SIFT image features. In: 9th International Conference on Industrial and Information Systems (ICIIS), pp. 1–6. IEEE (2014)
21. Zheng, J., Liu, Y., Ren, J., Zhu, T., Yan, Y., Yang, H.: Fusion of block and keypoints based approaches for effective copy-move image forgery detection. Multidimension. Syst. Signal Process. **27**(4), 989–1005 (2016)
22. He, Z., Lu, W., Sun, W., Huang, J.: Digital image splicing detection based on Markov features in DCT and DWT domain. Pattern Recogn. **45**(12), 4292–4299 (2012)
23. Alahmadi, A.A., Hussain, M., Aboalsamh, H.: Splicing image forgery detection based on DCT and local binary pattern. In: IEEE Global Conference on Signal and Information Processing. IEEE (2013)
24. Zhao, F., Shi, W., Qin, B., Liang, B.: Image forgery detection using segmentation and swarm intelligent algorithm. Wuhan Univ. J. Nat. Sci. **22**(2), 141–148 (2017)
25. Wo, Y., Yang, K., Han, G., Chen, H., Wu, W.: Copy–move forgery detection based on multi-radius PCET. IET Image Proc. **11**(2), 99–108 (2016)
26. Mahmood, T., Mehmood, Z., Shah, M., Khan, Z.: An efficient forensic technique for exposing region duplication forgery in digital images. Appl. Intell. **48**(7), 1791–1801 (2018)
27. Yang, F., Li, J., Lu, W., Weng, J.: Copy-move forgery detection based on hybrid features. Eng. Appl. Artif. Intell. **59**, 73–83 (2017)
28. Zhang, J., Zhao, Y., Su, Y.: A new approach merging Markov and DCT features for image splicing detection. In: IEEE International Conference on Intelligent Computing and Intelligent Systems, vol. 4, pp. 390–394 (2009)
29. Fan, Y., Carré, P., Fernandez-Maloigne, C.: Image splicing detection with local illumination estimation. In: IEEE International Conference on Image Processing, pp. 2940–2944 (2015)
30. Park, T.H., Han, J.G., Moon, Y.H., Eom, I.K.: Image splicing detection based on inter-scale 2D joint characteristic function moments in wavelet domain. EURASIP J. Image Video Process. **1**, 30 (2016)
31. Sundaram, A.M., Nandini, C.: ASRD: Algorithm for spliced region detection in digital image forensics. In: Computer Science On-line Conference. Springer, Cham (2017)
32. Zhu, N., Li, Z.: Blind image splicing detection via noise level function. Signal Process. Image Commun. **68**, 181–192 (2018)
33. Hayat, K., Qazi, T.: Forgery detection in digital images via discrete wavelet and discrete cosine transforms. Comput. Electr. Eng. **62**, 448–458 (2017)
34. Huynh-Kha, T., Le-Tien, T., Ha-Viet-Uyen, S., Huynh-Van, K., Luong, M.: A robust algorithm of forgery detection in copy-move and spliced images. Int. J. Adv. Comput. Sci. Appl. **7**(3) (2016)
35. Rao, Y., Ni, J.: A deep learning approach to detection of splicing and copy-move forgeries in images. In: IEEE International Workshop on Information Forensics and Security, pp. 1–6 (2016)

MSB Based Iris Recognition Using Multiple Feature Descriptors

Sunil S. Harakannanavar[1(✉)], C. R. Prashanth[2], and K. B. Raja[3]

[1] S. G. Balekundri Institute of Technology, Belagavi, Karnataka, India
sunilsh143@gmail.com
[2] Dr. Ambedkar Institute of Technology, Bangalore, Karnataka, India
[3] University Visvesvaraya College of Engineering, Bangalore, Karnataka, India

Abstract. Biometric refers to a science for analyzing the human characteristics such as physiological or behavioral patterns. Iris is a physiological trait, which is unique among all the biometric traits to recognize an individual effectively. In this paper, MSB based iris recognition based on Discrete Wavelet Transform, Independent Component Analysis and Binarized Statistical Image Features is proposed. The left and right region is extracted from eye images using morphological operations. Binary split is performed to divide the eight-bit binary of every pixel into four bit Least Significant Bits and four bit Most Significant Bits. The DWT is applied on four bit MSB to extract the iris features. Then ICA is applied on approximate sub band to extract the significant details of iris. The obtained features are then applied on BSIF to obtain the enhanced response with final features. Finally, generated features are then matched with test features using Euclidean distance classifier on CASIA v1.0 database to analyse the proposed iris model.

Keywords: Biometrics · Discrete wavelet transform · Independent component analysis · Iris · Euclidean distance

1 Introduction

Biometrics deals with the identification of individuals based on their physiological or behavioral characteristics. The characteristics of human having unique features like face, iris, retina, speech, palm print, hand geometry, signature, fingerprint, and so on [1]. Due to the presence of unique characteristics of iris, it is the most available reliable biometric system. In addition, iris has unique features including accuracy, uniqueness, stability, reliability and real-time access capability compared with other biometric patterns [2]. Iris patterns of human will not interchange among other biometric traits and remains constant over person's lifetime from human's birth to until death [3]. Iris is a thin circular part of an eye between pupil and sclera. Towards its uniqueness and stability, iris recognition is a reliable identification technique [8].

In this paper, iris recognition is performed based on the combination of Discrete Wavelet Transform (DWT), Independent Component Analysis (ICA) and Binarized Statistical Image Features (BSIF) are applied on four bit binary most significant bit details to obtain the final features. Finally, generated features are then matched with test

© Springer Nature Switzerland AG 2020
A. P. Pandian et al. (Eds.): ICICCS 2019, AISC 1039, pp. 615–623, 2020.
https://doi.org/10.1007/978-3-030-30465-2_68

features using Euclidean distance classifier on CASIA v1.0 database to analyse the proposed iris model. The rest of the paper is organized as follows: Sect. 2 deals with the related work of the existing systems on iris recognition. Section 3 provides the proposed iris model. Result analysis is described in Sects. 4 and 5 concludes the proposed work.

2 Related Work

Afreen et al. [1] applied Histogram Equalization and two dimensional Discrete Wavelet transform to extract the significant details of the iris images. The process of clustering is carried out by k-means algorithm. The test features and significant features are then classified using ED classifier on CASIA database. Jayshree et al. [2] adopted image wavelet transform to extract the significant details of the iris. The decomposed image consists of LL, LH, HL and HH band where the HH band contains higher energy level and further taken into the account for next level as an input sample to generate iris code. The obtained feature vectors are classified with test feature vectors using k-NN and ED classifier. Bhanu [3] adopted one dimension Gabor filter to extract the informative features of the iris images. The obtained features are matched with test image features using hamming distance technique to measure the accuracy of the model. Bansal et al. [4] explained an effective tool to determine Obstructive Lung Diseases which includes bronchitis, asthma or even lung cancer. The wavelet transform and Gabor filter are applied on segmented iris image to extract the significant coefficients. The extracted coefficients are then classified with the test coefficients using Support Vector Machine. Monika et al. [5] adopted Savitzky-Golay filter to extract the informative details of the iris image. The extracted informative details are stored in the form of feature vectors. The extracted coefficients are then matched with the test coefficients using Hamming Distance operator to determine the similarity of two iris images. Thirumurugan et al. [6] explained Canny detection and Hough transform to detect the boundaries of iris in the eye image. The deterministic patterns (coefficients) are extracted by applying iris image on discrete wavelet transform technique which decomposed into four sub-bands such as LL, LH, HH and HL bands. The extracted pattern coefficients are matched with the test template patterns using Hamming Distance operator to determine the similarity of two iris images. Abhineet et al. [7] applied Circular Hough transform to localize the iris and Daughman's rubber-sheet model to normalize the iris images. The two dimensional Discrete Cosine Transform are applied on the normalized iris image to extract the informative details of the images. The extracted informative details are matched with test details using Hamming Distance operator to determine the similarity of two iris images. Rita et al. [8] developed Ridge Energy Direction (RED) algorithm to extract the significant features (ridge patterns) of normalized iris images. The model using RED is performed on one-fourth of the total iris area and further applied filtering which records the largest value of the output and encoded with 1 bit to represent its identity as directional filter. Binsu et al. [9] applied HE to improve the quality of normalized iris texture. Canny edge detector was applied on image to extract the boundaries of iris from digital eye image. One dimensional energy signal was recorded by applying vertical projection approach on two dimensional images. DWT was applied on one dimensional image to extract the significant coefficients of the iris. Atul et al. [10] explained statistical feature technique to extract the features of iris images. The iris model was implemented based on the correlation between adjacent pixels. The coefficients generated from the statistical features such as radial direction and angular direction are then matched with test coefficients using Hamming Distance operator to determine the similarity of two iris images.

3   Proposed Methodlogy

In this section, iris recognition is performed based on the combination of DWT, ICA and BSIF are applied on four bit binary most significant bit details to obtain the final features. The proposed model is shown in Fig. 1. The model is tested on CASIA V 1.0 database for various combinations of Person inside Database (PID) and Person outside Database (POD).

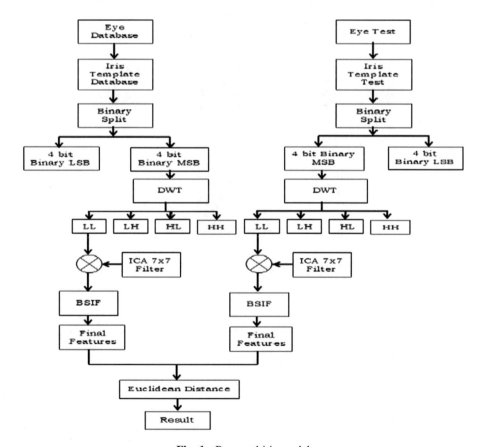

Fig. 1. Proposed iris model

3.1   CASIA Iris Database

The Chinese Academy of sciences Institute of Automation (CASIA V 1.0) Iris database is used to test performance of proposed iris model. The samples of iris images of one person are shown in Fig. 2.

Fig. 2. Eye image of CASIA database

3.2   Iris Extraction

Iris is circular part of an eye located between pupil and sclera. The iris part of left and right side of the pupil is considered having 40 pixel values on each side. Finally the left and right region is concatenated to form an iris image template as shown in Fig. 3.

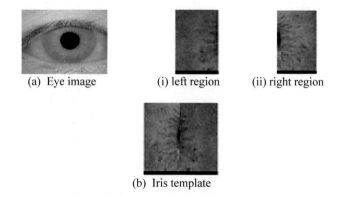

(a) Eye image   (i) left region   (ii) right region

(b) Iris template

Fig. 3. Extracted iris template from the CASIA database image

3.3   Binary Split

The decimal values of each pixel are converted into binary of eight bits. The eight-bit binary of every pixel is divided into four bit Least Significant Bits (LSB) and four bit Most Significant Bits (MSB) as shown in Fig. 4.

3.4   Discrete Wavelet Transform

In the proposed iris model, two dimensional DWT is applied on the four bit of MSB binary image to generate approximation LL band that has significant information of iris template and other detailed information such as detailed LH band having horizontal

(a) 4 bits of LSB   (b) 4 bits of MSB

Fig. 4. Binary split images

edge information, the detailed band HL has vertical edge information and HH band has information of diagonal edges of iris template. Figure 5 shows the approximate and detailed DWT decomposition bands of iris template.

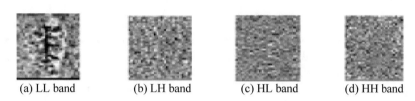

(a) LL band   (b) LH band   (c) HL band   (d) HH band

Fig. 5. DWT decomposition bands on iris template

3.5   ICA Filter Extraction

The ICA filter is used to select the appropriate images to extract the features. A non-linear filter is applied on each image to increase the high frequency component. Filters obtained from 7X7 windows with length 10 are used to filter the iris template to obtain binary images.

3.6   Binariezed Statistical Image Features (BSIF)

Binarization of an image is the method of filtering the image using ICA filter and applying proper threshold condition. In the proposed model 10-bits BSIF images are considered. The ICA has contained eight 7X7 windows with filter coefficients. The concatenation of each output bit value to single 10-bit frame is performed to obtain the binarized BSIF output of iris template as shown in Fig. 6.

Fig. 6. Binariezed output of BSIF

3.7   Euclidean Distance

Euclidian distance is used to match the final features of test images and generated features to identify a person using Eq. 1.

$$\text{Euclidean Distance} = \sum_{i=1}^{M}(P_i - q_i)^2 \tag{1}$$

Where, 'M' represents number of coefficients in a vector, P_i denotes the coefficient values of vectors in database and q_i denotes the values related to test images.

4   Experimental Results

The performance of the FAR, FRR and TSR are evaluated by creating the database of 6 Iris images of first 50 persons. The remaining one, iris image from 50 persons is considered as out of the database and helps for the calculation of FAR. For the different combinations of PID's with constant POD at 30, the percentage variations of optimum TSR, maximum TSR and EER are tabulated in Table 1. It is recorded that percentage OTSR decreases and percentage EER values increases with increase in PID's keeping POD constant.

Table 1. Performance parameters for CASIA database keeping POD constant

PID	POD	EER	OTSR	MTSR
20	30	12.5	87.5	95
40	30	12.5	87.5	97.5
50	30	16.67	84	98
60	30	18.35	81.67	95

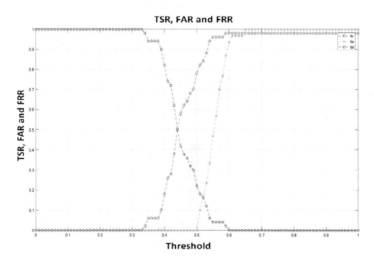

Fig. 7. Performance parameters plot for PID and POD of 50:30

The percentage of FAR, FRR and TSR with threshold for PID and POD of 50:30 is shown in Fig. 7. In the Proposed model it is observed that the value of FRR decreases with the threshold whereas the values of FAR and TSR are increased with the threshold. The percentage of TSR values of proposed iris model is 98 resulting EER value of 16.67 for PID and POD combination of 50:30.

For the different combinations of POD's with constant PID at 30, the percentage variations of optimum TSR, maximum TSR and EER are tabulated in Table 2. Hence it is recorded that percentage OTSR results constant of 90, whereas the percentage EER values results constant of 10, and the percentage maximum TSR results 96.67 for the combinations of 30:20, 30:30, 30:40 and 30: 60 respectively.

Table 2. Performance parameters for CASIA database keeping PID constant

PID	POD	EER	OTSR	MTSR
30	20	10	90	96.67
30	30	10	90	96.67
30	40	10	90	96.67
30	50	11	89	96.67
30	60	10	90	96.67

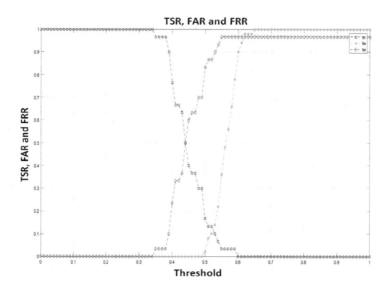

Fig. 8. Performance parameters plot for PID and POD of 30:50

The percentage of FAR, FRR and TSR with threshold for PID and POD of 30:50 is shown in Fig. 8. In the Proposed model it is observed that the value of FRR decreases with the threshold whereas the values of FAR and TSR are increased with the threshold. The percentage of TSR values of proposed iris model is 96.67 resulting EER value of 11 for PID and POD combination of 30:50.

5   Conclusion

MSB based iris recognition based on Discrete Wavelet Transform, Independent Component Analysis and Binarized Statistical Image Features is proposed. The left and right region is extracted from eye images using morphological operations. Binary split is performed to divide the eight-bit binary of every pixel into four bit least significant bits and four bit most significant bits. Now DWT is applied on four bit MSB to extract the iris features. Then ICA is applied on approximate sub band to extract the significant details of iris. The obtained features are then applied on BSIF to obtain the enhanced response with final features. The generated features are matched with test features using Euclidean distance classifier on CASIA V1.0 database to analyse the performance rate of proposed iris model.

References

1. Afifa Afreen, M., Diana Judith, I.: Iris recognition using hybrid technique, methods of moment and K means algorithm. In: International Conference on Advancements in Computing Technologies, vol. 4, no. 2, pp. 1–4 (2018)
2. Sali, J., Kadu, C.B.: Reliable human identification using iris as a biometric. Int. J. Sci. Eng. Manag. 1(8), 30–33 (2016)
3. Nagireddy, B.: Iris recognition using graphical user interface. Int. J. Adv. Res. Electr. Electron. Instrum. Eng. 7(1), 66–72 (2018)
4. Bansal, A., Agarwal, R., Sharma, R.K.: Iris images based pre-diagnostic tool to predict obstructive lung diseases. In: International Conference on Biomedical Research, pp. 517–522 (2018)
5. Singh, M., Sharma, S.: Iris recognition using Savitzky-Golay filter for better security outcomes. Int. J. Innovative Res. Comput. Commun. Eng. 6(6), 6329–6336 (2018)
6. Thirumurugan, P., Mohanbabu, G.: Iris recognition using wavelet transformation techniques. Int. J. Comput. Sci. Mob. Comput. 3(1), 75–83 (2014)
7. Kumar, A., Potnis, A., Pratap Singh, A.: Iris recognition and feature extraction in iris recognition system by employing 2D DCT. Int. Res. J. Eng. Technol. 3(12), 503–510 (2016)
8. Patil, R.B., Patodkar, N., Deshmukh, P.D.: Comparative performance analysis of feature extraction techniques of iris recognition. In: International Conference on Recent Advances in Computer Science, Engineering and Technology, pp. 22–26 (2017)
9. Kovoor, B.C., Supriya, M.H., Poulose Jacob, K.: Iris biometric recognition system employing canny operator. In: International Conference of Computer Science & Information Technology, pp. 65–74 (2013)
10. Bansal, A., Agarwal, R., Sharma, R.K.: Iris biometric recognition system employing canny operator. Int. J. Indian Acad. Sci. 41(5), 507–518 (2016)
11. Raja, K.B., Ragahavendra, R., Busch, C.: Scale-level score fusion of steered pyramid features for cross-spectral periocular verification. In: IEEE International Conference on Information Fusion, pp. 1–5 (2017)
12. Joshi, K., Agrawal, S.: An iris recognition based on robust intrusion detection. In: Annual IEEE India Conference, pp. 1–6 (2016)

13. Arunalatha, J.S., Rangaswamy, Y., Shaila, K., Raja, K.B., Anvekar, D., Venugopal, K.R., Iyengar, S.S., Patnaik, L.M.: Iris recognition using hybrid domain features. In: Annual IEEE India Conference, pp. 1–5 (2015)

14. Gale, A.G., Salankar, S.S.: Evolution of performance analysis of iris recognition system by using hybrid method of feature extraction and matching by hybrid classifier for iris recognition system. In: IEEE International Conference on Electrical, Electronics and Optimization Techniques, pp. 3259–3263 (2016)

Automatic Generation Control for Autonomous Hybrid Power System Using Single and Multi-objective Salp Swarm Algorithm

V. S. R. Pavan Kumar Neeli[(⊠)] and U. Salma

Department of EEE, GITAM (Deemed to be University),
Visakhapatnam, AP, India
pavanscholar123@gmail.com

Abstract. The main objective of the paper is to design a load frequency controller for Autonomous Hybrid power system by using Single and Multi objective soft computing technique. A novel controller such as Two degree of freedom PID (2DOFPID) controller has established a secondary controller for minimization of the frequency of oscillations. Recently proposed Salp swarm algorithm (SSA) technique was used to obtain the parameter of the controller. In this regard, to obtain the effectiveness of proposed technique, the performances are compared with other techniques such as Grasshopper optimization algorithm (GOA), Ant Lion optimizer (ALO) and Particle swarm optimization (PSO). The simulation studies (results) are carried out under five different sets of load disturbance which reveal the efficacy and superiority of Salp swarm algorithm based controller compared with other controllers.

Keywords: Salp swarm algorithm (SSA) · Ant Lion optimizer (ALO) · Grasshopper optimization algorithm (GOA) · Particle swarm optimization (PSO) · Two degree of freedom PID (2DOFPID) controller

1 Introduction

In this digital era, the electricity utilization is increasing at an unprecedented rate. Electricity generation using conventional sources such as coal combustion in thermal power stations, Hydro power stations are not a able to meet the emerging demands of the consumer's utilization. The raw materials used for generation are also decreasing gradually. In this scenario most of the generation units are accomplished the promising challenge of integration of generating unit with Non conventional based generating units. Wind energy and Solar energy sources which non harmful to the environment are picked as a renewable sources for the production of electricity. Though these two energy sources cannot able meet the entire demand but can be used for reducing the total electricity production from conventional generating units. The main aim of the Automatic generation control (AGC) was to establish balance between generations and demand maintaining the frequency of the whole system within acceptable range [4].

© Springer Nature Switzerland AG 2020
A. P. Pandian et al. (Eds.): ICICCS 2019, AISC 1039, pp. 624–636, 2020.
https://doi.org/10.1007/978-3-030-30465-2_69

Autonomous Hybrid power (HPS) system presented with Lee and wang [6] is a combination of Thermal generating unit which is integrated with Distributed generation (DG) resources. These DG resources consists of both conventional and Non conventional energy units (storage) such as Wind turbine generator (Wg), Solar PV system, Fuel cell with aqua electrolyser, Diesel engine (Dg) generator and energy storages unit such as Battery storage system [2]. Two different design objectives, the frequency deviations and Area control error (ACE) are considered for controller design which forms as a multi objective are combined and resulting to form as a single objective function.

The main contribution of the work are summarised below:

(1) To develop the model of Autonomous hybrid power system
(2) To obtain the parameters of proposed controller using different optimization techniques.
(3) Comparitive analysis is to be carried out under different loading conditions.

2   Modelling of the Power System

For simulation of the large scale systems, simplified models such as transfer function models are to be developed. Hence all generating units are developed as first order transfer function models. Therefore the total power obtained is the combination of power from thermal unit and power from the DG resources [3]. The output power of Distributed Generation system is given

$$\Delta P_{DG} = P_{Wg} + P_{Pv} + P_{Dg} + P_{Fc} - P_{Ae} \pm P_{Bss} \tag{1}$$

For small signal stability analysis, the generating units like Wind generator, Solar PV, Fuel cell with electrolyzer and Diesel generator can be modelled by the single order (first order) transfer functions with system gains value and time constants [12]. The simplified and linearized model of various generation systems is represented as

$$H_{Wg}(s) = \frac{K_{Wg}}{1 + T_{Wg}s} \tag{2}$$

$$H_{PV}(s) = \frac{K_{PV}}{1 + T_{PV}s} \tag{3}$$

$$H_{Fc}(s) = \frac{K_{Fc}}{1 + T_{Fc}s} \tag{4}$$

$$H_{Dg}(s) = \frac{K_{Dg}}{1 + T_{Dg}s} \tag{5}$$

$$H_{Ae}(s) = \frac{K_{Ae}}{1 + T_{Ae}s} \tag{6}$$

$$H_{Bss}(s) = \frac{K_{Bss}}{1 + T_{Bss}s} \tag{7}$$

The general block diagram of Autonomous hybrid power system (HPS) [1] shown below (Fig. 1).

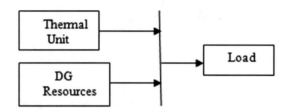

Fig. 1. Block diagram of power system

2.1   Two Degree of Freedom (2DOFPID) PID Controller

2DOFPID was a controller whose serial compensator is a PID element and feedback compensator is a PD element. Degrees of freedom of a control system is the number of closed-loops transfer function which can be adjustable independently. 2DOFPID controller is a two input and single output system [5] (Fig. 2).

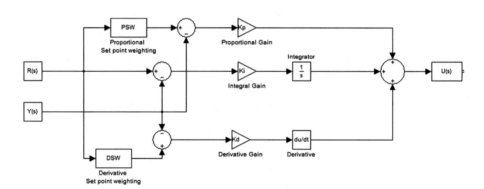

Fig. 2. Structure of 2DOFPID controller

R(s) and Y(s) and U(s) represent references signal, feedback from the measured system output with output signal respectively. Proportional, integral and derivative gains are represented respectively by Kp, Ki and Kd. PSW and DSW are the set point weights of proportional and derivative respectively. It evaluates a weighted difference signal depending on the proportional and derivative set point weighting for each of the three actions, that is, proportional, integral and derivative. Its generated output signal is determined by the difference in the reference and measured system output. General layout of 2DOFPID controller is shown as (Fig. 3).

The applied variations in components like generated powers (P_W; P_{sol}) are shown as

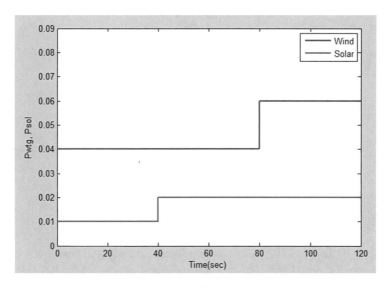

Fig. 3. Realization of the renewable power generation

3 Problem Formulation

Design of a controller depends upon well-defined objectives meeting the system requirements.

3.1 Single Objective Function

Each and every objective functions are having their own merits and demerits. Therefore in the present study is carried out with Integral of Square errors {ISE} is chosen with an desired objective functions for tuning and obtaining gains of 2DOFPID controller.

$$ISE = J = \int_0^T (\Delta f)^2 dt \tag{8}$$

3.2 Multi-Objective (MO) Function

Multi-Objective (MO) functions considered with combination of Integral of sum of squares with Incremental of Change in frequency deviations (Δf) and Area Control error (ACE). The final Objective function [11] is given below.

$$Min\, J = Min\,(J1 + J2) \tag{9}$$

Where J1 is given by

$$J_1 = \min \int_0^T (\Delta f^2)dt \tag{10}$$

And J2 is given by

$$J_2 = \min \int_0^T (ACE^2)dt \tag{11}$$

Where 'J' is minimized subjected to

$$K_P^{\min} \leq K_P \leq K_P^{\max} \qquad K_I^{\min} \leq K_I \leq K_I^{\max} \qquad K_D^{\min} \leq K_D \leq K_D^{\max}$$
$$PSW^{\min} \leq PSW \leq PSW^{\max} \qquad DSW^{\min} \leq DSW \leq DSW^{\max}$$

4   Soft Computing Technique

Salp Swarm Algorithm (SSA) was a recent Swarm intelligence algorithms [7] developed in year 2017 by Mirjalili. SSA is a population based method which explains the mimicking behaviour of Salp Swarms and their social interaction. The group of Salps called salp chains mathematically divide in to two groups: head salp is a leader and other are followers. Till now, the behaviour of salp swarm is not well conveyed, hence the researcher scholars consider the behavior of it to intensify their movement in seeking for food.

Steps followed in SSA:

1. Parameter initialization: The algorithm starts by initializing the parameters such as size in population N, no. of iterations t, and maximum iterations max_{iter}.
2. Initial Population: We generate initial population x_i, $i = \{1.....n\}$ randomly in the range of $[u, l]$ where u, l are upper and lower boundaries respectively.
3. Individuals Evaluations: Each individual (solution) in the population are evaluated by calculating its objective function value and the overall the best solution is assigned for F.
4. Exploration and exploitation: In order to balance between the Explorations and exploitations of the algorithm, we update the value of parameter c_1 given in the equation

$$c_1 = 2e^{-(\frac{4l}{L})^2} \tag{12}$$

Where l was the present iterations and L was the maximum no.of. Iteration
5. Position updation of solutions: The position of the leader solution and the other follower solutions are updated as given by

$$x_j^1 = \begin{cases} F_j + c_1((ub_j - lb_j)c_2 + lb_j \quad forc_3 \geq 0 \\ F_j - c_1((ub_j - lb_j)c_2 + lb_j \quad forc_3 < 0 \end{cases} \tag{13}$$

Where x_j^l is the leader position in j^{th} dimension and ub_j & lb_j are the max and min boundaries for j^{th} dimension and F_j is the food source position.
And

$$x_j^i = \frac{1}{2}(x_j^i + x_j^{i-1})\tag{14}$$

Where $i \geq 2$; x_j^i depicts the position of ith follower Salp at the jth dimensions.
6. Boundaries violations: If any solution violates the range of the search space during the update process, it returned back in the range of the problem.
7. Termination criteria: The number of iterations t is increased gradually untill it reaches to maximum iterations max_{iter} then the algorithm terminates search process and produces the overall best solution found so far.

5 Result and Analysis

This below session deals with the presentations of system dynamic response i.e. deviations in frequency(Δf) with various load uncertainities (ΔP_L) presented in Table 1 Simulations are carried out with Matlab 7.10.0 (2010a) software and Intel i3 processor, 4 GB ram based system.

The efficacy and superiority of proposed SSA technique is demonstrated by comparing its dynamic response with GOA [9], ALO [8] and PSO [10] techniques.

Table 1. Details of studied scenarios

Scenario	Simulation time	Ranges	Conditions
1.	120 s	$P_{Wtg} = \begin{cases} 0.04\,p.u., with & 0 \leq t < 80\,s \\ 0.06\,p.u., with & t \geq 80\,s \end{cases}$ $P_{Sol} - \begin{cases} 0.01\,p.u., with & 0 \leq t < 40\,s \\ 0.02\,p.u., with & t \geq 40\,s \end{cases}$ $P_L = \{0.06\,p.u., with \quad 0 \leq t \leq 120\,s$	Base Loading
2.	120 s	$P_{Wtg} = \begin{cases} 0.04\,p.u., with & 0 \leq t < 80\,s \\ 0.06\,p.u., with & t \geq 80\,s \end{cases}$ $P_{Sol} = \begin{cases} 0.01\,p.u., with & 0 \leq t < 40\,s \\ 0.02\,p.u., with & t \geq 40\,s \end{cases}$ $P_L = \begin{cases} 1.0\,p.u., with & 0 \leq t < 80\,s \\ 0.072\,p.u., with & t \geq 80\,s \end{cases}$	Increment of load by 20% at $t = 80$ s from Base load
3.	120 s	$P_{Wtg} = \begin{cases} 0.04\,p.u., with & 0 \leq t < 80\,s \\ 0.06\,p.u., with & t \geq 80\,s \end{cases}$ $P_{Sol} = \begin{cases} 0.01\,p.u., with & 0 \leq t < 40\,s \\ 0.02\,p.u., with & t \geq 40\,s \end{cases}$ $P_L = \begin{cases} 1.0\,p.u., with & 0 \leq t < 80\,s \\ 0.048\,p.u., with & t \geq 80\,s \end{cases}$	Decrement of load by 20% at $t = 80$ s from Base load

(continued)

Table 1. (*continued*)

Scenario	Simulation time	Ranges	Conditions
4.	120 s	$P_{Wtg} = \begin{cases} 0.04\,p.u., with & 0 \le t < 80\,s \\ 0.06\,p.u., with & t \ge 80\,s \end{cases}$ $P_{Sol} = \begin{cases} 0.01\,p.u., with & 0 \le t < 40\,s \\ 0.02\,p.u., with & t \ge 40\,s \end{cases}$ $P_L = \begin{cases} 0.06\,p.u.., with & 0 \le t < 40\,s \\ 0.072pu., with & 40 \le t < 80\,s \\ 0.048\,p.u., with & t \ge 80\,s \end{cases}$	Variation of Load by $\pm 20\%$ from Base load
5.	120 s	Corporates randomly variables of Load demand	Random loading

Scenario-1: Base loading Condition

In below scenario the dynamic performances of Hybrid power system(HPS) was investigated subjected to variations in wind, solar and load. As presented in Table 1 P_{wtg} is maintained at 0.04 p.u upto 80 s and increased to 0.06 p.u after 80 s. Similarly P_{Sol} is maintained at 0.01 p.u upto 40 s and increased to 0.02 p.u after 80 s and the load demand is 0.06p.u during the period $0 \le t \le 120$ s. The mismatches between the generations and load demand alleviated by the action of controllers. Figures 4(a) and 5 (a) reveals the comparative performance of frequency deviations of hybrid power system w.r.t different controllers and different soft computing techniques.

Scenario-2: Increment of Load by 20% at t = 80 s from Base Load

In this scenario sensitivity analysis of different controllers are performed to determine their robustness. As presented in Table 1 the variations in P_{wtg} and P_{Sol} are maintained similar in all scenarios while the load demand increased by 20% from base load at t = 80 s. Comparative responses of (Δf) obtained in the action of three controllers based on the SSA are shown Figs. 4(b) and 5(b).

Scenario-3: Decrement of Load by 20% at t = 80 s from Base Load

This scenario is similar to previous one but the only difference is the load demand is being decreased by 20% from base laod. Figure 4(c) and Fig. 5(c) presents the frequency deviations of the proposed system and the responses reveals the robustness of SSA optimized 2DOFPID controller compared with PI and PID.

Scenario-4: Variation of load by ±20% from Base Load

Another sensitivity analysis is performed to determine efficacy of proposed controller under the variation of wind energy, solar energy and load demand. In this scenario the load demand is maintained at 0.06 p.u up to 40 s and increased by 20% from nominal load and maintained upto 80 s and later the load demand is decreased to 20% from nominal load. Figures 4(d) and 5(d) presents the frequency deviations carried out during this sensitivity analysis.

Responses for Single Objective

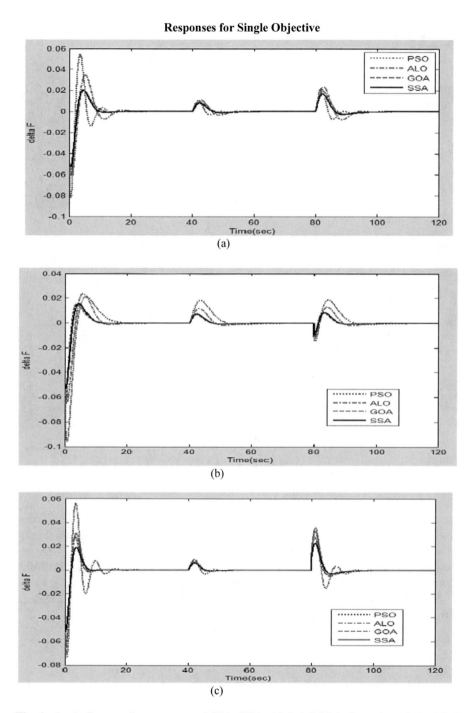

Fig. 4. (a–e): Comparative responses of SSA, GOA, ALO & PSO in Scenario 1, 2, 3, 4 & 5

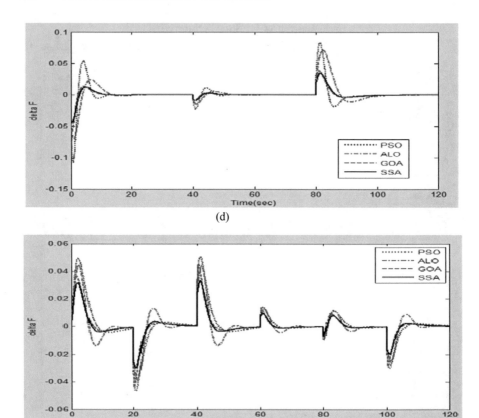

Fig. 4. (*continued*)

Scenario-5: Random Variation of Load

The supermacy analysis of the proposed controllers is carried out under random loading condition in this scenario. The dynamic performances are illustrated in Figs. 4(e) and 5 (e) with deviations in frequency due to different optimized controllers. All the test results in all scenarios confers that the responses of proposed SSA optimized 2DOFPID controller exhibits superior performance over other techniques implemented.

The robustness of SSA technique is being alleviated by comparing their responses with other techniques are presented in respective Figs. 4 and 5.

Responses of Multi-Objective

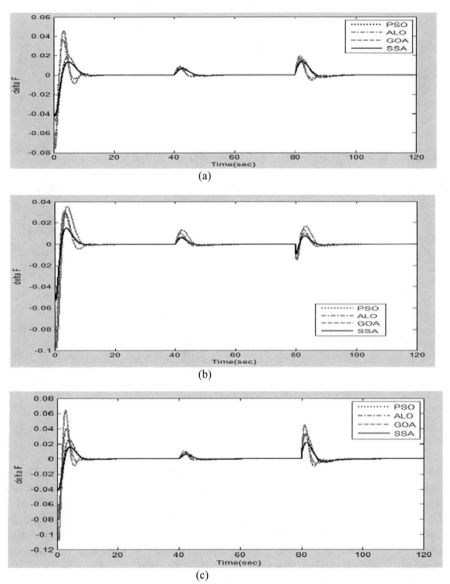

(a)

(b)

(c)

Fig. 5. (a–e): Comparative responses of SSA, GOA, ALO & PSO in Scenario 1, 2, 3, 4 & 5

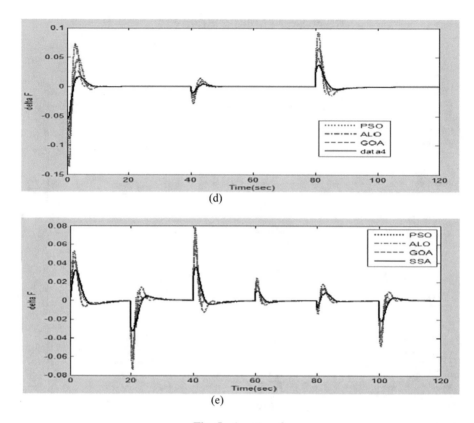

Fig. 5. (*continued*)

6 Conclusion

This research article focussed to propose an effective 2DOFPID controller as load frequency controller of Autonomous Hybrid power system (HPS) under various load un-certainities. To acquire the optimal gain values of proposed controllers, a novel Metaheuristic algorithm named as Salp swarm algorithm (SSA) through Single and Multi objective technique is implemented and the responses reveals that the proposed controller is more effective than PI and PID controller in reducing the frequency deviations. The time domain responses also confer the supremacy and efficacy of SSA technique in comparision with other techniques.

Finally it can concluded, the SSA optimized 2DOFPID controller is a better option for Autonomous Hybrid power system to enable automatic generation control in both Single and Multi objective conditions.

Table 2. Comparision of fitness value for single objective

Soft computing technique	Fitness value (ISE value with Single objective function)				
	Scenari1	Scenari2	Scenari3	Scenari4	Scenari5
PSO	0.01437	0.02150	0.01497	0.03586	0.02178
ALO	0.01180	0.01016	0.00960	0.03136	0.01674
GOA	0.00750	0.00503	0.00695	0.00788	0.01104
SSA	**0.00597**	**0.00447**	**0.00457**	**0.00664**	**0.00826**

Table 3. Comparision of fitness value for multi objective

Soft computing technique	Fitness value (ISE value with Multi objective function)				
	Scenari1	Scenari2	Scenari3	Scenari4	Scenari5
PSO	0.010630	0.01605	0.01931	0.03569	0.01794
ALO	0.008033	0.01068	0.01054	0.02168	0.01504
GOA	0.005446	0.005462	0.009813	0.01090	0.01213
SSA	**0.003831**	**0.003481**	**0.004203**	**0.006926**	**0.00788**

From the Tables 2 and 3, it is evident that the objectives of minimization of frequency deviations and area control error are achieved by using more significantly with SSA technique compared with other techniques.

References

1. Pandey, S.K., Mohanty, S.R., Kishor, N.: Frequency regulation in hybrid power systems using particle swarm optimization and linear matrix inequalities based robust controllerdesign. Int. J. Electr. Power Energy Syst. **63**, 887–900 (2014)
2. Das, D.C., Sinha, N., Roy, A.K.: Automatic generation control of an organic rankine cycle solar–thermal/wind–diesel hybrid energy system. Energy Technol. **2**(8), 721–731 (2014)
3. Lal, D.K., Barisal, A.K.: Load frequency control of AC microgrid interconnected thermal power system. In: IOP Conference Series: Materials Science and Engineering, vol. 225, no. 1, p. 012090. IOP Publishing (2017)
4. Shankar, G., Mukherjee, V.: Load frequency control of an autonomous hybrid power system by quasi-oppositional harmony search algorithm. Int. J. Electr. Power Energy Syst. **78**, 715–734 (2016)
5. Gaur, P., Bhowmik, N.S.D.: Impact assessment of vehicle-to-grid technology in LFC of multi-area solar-thermal power system. Int. J. Renew. Energy Res. (IJRER) **8**(3), 1580–1590 (2018)
6. Lee, D.-J., Wang, L.: Small-signal stability analysis of an autonomous hybrid renewable energy power generation/energy storage system part I: time-domain simulations. IEEE Trans. Energy Convers. **23**, 311–320 (2008)
7. Mirjalili, S., Gandomi, A.H., Mirjalili, S.Z., Saremi, S., Faris, H., Mirjalili, M.: Salp swarm algorithm: a bio-inspired optimizer for engineering design problems. Adv. Eng. Softw. **114**, 163–191 (2017)

8. Seyedali, M.: The ant lion optimizer. Adv. Eng. Softw. **83**, 80–98 (2015)
9. Saremi, S., Mirjalili, S., Lewis, A.: Grasshopper optimisation algorithm: theory and application. Adv. Eng. Softw. **105**, 30–47 (2017)
10. Eberhart, R., Kennedy, J.: Particle swarm optimization. In: Proceeding of the IEEE International Conference on Neural Networks, vol. 4, pp. 1942–1948 (1995)
11. Kalyanmoy, D.: Multi-objective genetic algorithms: problem difficulties and construction of test problems. Evol. Comput. **3**, 205–230 (1999)
12. Das, D.Ch., Roy, A.K., Sinha, N.: GA based frequency controller for solar thermal–diesel–wind hybrid energy generation/energy storage system. Int. J. Electr. Power Energy Syst. **43**(1), 262–279 (2012)

Investigation of the Temporal Evolution in Patient Networks Related to Chronic Diseases on Social Media

Sridevi Mutyala$^{(\boxtimes)}$ and B. R. Arunkumar

VTU Research Centre, Department of MCA,
BMS Institute of Technology and Management, Bengaluru, India
{sridevim, arunkumarbr}@bmsit.in

Abstract. The paradigm shift in the healthcare domain due to the advent of social media usage by healthcare seekers, practitioners and providers led to interdisciplinary research, which results in a completely new stream coined as Psychology Informatics. The study of structural dynamics of patient networks would reveal interesting results with respect to their social ties and the disease characteristics. The current work has attempted to study the temporal evolution in the patient networks related to chronic diseases viz. Alzheimer's and Diabetes on social media by using two important graph measures Entropy and Eigenvector centrality. Proctor controlled and transient patient networks are not under the purview of this research work.

Keywords: Patient networks · Social media · Entropy · Eigenvector centrality · Alzheimer's · Diabetes · Chronic diseases

1 Introduction

Social networks have gained importance in the recent years as they help to connect people across the globe to share their thoughts and experiences to overcome geographical isolation. Social media has brought drastic and unbelievable changes in the way people interact with each other. It is a fact that these days people are trying to interact more, digitally rather than directly to suffice their tendencies to maintain durable social relationships. Technology has changed this process, but it did not have a significant impact on the basic need to build supportive bonds with their counterparts [1]. When it comes to the context of health and healthcare, patients feel it comfortable to communicate with their peers online to seek informational and emotional support irrespective of their geographical location. Health related social networking increases awareness of healthcare facilities available and sensitizes the healthcare seekers about their need for healthcare [2]. Moreover, the patient networks discuss on patient-centric treatment in the scenario of life-style and chronic diseases like Alzheimer's and Diabetes which require prolonged care and customized treatment strategies. Besides, these

S. Mutyala—Research Scholar.
B. R. Arunkumar—Research Supervisor.

A. P. Pandian et al. (Eds.): ICICCS 2019, AISC 1039, pp. 637–644, 2020.
https://doi.org/10.1007/978-3-030-30465-2_70

persistent diseases can be managed and dealt with, by maintaining better social relationships which motivate the patients as well as the caregivers to stand strong with good emotional support that helps in prolonged survival. Viewing these relations from the network perspective, we can study the patterns of relational structures that focus on the characteristics of the network as a whole. The regularities and patterns in the interactions between the participants of the network gives rise to interesting structures which help us to study the social behavior of patients/caregivers dealing with certain diseases. This section is followed by the extensive review of relevant literature, description of the datasets, measures used for structural analysis of patient networks followed by discussion of experimental results and conclusion.

2   Literature Review

Social media provides us a treasure trove of data we never anticipated before and we can use this information to understand the pulse of the online community and bring drastic changes in population health and policy making in the healthcare domain. One-third of the 200 million daily Tweets and 55 million daily Facebook updates are health-related as per the survey made by University of Utah [3]. The usage of social media for healthcare information seeking and sharing is increasing day by day. During this process, the distribution of relational properties, understanding individual actions in the context of structured relationships or directly studying the structures themselves would unveil interesting results about the individual as well as group level behavior of the participants. The study of impact of the structure of network on the dynamics of the group and its influence on individuals within the group would open new avenues and insights into the healthcare domain.

Bronfenbenner (1943) and Jennings and Moreno (1945) were the first to make a quantitative study of the network methods that include tendencies towards mutuality of relationships, transitivity and structural balance. The authors of [4] presented a technique for analyzing a social network by using Shannon's entropy measure. Centrality measure entropies are utilized to perform the sensitivity analysis of the system employing the entropy changes of the participants in a social network.

In [5], a study was made on the entropy of weighted networks with the degree-based topological indices such as Zagreb index and Randic Index as weights. Topological indices serve as numerical parameters of a graph which characterize its structure or layout. A very important class of measure depends on Shannon's entropy to characterize and understand the graphs by deciding on their content of structural information. The research by [6] presented a graph-based entropy index which measures the diversity of events with respect to their distribution to the parts of a co-occurrence graph to detect signs of structural changes in the network that explains the potential dynamics of the participant behavior.

The work in [7] evaluated network measures, which are information-theoretic, on publication networks. The graph entropy measures discussed can be considered as measures of graph complexity that evaluate the structural complexity depending on the corresponding concept. In [8], the authors proposed a novel way to measure the graph entropy. They focused on the usage of event based entropy to find which are the influential nodes in a network. Tong et al. [9] introduced a new type of centrality measure that decomposes a graph into subgraphs and entropy calculation of neighbor nodes. It outperformed the famous measures including betweenness centrality, degree centrality, closeness centrality and Eigenvector centrality.

The literature reviewed so far discussed about the application of entropy as a graph measure to study complex networks. The application of this measure along with Eigenvector centrality for performing patient network analysis is a new approach in the current work. Patient network analysis can also be used to understand the process of change within a group over time which may lay ground for further analysis by the healthcare researchers. The social network approach thus has a different orientation in which structures, their effect, and evolution becomes the main focus [10].

Social network theories can be specified in terms of patterns of relations, representing a group or a social system as a whole. Approaches that do not have network measurements are not feasible to investigate theories about structural properties. In the current work, an attempt is made to study the patient groups on Facebook related to two chronic diseases namely Alzheimer's and Diabetes. These patient/caregiver networks are formed to discuss and share information related to specific diseases regarding the symptoms, diagnosis, treatment, diet, life style, progress of the disease and caretaking. The structural dynamics of these networks would be of interest to understand the social behavior of patients and their caregivers, and how they collaborate in a group to overcome geographical and emotional isolation. A comparative study of the measures viz. Average Degree, Entropy, and the Eigenvector centrality of the patient networks and their changes over one year period is visualized. This helps the healthcare researchers to understand the graph heterogeneity and the diversity of link distribution in the networks to study the social connections of the patients and caregivers. The correlation between the heterogeneity of the network and the percentage of influential nodes in the network is presented which exemplifies the relation between the entropy of a graph/network and the percentage of nodes with high Eigenvector centrality.

3 Description of the Datasets and Data Preprocessing

The data of the patient and caregiver networks related to two chronic diseases namely Alzheimer's and Diabetes was collected from Facebook by becoming a member of the above said groups. As Alzheimer's is a neuro-degenerative disease, the networks formed to discuss about it consist of caregivers as members rather than the patients directly involving in discussions. For each of the diseases considered, 3 patient support groups data over the period of 1 year taken quarterly was stored in the 'edgelist' format saved as .csv files. The data that is collected was sampled to maintain evenness among the data taken for analysis as the groups were actually differing in the total members in each group. The network is modeled in such a way that a member posting some content and the other members responding either by comment/reply or by like/emojis will have an edge between them. Redundant edges were removed and duplicate data was eliminated. Python programming language, equipped with large number of libraries to perform data analysis and visualization was used for experimenting with the datasets and obtain the results.

4 Measures Used for Structural Analysis of Patient Networks

4.1   Eigenvector Centrality

Eigenvector centrality Ev(x) is a measure of a vertex's influence in a network. This measure assigns corresponding scores to every vertex or node in the network based on the point that relationships with high-scoring vertices add more value to the score of the vertex being considered than the same number of connections to low-scoring vertices. For a node x, the Eigenvector centrality can be represented as [11]:

$$Ev(x) = \frac{1}{\lambda_{max(A)}} \cdot \sum_{i=1}^{n} a_{ix}.v_i \tag{1}$$

where $v = [v_1, v_2, \_\_\_, v_n]^T$ denotes an Eigenvector for the maximum Eigenvalue $\lambda_{max}(A)$ of the adjacency matrix named A.

4.2   Entropy

Rashevsky, Trucco and Mowshowitz defined a graphs topological information content based on their investigation on Shannon entropy of graphs. The *n* vertices of a given graph were partitioned into *k* classes of equivalence vertices based on their degree distribution. Then, the patitions were assigned a probability obtained as the number of vertices in this particular partition divided by the total number of vertices/nodes.

For a given graph G, the entropy can be defined as [12]:

$$H(G) = -\sum_{j=1}^{d_i} P_{ij} \log(P_{ij}) \tag{2}$$

where P_{ij} is the probability of a node belonging to a particular partition.

High entropy indicates that many vertices are equally important whereas low entropy indicate that only a few vertices are important. So, this Shannon's entropy measure can be used as a measure of structural characteristics of a network.

5 Experimental Results and Discussion

The entropy values of three groups for each disease (Alzheimer's and Diabetes) over one year period were calculated. The Eigenvector centrality of the nodes for each group over the same period was calculated which gives a node level measure. A threshold of 0.5 was fixed and all the nodes having Eigenvector centrality greater than 0.5 are considered influential in the network. Percentage of thus obtained influential nodes in the entire network is calculated. Now, the correlation between the average degree of the network, entropy and the Eigenvector centrality is studied to understand the structural properties of the network with respect to the diversity of the link distribution and the highly influential nodes in the networks. The extent of correlation between the measures being considered was depicted as shown in the Figs. 1, 2, and 3 for Alzheimer's support groups 1, 2 and 3, respectively.

Correlation graph between the three graph measures for Alzheimer's caregivers support group1

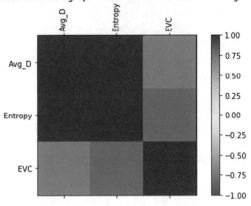

Fig. 1. Correlation graph for Alzheimer's Support Group 1

Correlation graph between the three graph measures for Alzheimer's caregivers support group2

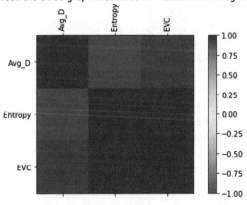

Fig. 2. Correlation graph for Alzheimer's Support Group 2

Correlation graph between the three graph measures for Alzheimer's caregivers support group3

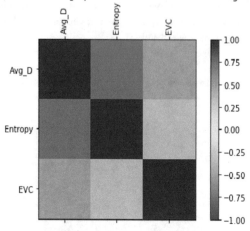

Fig. 3. Correlation graph for Alzheimer's Support Group 3

From the Fig. 1 visualizing the Alzheimer's support group 1, we can infer that Entropy and Eigenvector centrality are negatively correlated indicating that the diversity in the link distribution didn't positively affect the influentiality of the nodes in the network. This means that more number of diverse connections between the participants reduced the number of influential nodes over one year period indicating that many nodes are playing equally important role in the network.In all the three groups presented by Figs. 1, 2 and 3, average degree and the entropy are positively correlated strengthening the basic notion of entropy. The correlation between average degree and Eigenvector centrality varies from group to group indicating each patient group has a different structural behavior though they are formed with a common focal point. The same analysis can be applied to the results derived from the Diabetes groups also.

The Alzheimer's groups, as visualized in Fig. 4 show that the average degree and entropy go hand in hand in all the 3 groups over the one year period whereas, in except one group, the Eigenvector centrality varies irrespective of the average degree and entropy values. This is in contradiction to the Rich-Get-Richer phenomenon of the real-world networks as the influential nodes in these groups are not stably maintaining their influentiality and their percentage in the network kept changing. This can be related to the fact that Alzheimer's is a progressive disease [13] that gradually reduces the cognitive capabilities of the patient leading to final stage and thus the percentage of influential nodes in the network keeps changing as and when the so far influential participants (caregivers) exit from the group or remain passive or latent.

Temporal changes in the graph measures of the three Alzheimer's groups over one year duration

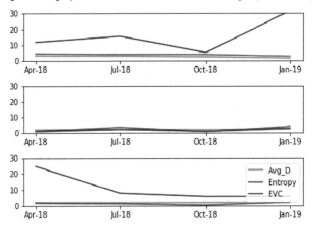

Fig. 4. Temporal changes in graph measures of Alzheimer's groups considered

The Diabetes support groups shown in Fig. 5, in contrast, exhibit a good synchronization and minimal variation between the graph measures indicating that most of the time, the influential nodes are stably maintaining their importance complying with the Rich-Get-Richer phenomenon of the real-world networks. This can be correlated to the characteristic of the disease as a manageable one with lifestyle changes and disciplined diet [14] thus resulting in the stability of the influential role played by the participants of the group.

Temporal changes in the graph measures of the three Diabetes groups over one year duration

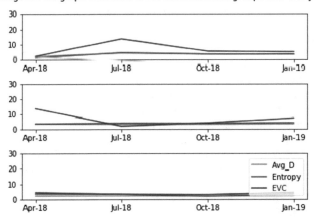

Fig. 5. Temporal changes in graph measures of Diabetes groups considered

6   Conclusion

The present work focuses on the structural analysis of the patient networks related to chronic diseases on social media. Two important measures namely Entropy and Eigenvector centrality are taken to understand how the structural dynamics of the patient networks and their social connections are related to the specific disease characteristics. Proctored patient networks are not considered for this research as they are formed to take up particular surveys and are transient. Seasonal disease networks were also not considered as they do not have well-defined patient groups on social media due to their inherent nature of short-term prevalence. Future enhancements of this work will head towards finding the highly collaborative patient networks related to chronic diseases on social media that would lay foundation for further analysis by the healthcare researchers.

References

1. Psychology Today – Social Life. https://www.psychologytoday.com/us/basics/social-life
2. Helzner, E.P., Scarmeas, N., Cosentino, S., Tang, M.X., Schupf, N., Stern, Y.: Survival in Alzheimer's disease: a multi-ethnic, population-based study in incident cases. Neurology **71**, 1489–1495 (2008). https://doi.org/10.1212/01.wnl.0000334278.11022.42
3. Algorithms for Innovation – Social Data Genius. https://uofuhealth.utah.edu/innovation/algorithm/2014/data/socialdatagenius.php
4. Ekrem, S., Selim, B.: Entropy based sensitivity analysis and visualization of social networks. In: IEEE/ACM International Conference on Advances in Social Networks Analysis and Mining (2012). ISBN: 978-1-4673-2497-7, https://doi.org/10.1109/asonam.2012.189
5. Ramin, K.: Entropy of weighted graphs with the degree-based topological indices as weights. In: MATCH Communications in Mathematical and in Computer Chemistry, vol. 76, pp. 69–80 (2016). ISSN 0340-6253
6. Yukio, O.: Graph-based entropy for detecting explanatory signs of changes in market. Rev. Socionetw. Strat. **12**, 183–203 (2018)
7. Andreas, H., et al.: On graph entropy measures for knowledge discovery from publication network data. In: 1st Cross-Domain Conference and Workshop on Availability, Reliability, and Security in Information Systems (CD – ARES). Lecture Notes in Computer Science. LNCS, vol. 8127, pp. 354–362. Springer (2013)
8. Jitesh, S., Jafar, A.: Discovering important nodes through graph entropy – the case of Enron email database. In: Proceedings of the 3rd International Workshop on Link Discovery, pp. 74–81 (2005). https://doi.org/10.1145/1134271.11342282
9. Tong, Q., Wei, S., Chang, Z.: How to identify the most powerful node in complex networks? A novel entropy centrality approach. Entropy **19** (2017). https://doi.org/10.3390/e19110614
10. Stanley, W., Katherine, F.: Social Network Analysis – Methods & Applications. Cambridge University Press, Cambridge (1994)
11. Bonacich, P., Lloyd, P.: Eigenvector-like measures of centrality for asymmetric relations. Soc. Netw. **23**, 191–201 (2001)
12. Bromiley, P.A., Thacker, N.A., Bouhova, T.E.: Shannon entropy, Renyi entropy, and information: Tina-memo No. 2004-004. The University of Manchester (2010)
13. Alzheimer's Disease Fact Sheet - National Institute on Aging, U.S. Department of Health and Human Services. https://www.nia.nih.gov/health/alzheimers-disease-fact-sheet
14. Diabetes Type 2 is Serious but Manageable – NIH MedlinePlus, vol. 9, pp. 12–13 (2014)

Automated Lighting Smart Parking Using Internet of Things

Rajasekhar Shastry, B. V. Ramana Murthy,
C. Kishor Kumar Reddy$^{(\boxtimes)}$, and P. R. Anisha

Department of Computer Science and Engineering,
Stanley College of Engineering & Technology for Women, Hyderabad, India
kishoar23@gmail.com

Abstract. As the field is changing over with the upgrades being made in Computer Technology, we are getting into the new innovation of omnipresence ruled with the guide of Internet of Things. IoT is making snappy upgrades and is hunting down each body's advantage. This inclining innovation is being employed in practically each subject: Engineering, Medical, Business, Services, Transportation, Administration, etc. With the blast in populace comes blast in assortment of vehicles on road which needs green and achievable Parking System-Parking Spaces which are Smart. Keeping in considerations the over-population issues, and vehicle leaving inconveniences we plan to format an adaptation that exhibits a Smart Parking System, the utilization of cloud to advise the chairman around opening, assuming any and furthermore, to demonstrate the opportunities accessible to the clients, informing them the whole scope of inhabitances and opening in the parking area. Going to the stopping masses in living arrangements, shopping retail establishments, work-places, etc. the stopping is inside the basement. This needs relentless use of lighting installations device. Lights exchanged on for the whole day isn't beyond question an exceptionally smart thought keeping up at the top of the priority list the present quality resources. Along these lines we give you the idea of programmed lights which switches at the lighting least complex while a vehicle's quality is detected. Subsequently consolidating Smart Parking with Automated Lighting is a productive, conceivable and helpful idea.

Keywords: Automated lighting · Internet of Things · Smart Parking

1 Introduction

With the development in worldwide people, comes a development being used of engines for transportation. This request stopping territories. Stopping issues are not surprising in the vast majority of the urban communities. It is truly chafing and a period ingesting enthusiasm for drivers to search for an empty parking spot in a spot or vehicle leaving zone without understanding the accessibility of parking spot. This issue pulled in vital ventures from industry parts to improve stopping incomes through age empowered arrangements. The charge for stopping expansion is normally expensive or incredibly high. Brilliant stopping makes utilization of different innovations to effectively deal with a Parking Area. Moving towards Smart City, Smart Parking is a

© Springer Nature Switzerland AG 2020
A. P. Pandian et al. (Eds.): ICICCS 2019, AISC 1039, pp. 645–652, 2020.
https://doi.org/10.1007/978-3-030-30465-2_71

wonderful occurrence wherein IoT and can be accurately utilized in our every day presence to offer remarkable contributions. The idea of making a Smart City is being done with the rise of the Internet of Things. Additionally in the present worldwide, it's miles basic to adapt to the strategies to reduce vitality utilization keeping up our vitality sources as a top priority. Consequently a Parking System with Reduced Power Consumption by Lights may turn out to be a systemized, legitimate and viable Parking System.

The proposed adaptation deals with Arduino Uno and NodeMCU ESP8266. Ultrasonic sensors experience the nearness of a car, and include the data the cloud. At that point the head is told about the whole amount of opening and inhabitances inside the part. This encourages to give the precise records around the openings to be had in the parking garage. Computerization of Lights inside the Parking Area is encouraged by methods for the utilization of the ultrasonic sensors that flip lights ON and OFF by methods for ascertaining the separation while the vehicle is in the front of the sensor. The assortment of Ultrasonic Sensor being 2 cm to 400 cm that is sufficient to degree remove with the goal that the arrangement of space might be respected. This accessibility is being posted on cloud. The cloud we're the utilization of here is IBM Cloud. The Node-RED buoy empowers us in growing a web webpage application for the Parking System. NodeMCU Board is the psyche of the Parking System while Arduino Uno cares for Light Automation. The - Smart Parking and Light Automation are being done on two distinct gatherings. The thought process being that is we need the notoriety of accessibility of the opening to be shown on net utility with the goal that the director can sit and uncover effectively. The Automation of lights is for preserving quality and to have savvy installments. Mechanization is wherein matters are associated in a way that they impart by means of each other without the need of a system. A straightforward code nourished inside the board with basic associations is sufficient for this Light Automation. Thus while those two thoughts are clubbed up it makes a Simple Parking Lot a Smart Parking Lot. This model is structured especially to stop in Residential Apartments, Shopping Malls, Hospitals and Offices in which stopping is in Cellar and requests persistent utilization of lights in manual operation. However, while Automation comes into utilization it permits in diminishing power admission.

2   Relevant Work

Antonio Carlos Bento in his paper "IoT: NodeMCU 12e X Arduino Uno, Results of a test and relative study" looks at the 2 microcontroller sheets NodeMCU and Arduino Uno-their upsides and downsides, etc. The most essential qualification being that Arduino Uno requires set up of late devices to associate with a network. NodeMCU anyway has direct association with Wi-Fi. NodeMCU being more valuable than Arduino-having numerous favors next to this like lower long, blast in preparing pace, running voltages, Larger RAM, Flash Memory, and so on. Anyway Arduino has nearly more prominent assortment of pins and hence does now not require Shield or Expansion Board for more associations Shrivastava et al. [1]. In "Separation Measurement of an Object or Obstacle through Ultrasound Sensors utilizing P89C51RD2" made utilization of Ultrasonic Sensors for separation estimation as it turned into the most economical answer and a Phillips P89C51RD2 microcontroller based absolutely gadget. They are flexible and valuable for measurement in air and submerged. They are utilized to ascertain remove indirectly from pace of sound in air and time taken for techniques to travel, hit the item and return came back to source [2].

Maher Hassan Kadhim inside the paper, "Arduino Smart Parking Manage System dependent on Ultrasonic Internet of Things (IoT) Technologies" proposes the Smart Parking engineering. The float of activity is enrollment, holding demand, digital book leaving after which leave vehicle. The proposed machine improves the execution by methods for bringing down the unfilled space in a stopping, and moreover sparing the fuel squandered in endeavoring to discover a stopping Revathi et al. [3]. Inside the paper "Brilliant Parking Systems and Sensors: A Survey" features a 3-layered usefulness of a Parking System. They are named as Sensing usefulness, Data Forwarding usefulness and the Upper-level Handling Data Storage, Processing and Client interfaces. Grouping of Parking Systems is done as Centralized helped seek, Opportunistically Assisted, etc. Different innovations being utilized in Parking Systems are Ultrasonic Sensors, Passive Infra-Red Sensors, Electromagnetic Sensors, etc. [4].

Yuvaraju et al. In his paper "IoT Based Vehicle Parking Place Detection the utilization of Arduino" gives a variant that utilizes Infrared Sensors related to the Arduino and refreshing information to the cloud utilizing GSM. Infrared Sensors are situated on openings which can be connected to Arduino Board. The information is refreshed to GSM which offers client the data about the parking spot accessibility. At last the client needs to make cost fundamentally dependent on schedule for which the opening transformed into involved [5].

3   Proposed Architecture

The proposed architecture consists basically of the following layers: The first layer of architecture is the sensing layer. In the sensing layer, we have 3 ultrasonic sensors (2 for Smart Parking and 1 for Light Automation). The microcontroller calculates the effective distance when the code is uploaded into the board which serves the purpose of the sensed data which is further pushed on the next layer. In the next layer, we use Wi-Fi or Mobile Hotspot for communication. In the next layer, we publish the data on the cloud using MQTT protocol because it is envisaged of providing better services, as the published data is automatically updated which is reflected on the web/desktop application we provide. Finally in the last layer the status of the parking slot is given on the Web Application. For the Light Automation, the microcontroller sends a logic 1 signal to the LED depending on the entry of the car and the LED switches ON else it remains OFF. The architecture is summarized in Fig. 1.

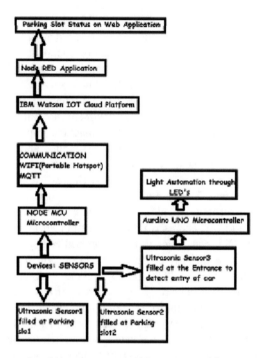

Fig. 1. Proposed architecture: smart parking system with automated lighting.

4   Implementation and Discussion

The connections for the demonstration are shown below in the Fig. 2:

Ultrasonic Sensor: It uses SONAR (SOund Navigation And Ranging). It has a transmitter and receiver within it. A square pulse is transmitted in the surrounding. When the pulse encounters an object, it gets reverted, which is received by receiver in sensor and is acknowledged at echo pin.

Working Principle of Ultrasonic Sensors: The time taken by the square pulse generated by transmitter to return back to the sensor, is calculated by the function pulseIn().

Distance = Speed of Sound (340 m/s)*(time/2). Here time calculated, is the time to travel from the sensor to obstacle and back to sensor (hence it is divided by 2).

We use 3 Ultrasonic sensors of which 2 sensors are used to smart parking system and 1 for the automated lighting system. In smart parking system, when the car is at a distance of 8 cm and more, the status shown is vacant. When the car is parked the parking slot, the distance is less than 8 cm, which shows, the status as occupied.

In automated lighting system, when the presence of car is sensed, the sensed data results in the computation of distance and the actuator used here is the LED which glows.

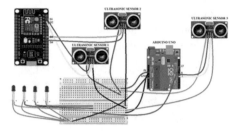

Fig. 2. IoT based smart parking system with automated lighting connections

LEDS: Whenever presence of a car is sensed the LEDs get switched ON for a while and then get switched OFF.

Ardulno Uno: The Arduino Uno Microcontroller runs the code for Automated Lighting. The distance is calculated indirectly from the Ultrasonic Sensor data (Sensor present at the entrance of Parking Area). Depending on this distance lights (LEDs) are switched ON and OFF.

NodeMCU ESP8266: The code for implementation is written in Arduino IDE where the code is written and uploaded. The code for Smart Parking System is uploaded in the NodeMCU board. Ultrasonic sensors are fixed on walls of the parking slot. Whenever a car is present in a parking slot, the distance between the car and the sensor is quite very small say maximum about 10 cm. Similarly when a car is not present the distance calculated in the code will be more than 10 cm and this implies that the slot is vacant. This distance is hence the parameter to determine the status of the parking slot and hence this is published on the Cloud-IBM Bluemix. The reason why the Arduino is not being used for implementing this is-

Arduino Uno does not have a built in Wi-Fi module and hence requires an external device to connect to a network which increases its cost. NodeMCU on the other hand has a built in Wi-Fi Module and is also inexpensive.

IBM Bluemix: It is a cloud platform developed by IBM. It is a cloud Platform as a service (PaaS). It provides a platform to develop IOT applications-the IBM Watson IoT platform and a very efficient programming tool- Node-RED. NodeMCU is connected to the IBM Cloud that sends and receives data (Fig. 3).

Fig. 3. Distance values being published on the IBM Cloud

Node-RED Flow Editor: It is a tool for visual programming. It is used to connect hardware devices and online services. It uses JavaScript and makes use of JSON objects. It provides different functionalities in the form of nodes. Through JSON parsing data (values of distances) is received from the device. Switch nodes are used to select the status of parking slots using the distances obtained and comparing them with a value of maximum distance given. The UI text node publishes the data on the web application when the flow is successfully deployed. Figure 4 gives the final view of the Node-RED flow for Smart Parking System.

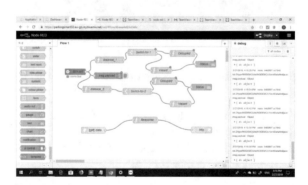

Fig. 4. The final view of node-RED flow editor for smart parking system

MQTT (Message Queuing Telemetry Transport): The MQTT protocol is based on Publisher-Subscriber model. It is preferred over http as in MQTT it is not necessary to make request whenever a change is made in the data unlike http. To manage connection and data between publisher-subscriber we require a Broker. The IBM Bluemix cloud is the Broker being used here. Web Application uses this protocol.

5   Conclusion and Future Scope

To make the research better and more efficient, we plan to add a Servo Motor at the entrance of the gate, along with an ultrasonic sensor. This sensor senses the presence of the car at the entrance, and will allow the servo motor to open only if any one or both the slots are vacant. When a car is at the entrance, and both slots are full, the servo motor doesn't function and will not allow the car to enter in the parking slot. This will reduce the time wasted for searching a slot when there is no empty slot available, and will direct the driver to go to another parking area.

Furthermore, we can add more lights and sensors to light the parking lot when the car is sensed in different places inside the lot. This is when we aim for large lots with more space and slots.

References

1. Bento, A.C.: IoT: NodeMCU 12e X Arduino Uno, Results of an experimental and comparative survey. Int. J. Adv. Res. Comput. Sci. Manage. Stud. **6**(1), 46–56 (2018)
2. Shrivastava, A.K., Verma, A., Singh, S.P.: Distance measurement of an object or obstacle by ultrasound sensors using P89C51RD2. Int. J. Comput. Theory Eng. **2**(1), 64–68 (2010)
3. Kadhim, M.H.: Arduino smart parking management system based on ultrasonic Internet of Things (IoT) technologies. Int. J. Eng. Technol. **7**, 494–501 (2018)
4. Revathi, G., Sarma Dhulipala, V.R.: Smart Parking Systems and Sensors: A Survey
5. Yuvaraju, M, Monika, M.: IoT based vehicle parking place detection using arduino. Int. J. Eng. Sci. Res. Technol., 536–542 (2017)
6. Gupta, A., Kulkarni, S., Jathar, V., Sharma, V., Jain, N.: Smart car parking management system using IoT. Am. J. Sci. Eng. Technol. **2**, 112–119 (2017)
7. Khanna, A., Anand, R.: IoT based smart parking system. In: International Conference on Internet of Things and Applications (IOTA), pp. 266–270, January 2016
8. Chinrungrueng, J., Sunantachaikul, U., Triamlumlerd, S.: Smart parking: an application of optical wireless sensor network. In: Institute of Electrical and Electronics Engineers (2007)
9. Lu, R., Lin, X., Zhu, H., Shen, X.(S.): SPARK: a new VANET-based smart parking scheme for large parking lots. In: Institute of Electrical and Electronics Engineers, pp. 1413–1421 (2009)
10. Thangam, E.C., Mohan, M., Ganesh, J., Sukesh, C.V.: Internet of Things (IoT) based smart parking reservation system using raspberry-pi. Int. J. Appl. Eng. Res. **13**(8), 5759–5765 (2018)
11. Subhash, R., Praveen, M., Dineshkumar, V., Vijayakumar, V., Ramachandiran, R., Shanmugam, M., Kalaipriyan, T., Raghav, R.S.: IoT based smart parking system. Int. J. Pure Appl. Math. **119**(14), 367–375 (2018)
12. Basavaraju, S.R.: Automatic Smart Parking System using Internet of Things(IOT). International Journal of Scientific and Research Publications **5**(12), 629–632 (2015)
13. Atif, Y., Ding, J., Jeusfeld, M.A.: Internet of Things approach to cloud-based smart car parking. In: The 7th International Conference on Emerging Ubiquitous Systems and Pervasive Networks (EUSPN), pp. 193–198 (2016)
14. KoKilavani, M., Malathi, A.: Smart street lighting system using IoT. Int. J. Adv. Res. Appl. Sci. Technol. **3**(11), 08–11 (2017)

15. Magan, T., Bahur, A., Ghatshile, S., Jagtap, P., Uplaonkar, D.: Smart lightning in street lights based on IOT. Int. J. Innov. Res. Comput. Commun. Eng. **5**(12), 16821 (2017)
16. Badgaiyan, C., Sehgal, P.: Smart street lighting system. Int. J. Sci. Res. **4**(7), 271–274 (2015)
17. Keni, P., Wajid, S.M., Ahmad, S.Z., Rahimunnisa, Shruthi, K.: Automated street lightning system using IoT. Int. J. Adv. Res. Ideas Innov. Technol. **4**(3), 1970–1973 (2018)

An Experimental Study Using Scale Invariant Feature Transform and Key-Point Extraction for Human Ear Recognition System

Subhranil Som[1][✉] and Renuka Mahajan[2][✉]

[1] Amity University Uttar Pradesh, Noida, India
subhranil.som@gmail.com, ssom@amity.edu
[2] Jaipuria Institute of Management, Noida, UP, India
renuka.mahajan@jaipuria.ac.in

Abstract. Abundant research has been done on the improvement of the security and trustworthiness of biometric systems. The aim of this paper is to demonstrate the image key-point extraction technique and establish its uniqueness for biometric identification. Ear features comes out to be one of the important biometric systems, which prove to have great potential, in identifying humans in the real world applications. In this work, key-point based matching and recognition is done using SIFT (Scale Invariant Feature Transform) technique. This approach extracts features from images of distinctive invariant. These images are utilized to perform consistent matching between various objects (ear). The key-points are invariant to image scale and hence can provide good matching over a wide range of images. The distinctive features have been matched correctly using the proposed technique and tested on a large database of ear images. This study helps in establishing that the experimental results show improvements in recognition accuracy.

Keywords: Scale Invariant Feature Transform (SIFT) · Laplacian of Gaussian (LoG) · Feature Points (FPs) · Difference of Gaussian (DoG) · Key Points (KPs) · Data set (DS) · Match Points (MPs)

1 Introduction

Biometric refers to identifying human by their physiological and behavioral traits, containing well defined characteristics for each individual [1–3]. Physiological traits depend upon the individuals' shape and hence, they vary from one person to another e.g. ear, face and fingerprints. Behavioral traits are linked to individuals' behavior e.g. handwriting, signatures etc. [4]. Biometric authentication is commonly used in several fields like commercial software engineering applications (electronic banking, investment and other financial transactions); government documents (identification cards, driving license, passports etc.) and forensic applications (identification of corpse, criminal investigation, terrorist identification, determination of parenthood and missing children) [4, 5] and lately in smartphone [6, 7]. Many biometric traits have already contributed in building an authentication system [8]. However, with the passage of time, they have shown certain pitfalls, which create a doubt on the existing

© Springer Nature Switzerland AG 2020
A. P. Pandian et al. (Eds.): ICICCS 2019, AISC 1039, pp. 653–665, 2020.
https://doi.org/10.1007/978-3-030-30465-2_72

authentication system. For example, a biometric trait such as face, which is unique for almost every person, but the patterns of face also tend to change with growing years. So this generates problem for the authentication systems which require continuous updating, which is quite unacceptable with respect to time and money. The aim of the paper is focused on the concept of building a reliable authentication system. Thus a new biometric trait (ear) can be considered for building an authentication system which will not require any rapid updating [9].

For a particular feature to be utilized for the purpose of identification, it should necessarily possess certain desirable properties [10]. An important property is the uniqueness of the feature [11]. If one reviews the literature of personal identification, then many researchers have claimed that the external ear could be a potential feature for individualization [12–15]. This is because the ear has many required properties such as fully developed at the time of birth, universality, stability and distinctiveness [16]. Ear biometrics has recently evolved as an important factor as far as authentication is concerned because of the fact that the shape of the human ear remains unchanged [11, 17, 18]. The structure of the ear is unaltered by facial expressions and hardly varies with age. According to [19], the ears are constantly fixed on the sides of the head. They are quite immune to hygiene issue, make-up (except jewelry), spectacles, beards and moustaches, unlike fingerprints. Further, an ear can be clearly acquired from a distance as it is larger compared to other features like retina, iris and fingerprints. Iris and retina measurements even lead to anxiety. So, the ear is not prone to covariate interference and particularly invariant to age as compared to other biometrics. Thus, biometric system of an ear may be viewed for security and identification system [20, 21].

Figure 1 shows the outward structure of a human ear (or pinna). It is formed by the helix, the antihelix, the lobule, the anti-helical fold, the tragus, the antitragus [19]. There are many valleys and ridges on the outer surface of the ear. Hence, the experimental images can be reduced to set of features and these features can be compared with the features set of others ear images, to determine its identity [22]. In fact, in a study, it has seen that the ear feature of identical twins also gives distinguishable features [11]. Thus variability between the ears is large enough to get a unique identification [23].

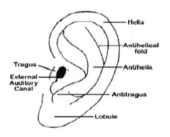

Fig. 1. Outward appearance of human ear [11]

1.1 Related Work

There has been extant research in utilizing ear as a characteristic for building a security system. The pioneer of using ear as a trait for security system was Iannarelli [12, 13, 16, 24]. He used over 10,000 ear images for examination to prove their uniqueness. He manually measured the distance between specific points of the ear. However, the drawback was to extract those key-points which will be invariant in different situations. Burg and Burger in 1997 built another system, which developed neighborhood graph, from the Vornoi diagram. It gives the curve segment and this curve is change prone with the recast of camera-to-ear or change in illumination. Hence, this again raises the issue of instability [25–27]. Further, Moreno et al. 1999 utilized the feature points of the external ear, the contours and the data obtained from the specific ear shape [33]. Chorus in 2005, proposed the geometric feature extraction method based on the number of pixel with same radius thus focusing on the curve avoiding the area outside the curve. The image matching was developed using a set of local interest points from Moravec work on stereo matching (using a corner detector) [32]. Harris and Stephens improved this Moravec detector (for efficient motion tracking and 3D structure from motion recovery) [25]. This Harris corner detector is used for many other image matching purposes [24, 25]. Harris also showed that it was rotation invariant, which means, on rotation of the image, the same corner was retrieved [31]. But, it suffered from scaling problem. It was not considered a good basis for matching images of various sizes, as it was extremely sensitive to variations in the image scale. Lowe 1999 extended the local feature approach to achieve scale invariance by incorporating a new local descriptor. It was comparatively less sensitive to local image distortions like 3D viewpoint change and had more distinctive features [14, 27].

1.2 Motivation

Previous researchers explored use of ears for biometric identification. It has been debated that both the shape as well as the appearance of the outer ear remains unique and unchanged in an individual during his lifetime. The review of previous researches show that so far the ear biometrics is still in nascent stage for commercial use. However, there exist some established methods of feature extraction from ear images [28]. In this paper, we discuss one of the technique called SIFT (Scale Invariant Feature Transform) for ear recognition (as done by [29]. Lowe 2004 first developed 'SIFT' as a method for image-based matching and recognition [13]. It is a method of finding distinctive and invariant features. This method is invariant to scaling, transformations in the image domain [13]. Experimentally, under real-world conditions, the SIFT has been found to be beneficial for image matching and object recognition [30]. But most of the researchers used their own databases. Moreover, the researchers' databases are too small leading to bad result in some cases. We conducted experiment on the large image set of IIT Delhi, using SIFT implementation.

The paper is organized as shown. Section 2 explains the proposed work. Section 3 describes the experimental results and discussions. The concluding remarks are shown in Sect. 4. Lastly, Sect. 5 gives pointers for future work.

2   Proposed Work

2.1   Images Used in Experimental Study

A data set of 792 processed ear images and 492 processed ear images are taken from the database of IIT-Delhi, as shown in Figs. 2 and 3. The processed image is of resolution approx. 50 × 180 pixels as per width x height basis and the raw set of images have resolution of approx. 272 × 204 pixels as per width x height [Abate et al. 2006], [Lowe 2004]. The experiment is performed on 100 images taken from the data set.

2.1.1   Some Raw Data

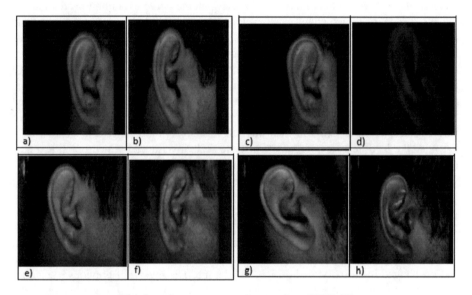

Fig. 2. a–h some raw ear images from IIT-Delhi

2.1.2   Some Processed Data

This is an experimental study performed over the data set. The ear data set has been passed through SIFT algorithm, which results in the extraction of invariant feature point of the ear image. A SIFT feature is a selected image region (also called key-point) by applying the Gaussian filter, in smoothing of the image.

2.2   Human Ear Recognition Experiment

Since there is no easy answer to decide what should be the σ value, the width of the mask to be used, we assumed many values like σ = 1, 2, 3… This is because we do not know at what scale the detail of the image will appear. Next, we combine the series of

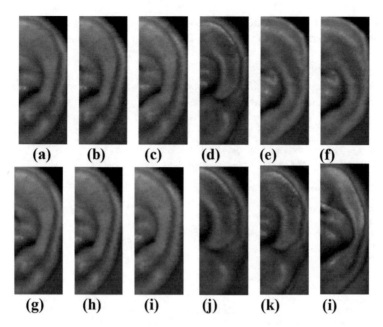

Fig. 3. (a–i) shows processed ear images from the database of IIT-Delhi

edge map which result from different σ value. Here, we are getting zero crossings, that is, the edges where intensity changes. So we scale is the sigma value of Gaussian. Typical value of σ = 1.6 but here the σ is taken as 0.3. We need to apply many scale, we can find the local maxima or minima, which is known as SIFT interest points or KPs (key points). The KPs are extracted by the SIFT detector and their description (location, orientation) are computed by the SIFT descriptor [4, 7]. A point in the scale space that can be treated as interest point is made the basis of SIFT detector [4]. In order to find a point and check whether it is interest point or not, 3 × 3 neighborhood of that point of that scale and also the 3X3 neighborhood of the scale above and below is considered. So if the centered point has LoG maxima or minima of other point, then it can be considered as the SIFT interest point. The descriptor computes the image gradient (a directional change in the image intensity/color). Here we focus on intensity. It is quite common to use the SIFT detector (i.e. computing the KPs without descriptors) independently or with the SIFT descriptor, as this work deals with finding only the key-point at the points where intensity changes. These KPs remain fixed with change in scale (similar images but the size may vary by multiplicative constant). The KPs are then used for matching purpose. KPs between two images are matched by identifying their nearest neighbors. The ratio of closest-distance to second-closest distance is considered. If it is much greater or too low than the decided threshold value (threshold value is 0.8), they are discarded. It rejects maximum of the false matches. The noise due to too short peaks are also rejected. On comparing the absolute value of the DoG scale space with the peak threshold 'tp', the peaks with values lower than the threshold are discarded. Too flat peaks that are generated by edges do not give stable features. Thus, such peaks identified are also discarded. A peak in the DoG scale space fixes two parameters of the key-point i.e. the position and the scale.

2.2.1   Sift Algorithm

This takes as an input an ear image and converts it into PGM, which is a text based image format for grey scale image.

```
[image, location] = sift(Image_input)
//This function reads an image and returns its feature points//
Input:    Image_input: the file name for the image.
Output:    Image_output: the image array in double format
I.        Load image
II.       Color images which will be converted to grayscale.
          image = rgb2gray(image_input);
III.      Size of the image
[rows, cols] = size (image);
IV.       Declare a Threshold value, sigma value and octave
Threshold = 0.8
Sigma = 0.3
Octave = 4
V.        Calculate Difference of Gaussian
For O=1 to Gs.0
DoG=Gs.octave{0}
(: , : ,s+1)-Gs.octave{0}(: ,: , s);    // octave is the set of sample of same image obtain by applying different σ value
and Gs.octave is the number of octave on which difference of Gaussian is applied.
VI.       Finding Feature points
          For O=1:Gs.0
          Sigma=1.6*2^(0-1)*(2^(1/s))^s
          current_DOG=DOG.octave {0}(: ,:,s);
          down_DOG= DOG.octave{0}(: ,: ,s-1);
          up_DOG =DOG.octave{0}(: ,: ,s+1);
VII.      Find Extremum
          IF extremum(1,1)
VIII.     Accurate Localization To Subpixel And Eliminate Some unsuitable Points
extremum= Localize_Eliminate(extremum, up_DOG,down_DOG,current_DOG,Thresh);
IX.       Show The Image With Feature Point
          Imshow(I)
          Plot(extremum)
```
This algorithm extracts the feature points from the image

2.2.2   Matching Algorithm

This function reads two images, finds their SIFT features, and displays lines connecting the matched key-points. It returns the number of matches displayed.

Steps of Matching

1. Before matching two images, the feature descriptors of the two images are obtained. Now for each descriptor in the first image, a match is selected in the second image.
2. For obtaining the match of the first key-point in the first image, it is compare with the other key-point of the second image. This is done for every key-point of the first image.
3. Step 2 is repeated for obtaining the matching key-points between the two images.
4. Finally, two images are appended side by side and a straight line is drawn between match key-point.

2.2.3   Resource: MATLAB (R2010b)
See Fig. 4.

2.3   Matlab Command for Sift

For finding the SIFT points for 002_2.bmp:
[image, descrips, locs] = sift('002_2.bmp');
For drawing the SIFT points for 002_2.bmp:
showkeys(image, locs);
For matching 002_2.bmp with 002_2.bmp
match('002_2.bmp','002_2.bmp');
For matching 002_2.bmp with 002_1.bmp
match('002_2.bmp','002_1.bmp');

3   Results and Discussions

3.1   Output and Discussions

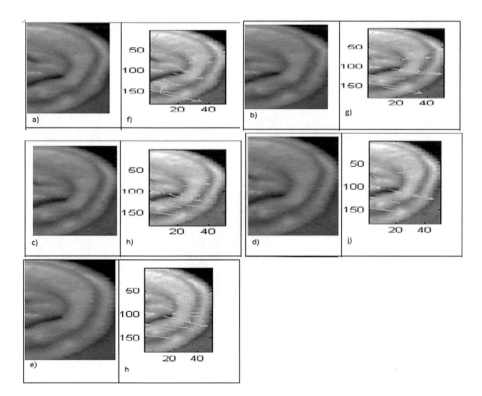

Fig. 4. Image a, b, c, d, and e shows input image and image f, g, h, i, j shows the corresponding output image.

Few images are taken from the data set (discussed in Sect. 2.1) to show the KPs over the ear image.

Image name	Number of SIFT points generated
002_2.bmp	37 keypoints
002_1.bmp	20 keypoints
001_1.bmp	15 keypoints
001_2.bmp	17 keypoints
001_3.bmp	14 keypoints
001_4.bmp	15 keypoints
001_5.bmp	16 keypoints

3.2   Match Point Between Two Different Ear Images After SIFT

To find the matching between two different ear images we follow (Figs. 5 and 6)

1. An array has been declared
2. Take the range of the image, we want find the match point
3. Display the number of match point

```
B []
for i = 1:n
for j = 1:n
M = matching (i,j);
B(i,j) = M;
end;
end
dlmwritre(output,B);
type output;
```

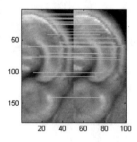

Fig. 5. Image a showing the match point between two similar ear

3.3   Match Points Between 3 × 3 Ear Images

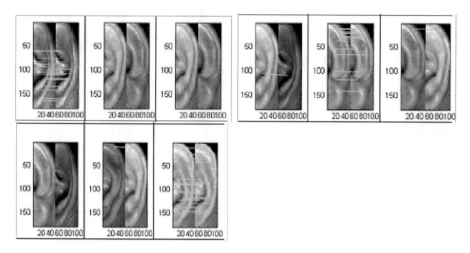

Fig. 6. (Figures are a, b, c, d, e, f, g, h, i) Image a, e, i showing the matching between the similar image and b, c, d, g, h showing matching between different images. Same image has large number of match point and different image have less number of match point.

Original Image	Image with SIFT point	Number of key points found	Original Image	Image with SIFT point	Number of key points found
001_1.bmp		15	001_2.bmp		17
001_3.bmp		14	001_4.bmp		15

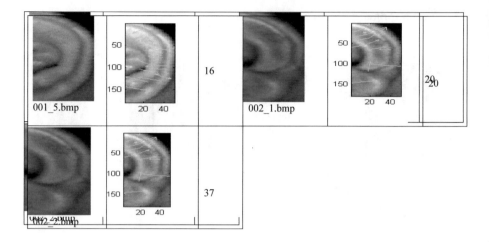

MATCHING

The matching is which is performed between same ear images, gives more match-points [Lowe 2004] (Fig. 7).

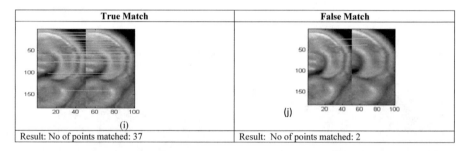

Fig. 7. Image i showing true match and image j showing false match

4 Conclusion

Ear biometric can thus be considered as one of the most reliable and invariant bio-metrics. Based on the stated objective, the system is built using reliable SIFT algorithm, which consequently extracts the features from the image. These features remain invariant to size and rotation of the ear. The DS is passed through the SIFT algorithm which extracts the invariant features. These invariant features are then matched with other feature points of some other images in the DS. The feature over the image is taken by considering a threshold value and intensity of each pixel along with its neighboring pixel are compared with that threshold value and if the pixel's intensity exceeds the threshold value, then it is considered as a feature which is not included in the feature set. Thus, we get several SIFT key-points over the image. After obtaining the feature points, it is necessary to match those features with the features set stored in the database. If the matching is done in between two similar images, then it returns high value of match points, but if the matching is done between two dissimilar images, then it returns either zero match points or negligible match points.

From several experiments conducted, we conclude that the SIFT algorithm gives reliable results for matching. Thus, if any trait (ear) which is unknown to the system wants to enter, then it will find no match to the existing data set of the system, thereby providing an alert to the system. The beauty of this technique is to find and filter out all those points in the image which will remain unaffected under any circumstances, thus contributing to a reliable authentication system.

5 Future Scope

This extraction technique is used to extract key-point, which is change in intensity, from the ear images that are invariant to change in scale. The Key-point extraction method used here requires an σ value, which is a scale factor. This work can be made more efficient in biometric identification by using more efficient value of σ. Further, an approach can be considered by rotating the ear image at different degree of orientation, to show that the image is rotation invariant. Thus, the use of ear biometrics for trustworthy identification system is presented and further development of this technique will constitute a reliable biometric approach.

References

1. Jain, K., Hong, L., Pankanti, S.: Biometric: promising frontiers for emerging identification market. Commun. ACM **2**, 91–98 (2002)
2. Delac, K., Grgic, M.: Electronics in marine. A survey of biometric recognition methods. In: 46th International Symposium on IEEE Conference Publications Proceedings, Elmar, pp. 184–193 (2004)
3. Li, S.Z.: Encyclopedia of Biometrics, 1st edn. Springer, Boston (2009)
4. Bhattacharyya, D., Ranjan, R., Alisherov, F.: Biometric authentication: a review. Int. J. u e Serv. Sci. Technol. **2**(3), 13–28 (2015)
5. Awasthi, R., Ingolikar, R.A.: A study of biometrics security system. Int. J. Innovation Res. Dev. **2**(4), 737–760 (2013)
6. Descartesbiometrics (2018). http://www.descartesbiometrics.com/ergo-app/
7. Oravec, M., Pavlovičová, J., Sopiak, D., Jirka, V., Loderer, M., Lehota, Ľ., Vodička, M., Fačkovec, M., Mihalik, M., Tomík, M., Gerát, J.: Mobile ear recognition application. In: 2016 International Conference on Systems, Signals and Image Processing (IWSSIP), pp. 1–4. IEEE (2016)
8. Masaud, K., Algabary, S., Omar, K., Md. Nordin, J., Abdullah, S.N.H.S.: A review paper on ear biometrics: models, algorithms and methods. Aust. J. Basic Appl. Sci. **7**(1), 411–421 (2013)
9. Izadi, M., Emadi, M.: A review on features extraction of two dimensional ear images and occlusion challenge. Int. J. Adv. Res. Electr. Electron. Instrum. Eng. 6938–6945 (2016)
10. Purkait, R.: Application of external ear in personal identification: a somatoscopic study in families. Ann. Forensic. Res. Anal. **2**(1), 1015 (2015)

11. Iannarelli, A.: Ear İdentification, Forensic İdentification Series. Paramount Publishing Company, Paramount (1989)
12. Choras, M.: Image feature extraction methods for ear biometrics in a survey. In: IEEE 6th International Conference on Computer Information Systems and Industrial Management Applications, CISIM 2007, pp. 261–265 (2007)
13. Lowe, D.G.: Distinctive image features from scale-invariant keypoints. Int. J. Comput. Vis., 1–28 (2004)
14. Abate, A., Nappi, M., Riccio, D., Stefano, R.: Ear recognition by means of rotation invariant descriptor. In: ICPR 2006 Proceedings of the 18th International Conference on Pattern Recognition, vol. 04, pp. 437–440. IEEE Computer Society Washington, DC (2006)
15. Abaza, A., Hebert, C., Harrison, M.A.F.: Fast learning ear detection for real time surveillance. In: IEEE Xplore in Conference: Biometrics: Theory Applications and Systems (BTAS) (2010). https://doi.org/10.1109/btas.2010.5634486
16. Pflug, A., Busch, C.: Ear biometrics: a survey of detection, feature extraction and recognition methods. Biometrics IET **1**(2), 114–129 (2012)
17. Narendira, K., Srinivasan, B.: Ear biometrics in human identification system. Int. J. Inf. Technol. Comput. Sci. **2**, 41–47 (2012). Published Online March 2012 in MECS. http://www.mecs-press.org/, https://doi.org/10.5815/ijitcs.2012.02.06
18. Singh, S., Singla, S.K.: A review on biometrics and ear recognition techniques. Int. J. Adv. Res. Comput. Sci. Softw. Eng. **3**(6), 1624–1630 (2013)
19. Hurley, D.J., Arbab-Zavar, B., Nixon, M.S.: Handbook of Biometrics, pp. 131–150 (2007)
20. Chen, H., Bhanu, B.: Shape model-based 3D ear detection from side face range images. In: Proceedings of IEEE Computer Society Conference on Computer Vision and Pattern Recognition, USA, p. 122 (2005)
21. Jain, K., Ross, A., Prabhakar, S.: An introduction to biometric recognition. IEEE Trans. Circuits Syst. Video Technol. Spec. Issue Image Video Based Biomet. **14**(1), 4–20 (2004)
22. Balakrishanan, G., Umamaheshwari: Human ear biometric authentication system. Int. J. Eng. Sci. Res. Technol., 542–546 (2014)
23. Burge, M., Burger, W., Jain, A.K., Bolle, R., Pankanti, S.: Ear Biometrics in Springer Biometrics: Personal Identification in Networked Society, pp. 273–286 (2013)
24. Jeges, E., Mate, L.: Model based human ear localization and feature extraction in world automation congress. In: 5th International Forum on Multimedia and Image Processing (IFMIP), vol. 1, no. 2, pp. 101–112 (2007)
25. Yuan, L., Mu, Z.: Ear recognition based on Gabor features and KFDA. Sci. World J. **2014**, 12 (2014). Article ID 702076
26. Burge, M., Burger, W.: Ear biometric in computer vision. In: Proceedings of ICPR 2000 IEEE., pp. 822–826 (2000)
27. Davesh, N., Sipi, D.: A survey paper on human identification using ear biometric. Int. J. Innovative Sci. Modern Eng. Blue Eyes Intell. Eng. Sci. Publ. **2**(10), 9–13 (2014)
28. Tiwari, S., Kumar, S., Kumar, S., Sinha, G.R.: Ear recognition for newborns. In: 2015 2nd International Conference on Computing for Sustainable Global Development (INDIACom), pp. 1989–1994. IEEE, March 2015
29. Anwar, A.S., Ghany, K.K.A., Elmahdy, H.: Human ear recognition using geometrical features extraction. Procedia Comput. Sci. **65**, 529–537 (2015)
30. Ghoualmi, L., Chikhi, S., Draa, A.: A SIFT-based feature level fusion of iris and ear biometrics. In: Multimodal Pattern Recognition of Social Signals in Human-Computer-Interaction, vol. 8869 of the series Lecture Notes in Computer Science, pp. 102–112 (2015)
31. Harris, C., Stephens, M.: A combined corner and edge detector. In: Fourth Alvey Vision Conference, Manchester, UK, pp. 147–151 (1988)

32. Moravec, H.: Obstacle avoidance and navigation in the real world by a seeing robot rover. Technical report CMU-RI-TR-3, Carnegie-Mellon University, Robotics Institute (1980)
33. Moreno, B., Sanchez, A.: On the use of outer ear images for personal identification in security applications. In: Proceedings IEEE 33rd Annual International Conference on Security Technology, pp. 469–476 (1999)

Feature Selection and Classification of Big Data Using MapReduce Framework

D. Renuka Devi$^{(\boxtimes)}$ and S. Sasikala

IDE, Department of Computer Science, University of Madras, Chennai, India
renukadevi.research@gmail.com

Abstract. The Feature selection (FS) plays an imperative role in Machine Learning (ML) but it is really demanding when we apply feature selection to voluminous data. The conventional FS methods are not competent in handling big datasets. This leads to the need of a technology that processes the data in parallel. MapReduce is a new programming framework used for processing massive data by using the "divide and conquer" approach. In this paper, a novel parallel BAT algorithm is proposed for feature selection of big datasets and finally classification is applied to the set of known classifiers. The proposed parallel FS technique is highly scalable for big datasets. The experimental results have shown improved efficacy of the proposed algorithm in terms of the accuracy and comparatively lesser execution time when the number of parallel nodes is increased.

Keywords: Feature selection · Machine Learning · Bigdata · Parallel · MapReduce

1 Introduction

With the advancement of technological developments, data are gathered from different sources such as websites, sensor, logs, and social media and so on which are collectively named big data sets [11]. At present, the evolving data brings along a big challenge to the conventional data mining processing methods [13] and thereby comes the need of a novel framework to store and process the large datasets.

The characteristics of big data are identified as (i) Volume, (ii) Variety, (iii) Velocity, (iv) Veracity, and (v) Value [1]. Volume is measured in terms of peta bytes or hexa bytes which is the accumulation of data from different sources. Variety means heterogeneity of data that includes text, image, audio, and video. Velocity refers to the rate at which the data are captured and veracity is the uncertainty in data collected.

The intention of FS is to build a simpler classification model with minimal number of features which should be free from redundancy and insignificant information. It is the significant preprocessing strategies in Machine Learning (ML) and Data mining that the advent of a big data era has elevated the challenges to higher bounds in FS [3].

Generally, FS methods are divided into the filter, wrapper, and embedded [4]. The conventional FS methods are replaced by the innovative methods where big datasets are processed in parallel [5]. This reduces the computational complexity with added

© Springer Nature Switzerland AG 2020
A. P. Pandian et al. (Eds.): ICICCS 2019, AISC 1039, pp. 666–673, 2020.
https://doi.org/10.1007/978-3-030-30465-2_73

accuracy. MapReduce is the programming model, where the input datasets are split into a number of smaller sets and distributed, processed in parallel.

In this paper, we propose a novel scalable parallel BAT algorithm for FS. The input features are divided into smaller chunks, and FS is applied in parallel. The features are combined to form a minimal relevant feature set for further classification by a set of classifiers. In this work, Feature Selection is applied to varied number of parallel nodes for performance comparison. The experimental results have shown higher accuracy and lesser computation time with the increase in parallel nodes, thereby proving the efficacy of the parallel processing.

2 Review Study

Reggiani et al. [12] proposed a MapReduce based FS for higher dimensional data. In this work, iterative MapReduce based, Minimum Redundancy Maximum Relevance algorithm is applied for FS. The scalability is experimented with a number of nodes 1, 2, 5, and 10. Judy et al. [5] suggested a FS model for big data set classification. The evolutionary MapReduce FS technique is used. Fuzzy Cognitive Maps (FCM) is being used for selecting the significant features parallel. The results shown the supremacy of this method in comparison with other FS.

Palma-Mendoza [7] proposed a distributed ReleifF-based feature selection of big datasets, based on spark computing model. The results of distributed computing are compared with the conventional FS method, which have shown the scalability of the proposed algorithm. Tsamardinos et al. [6] proposed a Parallel, Forward–Backward with Pruning (PFBP) algorithm. This model is evaluated on simulated datasets for scalability. The proposed algorithm is proven with improved scalability and accuracy than the other state-of-the-art FS methods.

Khalil et al. [2] recommended Distributed Whale Optimization Algorithm based on MapReduce for large scale optimization problems. Kečo et al. [8] suggested a parallel genetic algorithm for FS. It is experimented on GEMS data base. The meta heuristic algorithm is applied on MapReduce paradigm, followed by classification with a set of classifier methods ANNs (artificial neural networks) and SVMs (support vector machines).

3 Methodology

The proposed model is shown in the Fig. 1.

The methodology of the proposed work is divided into following phases

1. Preprocessing
2. Feature selection – MapReduce based FS
3. Classification.

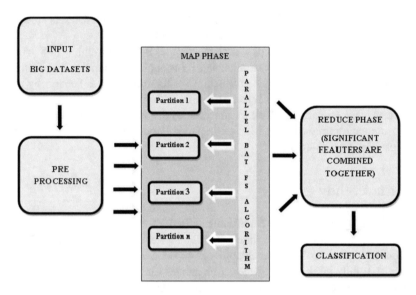

Fig. 1. Proposed model

3.1   Preprocessing

The data sets are to be preprocessed before FS and classification. The min-max normalization method [14] used is given in the Eq. (1). This method converts the given datasets to a value that vary between 0 and 1. The v' is the normalized value derived from feature v.

$$v' = \frac{v - min_A}{max_A - min_A} \left(new\_max_A - new\_min_A\right) + new\_min_A \tag{1}$$

3.2   Feature Selection – MapReduce Based FS

The big datasets passed as inputs are fragmented into smaller number of sets. MapReduce, a distributed computing model can efficiently tackle big data issues. In this work, a parallel FS-MR model is proposed. Generally, the big datasets are partitioned into a number of data sets of smaller in size where each computational node performs feature selection. The selected feature variables are then pooled into one feature vector in Reduce job. This parallel FS method provides scalability when working with outsized data sets.

A MapReduce job includes Map step and Reduce step wherein both steps include the transformation of input data element lists to output data element lists. Both the steps are executed consequently, thus reduce step always follow the completion of the map step. The algorithms that use MapReduce, scale up with the cluster size, and Execution Time (ET) is reduced by escalating the number of nodes. The design of the algorithm and the data layout are significant factors that influence the ET.

For every partition the parallel accelerated BAT algorithm is applied. BAT algorithm (BA) is the meta- heuristics optimization algorithm, which is applied to many problems. Profoundly, this algorithm stands on the echolocation nature of bats in finding its prey [15].

The standard BA involves,

1. Best features are found by the echolocation is alike to find the food or prey.
2. Every bat has velocity V_i at the position X_i with a frequency freq.
3. The loudness ranges from A_0 to a minimum constant value A_{min}.

In general, the frequency freq in a range $[freq_{min}, freq_{max}]$ relate to wavelengths $[\lambda_{min}, \lambda_{max}]$. Define the rules how feature position fp_i and velocities ve_i in a d-dimensional search space are updated at time step t are given by the following equations.

$$freq_i = freq_{min} + (freq_{max} - freq_{min})\beta \tag{2}$$

$$ve_i^t = ve_i^{t-1} + (fp_i^t - fp_*)freq_i \tag{3}$$

$$fp_i^t = fp_i^{t-1} + ve_i^t \tag{4}$$

where $\beta \in [0, 1]$ is a random vector drawn from a uniform distribution.

3.3   Classification

The significant features selected are finally combined for further classification phase. The classification algorithm is tested on training datasets, subsequently the test data is applied to classify the given datasets into positive or negative class. The features are classified by the set of classifiers SVM, Naive Bayes (NB), and Hoeffeding Tree (HD).

3.3.1   SVM

SVM is used for both classification and regression. SVM is based on the concept of decision boundaries that are defined using decision planes. A decision plane (hyperplane) is used to sepraret the class members belongs to the different class labels. The classification is perfomed by finding the hyperplane that clearly segeregates the two classes with the help of support vectors. SVMs train the classifier function using pre-labeled data using a class of supervised machine learning algorithms [9].

3.3.2   NB

Fundamentally, a NB classifier is a basic classifier established on conditional probability. The classifier that applies the Bayes' theorem with naïve i.e., strong independence assumptions between features. Based on NB, Gaussian Naive Bayes is used for classification based on the binomial (normal) distribution of data.

The Bayes Theorem is [10]:

$$P(H \mid X) = P(X \mid H)\,P(H)\,P(X) \tag{5}$$

Where,

H - Hypotheis
X - Data Set belongs the class C
P(H|X) - Probability of H on X
P(X|H) - Probability of X on H.

3.3.3 HD

The Hoeffding tree learners incrementally as the features arrive in order. The advantage of incremental approach is to update the classification model without recreating the model once again. This approach is very well suited for data streams. It incrementally constructs a decision tree based on the theoretical guarantees of the Hoeffding bound (or additive Chernoff bound) and an analysis is done on the tree. Hoeffding bounds are validated for a higher level of confidence is given in the Eq. (6).

r = real-valued random variable, with range R
n = number of independent observations have been made
$\bar{r}$ = mean value computed from n independent observations

Hoeffding bound states that with probability $1 - \delta$, the true mean of the variable is at least $\bar{r} - \epsilon$. And ϵ is given by the following formula:

$$\epsilon = \sqrt{R2ln(1/\delta)/2n} \tag{6}$$

4 Experimental Results

The proposed methodology is implemented with Madelon dataset from the UCI machine learning repository. The data set characteristics are shown in Table 1.

Table 1. Data set

Dataset	No of Instances	No of Attributes
Madelon	4400	500

The performance results are measured in terms of Accuracy, Precision, Recall, F-measure, and Processing time. The computational results are listed in the Table 2.

Table 2. Performance results

FS With Classifier/ number of nodes		Precision	Recall	F-measure	Accuracy	Processing Time (seconds)
NB	8	80.76	80.76	80.76	80	21.02
	4	80.36	80.16	80.7	80	27.02
	2	80	80.6	80	79.8	33.02
SVM	8	80.76	100	89.36	90.0	27.89
	4	80.76	100	89	90.0	33.89
	2	80.6	100	89.36	90.0	38.89
HD	8	96.15	83.33	89.28	88.8	12.98
	4	96	83	89.8	88.0	18.98
	2	96.5	83.36	89.18	88.1	22.98

With the increase in the number of nodes, there is a considerable decrease in the execution time is represented in the graph Fig. 2.

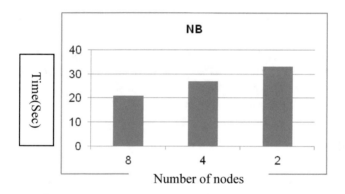

Fig. 2. FS with NB vs number of nodes in MapReduce framework

The Parallel FS with SVM classifier is represented in Fig. 3

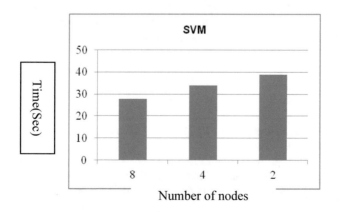

Fig. 3. FS with SVM vs number of nodes in MapReduce framework

The Parallel FS with HD classifier is represented in Fig. 4.

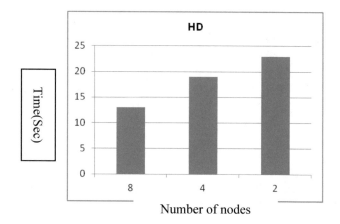

Fig. 4. FS with HD vs number of nodes in MapReduce framework

With inferences from the above graph, the scalability is optimized with an increase in the number of nodes. The implementation work shows improved efficacy in scalability and accuracy. The classification is accomplished through the classifiers NB, SVM, and HD, with the accuracy percentage of 80, 90, 88.8 respectively.

5 Conclusion

Distributed computing is indeed necessary in the case of handling data which are escalating per second across the globe. Thus, these open issues lead to the parallel programming development, wherein the entire processing is carried out by splitting and executing in parallel across the number of nodes. In this work, we recommended a framework for optimal FS of immense datasets using scalable MapReduce based BAT FS algorithm followed by classification. This methodology is extremely adaptable for large dimensional data and leverages the effectiveness of parallel algorithms. The experimental outcome has revealed the improved efficacy in terms of scalability and accuracy.

References

1. Gill, S.S., Buyya, R.: Bio-inspired algorithms for big data analytics: a survey. Taxon. Open Chall., 1–17 (2019)
2. Khalil, Y., Alshayeji, M., Ahmad, I.: Distributed whale optimization algorithm based on MapReduce. Concurr. Comput.: Pract. Exp. **31**(1), e4872 (2019)
3. Li, J., et al.: Feature selection: a data perspective. ACM Comput. Surv. (CSUR) **50**(6), 94 (2018)

4. Rodriguez-Galiano, V.F., et al.: Feature selection approaches for predictive modelling of groundwater nitrate pollution: An evaluation of filters, embedded and wrapper methods. Sci. Total Environ. **624**, 661–672 (2018)

5. Judy, M.V., Soman, G.: Parallel fuzzy cognitive map using evolutionary feature reduction for big data classification problem. In: Annual Convention of the Computer Society of India. Springer, Singapore (2018)

6. Tsamardinos, I., et al.: A greedy feature selection algorithm for Big Data of high dimensionality. Mach. Learn. **108**(2), 149–202 (2019)

7. Palma-Mendoza, R.-J., Rodriguez, D., De-Marcos, L.: Distributed ReliefF-based feature selection in Spark. Knowl. Inform. Syst. **57**, 1–20 (2018)

8. Kečo, D., Subasi, A., Kevric, J.: Cloud computing-based parallel genetic algorithm for gene selection in cancer classification. Neural Comput. Appl. **30**(5), 1601–1610 (2018)

9. Ghaddar, B., Naoum-Sawaya, J.: High dimensional data classification and feature selection using support vector machines. Eur. J. Oper. Res. **265**(3), 993–1004 (2018)

10. Bista, S., Chitrakar, R.: DDoS attack detection using heuristics clustering algorithm and naïve bayes classification (2018)

11. Sasikala, S., Renuka Devi, D.: A review of traditional and swarm search based feature selection algorithms for handling data stream classification. In: Third International Conference on Sensing, Signal Processing and Security (ICSSS). IEEE (2017)

12. Reggiani, C., Le Borgne, Y.A., Bontempi, G.: Feature selection in high-dimensional dataset using MapReduce. In: Benelux Conference on Artificial Intelligence. Springer, Cham (2017)

13. Wu, X., Zhu, X., Wu, G.-Q., Ding, W.: Data mining with big data. IEEE Trans. Knowl. Data Eng. **26**(1), 97–107 (2014)

14. Liu, Z.: A method of SVM with normalization in intrusion detection. Proc. Environ. Sci. **11**, 256–262 (2011)

15. Yang., X.-S.: A new metaheuristic bat-inspired algorithm. In: Nature inspired Cooperative Strategies for Optimization (NICSO 2010), pp. 65–74. Springer, Heidelberg (2010)

Apprenticeship Learning Based Load Balancing Technique for Cloud Environment

S. P. Vatsalya[1]($\boxtimes$), N. Vidhya Sree[1]($\boxtimes$), C. M. Chethan Malode[2]($\boxtimes$),
and K. Bhargavi[3]($\boxtimes$)

[1] Department of Electrical and Electronics Engineering,
Siddaganga Institute of Technology, Tumakuru, Karnataka, India
vatsalyavidc@gmail.com, vidhyasreen@gmail.com
[2] Department of Information Science and Engineering,
Siddaganga Institute of Technology, Tumakuru, Karnataka, India
malodechethan@gmail.com
[3] Department of Computer Science and Engineering,
Siddaganga Institute of Technology, Tumakuru, Karnataka, India
bhargavi.tumkur@gmail.com

Abstract. Cloud Computing is accessing and handling data and documents from the internet rather than from any individual computer hard drive. The issues faced by cloud computing are security, privacy, vendor lock-in, server downtime, network connectivity, dependency, vulnerability to attacks, load balancing, etc. Load balancing in cloud computing is one of the important issues as huge amount of load need to be efficiently distributed among the servers. The existing approaches to address load balancing issue are throttled technology, active clustering, central policy for virtual machine, round robin technology, max-min min-min, fuzzy monitoring, honeybee foraging behavior, reinforcement learning, etc. The primary drawbacks of above-mentioned approaches towards load balancing are lowered throughput, high migration rate, overloading and under-loading of resources. This paper proposes a novel architecture which applies apprenticeship learning for load balancing in the cloud. Its performance is found to be good with respect to parameters like response time, accuracy, learning rate and speed.

Keywords: Cloud-computing · Load balancing · Apprenticeship learning · Learning rate · Accuracy · Response time

1 Introduction

Cloud computing handles data and documents from the internet rather than an individual computer's hard drive. The main services of cloud computing are IAAS (Infrastructure as a service), SAAS (Software as a service) and PAAS (Platform as a service) [1]. Several applications are launched in the cloud like Apple icloud, CloudSponge, ShareThis, Dropbox, Google calendar, Google drive, War Light, etc.

Cloud computing comes up with many advantages like more economical, always accessible, flexibility, scalability, all over functioning, automated updates on software, enhanced collaboration, control on documents, easily manageable, etc. But there are

© Springer Nature Switzerland AG 2020
A. P. Pandian et al. (Eds.): ICICCS 2019, AISC 1039, pp. 674–681, 2020.
https://doi.org/10.1007/978-3-030-30465-2_74

certain issues faced by cloud computing which includes security, privacy, vendor lock-in server, downtime, network connectivity dependency, vulnerability to attacks, load balancing, scheduling, resource management, etc. [2].

Since the cloud has many users, the load has to be properly distributed among the servers in a parallel manner else it affects the performance of cloud computing. There are many approaches to overcome this issue such as throttled technology, active clustering, central policy for a virtual machine, round robin technology, max-min and min-min method, fuzzy monitoring, honeybee foraging behavior, reinforcement learning, etc. [3].

The main drawback of the above-mentioned approaches is there is a chance of improper allocation of workload due to the random distribution of work to any server without knowing its internal features (storage, speed, response time, memory, bandwidth, etc.). There by the rate of under loading or overloading the machine is high.

So in this paper, apprenticeship learning technique is used which learns by monitoring the specialist hence the work of efficient load balancing terminates in less number of trails and outcome will be close to the expert performance [4].

2   Organisation of the Paper

The rest of the paper is organized as follows: Sect. 3 contains some of the related work, Sect. 4 describes some of the important definitions, Sect. 5 take up proposed work and explains the architecture in detail, Sect. 6 describes the results obtained and finally Sect. 7 gives a conclusion.

3   Related Work

Load balancing in cloud computing is discussed in [5]. In this paper, the load is randomly distributed among virtual machines hence there is a chance of misallocation of the load and according to the requirement of load virtual machine and physical machine cannot be remapped. Hence it does not effectively distribute the load to all machines for various inputs.

Cloud computing resources provisioning research based on reinforcement learning is discussed in [6]. Here, resource management is achieved by mainly concentrating on 3 types of imbalances i.e., imbalance in the needs of applications, imbalance in the application time, imbalance in the distribution of applications. But they haven't discussed the internal features of the virtual machine and physical machine like memory, speed, accuracy, response time, by which resources will be wasted to some extent.

Dynamic selection of virtual machine in cloud data centres using reinforcement learning approach is discussed in [7]. In this paper, they have discussed only the selection of virtual machine whereas the selection of physical machines is equally necessary for load balancing and the technology implemented here for load balancing is reinforcement learning where it takes more time to execute.

Inverse reinforcement learning is discussed in [8]. In this paper, they have made a survey regarding the basics of inverse reinforcement learning but they have not explained the application of it, as a result it lacks practical application part.

4   Definitions

In this section technical terminologies used in the paper are defined.

Virtual Machine (VM): It is a device composed of user applications and is supported by a physical server which runs an operating system.

Physical Machine (PM): It is a device consisting of several virtual machines on a single hardware device (computer) which permits to access many operating systems and assign computing resources to VM as required.

Apprenticeship Learning Agent (ALA): It is an agent which consists of a set of best policies which are acquired by observing specialist.

Quality of Service (QoS): It is the measure of the capacity of the network to give services to users and applications with respect to QoS parameters like throughput, delay, latency, jitter, etc.

5   Proposed Work

The architecture for apprenticeship learning based load balancing technique for cloud environment is shown in Fig. 1. The proposed architecture is divided into 2 modules, they are apprenticeship learning agent for physical machine load balancing (ALA-PLB) and apprenticeship learning agent for virtual machine load balancing (ALA-VLB).

ALA-PLB
The prime function of ALA-PLB is to accept the inputs and select the suitable PM according to the requirements of input. The end output of this module is, a task done with maximum QoS. It consists of a set of predefined best policies initialized as π. The quality function (Q) with the state (S) and action (a) is selected with a suitable policy which meets the requirements of the task given by the client. The PM is selected based on the ratio of accuracy of the current required PM (ACC(c)), accuracy of utilized PM (ACC(u)) and response time (RT) such that the ratio of the current required PM should be greater than the utilized PM as shown in Eq. 1.

$$\left[\frac{\text{ACC(c)}}{\text{RT}}\right]PM > \left[\frac{\text{ACC(u)}}{\text{RT}}\right]PM \tag{1}$$

ALA-VLB
The ALA-VLB accepts the inputs from ALA-PLB and select the appropriate VM according to the requirements of the PM and completes the task thereby returns the output to corresponding ALA-PLB. The working of ALA-VLB is given in the algorithm of ALA-VLB. The VM which is selected based on the ratio of the speed of current required VM (SPEED(c)), speed of utilized VM (SPEED(u)) and memory (MEM) such that the ratio of the current required VM should be greater than the utilized VM as shown in Eq. 2.

$$\left[\frac{\text{SPEED(c)}}{\text{MEM}}\right]\text{VM} > \left[\frac{\text{SPEED(u)}}{\text{MEM}}\right]\text{VM} \tag{2}$$

Algorithm for ALA-PLB:

Input: Set of tasks $T = \{T_1, T_2, T_3 \ldots T_n \}$
Output: Task done with maximum QoS (T)
Begin

    Initialize Q (S, a, PM) according to the task given by a set of policies $\pi = \{\pi_1, \pi_2, \ldots \ldots \pi_n\}$

    For every task given **do**

        For every state of PM **do**

            Choose PM based on QoS

            $\left[\frac{\text{ACC(c)}}{\text{RT}}\right]PM > \left[\frac{\text{ACC(u)}}{\text{RT}}\right]PM$

            Execute action a_t of a state S_t by a selected PM

            Output is compared with the best policy

            Observe S_{t+1}

            Reward r_t is collected

            $Q (S_t, a_t) = Q (S_t, a_t) + [\alpha\, r_t + \beta \max Q (S_{t+1}, a_{t+1}) - Q(S_t)]$

        End for

    End for

    The given task is done with maximum QoS (T) = {QoS (t_1), QoS (t_2), QoS (t_3) .... QoS (t_n)}

End

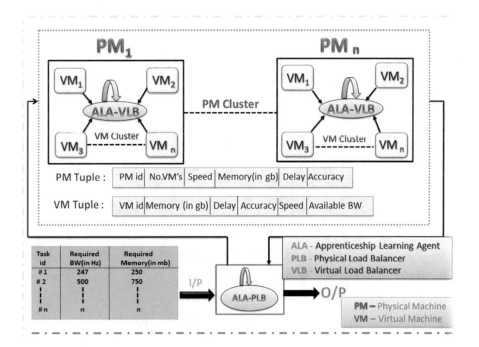

Fig. 1. Architecture for apprenticeship learning based load balancing technique for cloud environment.

Algorithm for ALA-VLB:

Input: Set of tasks T = {T₁,T₂,T₃... ....Tₙ }
Output: Task done with maximum QoS (T)
Begin
    Initialize
    Q (S, a, VM) according to task given by set of policies $\pi = \{\pi_1,\pi_2,\quad \pi_n\}$
    Q (S,a,VM)={Q (S₁,a₁,VM₁),Q (S₂,a₂,VM₂),...... Q (Sₙ , aₙ, VMₙ)}
    For every task given **do**
        For every state of VM **do**
            Choose VM based on QoS
$$\left[\frac{SPEED(c)}{MEM}\right] VM > \left[\frac{SPEED(u)}{MEM}\right] VM$$
            Execute action aₜ of a state Sₜ by a selected VM Output is compared with the best policy
            Observe sₜ₊₁
            Reward rₜ is collected
            Q (Sₜ ,aₜ) = Q (Sₜ,aₜ) + [α rₜ+β max Q (Sₜ₊₁,aₜ₊₁) - Q (Sₜ)]
        End for
    End for
    The given task is done with maximum QoS (T)={QoS (t₁), QoS (t₂), QoS (t₃) ....QoS (tₙ)}
End

---

6 Result

In this section, the performance of the RL (Reinforcement Learning) and AL (Apprenticeship Learning) is evaluated with respect to various performance parameters like accuracy, efficiency, and time delay.

6.1 Trend Towards RL and AL Over a Few Years

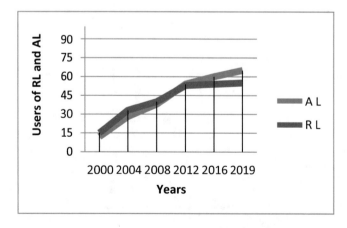

Fig. 2. Years versus users of RL and AL.

The graph of years versus users of AL and RL is shown in Fig. 2. It is observed from the graph that in the earlier 2000, the number of users of RL and AL was minimal, at the time of 2004–2012 number of users started to rise and stayed constant. Overall in the year of 2019 number of users for reinforcement learning reached up to 70% of total users but the users for apprenticeship learning technique are still increasing and have reached 90% of total users due to the improved version which provides higher accuracy, low response time and less erroneous.

6.2   Learning Rate

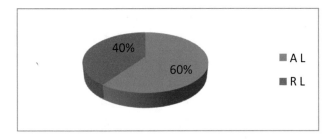

Fig. 3.   Learning rate of AL and RL

The graph of learning rate of apprenticeship learning and reinforcement learning is shown in Fig. 3. The speed of learning is more for apprenticeship learning technique compared to reinforcement learning because the apprenticeship learning technique performs the given task by observing the expert while reinforcement learning performs a set of tasks without any specialist.

6.3   Accuracy

The graph of RL and AL versus accuracy is shown in Fig. 4. Since in apprenticeship learning technique, work ends in less number of iterations, as a result, its outcome will be close to the expert performance, its accuracy will be more compared to reinforcement learning.

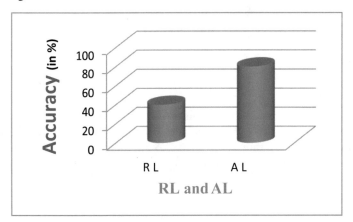

Fig. 4.   The graph of AL and RL versus accuracy

6.4   Response Time

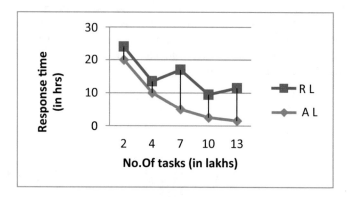

Fig. 5. The graph of number of tasks versus response time

The graph of number of tasks versus response time is shown in Fig. 5. Due to the non-erroneous nature of apprenticeship learning the reward gathering will be fast and accurate so response time will be less as compared to reinforcement learning where an error will be more because of the dynamic performance of tasks.

6.5   Error Identification

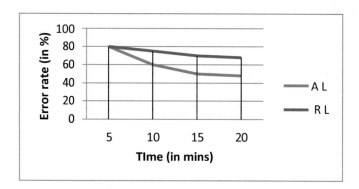

Fig. 6. The graph of time versus error rate

The graph of time versus error rate is shown in Fig. 6. Since apprenticeship learning technique conducts the duty with best policies which are obtained by monitoring expert i.e. minimum exploration time as compared to reinforcement learning, as it performs the given task abruptly without guidance from an expert.

7   Conclusion

The paper presents a unique architecture using apprenticeship learning for load balancing in cloud environment. The performance of the proposed architecture is found to be good with respect to parameters like accuracy, learning rate, and response time. In future, the proposed architecture can be improvised to do efficient scheduling and resource management operations in cloud.

References

1. Kaur, A., Luthra, P.: A review on load balancing in cloud environment. Int. J. Comput. Technol. **17**, 7120–7125 (2018)
2. Rahman, A., Sarfraz, S., Shoaib, U., Abbas, G., Sattar, M.A.: Cloud based E-learning, security threats and security measures. Indian J. Sci. Technol. **9** (2016)
3. Mishra, N.K., Mishra, N.: Load balancing techniques: need, objectives and major challenges in cloud computing - a systematic review. Int. J. Comput. Appl. **131**, 148–154 (2015)
4. Zhifei, S., Joo, E.M.: A review of inverse reinforcement learning theory and recent advances. In: Congress on Evolutionary Computation. IEEE (2012)
5. Mesbahi, M., Rahmani, A.M.: Load balancing in cloud computing: a state of the art survey. Int. J. Mod. Educ. Comput. Sci. **8**, 64 (2016)
6. Peng, Z., Cui, D., Zuo, J., Lin, W.: Research on cloud computing resources provisioning based on reinforecement learning. Math. Probl. Eng. **2015**, 12 (2015)
7. Duggan, M., Flesk, K., Duggan, J., Howley, E., Barrett, E.: A reinforcement learning approach for dynamic selection of virtual machines in cloud data centres. In: Sixth International Conference on Innovative Computing Technology. IEEE (2016)
8. Abbeel, P., Ng, A.Y.: Algorithms for inverse reinforcement learning. In: International Conference on Machine Learning, vol. 1 (2000)

Distance Measurement for Self-driving Vehicles Using Data Fusion and Machine Learning

S. Sreekuttan$^{(\boxtimes)}$ and S. Adarsh

Department of Electronics and Communication Engineering, Amrita School of
Engineering, Coimbatore, Amrita Vishwa Vidyapeetham, Coimbatore, India
ssreekuttan33@gmail.com, s_adarsh@cb.amrita.edu

Abstract. For certain mobile robots and self driving vehicles, accurate measurement of distance ahead of them is indispensable. Several sensors are utilized to achieve this. This work is on fusing the obtained data from two sensors namely Leddar M-16 and RPLidar 360. A comparison of the accuracy of the distance measured from the vehicle to obstacle using Leddar M-16 and RPLidar 360 is being done. Also these results are being compared to the improved accuracy of the resultant from both sensors after the data is fused together to produce a different set of values. Analysis on the data is done using a tool named Weka. Test bed and experiments were designed for collection of data. A machine learning technique, linear regression is used for improving accuracy of the measurement.

Keywords: Machine learning · Multi sensor data fusion · Linear regression · Self driving vehicle · Lidar

1 Introduction

Driving is more of a behavior than a rule based operation as far as normal human drivers are concerned. The drivers knowingly or unknowingly keep track of the distance ahead of him. He monitors the distance between vehicle and the obstacles on his path using his eyes. Driverless vehicles and mobile robots are no different in the aspect of distance measurement. These vehicles also need constant monitoring of the distance to the obstacle ahead of it. Sensors such as cameras, RADARs and LIDARs are most commonly used to achieve this. In this work, an RPLidar 360 and a Leddar M-16 are used to serve the purpose. An experiment setup is designed for the purpose of data collection from these sensors. The accuracy of measurements by these sensors is calculated by actual comparison with ground truth data from the experimental setup. The process of fusing data from these sensors is used to improve the accuracy of distance measured. The core idea is to identify the errors over a range of distance and compensate it using corrective measures with the help of machine learning techniques to achieve better accuracy.

© Springer Nature Switzerland AG 2020
A. P. Pandian et al. (Eds.): ICICCS 2019, AISC 1039, pp. 682–689, 2020.
https://doi.org/10.1007/978-3-030-30465-2_75

2 Related Works

Some works related to the sensors namely Leddar M-16 and RPLidar 360 are discussed in this section. The work by Adarsh et al. [10] and Adhvaryu et al. [9] is on the usage of fuzzy methods used to improve the performance of the sensors like RPLidar and ultrasonic sensors. The work by Mimeault et al. [3] briefs the usability of Leddar Tech sensors for automobiles which is also indicative of the usability of the Leddar sensors for mobile robots. Almahzumy et al. [6] Discusses the applicability of RPLidar sensor for mobile robots. Kim et al. [7] discusses the use of RPLidar as a sensor of unmanned vehicle where it is used in wall avoidance which is more like a path finding autonomous vehicle. Fuchun et al. [1] discusses in detail about the use of lidar sensor for localisation for a mobile robot. Thereby it is clear that the use of RPLidar 360 and Leddar M-16 is a practicable option for both mobile robots and some of the specific purpose self driving vehicles.

3 Experiment Setup

Fig. 1. Experimental setup with planar surface obstacle

The test bed for the experiment was designed to perform the data collection using both Leddar M-16 and RPLidar 360. Initially the sensors were fixed onto an electric buggy as shown in Figs. 1 and 2. Then the electric buggy is parked at a place where the ground surface is plane and also there is no obstacle ahead of the vehicle for at least twenty metre.

- **Test Bed:** The test bed is arranged in such a way that, the buggy is placed on floor. From the buggy, on the floor a perpendicular line is marked. That line is marked with distances with an interval of 10 cm. The range of the distance marked is from 0 cm to 2000 cm. Obstacle is a plane surface cardboard held vertical as seen in Fig. 2. Electric buggy and the obstacle are placed face to face. Buggy is fixed to a spot whereas the obstacle is easily movable.

Fig. 2. Test bed: buggy with sensors facing the obstacle

- **Sensors:** The sensors used are RPLidar 360 and Leddar M-16. Both of the sensors are mounted on the electric buggy as shown in Fig. 4. Leddar M-16 is powered visa adapter from a UPS connected in the vehicle. RPLidar 360 is powered by the data acquisition system that is the laptop.
- **Acquisition of data:** Raw data from RPLidar 360 is collected using Python 3.6.7 code in Ubuntu OS. The data from Leddar M-16 is collected using Leddar M-16 configurator in OS Windows 10.
- **Ground truth data:** The data that is used as the reference is the markings on the ground which was made with the help of measuring tape and drawing tools.
- **The data collection procedure:** The obstacle is initially placed at 10 cm distance marking on the ground. It is 10 cm away from the vehicle. The distance measured using Leddar M-16 and RPLidar 360 is recorded manually. Then the obstacle is moved to 20 cm marking and the process is repeated. After each 10 cm marking, the measurement is made and values are recorded. This process repeats till 20 m. The whole process starting from 10 cm to 20 m is repeated thrice.
- **Averaging:** The obtained measurements are three repetitions of same readings from Leddar M-16 and RPLidar 360. So for each reference value, the corresponding measurements are averaged out and this averaged value set and the ground truth data together is used for analysis. The ground truth data is mentioned as Actual distance.
- **File Format:** The file format in which the data is saved is .csv

The obtained .csv file is used for data fusion and analysis. From analysis, new equations are obtained which is further tested with the test data.

4   Data Fusion and Analysis

The obtained .csv file is used for analysis of the data so as to measure accuracy, do the fusing process of data and compare the accuracy resulted in each case. When it comes to cases, three cases are considered here. First case is measurement using Leddar M-16 only. Second case is measurement using RPLidar 360 only and the third and final case considered is that of the fusion of both RPLidar 360 and Leddar M-16 data.

Linear regression is applied to the RPLidar 360 set of data, Leddar M-16 set of data and the combined data. The .csv file was processed with linear regression using the software WEKA 3.8. Details related to the linear regression applied is shown in the following figure (Fig. 3).

```
=== Run information ===

Scheme:       weka.classifiers.functions.LinearRegression -S 0 -R 1.0E-8 -num-decimal-places 4
Relation:     LidarLeddar1
Instances:    60
Attributes:   3
              Actual Distance
              Leddar
              Lidar
Test mode:    10-fold cross-validation

=== Classifier model (full training set) ===


Linear Regression Model

Actual Distance =

      0.5818 * Leddar +
      0.4316 * Lidar +
   -203.9325

Time taken to build model: 0.02 seconds
```

Fig. 3. Details of the regression

For the regression, 10 fold cross-validation was used. The training and testing data were split as 66% of data as training data and the remaining as testing data. The following figure shows a snapshot of data before and after regression and the corresponding errors.

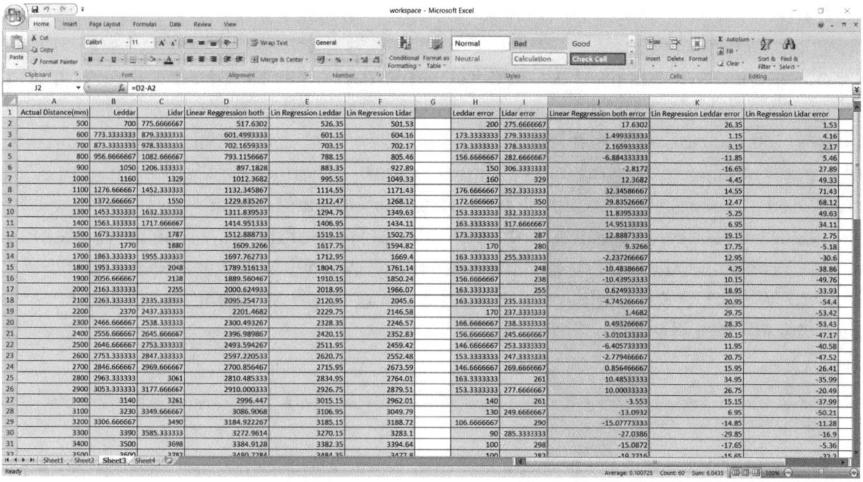

Actual Distance(mm)	Leddar	Lidar	Linear Reggression both	Lin Regression Leddar	Lin Regression Lidar	Leddar error	Lidar error	Linear Reggression both error	Lin Regression Leddar error	Lin Regression Lidar error
500	700	775.6666667	517.6302	526.35	501.53	200	275.6666667	17.6302	26.35	1.53
600	773.3333333	879.3333333	601.4993333	601.15	604.16	173.3333333	279.3333333	1.499333333	1.15	4.16
700	873.3333333	978.3333333	702.1659333	703.15	702.17	173.3333333	278.3333333	2.165933333	3.15	2.17
800	956.6666667	1082.666667	793.1156667	788.15	805.46	156.6666667	282.6666667	-6.884333333	-11.85	5.46
900	1050	1206.333333	897.1828	883.35	927.89	150	306.3333333	-2.8172	-16.65	27.89
1000	1160	1329	1012.3682	995.55	1049.33	160	329	12.3682	-4.45	49.33
1100	1276.666667	1452.333333	1132.345867	1114.55	1171.43	176.6666667	352.3333333	32.34586667	14.55	71.43
1200	1372.666667	1550	1229.835267	1212.47	1268.12	172.6666667	350	29.83526667	12.47	68.12
1300	1453.333333	1632.333333	1311.839533	1294.75	1349.63	153.3333333	332.3333333	11.83953333	-5.25	49.63
1400	1563.333333	1717.666667	1414.951333	1406.95	1434.11	163.3333333	317.6666667	14.95133333	6.95	34.11
1500	1673.333333	1787	1512.888733	1519.15	1502.75	173.3333333	287	12.88873333	19.15	2.75
1600	1770	1880	1609.3266	1617.75	1594.82	170	280	9.3266	17.75	-5.18
1700	1863.333333	1955.333333	1697.762733	1712.95	1669.4	163.3333333	255.3333333	-2.237266667	12.95	-30.6
1800	1953.333333	2048	1789.516133	1804.75	1761.14	153.3333333	248	-10.48386667	4.75	-38.86
1900	2056.666667	2138	1889.560467	1910.15	1850.24	156.6666667	238	-10.43953333	10.15	-49.76
2000	2163.333333	2255	2000.624933	2018.95	1966.07	163.3333333	255	0.624933333	18.95	-33.93
2100	2263.333333	2335.333333	2095.254733	2120.95	2045.6	163.3333333	235.3333333	-4.745266667	20.95	-54.4
2200	2370	2437.333333	2201.4682	2229.75	2146.58	170	237.3333333	1.4682	29.75	-53.42
2300	2466.666667	2538.333333	2300.493267	2328.35	2246.57	166.6666667	238.3333333	0.493266667	28.35	-53.43
2400	2556.666667	2645.666667	2396.989867	2420.15	2352.83	156.6666667	245.6666667	-3.010133333	20.15	-47.17
2500	2646.666667	2753.333333	2493.594267	2511.95	2459.42	146.6666667	253.3333333	-6.405733333	11.95	-40.58
2600	2753.333333	2847.333333	2597.220533	2620.75	2552.48	153.3333333	247.3333333	-2.779466667	20.75	-47.52
2700	2846.666667	2969.666667	2700.856467	2715.95	2673.59	146.6666667	269.6666667	0.856466667	15.95	-26.41
2800	2963.333333	3061	2810.485333	2834.95	2764.01	163.3333333	261	10.48533333	34.95	-35.99
2900	3053.333333	3177.666667	2910.000333	2926.75	2879.51	153.3333333	277.6666667	10.00033333	26.75	-20.49
3000	3140	3261	2996.447	3015.15	2962.01	140	261	-3.553	15.15	-37.99
3100	3230	3349.666667	3086.9068	3106.95	3049.79	130	249.6666667	-13.0932	6.95	-50.21
3200	3306.666667	3490	3184.922267	3185.15	3188.72	106.6666667	290	-15.07773333	-14.85	-11.28
3300	3390	3585.333333	3272.9614	3270.15	3283.1	90	285.3333333	-27.0386	-29.85	-16.9
3400	3500	3698	3384.9128	3382.35	3394.64	100	298	-15.0872	-17.65	-5.36

Fig. 4. Snapshot of data before and after linear regression and the corresponding errors

The following table shows the obtained relationships after usage of linear regression. Here Actual Distance refers to the ground truth data, Lidar refers to the RPLidar 360 measured data and Leddar M-16 refers to measurement data from Leddar M-16.

It is found that if both RPLidar 360 and Leddar M-16 data are fused together, the accuracy is drastically improved. Also that, with a simplistic mathematical process such as linear regression, the effort on processing to reduce error is very less. So it is advantageous to use Linear Regression with the data of both the sensors.

The limitation with this work: the range in which the RPLidar 360 works is just 6 m which is very small compared to the 100 m range of Leddar M-16. The fusion happens only in the 6 m range. The remaining measurements will have to rely fully upon the values provided by single sensor that is Leddar M-16.

The following relationships were obtained:

When only Leddar M-16 is considered and when linear regression is applied, the following relationship.

$$\text{Actual Distance (mm)} = 1.02 * \text{Leddar} - 187.65 \tag{1}$$

When only RPLidar 360 is considered and when linear regression is applied, the following relationship.

$$\text{Actual Distance (mm)} = 0.99 * \text{Lidar} - 266.38 \tag{2}$$

When both Leddar M-16 and RPLidar 360 is considered and when linear regression is applied, the following relationship. This is the result of fusion of data from both the sensors.

$$\text{Actual Distance (mm)} = 0.6865 * \text{Leddar} + 0.3234 * \text{Lidar} - 213.7705 \tag{3}$$

5   Results

The obtained equations from linear regression were tested on a different set of test data. This time, the following results were obtained. The tool named Weka 3.8 was used to do analysis and to get results. The results show a clear improvement in performance upon the use of both Leddar M-16 and RPLidar 360 over the use of single sensor individually.

The following table throws light on some of the results obtained after testing (Table 1).

Table 1. Results of different equations obtained after regression and used for testing

Leddar M-16 Only	RPLidar 360 only	Both Leddar M-16 and RPLidar 360
Actual Distance(mm) =1.02*Leddar M-16 - 187.65	Actual Distance(mm) =0.99*Lidar - 266.38	Actual Distance(mm) = 0.6865*Leddar M-16 + 0.3234*Lidar - 213.7705
Time taken to build model: 0 seconds === Summary === Correlation coefficient 0.9789 Mean absolute error 17.223 Root mean squared error 20.1821 Relative absolute error 1.1348 % Root relative squared error 1.1531 % Total Number of Instances 60	Time taken to build model: 0 seconds === Summary === Correlation coefficient 0.9878 Mean absolute error 26.1776 Root mean squared error 33.139 Relative absolute error 1.7247 % Root relative squared error 1.8933 % Total Number of Instances 60	Time taken to build model: 0 seconds === Summary === Correlation coefficient 0.9989 Mean absolute error 11.4572 Root mean squared error 15.0489 Relative absolute error 0.7494 % Root relative squared error 0.8517 % Total Number of Instances 60

The following figure shows the plot of actual and predicted values of the distances measured using both sensors after linear regression is applied to the obtained data as per Eq. (3) (Fig. 5).

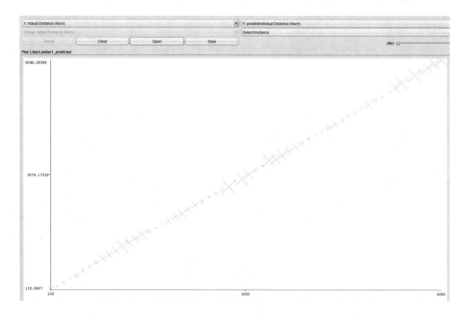

Fig. 5. Linear regression on the data from both sensors giving the combined results

This result shows a mean absolute error of 11.4572 and root mean squared error of 15.0489 after data fusion. This error is lesser than the errors obtained when only single sensor is used. Thus an improvement in accuracy is obtained by implementing fusion of data from both sensors.

6   Conclusion

The results shown in the previous section clearly indicates a better accuracy for the third equation. This portrays the fact that the results obtained after fusing the data of both the senors together using linear regression gives a much better accuracy in the measurement of distance. This resultant measurement system can be used as a part of the emergency braking systems and path planning systems for self diving cars are autonomous mobile robots.

References

1. Fuchun, L., Yifeng, C., Yunze, L. Research on indoor robot SLAM of RBPF improved with geometrical characteristic localization. In: Proceedings of the 29th Chinese Control and Decision Conference, CCDC 2017, Art. no. 7979080, pp. 3325–3330 (2017)
2. Garrote, L., Rosa, J., Paulo, J., Premebida, C., Peixoto, P., Nunes, U.J.: 3D point cloud downsampling for 2D indoor scene modelling in mobile robotics. In: 2017 IEEE International Conference on Autonomous Robot Systems and Competitions, ICARSC 2017, Art. no. 7964080, pp. 228–233 (2017)
3. Mimeault, Y., Gidel, S., Tremblay, M.L.: M-16TM technology for efficient automatic vehicle classification sensor. In: 19th Intelligent Transport Systems World Congress, ITS 2012, pp. AM-00129 (2012)
4. Panich, S.: Object detection with laser scanner for navigation system. Far East J. Electron. Commun. 15(2), 111–120 (2015). https://doi.org/10.17654/FJECDec2015_111_120
5. Palacios, O.F.G., Salah, S.H.: Mapping marsian caves in 2D with a small exploratory robot. Paper presented at the Proceedings of the 2017 IEEE 24th International Congress on Electronics, Electrical Engineering and Computing, INTERCON 2017 (2017). https://doi.org/10.1109/intercon.2017.8079721
6. Almahzumy, H., Ardilla, F., Marta, B.S.: Board PSoC library design and evaluation for mobile robot implementation. Paper presented at the Proceedings IES-ETA 2017 - International Electronics Symposium on Engineering Technology and Applications, December 2017, pp. 257–262 (2017). https://doi.org/10.1109/elecsym.2017.8240413
7. Kim, M., Jang, J.: A study on avoidance technique for autonomous avoidance of unmanned vehicle. J. Eng. Appl. Sci. 12(11), 2861–2866 (2017). https://doi.org/10.3923/jeasci.2017.2861.2866
8. Qureshi, H.S., Glasmachers, T., Wiczorek, R.: User-centered development of a pedestrian assistance system using end-to-end learning. Paper presented at the Proceedings - 17th IEEE International Conference on Machine Learning and Applications, ICMLA 2018, pp. 808–813 (2019). https://doi.org/10.1109/icmla.2018.00129
9. Adhvaryu, A.D., Adarsh, S., Ramchandran, K.I.: Design of fuzzy based intelligent controller for autonomous mobile robot navigation. Paper presented at the 2017 International Conference on Advances in Computing, Communications and Informatics, ICACCI 2017, January 2017, pp. 841–846 (2017). https://doi.org/10.1109/icacci.2017.8125946
10. Adarsh, S., Ramachandran, K.I.: Design of sensor data fusion algorithm for mobile robot navigation using ANFIS and its analysis across the membership functions. Autom. Control Comput. Sci. 52(5), 382–391 (2018). https://doi.org/10.3103/S0146411618050036
11. Wang, H., Lin, Y., Wang, Z., Yao, Y., Zhang, Y., Wu, L.: Validation of a low-cost 2D laser scanner in development of a more-affordable mobile terrestrial proximal sensing system for 3D plant structure phenotyping in indoor environment. Comput. Electron. Agric. 140, 180–189 (2017). https://doi.org/10.1016/j.compag.2017.06.002
12. Wu, M., Ma, H., Fu, M., Yang, C.: Particle filter based simultaneous localization and mapping using landmarks with RPLidar (2015). https://doi.org/10.1007/978-3-319-22879-2_54

An Efficient Finger Vein Image Enhancement and Pattern Extraction Using CLAHE and Repeated Line Tracking Algorithm

Thenmozhi Ganesan$^{(\boxtimes)}$, Anandha Jothi Rajendran,
and Palanisamy Vellaiyan

Department of Computer Applications, Alagappa University, Karaikudi, India
thenmozhiganesan23@gmail.com,
ranandhajothi12@gmail.com, vpazhanisamy@yahoo.co.in

Abstract. Finger vein recognition is documented as one of the effectual biometric technique today for person identification because of its following
advantages: Contactless sensor, low cost, living body authentication and complex security. Pattern extraction is an essential process that is used to extract the
cleared specific vein patterns from the spurious finger vein image. In this paper,
the repeated line tracking algorithm is endeavored to extort the vein samples
adequately. In order to attain the improved extraction consequence, the image is
enhanced by using Contrast Limited Adaptive Histogram Equalization
(CLAHE) before employing the extraction progress. For this purpose we are
taking a publicly available dataset namely UTFVP. Subsequently a performance
comparison is given for our proposed work in the aspect of some quality
measures as PSNR, MSE and SSIM values for before and after processing.
Experimental result shows that the simplicity, reliability and sturdiness of the
proposed pattern extraction method.

Keywords: Image enhancement · Contrast limited adaptive histogram
equalization · Image segmentation · Feature extraction · Repeated line tracking

1 Introduction

Nowadays, biometric technology attains a major position in individual person identification. This system verifies the user's identity by using their physical or behavioral
biometric characteristics. Physical characteristics including the attributes such as iris,
finger print, face and vein whereas voice, signature and key stroke pattern and gait are
known as behavioral characteristics [1, 2]. Unlike traditional authentication methods
such as Personal Identification Number (PIN) and passwords, biometric based
authentication provides enriched security due to its robustness and also it eliminates the
forgetting issues of ancient security systems [3].

Vein identification technology is a significant way for human recognition. Finger
vein, palm vein, hand vein are three categories of vein technology. Among these, finger
vein based recognition is broadly used for its efficacious features such as uniqueness,
collectability, permanence, and universality. Undoubtedly, it offers high-level security
because finger veins are internal features of the human body and it is too hard to

© Springer Nature Switzerland AG 2020
A. P. Pandian et al. (Eds.): ICICCS 2019, AISC 1039, pp. 690–700, 2020.
https://doi.org/10.1007/978-3-030-30465-2_76

replicate or forge. Researchers proved that vein patterns are identical even between twins. The credibility of the finger vein authentication is quite high [4].

Finger vein images are usually grabbed using infrared based light transmission or reflection system. Near Infra-Red (NIR) light beam illuminate the finger and the camera placed under the finger is captured the vein image. During this illumination process NIR light source is engrossed by the hemoglobin so that the patterns of vein are visible as darker than the encompassing tissues [5]. The image which is captured in this way having several problems for instance low contrast, improper intensity and noise and it is rarely in good quality. This will reduce the accuracy and recognition rate.

To overcome these problems some image processing steps are needed. After enhancing the image, an efficient pattern extraction method is practiced to excerpt the obligatory vein features. Features of finger vein are extracted generally by the subsequent approaches: Subspace learning-based approach [6], Statistical-based techniques [7], Local invariant-based methods [8] and vessel extraction [1]. In vessel extraction technique the vein patterns are considered the same as dark lines and extracted as line-like formation. Appearance-based methods are used in subspace-based feature extraction method. Statistical-based approaches exploit the local or global statistical information. Local invariant-based methods are encouraged by the visions of computers.

In this work we are attempting CLAHE based finger vein enhancement for solving the contrast variation issues in acquired image. Further, vein samples are extracted by using repeated line tracking algorithm. The content of this paper is formed as follows: Sect. 2 discuss the related works carried out. Section 3 portrays the proposed methodology. Section 4 illustrates performance evaluation of this work. Experimental results are shown in Sect. 5 and the paper is concluded in Sect. 6.

2 Related Work

Extraction of required vein patterns from the vein image is a most prominent step of finger vein recognition. There are several techniques and filters are used in past for this purpose. In near infrared vein pattern based personal identification system [9] Kono used the background-reduction filter for image enhancement and normalized cross correlation method to indentify the similarity between two image patterns. The repeated line tracking algorithm [10] is proposed to extract the patterns of vein and afterward the accurateness of the extraction method is described and showed that how robustly the proposed algorithm worked against the irregular shadings in the captured images. Maximum curvature method is used in [11] to extract the features of finger vein image. In this method, the center lines of the vein patterns are highlighted. Gabor and wide line detector are worked [12] as a neighborhood region and if the current pixel is marked as vein point then the neighbors achieve the higher gray values.

In [13] different feature extraction algorithm namely line tracking which has random starting positions for vein pattern extraction and also having special segments such as pre-processing, image normalization and post-processing before applying extraction algorithm. An efficient algorithm is used to extract the vein regions according to the edges of finger image characteristics that proficiently extract desired regions without unwanted background in [14]. New vein extraction algorithm based on anatomy structure analysis [15] along with integrated matching approach is proposed which can extract the patterns using cross-sectional character. Modified Finite Radon Transform feature extraction method [16] is used to extract the local invariant features and it is an influential method to handle the structure of vein patterns for illumination variation. The competitive Gabor [17] response method for feature extraction is grouping a Gabor magnitude and competitive Gabor orientation. An adaptive Niblack threshold segmentation algorithm [18] for finger vein feature extraction is proposed along with adaptive histogram equalization filter.

3   Proposed Methodology for Finger Vein Feature Extraction

See (Fig. 1).

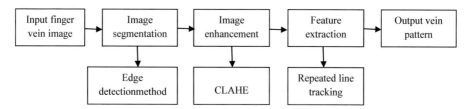

Fig. 1. Block diagram of pre-processing and feature extraction

3.1   Pre-processing

Image Segmentation: Initial step of the proposed work is image segmentation. The partitioning of image into number of regions or pixels is known as image segmentation. In other words, it is the partition of image into several categories that involve each picture element in the image. To facilitate representation of image into some meaningful format is the main intention of image segmentation process. The following are the important qualities of fine image segmentation [19].

- In order to form the connected region, same category pixels have similar grayscale of multivariate values.
- Pixels in the different categories or neighboring pixels have dissimilar values.

Edge Detection Based Method: Edge detection based image segmentation method is applied to achieve the border line of an object within the image. It is based on the intensity value which is changing rapidly in image since a sole intensity value cannot give all information about edges [19]. All the edges of the images are detected initially afterward they are connected to create the object limits in order to subdivide the desired sections in this edge detection based method. Gray histograms method and Gradient based method are the two edge based detection techniques. The edges are detected by using the techniques such as sobel operator, Robert's operator and canny operator. The result of this method is usually a binary image. We are using sobel operator for edge detection and its consist of 3 * 3 convolution kernels pair in which each kernel is similar to another and rotated for 90° as shown in Fig. 2.

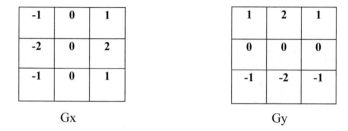

Gx   Gy

Fig. 2. The convolution kernel of sobel

These kernels are individually applied in key image to create individual gradient component measurements Gx and Gy. The gradient magnitude is found by combining these components. The gradient magnitude is shown in Eq. (1).

$$|G| = \sqrt{Gx^2 + Gy^2} \tag{1}$$

This magnitude can also be computed by the Eq. (2),

$$|G| = |Gx| + |Gy| \tag{2}$$

The orientation angle of an extracted edge is given in Eq. (3),

$$\theta = arc \; \tan(Gy/Gx) \tag{3}$$

In the image the contrast direction ranging from black to white is denoted as θ. Figure 3 shows the outcome of edge detection process.

Image Enhancement: The enrichment of image quality by using some filters and algorithms is known as image enhancement. It is used to get better contrasted and noiseless image.

Contrast Limited Adaptive Histogram Equalization (CLAHE): CLAHE is a customized form of adaptive histogram equalization in which the enrichment task is practiced on territory pixels and the conversion function is determined. It is used to automatically adjust the contrast variations such as low contrast and high contrast in image [20]. In this work, CLAHE is applied for the detected edges of input image to improve the trait of required region. CLAHE is used where the necessity of brightness is high. The images are partitioned into related regions and then equalization is found for each region. The quality of the CLAHE enhancement image is controlled by the key

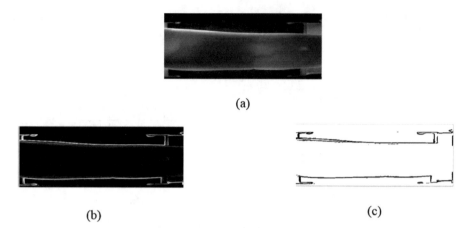

(a)

(b)   (c)

Fig. 3. Edge detection of finger vein image (a) Original (b) Sobel gradient (c) Sobel edges

parameters such as blocks size and clip limit. The basic algorithm for enhancing the region of individual image using CLAHE is

Step-1: Partition the input image into number of contextual region which are equal in size with 8×8 blocks and each region is corresponding to the 64 pixel neighborhood.

Step-2: Compute the intensity histogram of every region.

Step-3: Set the clip limits for histograms clipping. Clip limit is a threshold parameter which is used to alter the image contrast because higher clip limit raising the local image contrast so that it should be a minimum value.

Step-4: Modify the histogram by choosing the transformation function.

Step-5: Each histogram is transformed into the height of the selected clip limit. The mathematical representation of gray level transformation for standard CLAHE with uniform distribution is given in Eq. (4)

$$g = [g_{max} - g_{min}] * P(f) + g_{min} \tag{4}$$

g represents computed pel value, g_{min} represents the minimum pel value, g_{max} represents the maximum pel value and P (f) represents cumulative probability distribution.

For exponential gray level distribution,

$$g = g_{min} - \left[\frac{1}{\alpha}\right] * In[1 - P(f)] \tag{5}$$

Here α represents the clip parameter. Instead of operating a whole image, CLAHE works on small region of image named as tiles. The CDF of Rayleigh distribution is shown in Eq. (6)

$$y = P(f(x/b)) = \int\limits_{0}^{x} \frac{x}{b^2} e^{-\left(\frac{-x^2}{2b^2}\right)} \tag{6}$$

Step-6: By using bilinear interpolation the neighborhood tiles were combined. According to the modified histograms the gray values of the image were altered. Figures 4 and 5 shows the input and CLAHE enhanced images respectively.

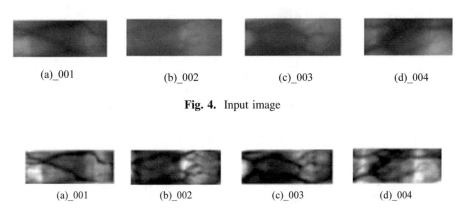

(a)_001   (b)_002   (c)_003   (d)_004

Fig. 4. Input image

(a)_001   (b)_002   (c)_003   (d)_004

Fig. 5. CLAHE enhanced vein image

3.2   Pattern Extraction

Conversion of the input data into desired patterns is recognized as pattern extraction. It includes the extraction of length, shape and width of the vein pattern. Repeated line tracking algorithm is attempted here to extract the patterns.

Repeated Line Tracking Algorithm: Repeated line tracking offers the potential outcome in finger vein based recognition. It initiates erratically on various portions. This can be carried out by moving the direction of vein patterns for each picture element to hit upon the dark line [10]. Veins in the image is tracked by selected the path in accordance with the previously defined possibilities in both coordination. The initial point labeled as seed is indiscriminately chosen. The entire progression is repetitively performed for finite times [21].

Repeated Trial of Dark Line Tracking: An intersected outline of a vein come across valley-like structure which shown in Fig. 6. The valley remains measurable but the depth of the valley has been varied due to its image shading. The tracking operation is started from any point in grabbed image. The place of existing picture element termed as current tracking point is moved pel by pel along the dark lines. Depth of the intersected outline is tested over the present tracking point [21].

Repeated Line Tracking for Pattern Extraction-Steps

Step-1: Determine the seed point for tracking and moving direction elements.
Step-2: Spot the darkened line track and the progress of such point.
Step-3: Update and track the count of points that is placed in locus space.
Step-4: Repeat step-1 to step-3 executions for X times.
Step-5: Achieve vein patterns from the locus space.

4   Performance Evaluation

4.1   Quality Measures

After the completion of pre-processing step, the quality assessment of enhanced vein image has been calculated by following measures: MSE, PSNR and SSIM. These proceedings are universally accepted and used to estimate the effectiveness of the enhancement [22].

4.2   Mean Square Error (MSE)

The cumulative squared error among input and output image is called as Mean Square Error and computed by using Eq. (7)

$$MSE = \frac{1}{MN} \sum_{i=1}^{M} \sum_{j=1}^{N} [M_1(i,j) - M_2(i,j)]^2 \tag{7}$$

$M_1(i, j)$ and $M_2(i, j)$ represent the input and filtered image matrix respectively. M * N indicates the image size.

4.3   Peak Signal to Noise Ratio (PSNR)

It is a statistical measure of image characteristic which is depends on the pixels difference among the input and resultant images. The main objective of the SNR is to measure quality of resultant image and evaluate it with input image. PSNR value was calculated by using Eq. (8)

$$PSNR = 10 * log_{10} \frac{R^2}{MSE} \tag{8}$$

R indicates highest value of pel present in image. For an 8-bit pixel image R = 255.

4.4   Structural Similarity Index Measure (SSIM)

SSIM based on three computations terms namely the luminance term, the structural term and the contrast term. The overall SSIM index is a combination of these separate terms. The structural similarity between input image x and enhanced image y was calculated by the following Eq. (9).

$$SSIM(x, y) = \frac{(2\mu_x\mu_y + c_1 L)(2\sigma_{xy} + c_2 L)}{(\mu_{x^2} + \mu_{y^2} + c_1 L)(\sigma_{x^2} + \sigma_{y^2} + c_2 L)} \tag{9}$$

Where μ_x, μ_y are the local means and σ_x^2 and σ_y^2 are standard deviations of x and y, σ_x, σ_y are the cross-covariance of x and y. L is the utmost possible value of x and C1, C2 are adaptable constants.

5   Experimental Results

UTFVP is a finger vein database that encloses a vein image of index, middle and ring finger for both left and right hand. Sobel filter is worked to extract the boundaries from the input vein image. The detected edges are then tested with the Contrast Limited Adaptive Histogram Equalization filter. It reduced the amplification of noise and adjusts the image contrast variations. CLAHE enhanced the vein image quality and furnished natural look to the image. We are applied and tested the proposed technique on hundred finger vein image in UTFVP database. Finally show the performance for selected ten right index finger vein images from 001 to 010 in database in accordance with the parameters PSNR, MSE and SSIM which is shown in Table 1. SSIM of vein image after enhancement is shown and it seems to moderately good. The proposed enhancement method provides the visually cleared image and after applying CLAHE the vessel patterns in the vein is appeared darker that helps to extract the features robustly and easily. We achieved the desired extraction outcome and accuracy visually. The comparison is also shown as graph in Figs. 8 and 9. Repeated line tracking algorithm is finally applied on filtered images. This algorithm effectively extracted the vein patterns and the extracted patterns are shown in Fig. 7.

Table 1. Comparison of MSE and PSNR values before and after using CLAHE and SSIM values after using CLAHE

Finger vein image	Before		After		SSIM
	MSE	PSNR	MSE	PSNR	0.50
001	2541	28.33	1071	60.51	0.72
002	1802	29.23	1120	62.34	0.57
003	3122	29.41	2971	60.24	0.81
004	2143	28.28	1541	61.14	0.53
005	2140	28.93	1340	62.71	0.59
006	2481	27.04	1513	60.31	0.46
007	2609	26.43	1070	59.80	0.64
008	3610	27.97	1879	60.51	0.79
009	2501	26.47	1501	59.15	0.80
010	2430	27.44	1081	60.70	0.50

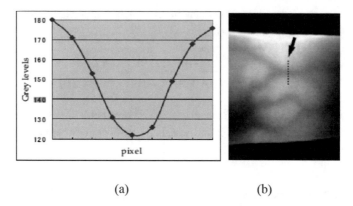

(a)   (b)

Fig. 6. Intersected outline of a vein (a) Intersected outline (b) Area of a intersection

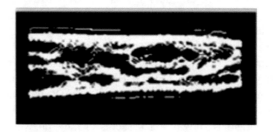

Fig. 7. Resultant RLT feature extraction image

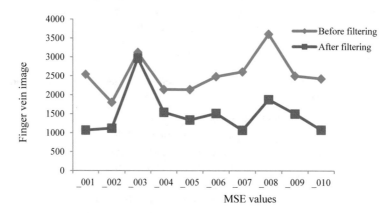

Fig. 8. Comparision of MSE after and before using CLAHE

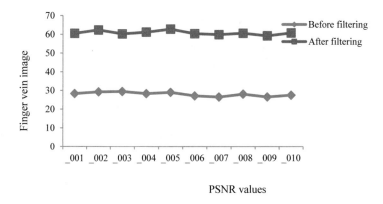

Fig. 9. Comparision of PSNR after and before using CLAHE

6   Conclusion

We conclude the paper of finger vein pattern based feature extraction method. Edge detection, filtering and vein pattern extraction were effectively applied on vein image and implemented in MATLAB. Finger vein pattern based feature extraction has been successfully performed by applying CLAHE based image enhancement and repeated line tracking algorithm based pattern extraction. It provided visually efficient performance and for our future work, it will help to get better matching accuracy of finger vein identification system.

Acknowledgement. This article has been written with the financial support of RUSA-Phase 2.0 grand sanctioned vide Letter No. F.24-51/2014-U. Policy (TNMulti-Gen). Dept. of Edn. Govt. of india, Dt. o9.10.2018.

References

1. Kumar, A., Zhou, Y.B.: Human identification using finger images. IEEE Trans. Image Process. **21**(4), 2228–2244 (2012)
2. Anandha Jothi, R., Palanisamy, V., Nithyapriya, J.: Evaluation of fingerprint minutiae on ridge structure using Gabor and closed hull filter. In: Computational Vision and Bio Inspired Computing. Springer, Heidelberg (2018)
3. Anandha Jothi, R., Palanisamy, V.: Analysis of fingerprint minutiae extraction and matching an algorithm. Int. J. Adv. Res. Trends Eng. Technol. **3**(20), 398–410 (2016)
4. Yang, J., Li, X.: Efficient finger vein localization and recognition. In: 20th International Conference on Pattern Recognition (ICPR), pp. 1148–1151. IEEE (2010)
5. Khanam, R., Khan, R., Ranjan, R.: Analysis of finger vein feature extraction and recognition using DA and KNN methods. In: Amity International Conference on Artificial Intelligence (AICAI), pp. 477–483 (2019)

6. Liu, Z.: Finger vein recognition with manifold learning. J. Netw. Comput. Appl. **33**(3), 275–282 (2010)
7. Lee, E.C., Lee, H.C., Park, K.R.: Finger vein recognition using minutia-based alignment and local binary pattern-based feature extraction. Int. J. Imaging Syst. Technol. **19**(3), 179–186 (2008)
8. Peng, J., Wang, N., Li, Q., Niu, X.: Finger-vein verification using gabor filter and sift feature matching. In: Eighth International Conference on Intelligent Information Hiding and Multimedia Signal Processing, pp. 45–48 (2012)
9. Kono, M., Ueki, H., Umemura, S.: A new method for the identification of individuals by using of vein pattern matching of a finger. In: Proceedings of the 5th Symposium on Pattern Measurement, pp. 9–12 (2000)
10. Miura, N., Nagasaka, A.: Feature extraction of finger-vein pattern based on repeated line tracking and its application to personal identification. Mach. Vis. Appl. **15**(4), 194–203 (2004)
11. Miura, N., Nagasaka, A., Miyatake, T.: Extraction of finger-vein patterns using maximum curvature points in image profiles. IEICE Trans. Inf. Syst. **90**(8), 1185–1194 (2007)
12. Huang, B., Dai, Y., Li, R., Tang, D., Li, W.: Finger-vein authentication based on wide line detector and pattern normalization. In: Proceedings 20th International Conference Pattern Recognition (ICPR), pp. 1269–1272 (2010)
13. Bhagyashree, B., Ramesh, K.: Extraction of segmented vein patterns using repeated line tracking algorithm. In: 3rd International Conference on Sensing, Signal Processing and Security (ICSSS), pp. 89–92 (2017)
14. Guo, Q., Qiao, B.: Research on the finger vein image capture and finger edge extraction. In: IEEE International Conference on Mechatronics and Automation (ICMA), pp. 275–279 (2017)
15. Yang, L., Yang, G., Yin, Y., Xi, X.: Finger vein recognition with anatomy structure analysis. IEEE Trans. Circ. Syst. Video Technol. **28**(8), 1892–1905 (2018)
16. Van, H.T., Thai, T.T., Le, T.H.: Robust finger vein identification base on discriminant orientation feature. In: Seventh International Conference on Knowledge and Systems Engineering, pp. 348–353 (2015)
17. Lu, Y.: Finger vein recognition using histogram of competitive Gabor responses. In: 22nd International Conference on Pattern Recognition (ICPR), pp. 25–43 (2014)
18. Liu, C.: A new finger vein feature extraction algorithm. In: 6th IEEE International Congress on Image and Signal Processing (CISP), pp. 395–399 (2013)
19. Dilpreet, K., Yadwinder, K.: Various image segmentation techniques: a review. Int. J. Comput. Sci. Mobile Comput. **3**(5), 809–814 (2014)
20. Brij, B.S., Shailendra, P.: Efficient medical image enhancement using CLAHE enhancement and wavelet fusion. Int. J. Comput. Appl. **167**(5), 0975–8887 (2017)
21. Manpreet, K., Geetanjali, B.: Finger vein detection using repeated line tracking, even gabor and multilinear discriminant analysis (MDA). Int. J. Comput. Sci. Inf. Technol. **6**(4), 3280–3284 (2015)
22. Anandha Jothi, R., Palanisamy, V.: Performance enhancement of minutiae extraction using frequency and spatial domain filters. Int. J. Pure Appl. Math. **118**(7), 647–654 (2018)

Simulation of Signal Generation and Measuring Circuit and Real Time IoT Based Electrical Bio Impedance Cardiac Monitoring System

Sheeba Santhosh[1,2], A. Vimala Juliet[3], and G. Hari Krishnan[4(✉)]

[1] Sathyabama Institute of Science and Technology, Chennai, India
[2] Vel Tech Rangarajan Dr. Sagunthala R&D
Institute of Science and Technology, Chennai, India
[3] Department of ICE, SRM Institute of Science and Technology, Chennai, India
[4] Department of EEE, Sree Vidyanikethan Engineering College, Tirupati, India
haris_eee@yahoo.com

Abstract. IoT plays a vital role in critical patient care remote monitoring anywhere around the globe. Cardiac patients in remote areas in specific aged patients have to visit physician for diagnosing their health state regularly. In our proposed system critical cardiac patient heath was diagnosed using electrical bioimpedance technique and the same diagnosed impedance cardiograph signal was make available in live to the physician through IoT. Before designing hardware, simulation circuits are designed to obtain the required specifications in order to avoid hardware damage during testing.

Keywords: Internet of Things · Bio-impedance · Diagnosis · Impedance cardiograph · Proteus Software · Interfacing · Arduino Uno · Node MCU

1 Introduction

In present scenario, cardiac failure has become one of the major causes for the increased mortality in most of the country. Many studies have been carried out to investigate the causes and effects of cardiac diseases [1–4]. The proper usage of electrocardiograph has helped physicians to diagnose most of the cardiac related ailments. Studies shows measurement of ECG signal is generally associated with various interferences causing discrepancy in the measurement system. Several methods were proposed to reduce these interferences in the ECG signal to provide accurate measurement. One such technique is the use of bio-impedance signal for cardiac measurement.

Bio-impedance basically refers to the electrical resistance provided by the tissue to the flow of current. The bio-impedance signal represents integrated measurement of various physiological changes corresponding to a system. It is a noninvasive method used to measure various physiological changes such as blood flow, body composition, etc. It is used in various medical applications such as measurement of respiration rate, measuring sweat activity, measurement of bone thickness and so on. It is an appropriate

© Springer Nature Switzerland AG 2020
A. P. Pandian et al. (Eds.): ICICCS 2019, AISC 1039, pp. 701–706, 2020.
https://doi.org/10.1007/978-3-030-30465-2_77

tool to the development of devices to measure cardiac output and cardiac blood volume changes [5]. It provides information about stroke volume, cardiac output, heart rate. Measurements of bio-impedance can be performed by placing surface electrodes on the chest or limbs after skin preparation [6]. Changes in the bio-impedance can be observed by passing a high frequency, low amplitude and constant magnitude current to human body. Bio-impedance measurement is basically represents passive electrical properties. The charge carriers in tissue are the ions which move outside and inside the cells. When an alternating current is applied, electrons and ions transition takes place at the electrode [8]. The current flow due to movement of ions causes changes in the biological material. The cell membrane which separates the intracellular and extracellular fluid has a very low conductance and behaves like a parallel plate capacitor with dielectric between the two regions.

Thus, biological materials are capacitive in nature [9]. At low frequencies (less than 10 kHz), almost all the current flows through extracellular space only and so the total impedance is largely resistive. At higher frequencies the current enters the intracellular space which makes overall impedance is lower [7]. Most changes between normal and pathological tissue occurs in the frequency range of 100 Hz–10M. The impedance of a biological tissue decreases as the frequency increases. When an external power source is applied to a biological tissue, ionic polarization takes place [10, 11]. It produces displacement of the positive ions with respect to the negative ions. The frequency of the applied signal determines the value of impedance measured and also depends on the time taken for the ions to change their position in the material. In general, bio-impedance method is non invasive, inexpensive and safe to be used without causing any risk to the patient [12].

The proposed system work flow starts with design of signal generation circuit to generate different waveform, Voltage and Frequency measuring circuit design using ISIS Proteus Software with in build Arduino Uno package. Hardware implementation of the system after system parameters are achieved by proper tuning of the devices in simulation circuit. Hardware circuit was interfaced with IoT platform using NodeMcu device.

2 Methodologies

Any critical Bio-impedance measuring system requires the impact of the applied electrical current signal on the tissue through surface electrodes. To have impact on tissues, circuit to generate electrical signal with required frequency and voltage is required. The generated electrical signal has to be applied on the surface of the human body through surface electrodes. Generally adhesive electrodes are most suited for bio-impedance purpose [10]. ISIS Proteus Software by Labcenter Electronics was used to simulate various circuits used for impedance based monitoring system.

  (i) Variable Frequency & Voltage signal generator design.
 (ii) Arduino based circuit for measure and display generated signal.
(iii) LCD Interface with Arduino Uno.
(iv) Circuit to acquire the generated signal using Arduino Uno.

Bio-impedance measurement system design requires four major sub circuits for sending the signal, acquiring the signal using electrodes, Sensing the response and Measuring the response signal. Signal generator circuit was simulated using 3 different means as using ICL8038 IC, the second method using IC741 and the third method using IC555 to control the frequency and voltage. ICL8038 IC is a specific signal generator IC which operates on voltage range between 8 V to 10 V. It generates sine, squire and triangular waveforms with voltage nearly equal to the supply voltage and frequency ranges from 0.001 Hz to 300 kHz.

Frequency variation can be obtained by placing variable capacitance between the pins 4 and 5. Triangular wave can be acquired at pin 3, square wave can be acquired at pin 8 and sine wave at pin 12. Maximum current range of the generated waveforms is around 25 mA based on the load applied. Arduino Uno is preferred due to its in built Analogue as well as digital pins to connect the external circuitry. PWM pins in Arduino are used to measuring the frequency of the applied signal. Arduino board is interfaced to 16×2 Matrix LCD display to shown the input signal frequency and voltage. Also virtual monitor in arduino software is used to obtain the instantaneous variation the input circuit.

The simulated circuit for Variable Frequency & Voltage signal generator design. Arduino based circuit for measure and display generated signal. Arduino Uno LCD Interfacing Circuit acquires the generated signal using Arduino Uno is as shown in Fig. 1. The output waveforms can be viewed using inbuilt Scope in Proteus software. Figure 1 shown the arduino interfacing with LCD and input signal generator circuit with display of voltage and frequency in LCD and display on input voltage in Arduino inbuilt virtual monitor.

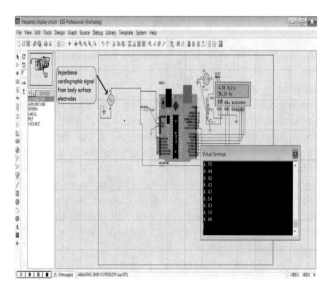

Fig. 1. Arduino based voltage and frequency measurement of simulated impedance signal

3   Results and Discussion

Once the main circuit as shown in Fig. 2 was simulated in Proteus with Arduino, signal generator, Scope, LCD and Virtual Monitor as major components. Validation of the signal generator for proper selection of Voltage and frequency was done by varying the capacitor. It is observed that if supply to the signal generator is less than 5 V output waveforms are distorted.

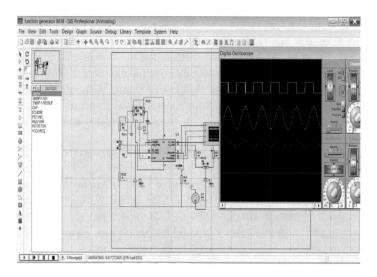

Fig. 2. Signal generator simulation with variable voltage and frequency.

Based on the analysis the variation in different waveforms sine square and triangular are studies as shown in Fig. 3 and sine wave is more suitable for the bio-impedance measuring systems.

Based on the simulation circuit designed using ISIS Proteus for signal generation and monitoring hardware circuit was implemented. NodeMCU was interfaced with arduino uno for connecting hardware to Internet. After interfacing wifi module with arduino uno and implementing the required code to connect wifi for internet and accessing data through Thinkspeak website.

In think speak fields were created and the API key has been generated for API request. The same key was copies and passed on the arduino program. Also the real time web access link for the generated field was copied from the website to program. After successful compilation and debugging impedance cardiac signal was visualized on the Thinkspeak website as shown in Fig. 4.

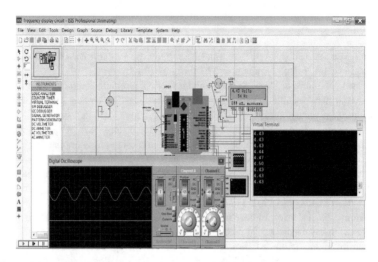

Fig. 3. Arduino based voltage and frequency measuring system showing input and output.

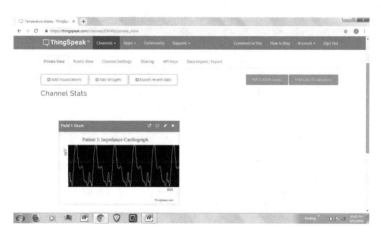

Fig. 4. Real time bioimpedance signal monitoring on Thinkspeak.

4   Conclusion

System for critical cardiac patient heath diagnosing using electrical bioimpedance technique was designed. Using arduino software, arduino, Node MCU and Thinkspeak the same diagnosed impedance cardiograph signal was make available in live to the physician. By implementing the IoT based Cardiac bioimpedance monitoring system patients at remote area can cable to easily diagnose by the physicians. Also same data can be accessed anywhere around the globe if needed.

References

1. Yan, J.Y., Lu, Y., Liu, J., Wu, X.Y.: Intelligent diagnosis of cardiovascular diseases utilizing ECG signals. Int. J. Inf. Acquisition **07**(2), 81–97 (2010)
2. Kumaravel, N., Sridhar, K.S., Nithiyanandam, N.: Automatic diagnosis of heart diseases using neural network. In: 1996 Biomedical Engineering Conference, Dayton, pp. 319–322 (1996)
3. Hong, B., Kai, J., Ren, Y., Han, J., Zou, Z., Ahn, C.H., Kang, K.A.: Highly sensitive rapid, reliable, and automatic cardiovascular disease diagnosis with nanoparticle fluorescence enhancer and MEMS. Adv. Exp. Med. Biol. **614**, 265–273 (2008)
4. Scherhag, A.W., et al.: Continuous measurement of hemodynamic alterations during pharmacologic cardiovascular stress using automated impedance cardiography. J. Clin. Pharmacol. **37**, 21S–28S (2013)
5. Palls-Areny, R., Gonzlez Landaeta, R. Casas, O.: Heart rate detection from plantar bio-impedance measurements. In: 2006 Annual International Conference of the IEEE Engineering in Medicine and Biology Society (EMBC), pp. 5113–5116. IEEE (2006)
6. Wang, J.-J., Hu, W.-C., Kao, T., Liu, C.-P., Lin, S.-K.: Development of forearm impedance plethysmography for minimally invasive monitoring of cardiac pumping function. J. Biomed. Sci. Eng. **4**, 122–129 (2011)
7. ChaoShi, R., Yan, W., Juan, D.: The application study of electrical impedance tomography. Chin. J. Med. Instrum. **31**(4), 235–238 (2007)
8. Hari Krishnan, G., Ananda Natarajan, R., Nanda, A.: Comparative study of rheumatoid arthritis diagnosis using two different methods. BPJ **7**(1), 379–382 (2014)
9. Cai, S., Yang, S., Zheng, F., Lu, M., Wu, Y., Krishnan, S.: Knee joint vibration signal analysis with matching pursuit decomposition and dynamic weighted classifier fusion. Comput. Math. Methods Med. **2013**, 1–11 (2013)
10. Tanaka, N.I., Miyatani, M., Masuo, Y., Fukunaga, T., Kanehisa, H.: Applicability of a segmental bioelectrical impedance analysis for predicting the whole body skeletal muscle volume. J. Appl. Physiol. **103**, 1688–1695 (2007)
11. Hari Krishnan, G., Ananda Natarajan, R., Nanda, A.: Impact of upper limb joint fluid variation on inflammatory diseases diagnosis. JEET **9**(6), 2114–2117 (2014)
12. Hari Krishnan, G., Ananda Natarajan, R., Nanda, A.: Synovial fluid density measurement for diagnosis of arthritis. BPJ **7**(1), 221–224 (2014)

Robotic Hand-Eye System Using Machine Learning

Saumil Jariwala[✉], Omkar Chaubal, Roshan Singh,
and Kanchan Bakade

Mukesh Patel School of Technology Management and Engineering (Mumbai
Campus), SVKM'S NMIMS University, Mumbai 400056, Maharashtra, India
saumiljariwala97@gmail.com, ochaubal@gmail.com,
roshansingh0202@gmail.com, kanchan.bakade@nmims.edu

Abstract. Robotics is one of the most potential scientific domains. Functions similar to a human arm can be performed by a robotic arm which acts as a robotic manipulator. The most essential feature of a robotic manipulator is the number of joints which enables it to move in a particular direction. The model proposed various applications of robotic arm using machine learning. The model contains a camera that captures the images which is used for either sorting or recognition. Based on these images the robotic arm can differentiate between objects. Raspberry Pi B+ which is an openly available Linux board and Arduino are used for the processing. Raspberry Pi and Arduino are used along with wireless module where Raspberry Pi acts as the brain of robotic arm. It processes the data by using machine learning and Arduino sends the signals to all the motors of the arm thus controlling its movements as well as speed. Inception classifier is used for classification of various objects.

Keywords: Gesture controlled arm · Robotic arm · Robotic mechanisms · Robot control · Inception classifier

1 Introduction

Robotics is the science of employing machines to undertake jobs which may be monotonous, difficult and risky to human life. An Industrial Robotic arm is an automatically programmable multi-functional manipulator capable of moving in one or more axes to perform certain job by using its end effectors. The most widespread use of a Robotic Arm is to multiply force. These are broadly used in Heavy lifting Industries, Mining Industries etc. The existing methods of control of a Robotic Arm include the usage of Biomedical Signals such as EMG or some external sensors such as accelerometers and gyroscope. For an autonomous system, specific programs are written so that the arm would perform certain tasks according to the user's requirement. Employing various techniques like image recognition and machine learning ensures minimum human intervention and high accuracy. Various open source software tools like tensorflow are now available for the end user. These software make it easy for the user to detect and classify different objects. A database of the objects to be classified has to be created. This paper elaborates on the implementation of a robotic manipulator

© Springer Nature Switzerland AG 2020
A. P. Pandian et al. (Eds.): ICICCS 2019, AISC 1039, pp. 707–715, 2020.
https://doi.org/10.1007/978-3-030-30465-2_78

which can perform actions similar to the human arm. The Raspberry Pi camera captures and sends the data to the Raspberry Pi board. It then processes the data and sends the signals to the Arduino which in turn sends the signals to the Motor Drivers thus controlling its movement and speed. The robotic arm can be used for multiple applications. It is equipped with five stepper motors and drivers thus giving the robot five degrees of freedom. Higher degrees of freedom makes it easier for the robot to acquire the object. Gripper is used for enhancing the arms movement.

1.1 Types of Robotic Arm

The extensive majority of robots available today exhibits one of the four rudimentary configurations which are polar, cartesian, cylindrical and joint-arm configuration. A standard industrial robot is equipped with four to six degrees of freedom. The four basic robot structures are shown Fig. 1 polar, Fig. 2 Cylindrical, Fig. 3 Cartesian, Fig. 4 Articulated arm.

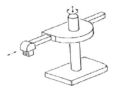

Fig. 1. Polar structure

Fig. 2. Cylindrical structure

Fig. 3. Cartesian structure

Fig. 4. Articulated structure

2 Literature Survey

Factories and Manufacturing Hubs require novel approaches to place mechatronics and robotics assisted systems to minimize the human labor and working hazard at the workplace. Robotics and Automation is extensively used in the field of Computer based, Electronic and Mechanical systems in operation and control production. [1] presents a robotic arm that employs a image sensor that captures the images which is used to process data for recognition or sorting. Object detection method is executed using various types of image processing techniques like contour detection and border extraction. Image Processing based Robotic arm with accelerometer are implemented in factories to do repetitive tasks such as placing the things to and from conveyor to other places [2]. An accelerometer based three joint automatic robotic manipulator is presented in [2]. The robotic arm is equipped with sensors to detect the obstacles in the work space. Microcontrollers like Arduino and Raspberry Pi are used to process the data captured by the camera. [3] discusses the application of Arduino for computer based semi-intuitive robotic arm. The programming feature allows a robot to be used for a variety of industrial applications. These applications may include moving or picking up an object. MEMS Sensor can also used to control the movements of the robotic arm. [4] provides four different gestures for controlling the robots using MEMS Sensor. [5] analyses the application of robotic vision system using synergetic pattern recognition for workpiece detection and manipulation. A pick and place robotic arm with 4-DOF employing haptic technology is discussed in [6], where haptic robotic arm designed to pick the objects of specific weight and place them in a desired location. Human-Machine communication interface between a Leap Motion controller and 6-DOF robotic arm is discussed in [7]. An algorithm based on ANN with back propagation for recognizing human arm gestures is proposed in [8].

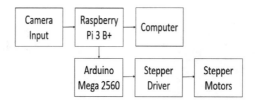

Fig. 5. Block diagram of the system

The block diagram of the system is shown in Fig. 5. The Raspberry Pi B+ module is equipped with a tensorflow which is employed for object recognition. It uses inception classifier for classifying the objects. Spyder IDE was used for training the tensorflow module, which was done using python language. The program was then transferred into Raspberry Pi module. Based on this algorithm the module takes the decision on which object is present in front of the robotic arm. Arduino Mega is connected to the motors with the stepper drivers. The Raspberry Pi is the brain of the robotic arm where all the decisions are been taken. With the help of machine learning algorithm and the input from the camera as well as the transmitter, it performs the required task. The image captured from the raspberry pi camera is matched with the database and the required task is performed. It sends the signal to the Arduino Mega with the help of serial communication. The Arduino Mega is connected to the stepper motors and controls them with the help of stepper motor drivers which are capable of handling high ampere motors. According to the decision taken by the raspberry pi, the Arduino mega moves the motors which are required. But in the automatic mode it follows the path defined previously and accordingly places the detected object at the required place. The main benefit is soft automation so that it is easy to program the robot for the required task.

3   System Design

The system is based on two microcontrollers, Arduino Mega and Raspberry Pi, six stepper motors and two stepper drivers connected to the motors.

3.1   Microcontrollers

The Raspberry Pi B+ shown in Fig. 6 is equipped with a five megapixel camera. The camera constantly feeds the input to the Raspberry Pi. Tensorflow is installed in the Raspberry Pi module for image recognition. Since the Raspberry Pi is more flexible for software applications it us used. Python is used for programming the tensorflow to employ the image recognition tool. Raspberry Pi B+ has 1.2 GHz CPU. BCM2837 Broadcom chip is used in Raspberry Pi 3. Quad-core ARM cortex A53 cluster is used in this chip. Raspberry Pi 3B+ contains a 1 GB RAM operating at a frequency of 900 MHz.

Fig. 6. Raspberry Pi

Fig. 7. Arduino Mega

Arduino Mega 2560 is shown in Fig. 7. The communication between Arduino Mega and Raspberry Pi is achieved with the help of UART. Arduino Mega 2560 contains 54 digital Input/Output pins and 16 analog inputs. Since it has more number of pins, Arduino Mega can be connected to more motors, thus enabling higher degree of freedom. It is based on a 16 MHz oscillator and has a operating voltage of 5 V.

3.2   Motors and Motor Drivers

Six stepper motors are employed in the articulated robotic arm. Two Stepper motor Tb6600 are used and four a4988 stepper motors are used. Figure 8 shows the motors used in the system. The six axis Robotic arm has been 3D printed and the material used for the 3D printing is ABS which is a strong long lasting plastic substance. Honeycomb printing pattern is used for printing the robotic arm as it has a better strength structure which can handle high weights without the problem of breakage. The motors used in the robotic arm are bipolar stepper motors which are having high torque and there are total 7 stepper motors in the robotic arm. The base motors are of high amps as they require more power to handle the weight. They are being controlled by the Arduino with the help of stepper drivers which are capable of handling 4A stepper motors. All the motors have been subjected to stress tests and the results were compared to the industrial standard. Motors are driven by drivers shown in Fig. 9.

Fig. 8. Stepper motors   **Fig. 9.** Stepper driver TB6600

3.3 Raspberry Pi Camera Module

The camera module shown in Fig. 10. is accompanied along with Raspberry Pi 3 B+ . It is connected to Raspberry Pi via a small socket present on the board. The Csi interface is employed for interfacing the microcontroller to the camera. The resolution of the camera is five megapixel and it is equipped with focus lens. The camera captures static image with capacity up to 2592×1944 pixels, and also supports $640 \times 480p$ 60,720p 60 and 1080p 30 video.

Fig. 10. Raspberry Pi camera

4 Methodology

The foundation of this project is based on the ability of the robotic arm to sort various objects by recognizing them with the help of a five megapixel camera equipped with Raspberry Pi b3+. The process begins when the object enters the work space, The camera sends the live feed to the microcontroller which is equipped with a software that detects the object and recognizes it. In the following implementation, the program is trained to recognize two different chocolate based on the color and design of their wrappers. The resulting outcome was communicated to Arduino via UART. The flowchart shown below in Fig. 11 illustrates the methodology of the project implemented.

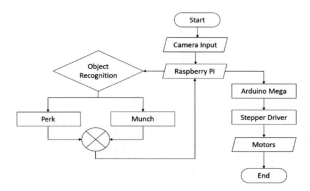

Fig. 11. Flow chart

- The Raspberry Pi B+ is equipped with a 5 megapixel camera, which is pointed towards the workspace of the robotic arm where the object is present.
- The Raspberry Pi B+ has tensorflow installed which is used for image detection and recognition.
- Raspberry Pi camera constantly feeds the input to the microcontroller where the image processing takes place. Inception algorithm is employed for recognizing various objects after creating their database.
- RaspberryPi communicates with the Arduino Mega via UART.
- Arduino Mega is connected to various stepper drivers which are in turn connected to stepper motors.
- The robotic arm is programmed to follow a predetermined based on the input from Raspberry Pi (Fig. 12).

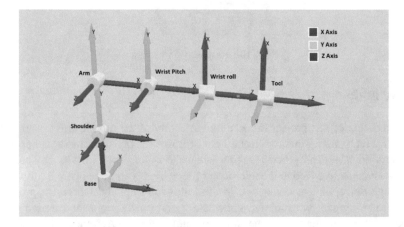

Fig. 12. Link coordinate system diagram of 5 axis robot

5   Result

A system is gauged based on its accuracy and response time. Trials were conducted to determine the accuracy of the project for sorting two chocolates namely 'Perk' and 'Munch'. Data set was created to train the system to identify and recognize the aforementioned chocolates. During the trials it was observed that the accuracy at the beginning stages was low and it increased as the number of trials conducted were increased. The Table 1 below depicts the results obtained:

Table 1. Results table

Sr. No	Number of iteration	Training stage		Object recognition			Accuracy
		Name of item	Data set quantity	Name of item	Trail	Correctly recognized	
1	500	Perk	20	Perk	60	51	85%
		Munch	20	Munch	60	49	81.66%
2	1000	Perk	45	Perk	60	58	96.6%
		Munch	45	Munch	60	56	93.3%

In the first 500 iterations, the data set consisted of 20 images of each Perk and Munch. It was observed that out of 60 trials conducted, the system could correctly recognize and sort 51 and 49 chocolates of 'Perk' and 'Munch' respectively resulting in an accuracy of 85% and 81.66% as shown in Fig. 13. During the next 1000 iterations the accuracy increased to 96.6% and 93.3% for 'Perk' and 'Munch' respectively which is shown in Fig. 14. Thus as the trials proceed further the accuracy increases. The graphs shown below illustrates the accuracy percentage of the system in the first stage with 500 iterations and second stage with 1000 iterations.

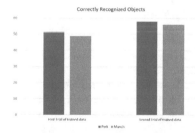

Fig. 13. Correctly recognised objects

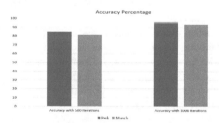

Fig. 14. Accuracy graph

6   Limitation

During testing of the robot, certain limitations were observed for the proposed model, which are as follows:

1. The object should be positioned in a proper way for the robotic arm to be able to pick it up. The object should be lying still. This robot is not applicable for picking up objects on a conveyor belt.
2. If there is an obstacle which breaks the line of sight of the receiver, the receiver won't be ble to detect the location of the transmitter, hence robotic arm may not be able to reach the object.
3. It can only move 180° i.e. it can only cover the area with an angle of 180° in the forward direction.
4. A constant power source is required as 9 V batteries get drained easily due to the presence of many sensors.

7   Conclusion and Future Scope

The robotic arm used for sorting of different objects is implemented successfully. The algorithm accurately identifies the object and the robotic arm picks it up only to be sorted according to the predetermined instructions. This concept can be incorporated to achieve the goal of minimizing human labor and increasing the productivity. Sensors can be added to the robotic to detect and it can be programmed to overcome the obstacles. Additionally, truly autonomous machines can be created which perform their function and are able to recharge without the need of any human intervention.

References

1. Dragusu, M., Mihalache, A.N., Solea, R.: Practical applications for robotic arms using image processxing. In: 2012 16th International Conference on System Theory, Control and Computing (ICSTCC), Sinaia, 2012, pp. 1–6 (2012)
2. Bhavsar, R., Jadav, H., Soni, K., Chatrola, K., Patel, V.: Image processing based robotic arm with accelerometer. Int. J. Comput. Sci. Inf. Technol. Res. **3**, 93–95 (2015)
3. Badrinath, S., Vinay, P.B., Hegde, P.: Computer vision based semi-intuitive robotic arm. In: 2016 2nd International Conference on Advances in Electrical, Electronics, Information, Communication and Bio-Informatics (AEEICB), Chennai, 2016, pp. 563–567 (2016)
4. Patil, A., Kadiri, H., Wani, A.: IOT based remote access human control robot using MEMS sensor. Int. J. Comput. Sci. Mob. Comput. **5**(3), 816–826 (2016)
5. Wang, F.-Y., Levert, P., Bing, P.: A robotic vision system for object identification and manipulation using synergetic pattern recognition. Robot. Comput. Integr. Manuf. **10**(6), 445–459 (1993)
6. Omijeh, B.O., Uhunmwangho, R., Ehikhamenle, M.: Design analysis of a remote controlled pick and place robotic vehicle. Int. J. Eng. Res. Dev. **10**(5), 57–68 (2014)

7. Bassily, D., Georgoulas, C., Güttler, J., Linner, T., Bock, T.: Intuitive and adaptive robotic arm manipulation using the leap motion controller. In: International Symposium on Robotics (ISR 2014) (2014)
8. Szabó, R., Gontean, A.: Industrial robotic automation with Raspberry Pi using image processing. In: 2016 International Conference on Applied Electronics (AE), Pilsen, 2016, pp. 265–268 (2016)
9. Gupta, G.S., Mukhopadhyay, S.C., Finnie, M.: WiFi based control of a robotic arm with remote vision. In: IEEE Conference Publications, 2009, pp. 557–562 (2009)
10. Bhuyan, A.I., Mallick, T.C.: Gyro-accelerometer based control of a robotic Arm using AVR microcontroller. In: 2014 9th International Forum on Strategic Technology (IFOST), Cox's Bazar, 2014, pp. 409–413 (2014)
11. Yenorkar, R., Chaskar, U.M.: GUI based pick and place robotic arm for multipurpose industrial applications. In: 2018 Second International Conference on Intelligent Computing and Control Systems (ICICCS), Madurai, India, 2018, pp. 200–203 (2018)
12. Patidar, V., Mishra, A., Tiwari, R.: Robotic gripper arm system with effective working envelope. In: 2018 Second International Conference on Intelligent Computing and Control Systems (ICICCS), Madurai, India, 2018, pp. 1061–1065 (2018)

A Robust Q-Learning and Differential Evolution Based Policy Framework for Key Frame Extraction

Sudipta Rudra and Senthil Kumar Thangavel[✉]

Department of Computer Science and Engineering, Amrita School
of Engineering, Coimbatore, Amrita Vishwa Vidyapeetham, Coimbatore, India
cb.en.p2csel7029@cb.students.amrita.edu,
t_senthilkumar@cb.amrita.edu

Abstract. With the recent development in multimedia technologies, in rapid conjunction with the increase of the volume of digital video data through internet and web technologies. For this purpose solely, content based video retrieval (CBVR) has become a wide and vast area of research throughout the last decade. The objective of this thesis is to present applications for temporal video frame analysis based on performance evaluation of key frames and video sequence from the extracted key frames retrieval based on different mathematical models. In this work, through performance analysis, we extracted the key frames from a video into its constituent units. This is achieved by identifying transitions between adjacent temporal features. The proposed algorithm aims to extract the key frames based on the validation measures and cross mutation function through the modified differential evaluation algorithm. Given the size of the vector containing image pixels, it can be modeled by a parameter based cross evaluation function of the parent vector. The proposed system, designed for extracting key frames, has led to reliable algorithm achieving high performance for object re-identification. In addition, the high computational time allows for key frame analysis in real time. In our research, we opted for a global method based on local optimization. The proposed methodology is being validated against various state of the art key frame extraction algorithm which proves this methodology as reliable and faster complexion process for object re-identification.

1 Introduction

A diversified summary of key frame selection method have been furnished throughout the last decade. It have been recovered safe that such methods usually have three forms, viz. shot boundary detection as a pre-processing form, primary form of key frame perception such as optical, functional, sound and textual features that are selected in each exposure, and then prepared and analyzed with machine learning and deep learning based performance, and the last form is post-processing form placed in dismissal of match if they happen in the consequent sequence of key exposures. Computation method and gettable trial video- recording collections also been determined. Video recording information has been inflated quickly due to fast improvement of

© Springer Nature Switzerland AG 2020
A. P. Pandian et al. (Eds.): ICICCS 2019, AISC 1039, pp. 716–728, 2020.
https://doi.org/10.1007/978-3-030-30465-2_79

digital video recording seizure and editing technology. Key exposure selection is a mighty way that utilize video-recording content by extracting a rigid of compact key exposures to correspond video-recording sequences. Most of the existing key exposures selection methods are not fit to an object re-identification problem, as they does not fit particular requisites. Here, key exposure selection methods are classified into three types, based on shot boundary, genetic algorithm and optimization based techniques and differential process based where our methodology exist. Nowadays, differential process based performance are generally theoretical in video-recording sequence and volume analysis explorer. Through these performance, key exposure selection is commonly framed in a representative parameter based procedure that split one video recording snap into a various frame segments and then one or various exposure are selected according to lower or higher level features. Usually some performances in closed area generally are unfit and dependable for different formatting of video recording sequences generated from the IOT devices and real time video acquisition devices. Accomplishing a purposeful key exposure is a crucial problem in respective multimedia technologies. Concentration of the activity is by corresponding the video-recording content competently also by accelerating from the extracted key frames. In this thesis, a local improvement based performance is constituted. Initially, the key exposure is characterized for video-recording sequence analysis. Then a key exposure selection rule supported on qualified version of differential evolution algorithm which has self-threshold selection and updation process based on global parameter search. This motivated us to choose this algorithm as a reliable methodology for object re-identification research problem. Nowadays digital video-recording is an upcoming factor in several multimedia system and web based utilization. A video recording sequence is a serial occurrence of the continuous visual images. The fast growths in network technologies have enhanced the usage of video-recording. A video-recording sequence is combined of multiple exposure at a exposure rate of 26 exposures per second (fps) that a person eye can't realize the separation and irregularity in the video-recording. This modification throughout the period of time got many future investigators from a newer technologies which intent in amending the effectiveness and prompt usage of videos. Outdated volume of video-recording data is tedious to many utilizations, and therefore, there is need to the process, which can be utilized for profit view and continuous exposure from the video-recording without looking the whole video. Key frame of a video furnishes the more reliable and compact information of the video-recording. They are also known as representative exposures, R-exposures, snap-picture conceptional otherwise unchanging storyboard. Key frames serves as the basic element for various undertaking like video tagging, video- report, inquiring, perceptive, chapter titles in DVD's based storage mediums and it is also being used in many compact utilizations such as video surveillance, medical video-recording, submerged video, web browsing video, athletics and news programmer, along with interior and exterior videos etc. Early approaches prefers to select key frames in selected or through unvarying choice of the video-recording exposures at pre-distinct intervals by the user. Traditional summarization techniques provide heavy static video visualization or accumulated video cover which are not properly reliable from extracting out inferences based on the features levels which comprises of both high level and low level features which comprises of texture, shape, size of the object in the video. Given a

VIDEO/IMAGE dataset divided into classes and subclasses. The objective of the project development would be towards:

1. Given a input data set, extract and model the spatial and temporal key frames based on the high level and low level optimized features.
2. Transformation of all the spatial and temporal features into a localized label set that helps for object re-identification.

2   Literature Survey and Related Works

This Survey iterates the few of the enquiry activity in the literature covering the key exposure selection problem of video-recording data. In [1], authors have disposed a key exposure selection algorithm which mechanism based on "shots" and exposure prepared in a compressible video-recording. Video-recording is segmental into "Shots" that contains uninterrupted series of actions. Various exposures are controlled in the snap from which key exposures are selected. From [1] an improved histogram algorithm is used to compute frame difference. compressible video of MPEG type is seized which classifies frames into I frames, P frames and B frames. After shot segmentation in the MPEG video stream, characteristics of I frames, P frames and B frames are used for key frame extraction [2].

An key frame extraction algorithm based on optimized key frame difference is proposed in [3]. The variation in adjacent frames are used to find the inter-frame similarity, initially. The suitable key frames are finally acquired after optimization. The complex calculation involved in the process of key frame extraction is avoided in this algorithm. Also, key frame repetitiveness is reduced through this algorithm. In [5] the author uses improved optimization of frame difference. In this approach, the number of key frames 'k' is set in advance and the distance between the current and the next frame is calculated using a formula. This is done from the first frame and thus they calculate inter-frame difference. Inter-frame difference calculates all the distance of the frames and gets the global maximum and minimum value of inter frame difference. After this they calculate the middle value and then they remove all local extreme points, and the leftover extreme points are defined as L. If k < L, then first 'k' frames are selected as key frames. In spite of the creation many formulation for solving key frame extraction problem [6–8], this paper proposes a simple algorithmic framework which combine DE algorithm with local improvisation approach. The technical contribution of this approach and its performance are well proved with different videos. The DE algorithm is similar to other EAs functionally. It performs iterative execution of mutation-crossover-selection process cycle over the set of campaigner in the existing population. However, the mutation strategy used in DE is specific in its nature. It is termed as differential mutation. The DE algorithm starts its search with a randomly initialized set of problem dependent candidate solutions. This set of candidates is known as the initial population (IP). Now, the IP experience a reproduction cycle for a predefined number of generations. At each generation, for each of the campaigner in the population [9], a new child is produced performing differential mutation followed by the crossover operation. After producing a child for a campaigner solution, they are compared by their fitness (measured by the problem specific objective function), a selection process is carried out to resolve the person for next generation. Thus, at the end of each generation a new population with the survivors is generated. This process will be continued until meeting a user defined stopping criteria. The DE algorithm is widely used for variety of complex optimization problems. The performance DE algorithm is still being improved by the researchers [10]. The theoretical investigation of DE algorithm is still an open challenge problem [9–11]. Many study papers are reported in the literature for the control parameters of DE algorithm [10].

3 Methodology

Key exposure selection is a built-in part of video analytics. The selected key exposures are being utilized for video-recording report along with content recovery. Here subsist several procedural formulation for resolution and examining key exposure selection trouble in video-recording calculus. The concentration of thesis is to widen the plan of action of integrating Evolutionary Computation methodological analysis with a self-adaptive technique based key frame extraction approach. With two other received approaches. The received approaches reasoned in this report for analysis are SSIM (Structural Similarity Index Method) Method, Entropy Method and Euclidean Distance method, PSNR method. This report also gives a new conceptualization for key exposure selection by combining the local improvisation method with Differential Evolution algorithm. The proposed approach is analyzed with all the active approaches by its preciseness rate and request rate. It is found from the examination that the proposed approach exceed other approaches. The consequence and discussion related to this experimental study are presented in this thesis.

Quality evolution mainly, does not form the usage of mutation operator that look on same conditional probability distribution function, but makes a new cheerful optimization based operator which depends upon the differences between randomly selected cross mutated pairs of individuals. The basic DE conferred above is just a model and it can be revised in assorted ways to gear the particularity of each different problem. If in one hand this large number of outlook can be seen as an advantage, on the other hand, its use by the uninitiate person can be hard. Even for knowledgeable users, the capable configuration of the algorithm usually demands a lot of time doing prelim experiments and/or; info about the problem (e.g. multi modality, separability) which is not always accessible. In this context, what is left to the user is the use of a standard configuration what can lead to a miserable execution of the algorithm. Therefore, to turn DE into an easy-to-use broad optimization tool some self-configuration quality must be provided. A population of nominee solutions randomly created within the domain region of the problem, usually described as:

$$X = \left\{ x \in R^D : x_{min} \leq x \leq x_{max} \right\} \qquad (1)$$

Where and are respectively the lower and upper limits of each variable and D is the problem dimension, i.e., the number of variables in the problem. In this paper, represents the respective with index i $\in$ [1, NP] in the population, during the generation g.

$j = 1,\ldots,$ D represents the design parameter index. A given individual is then represented by:

$$x_{g,i} = x_{g,1}, x_{g,2}, \ldots\ldots\ldots\ldots\ldots x_{g,N} > \tag{2}$$

After formatting, for each individual in the population a new individual, called mutant, is bring forth by a mutation operator. In Differential Evolution mutation is based on the difference between individuals randomly chosen from the current population as follows,

$$V_{g,i} = x_{g,1} + F\left(x_{g,2} - x_{g,1}\right) \tag{3}$$

After mutation a trial vector is produced through recombination of and. In the basic DE algorithm, the discrete recombination with probability CR is used, as we can see in the following scheme:

$$u_{g,i} = \begin{cases} v_{g,i}, & \text{if } U_{[0,1]} \leq F, CR \\ x_{g,i} & \text{otherwise} \end{cases} \tag{4}$$

Where, U [0, 1] represents the sampling of a random variable with uniform distribution in the interval [0, 1]. In this way, F and CR represent control parameters of the algorithm.

Finally, the trial vector competes with the current solution based on their objective function evaluations. The trial vector replaces the current solution if it is better than the current solution or if it has the same value of objective function. This process can be described by the equation below:

$$x_{g+1,i} = \begin{matrix} u \\ f([g,i]) \leq f\left(x_{g,i}\right) \\ \{ u_{g,i}, \text{ if } x_{g,i} \text{ otherwise} \end{matrix} \tag{5}$$

3.1   Configuration of Operators of DE

Q-learning: In this formulation the problem of allotment of operators is seen as a confirmation learning problem. In the regular model of reinforcement learning an agent interacts with the environment and receives information about its current state. Based on that information, the agent takes an action to be executed. This activity alters the situation which in turn twist a feedback sign to the agent. Through this sign, which can be a reward or a punishment, the agent "learns" how to conduct. For the function allocation problem the DE is seen as an agent which has to learn what function to use at each step. It act with the population of individuals by means of the operators (mutation + crossover) and the feedback sign is generated by evaluating the population. If the used operators generate good individuals they must be reinforced. One of the most used algorithms in support learning is called Q-learning. In this thesis we use a individual state version of Q-learning. Fundamentally, the agent starts playing an action and evaluates the outcomes of this action by means of the received reward r as follows:

$$Q(a_t) \leftarrow \alpha(r + \gamma \max_{a_{t+1}} Q(a_{t+1}) + (1 - \tau)Q(a_t) \tag{6}$$

$Q(a_t)$ is the reward expected by the execution of action $\gamma \max_{a_{t+1}} Q(a_{t+1})$ is an estimate of the maximum future reward and γ is a discount factor applied to it. Trying all actions repeatedly the algorithm learns the best execution policy based on the long term rewards and not only on the immediate ones.

4   Experiment and Results

The dataset being collected was a set of 5 video sequences of movable chairs and collected with the cps camera with optical zoom of 20 x, bandwidth of 54 dB and SNR of 20 dB along with resolution of 30 x. the five video sequences had a bit rate of 920 kbps each along with data rate of 800 kbps each. The frame width and height of the video sequences were of 640 * 480 pixels. This data set was being designed in order to compute on real time environment for the key frame extraction based applications.

Table 1. The below table describes the frame rate, frame height, data rate, bit rate, frame rate of input videos.

Input	Frame width	Frame height	Data rate	Bitrate	Frame rate
Video_1	640	480	800 kbps	920 kbps	30 fps
Video_2	640	480	800 kbps	920 kbps	30 fps
Video_3	640	480	800 kbps	920 kbps	30 fps
Video_4	640	480	800 kbps	920 kbps	30 fps
Video_5	640	480	800 kbps	920 kbps	30 fps

The above table represents the tabular view of the description and characteristics of the video set. This video sets have a property a continuous and symmetric frames which have been explained in the result below (Table 2).

Table 2. Monitoring of video data set

	Test the randomized number of key frames	Actual amount of key frames	Proposed performance to select the random number of key frames	Time (in seconds)
Video_1	500	274	248	67
Video_2	316	165	132	32
Video_3	274	212	184	43
Video_4	157	110	83	22
Video_5	357	249	217	54

The above table being presented reflects the results in order to test precision and recall values. Each video sequences were performed though various experiments and values were recorded. First column in the array of table represents the randomized number of key frames that can be generated using any method. Whereas the second column in the array of table represents the actual amount of key frames which were extracted using a simple differential evolution algorithm. From which it was observed that the results being derived were of higher deviation than total number of frames as many asymmetric and discontinuous frames were being deleted. Key frames as mentioned in the introduction are the frames which describes the information along with spacial and feature of the video.

Final column in a table represents the actual number key frames being extracted with the help of the proposed methodology. The frequency of key frames extracted deviated highly from the total number of key frames. Here the key frame being extracted represents the actual features that are context semantics of the video sequence.

The below Table 3 represents the validation of the proposed approach through the comparison with 3 category based algorithms. Where the first category consists of general key frame based extraction method and the second category represents the genetic algorithm optimization based parametric methods. It is being observed that the histogram based key frame extraction method achieves higher precision values than the proposed method and other algorithms which have a precision rate for 51.23%. Similarly, for video_2 it was observed that it achieved precision rate of 49.37% than 48.17% of the proposed algorithm. But for the video _5 it achieved more precision rate of 49.01%. As per the experiments conducted on the input. But for the video sequence 2, the traditional DE algorithm achieves more precision rate than the proposed algorithm. In some cases the DE algorithm proves out to be more robust and efficient than the GA/optimization parametric based and general key frame extraction of the video sequences. But from the outputs, it was precisely observed that still proposed algorithm stands better in terms of contrast, symmetricity and continuous frames which can be stated the key frames preserving the features of the video recorder. The formula for the precision is being below:

$$Precision = \frac{Detect\ the\ actual\ number\ of\ key\ frames}{total\ number\ of\ key\ frames} \times 100$$

Table 3. The below contains the validation result of precision value of the proposed algorithm with various key frame extraction algorithms.

Video shots	Category 1 (general key frame extraction dataset)			Category 2 (GA/optimization parameters based)				Category-3 (DE)	
	Histogram based	Shot boundary	Mutual information method	Entropy based	PSNR based	Euclidean based	SSIM based	Traditional DE	Proposed approach
Video_1	51.23%	46.32%	47.35%	45.84%	48.32%	47.24%	46.74%	47.32%	**48.48%**
Video_2	49.37%	47.25%	46.21%	47.75%	48.21%	48.27%	47.12%	48.48%	**48.17%**
Video_3	48.35%	47.19%	45.33%	47.10%	45.23%	46.72%	48.65%	49.97%	**49.42%**
Video_4	48.24%	47.96%	46.29%	48.45%	46.24%	47.32%	47.93%	48.84%	**48.92%**
Video_5	47.93%	47.12%	47.25%	48.26%	47.91%	47.49%	48.01%	49.26%	**49.01%**

The formula being presented was used to derive the results of key frame extraction method and was recorded for relation with other verbalize of the art algorithms being showed in Table 1. Precision measure is being used here as a validation measure which detects the false positive frames to be extracted out from the video sequences.

Recall was used as another measure of validation to test the proposed algorithm working and complete the evaluation process for the proposed methodology. The formula used for calculating recall is presented below:

$$\text{Recall} = \frac{Detect\ the\ actual\ number\ of\ key\ frames}{actual\ number\ of\ key\ frames} \times 100$$

Recall is being stated as measure to check the percentage of actual key frames extracted from the entire amount of key frames. Higher the recall value, higher is the percentage of actual key frames extracted which contains both high level and mid – level features. Such measures are the statistical based validation measures that are often used for performance evaluation. It has proved to be used as a measure in the terms of key frame extraction as it describes the inheritance of actual key frames along with it. More the precision rate and lesser the rate of recall, the extraction of genuine key frames are accurate otherwise the extraction process more false positive of extracted frames which are not visible from the common eye. Below the Table 4 represents the recall percentage obtained by the proposed algorithm which also shows the relation of the state of the art algorithms with methodology. It is being observed that for video_1, the proposed methodology and PSNR based optimization algorithm performs better than the rest of the methodologies. The proposed methodology has higher compared value in case of video_2 means the 87% of the key frames extracted from the entire amount of frames are accurate and are homogeneous. The extraction are homogeneous and higher in case of traditional DE which performs better for video_3 having a recall rate of 88.17%. In case of video_4 the proposed methodology has higher recall rate than the other key frame extraction algorithms where as, the DE algorithm has higher performance as well as recall rate for video_5 with measure of 86.17%.

Table 4. The below contains the validation result of recall value of the proposed algorithm with various key frame extraction algorithms

Video shots	Category 1 (general key frame extraction dataset)			Category 2 (GA/optimization)				Category-3 (DE)	
	Histogram based	Shot boundary	Mutual information method	Entropy based	PSNR based	Euclidean based	SSIM based	Traditional DE	Proposed approach
Video_1	86.56%	85.27%	87.49%	86.94%	87.81%	85.42%	84.48%	86.39%	**87.86%**
Video_2	83.4%	82.91%	84.33%	83.21%	84.74%	84.16%	83.12%	83.47%	**87.13%**
Video_3	85.41%	82.49%	86.67%	87.56%	88.28%	87.86%	86.31%	88.95%	**88.17%**
Video_4	84.76%	83.52%	83.16%	85.47%	84.97%	85.32%	85.98%	85.32%	**86.04%**
Video_5	85.23%	83.79%	84.85%	86.12%	85.31%	86.64%	85.97%	86.71%	**86.13%**

The below table displays the number of frames extracted from the video sequences each describing their precision and recall rate. The proposed algorithm extracts recall value of 88.17% from video sequence_3. It is observed that the stability of key frames extracted from the video sequences depicts their characteristics of movement, anomalies which mainly concentrates on the continuous and symmetric frames being extracted. From the Table 5, it depicts the video sequence_5 has more continuous and symmetric frames at a constant rate during first to fifth time interval. But still it cannot be said based on the ranges of the values, depending on the object characteristics, its key movements, texture appearance it can be evaluated. This values derived are for the purpose of validation and analysis of the working and development of key frame extraction that can be suitable for object re-identification problem.

Table 5. The below table shows the key frames extracted based on the validation measures for the proposed algorithm.

Input	Key frames extracted	Precision and recall value
Video_1	[9, 47, 58, 71, 123, 332, 370, 396, 452, 467]	(87.86%, 48.48%)
Video_2	[8, 29, 34, 40, 58, 73, 76, 82, 98, 100]	(87.13%, 48.17%)
Video_3	[12, 16, 19, 31, 41, 51, 56, 62, 67, 98]	(88.17%, 49.42%)
Video_4	[1, 4, 13, 15, 23, 30, 35, 48, 56, 99]	(86.04%, 48.92%)
Video_5	[11, 13, 14, 16, 21, 25, 30, 43, 69, 99]	(86.13%, 49.01%)

4.1   Results

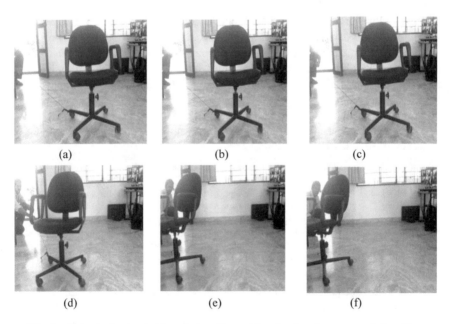

(a)   (b)   (c)

(d)   (e)   (f)

Fig. 1. Output of extracted key frames from video_1 using the proposed algorithm

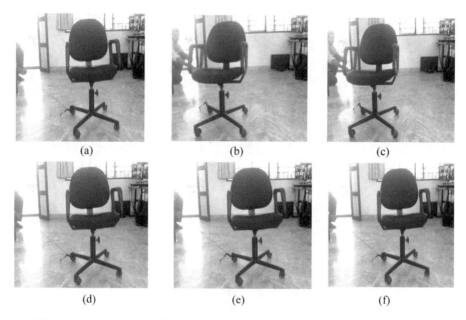

Fig. 2. Output of extracted key frames from video_2 using the proposed algorithm

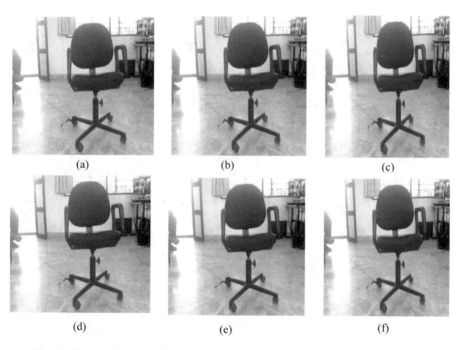

Fig. 3. Output of extracted key frames from video_3 using the proposed algorithm

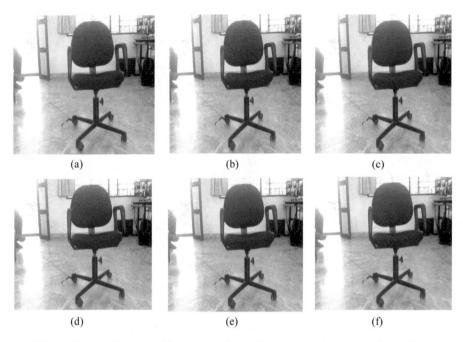

(a)   (b)   (c)

(d)   (e)   (f)

Fig. 4. Output of extracted key frames from video_4 using the proposed algorithm

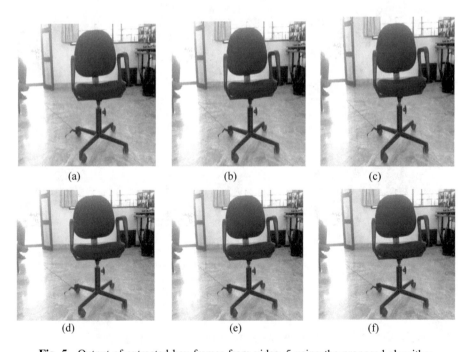

(a)   (b)   (c)

(d)   (e)   (f)

Fig. 5. Output of extracted key frames from video_5 using the proposed algorithm

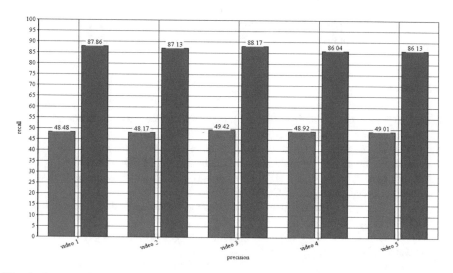

Fig. 6. Output of visualization of key frames from video sequences using the proposed algorithm.

From Fig. 6, it is being observed that in measure of precision rate the extracted key frames maintain a consistency rate at around 48% level to 50% level. Though the video sequence 4 and video sequence 5 achieves the precision rate of approximate 49%. While comparison with the state of other performance, this proposed methodology achieves better in terms of extracting key frames based on PSNR, SSIM, Euclidian distance and entropy based methods. Whereas, in terms of recall rate some algorithms outperform our proposed algorithm such as histogram based method. But it was observed that in terms of performance for stability of the continuous frames depicting the change in action and behavior of the object, the proposed algorithm maintains a good stability rate (Figs. 1, 2, 3, 4 and 5).

5   Conclusion

A key frame extraction performing based on self-adaption with Q-learning process with low level characteristic is being presented for video object determination problem. Precisely, the normal differential evolution algorithm is used to infuse accurate key frames to screen up the primary content for the entire video recording sequence. Initially, an secondary sequence is derived established on structural similarity variation in adjacent frames from original sequence and the peak signal to noise ratio. Then, the terminal key frame sequence is derived aside learning from the cognition self- adaptive process that embeds the parameters in to the search space to find out the best of total key frames providing important inferences and characteristics of the object from the alternative sequence. It can be viewed as a convex optimization problem using local parameters globally throughout the search space. Proved with respective video recorder datasets with identical content, info, and resolutions, but with different object activities.

It is shown that the planned method have advantages in procedure quality in cost of exactness, request rate, strengthens on formats, video resolution, and so on.

References

1. Li, Q., Qiu, Z., Yao, T., Mei, T., Rui, Y., Luo, J.: Action recognition by learning deep multi-granular spatio-temporal video representation. In: ICMR (2016)
2. Li, Y., Yao, T., Mei, T., Chao, H., Rui, Y.: Share-and-chat: achieving human-level video commenting by search and multi-view embedding. In: ACM MM (2016)
3. Pan, Y., Mei, T., Yao, T., Li, H., Rui, Y.: Jointly modeling embedding and translation to bridge video and language. In: CVPR (2016)
4. Pan, Y., Yao, T., Li, H., Mei, T.: Video captioning with transferred semantic attributes. In: CVPR (2017)
5. Qiu, Z., Yao, T., Mei, T.: Deep quantization: encoding convolutional activations with deep generative model. In: CVPR (2017)
6. Qiu, Z., Yao, T., Mei, T.: Learning spatio-temporal representation with pseudo-3D residual networks. In: ICCV (2017)
7. Xu, J., Mei, T., Yao, T., Rui, Y.: MSR-VTT: a large video description dataset for bridging video and language. In: CVPR (2016)
8. Yao, T., Pan, Y., Li, Y., Mei, T.: Incorporating copying mechanism in image captioning for learning novel objects. In: CVPR (2017)
9. Yao, T., Pan, Y., Li, Y., Qiu, Z., Mei, T.: Boosting image captioning with attributes. In: ICCV (2017)
10. Zhang, J.M., Liu, H.Y., et al.: Key frame extraction based on improved ant algorithm and agglomerative. Comput. Eng. Appl. **49**(3), 222–225 (2013)
11. Guo, X., Shi, F.: Quick extracting key frames from compressed video. In: Proceedings of the 2nd International Conference on Computer Engineering and Technology, ICCET-2010, Chengdu, China, vol. 4, pp. 163–165, April 2010

Author Index

© Springer Nature Switzerland AG 2020
A. P. Pandian et al. (Eds.): ICICCS 2019, AISC 1039, pp. 729–732, 2020.
https://doi.org/10.1007/978-3-030-30465-2

Printed in the United States
By Bookmasters